MW01502716

Edited by
Jeroen Kool
Wilfried M.A. Niessen

**Analyzing Biomolecular Interactions
by Mass Spectrometry**

Related Titles

Hillenkamp, F., Peter-Katalinic, J. (eds.)

MALDI MS

A Practical Guide to Instrumentation, Methods and Applications

Second Edition
2014
Print ISBN: 978-3-527-33331-8,
also available in digital formats

Przybylski, M.

Biopolymer Mass Spectrometry

Methods, Ion Chemistry, Bioanalytical Applications

2013
Print ISBN: 978-3-527-32955-7

Cannataro, M., Guzzi, P.H.

Data Management of Protein Interaction Networks

2012
Print ISBN: 978-0-470-77040-5,
also available in digital formats

Budzikiewicz, H., Schäfer, M.

Massenspektrometrie

Eine Einführung

Sechste Auflage
2012
Print ISBN: 978-3-527-32911-3,
also available in digital formats

Schalley, C.A. (ed.)

Analytical Methods in Supramolecular Chemistry

Second Edition
2012
Print ISBN: 978-3-527-32982-3,
also available in digital formats

Edited by
Jeroen Kool and Wilfried M.A. Niessen

Analyzing Biomolecular Interactions by Mass Spectrometry

WILEY-VCH

Verlag GmbH & Co. KGaA

The Editors

Dr. Jeroen Kool
VU University Amsterdam
Faculty of Science
Amsterdam Institute for Molecules
Medicines and Systems
Division of BioAnalytical Chemistry/
BioMolecular Analysis
De Boelelaan 1083
1081 HV Amsterdam
The Netherlands

Prof. Dr. Wilfried M.A. Niessen
hyphen MassSpec
de Wetstraat 8
2332 XT Leiden
The Netherlands

and

VU University Amsterdam
Faculty of Science
Amsterdam Institute for Molecules
Medicines and Systems
Division of BioAnalytical Chemistry/
BioMolecular Analysis
De Boelelaan 1083
1081 HV Amsterdam
The Netherlands

■ All books published by **Wiley-VCH** are carefully produced. Nevertheless, authors, editors, and publisher do not warrant the information contained in these books, including this book, to be free of errors. Readers are advised to keep in mind that statements, data, illustrations, procedural details or other items may inadvertently be inaccurate.

Library of Congress Card No.: applied for

British Library Cataloguing-in-Publication Data
A catalogue record for this book is available from the British Library.

Bibliographic information published by the Deutsche Nationalbibliothek
The Deutsche Nationalbibliothek lists this publication in the Deutsche Nationalbibliografie; detailed bibliographic data are available on the Internet at <http://dnb.d-nb.de>.

© 2015 Wiley-VCH Verlag GmbH & Co. KGaA, Boschstr. 12, 69469 Weinheim, Germany

Print ISBN: 978-3-527-33464-3
ePDF ISBN: 978-3-527-67342-1
ePub ISBN: 978-3-527-67341-4
Mobi ISBN: 978-3-527-67340-7
oBook ISBN: 978-3-527-67339-1

Cover Design Bluesea Design, McLeese Lake, Canada
Typesetting Laserwords Private Limited, Chennai, India
Printing and Binding Markono Print Media Pte Ltd., Singapore

Printed on acid-free paper

Contents

List of Contributors *XIII*
Preface *XVII*
Abbreviations *XIX*

1	**Introduction to Mass Spectrometry, a Tutorial** *1*	
	Wilfried M.A. Niessen and David Falck	
1.1	Introduction *1*	
1.2	Figures of Merit *1*	
1.2.1	Introduction *1*	
1.2.2	Resolution *2*	
1.2.3	Mass Accuracy *4*	
1.2.4	General Data Acquisition in MS *5*	
1.3	Analyte Ionization *6*	
1.3.1	Introduction *6*	
1.3.2	Electrospray Ionization *8*	
1.3.3	Matrix-Assisted Laser Desorption Ionization *10*	
1.3.4	Other Ionization Methods *10*	
1.3.5	Solvent and Sample Compatibility Issues *11*	
1.4	Mass Analyzer Building Blocks *12*	
1.4.1	Introduction *12*	
1.4.2	Quadrupole Mass Analyzer *13*	
1.4.3	Ion-Trap Mass Analyzer *13*	
1.4.4	Time-of-Flight Mass Analyzer *15*	
1.4.5	Fourier Transform Ion Cyclotron Resonance Mass Spectrometer *16*	
1.4.6	Orbitrap Mass Analyzer *17*	
1.4.7	Ion Detection *18*	
1.5	Tandem Mass Spectrometry *18*	
1.5.1	Introduction: "Tandem-in-Time" and "Tandem-in-Space" *18*	
1.5.2	Ion Dissociation Techniques *20*	
1.5.3	Tandem Quadrupole MS–MS Instruments *21*	
1.5.4	Ion-Trap MS^n Instruments *23*	
1.5.5	Tandem TOF (TOF–TOF) Instruments *23*	
1.5.6	Hybrid Instruments (Q–TOF, Q–LIT, IT–TOF) *24*	

1.5.7   MS–MS and MSn in FT-ICR-MS   *26*
1.5.8   Orbitrap-Based Hybrid Systems   *27*
1.5.9   Ion-Mobility Spectrometry–Mass Spectrometry   *28*
1.6   Data Interpretation and Analytical Strategies   *30*
1.6.1   Data Acquisition in MS Revisited   *30*
1.6.2   Quantitative Bioanalysis and Residue Analysis   *31*
1.6.3   Identification of Small-Molecule "Known Unknowns"   *32*
1.6.4   Identification of Drug Metabolites   *33*
1.6.5   Protein Molecular Weight Determination   *37*
1.6.6   Peptide Fragmentation and Sequencing   *38*
1.6.7   General Proteomics Strategies: Top-Down, Middle-Down, Bottom-Up   *39*
1.7   Conclusion and Perspectives   *43*
               References   *43*

Part I   Direct MS Based Affinity Techniques   *55*

2   Studying Protein–Protein Interactions by Combining Native Mass Spectrometry and Chemical Cross-Linking   *57*
               Michal Sharon and Andrea Sinz
2.1   Introduction   *57*
2.2   Protein Analysis by Mass Spectrometry   *58*
2.3   Native MS   *59*
2.3.1   Instrumentation for High-mass ion Detection   *60*
2.3.2   Defining the Exact Mass of the Composing Subunits   *60*
2.3.3   Analyzing the Intact Complex   *61*
2.4   Chemical Cross-linking MS   *64*
2.4.1   Types of Cross-linkers   *64*
2.4.2   MS/MS Cleavable Cross-linkers   *66*
2.4.3   Data Analysis   *68*
2.5   Value of Combining Native MS with Chemical Cross-linking MS   *68*
2.6   Regulating the Giant   *69*
2.7   Capturing Transient Interactions   *70*
2.8   An Integrative Approach for Obtaining Low-Resolution Structures of Native Protein Complexes   *72*
2.9   Future Directions   *73*
               References   *74*

3   Native Mass Spectrometry Approaches Using Ion Mobility-Mass Spectrometry   *81*
               Frederik Lermyte, Esther Marie Martin, Albert Konijnenberg, Filip Lemière, and Frank Sobott
3.1   Introduction   *81*
3.2   Sample Preparation   *82*
3.3   Electrospray Ionization   *84*

3.4   Mass Analyzers and Tandem MS Approaches   *88*
3.5   Ion Mobility   *90*
3.6   Data Processing   *95*
3.7   Challenges and Future Perspectives   *98*
           References   *102*

Part II   LC–MS Based with Indirect Assays   *109*

**4   Methodologies for Effect-Directed Analysis: Environmental
            Applications, Food Analysis, and Drug Discovery**   *111*
           Willem Jonker, Marja Lamoree, Corine J. Houtman, and Jeroen Kool
4.1   Introduction   *111*
4.2   Principle of Traditional Effect-Directed Analysis   *113*
4.3   Sample Preparation   *113*
4.3.1   Environmental Analysis   *113*
4.3.1.1   Aqueous Samples   *113*
4.3.1.2   Biological Samples   *116*
4.3.1.3   Sediment, Soil, Suspended Matter   *120*
4.3.2   Food Analysis   *121*
4.3.2.1   Chromatography-Olfactometry   *122*
4.3.2.2   Functional Foods   *123*
4.3.3   Drug Discovery   *124*
4.3.3.1   Combinatorial Libraries and Natural Extracts   *124*
4.4   Fractionation for Bioassay Testing   *126*
4.4.1   Environmental Analysis   *126*
4.4.2   Food Analysis   *130*
4.4.3   Drug Discovery   *131*
4.5   Miscellaneous Approaches   *133*
4.6   Bioassay Testing   *136*
4.6.1   Environmental Analysis   *136*
4.6.2   Food Analysis   *140*
4.6.3   Drug Discovery   *140*
4.7   Identification and Confirmation Process   *141*
4.7.1   Instrumentation   *141*
4.7.1.1   Gas Chromatography–Mass Spectrometry   *141*
4.7.1.2   Liquid Chromatography–Mass Spectrometry   *142*
4.7.2   Data Analysis   *143*
4.7.2.1   Deconvolution   *144*
4.7.2.2   Database Spectrum Searching   *145*
4.7.2.3   Database Searching on Molecular Formula   *145*
4.7.2.4   In Silico Tools   *145*
4.7.2.5   Use of Physicochemical Properties   *146*
4.7.2.6   Applications in Nontarget Screening and EDA   *146*
4.8   Conclusion and Perspectives   *148*
           References   *149*

Contents

5   **MS Binding Assays**   *165*
        Georg Höfner and Klaus T. Wanner
5.1   Introduction   *165*
5.2   MS Binding Assays – Strategy   *167*
5.2.1   Analogies and Differences Compared to Radioligand Binding
         Assays   *167*
5.2.2   Fundamental Assay Considerations   *169*
5.2.3   Fundamental Analytical Considerations   *170*
5.3   Application of MS Binding Assays   *171*
5.3.1   MS Binding Assays for the GABA Transporter GAT1   *171*
5.3.1.1 GABA Transporters   *171*
5.3.1.2 MS Binding Assays for GAT1 – General Setup   *172*
5.3.1.3 MS Binding Assays for GAT1 – Speeding up Chromatography   *177*
5.3.1.4 MS Binding Assays for GAT1 with MALDI-MS/MS for
         Quantitation   *179*
5.3.1.5 Library Screening by Means of MS Binding Assays   *181*
5.3.2   MS Binding Assays for the Serotonin Transporter   *183*
5.3.2.1 Serotonin Transporter   *183*
5.3.2.2 MS Binding Assays for SERT – General Setup   *184*
5.3.3   MS Binding Assays Based on the Quantitation of the Nonbound
         Marker   *187*
5.3.3.1 Competitive MS Binding Assays for Dopamine D1 and D2
         Receptors   *188*
5.3.4   Other Examples Following the Concept of MS Binding Assays   *189*
5.4   Summary and Perspectives   *191*
         Acknowledgments   *192*
         References   *192*

6   **Metabolic Profiling Approaches for the Identification of Bioactive
         Metabolites in Plants**   *199*
         Emily Pipan and Angela I. Calderón
6.1   Introduction to Plant Metabolic Profiling   *199*
6.2   Sample Collection and Processing   *200*
6.3   Hyphenated Techniques   *203*
6.3.1   Liquid Chromatography–Mass Spectrometry   *203*
6.3.2   Gas Chromatography–Mass Spectrometry   *206*
6.3.3   Capillary Electrophoresis–Mass Spectrometry   *207*
6.4   Mass Spectrometry   *207*
6.4.1   Time of Flight   *208*
6.4.2   Quadrupole Mass Filter   *208*
6.4.3   Ion Traps (Orbitrap and Linear Quadrupole (LTQ))   *209*
6.4.4   Fourier Transform Mass Spectrometry   *210*
6.4.5   Ion Mobility Mass Spectrometry   *210*
6.5   Mass Spectrometric Imaging   *210*
6.5.1   MALDI-MS   *211*

6.5.2   SIMS-MS   *212*
6.5.3   DESI-MS   *212*
6.5.4   LAESI-MS   *213*
6.5.5   LDI-MS and Others for Imaging   *213*
6.6   Data Analysis   *214*
6.6.1   Data Processing   *214*
6.6.2   Data Analysis Methods   *214*
6.6.3   Databases   *215*
6.7   Future Perspectives   *216*
          References   *216*

**7   Antivenomics: A Proteomics Tool for Studying the Immunoreactivity
          of Antivenoms   *227***
          *Juan J. Calvete, José María Gutiérrez, Libia Sanz, Davinia Pla, and
          Bruno Lomonte*
7.1   Introduction   *227*
7.2   Challenge of Fighting Human Envenoming by Snakebites   *227*
7.3   Toolbox for Studying the Immunological Profile of Antivenoms   *228*
7.4   First-Generation Antivenomics   *229*
7.5   Snake Venomics   *230*
7.6   Second-Generation Antivenomics   *232*
7.7   Concluding Remarks   *236*
          Acknowledgments   *236*
          References   *236*

Part III   Direct Pre- and On-Column Coupled Techniques   *241*

**8   Frontal and Zonal Affinity Chromatography Coupled to Mass
          Spectrometry   *243***
          Nagendra S. Singh, Zhenjing Jiang, and Ruin Moaddel
8.1   Introduction   *243*
8.2   Frontal Affinity Chromatography   *244*
8.3   Staircase Method   *247*
8.4   Simultaneous Frontal Analysis of a Complex Mixture   *249*
8.5   Multiprotein Stationary Phase   *252*
8.6   Zonal Chromatography   *253*
8.7   Nonlinear Chromatography   *260*
          Acknowledgments   *265*
          References   *265*

**9   Online Affinity Assessment and Immunoaffinity Sample Pretreatment
          in Capillary Electrophoresis–Mass Spectrometry   *271***
          Rob Haselberg and Govert W. Somsen
9.1   Introduction   *271*
9.2   Capillary Electrophoresis   *272*

9.3 Affinity Capillary Electrophoresis *276*
9.3.1 Dynamic Equilibrium ACE (Fast Complexation Kinetics) *276*
9.3.2 Pre-Equilibrium ACE (Slow Complexation Kinetics) *279*
9.3.3 Kinetic ACE (Intermediate Complexation Kinetics) *280*
9.4 Immunoaffinity Capillary Electrophoresis *281*
9.5 Capillary Electrophoresis – Mass Spectrometry *283*
9.5.1 General Requirements for Effective CE – MS Coupling *283*
9.5.2 Specific Requirements for ACE – MS and IA-CE-MS *284*
9.6 Application of ACE – MS *286*
9.7 Applications of IA-CE – MS *292*
9.8 Conclusions *295*
References *296*

10 **Label-Free Biosensor Affinity Analysis Coupled to Mass Spectrometry** *299*
David Bonnel, Dora Mehn, and Gerardo R. Marchesini
10.1 Introduction to MS-Coupled Biosensor Platforms *299*
10.2 Strategies for Coupling Label-Free Analysis with Mass Spectrometry *301*
10.2.1 On-Chip Approaches *301*
10.2.1.1 SPR-MALDI-MS *301*
10.2.1.2 SPR-LDI-MS *304*
10.2.2 Off-Chip Configurations *305*
10.2.2.1 ESI-MS *305*
10.2.2.2 In Parallel Approach *305*
10.2.3 Chip Capture and Release Chromatography – Electrospray-MS *306*
10.3 New Sensor and MS Platforms, Opportunities for Integration *307*
10.3.1 Imaging Nanoplasmonics *307*
10.3.2 Evanescent Wave Silicon Waveguides *308*
10.3.3 New Trends in MS Matrix-Free Ion Sources *309*
10.3.4 Tag-Mass *310*
10.3.5 Integration *310*
References *310*

Part IV **Direct Post Column Coupled Affinity Techniques** *317*

11 **High-Resolution Screening: Post-Column Continuous-Flow Bioassays** *319*
David Falck, Wilfried M.A. Niessen, and Jeroen Kool
11.1 Introduction *319*
11.1.1 Variants of On-line Post-Column Assays Using Mass Spectrometry *321*
11.1.1.1 Mass Spectrometry as Readout for Biological Assays *323*
11.1.1.2 Structure Elucidation by Mass Spectrometry *327*
11.1.2 Targets and Analytes *328*

11.1.2.1   Targets   *328*
11.1.2.2   Analytes and Samples   *330*
11.2   The High-Resolution Screening Platform   *330*
11.2.1   Separation   *330*
11.2.2   Flow Splitting   *334*
11.2.3   Bioassay   *336*
11.2.4   MS Detection   *340*
11.3   Data Analysis   *342*
11.3.1   Differences between HRS and HTS   *342*
11.3.1.1   Influence of Shorter Incubation Times   *342*
11.3.1.2   The Assay Signal   *344*
11.3.1.3   Dilution Calculations   *346*
11.3.1.4   MS Structure Elucidation   *347*
11.3.1.5   Structure – Affinity Matching   *349*
11.3.2   Validation   *350*
11.4   Conclusions and Perspectives   *353*
11.4.1   The Relation of On-line Post-Column Assays to Other Formats   *353*
11.4.2   Trends in High-Resolution Screening   *354*
11.4.3   Conclusions   *357*
            References   *358*

12   **Conclusions**   *365*
            Jeroen Kool

            Index   *373*

List of Contributors

David Bonnel
ImaBiotech
Parc Eurasanté
885 av. Eugène Avinée
59120 Loos
France

Angela I. Calderón
Auburn University
Harrison School of Pharmacy
Department of Drug Discovery
and Development
4306 Walker Building
Auburn
AL 36849
USA

Juan J. Calvete
Institut de Biomedicina de
València-CSIC
C/Jaume Roig, 11
46010 València
Spain

David Falck
VU University Amsterdam
Faculty of Science
Amsterdam Institute for
Molecules Medicines and
Systems
De Boelelaan 1083
1081 HV Amsterdam
The Netherlands

and

Leiden University Medical
Center (LUMC)
Center for Proteomics and
Metabolomics
Division of Glycomics and
Glycoproteomics
Albinusdreef 2
2300RC Leiden
The Netherlands

José María Gutiérrez
Universidad de Costa Rica
Instituto Clodomiro Picado
Facultad de Microbiología
San José
Costa Rica

Rob Haselberg
VU University Amsterdam
Faculty of Science
Amsterdam Institute for
Molecules Medicines and
Systems
De Boelelaan 1083
1081 HV Amsterdam
The Netherlands

Georg Höfner
Ludwig-Maximilians-Universität
Department für Pharmazie
Butenandtstr. 7
81377 München
Germany

Corine Houtman
VU University Amsterdam
Faculty of Earth and Life
Sciences
Institute for Environmental
Studies
De Boelelaan 1087
1081 HV Amsterdam
The Netherlands

Zhenjing Jiang
Jinan University
Department of Pharmacy and
Guangdong Province Key
Laboratory of Pharmacodynamic
Constituents of Traditional
Chinese Medicine and New Drug
Research
Guangzhou 510632
China

Willem Jonker
VU University Amsterdam
Faculty of Science
Amsterdam Institute for
Molecules Medicines and
Systems
Division of BioAnalytical
Chemistry/BioMolecular
Analysis
De Boelelaan 1083
1081 HV Amsterdam
The Netherlands

Albert Konijnenberg
University of Antwerpen
Department of Chemistry
Biomolecular and Analytical
Mass Spectrometry
Groenenborgerlaan 171
2020 Antwerpen
Belgium

Jeroen Kool
VU University Amsterdam
Faculty of Science
Amsterdam Institute for
Molecules Medicines and
Systems
De Boelelaan 1083
1081 HV Amsterdam
The Netherlands

Marja Lamoree
VU University Amsterdam
Faculty of Earth and Life Sciences
Institute for Environmental
Studies
De Boelelaan 1087
1081 HV Amsterdam
The Netherlands

Filip Lemière
University of Antwerpen
Department of Chemistry
Biomolecular and Analytical
Mass Spectrometry
Groenenborgerlaan 171
2020 Antwerpen
Belgium

Frederik Lermyte
University of Antwerpen
Department of Chemistry
Biomolecular and Analytical
Mass Spectrometry
Groenenborgerlaan 171
2020 Antwerpen
Belgium

Bruno Lomonte
Universidad de Costa Rica
Instituto Clodomiro Picado
Facultad de Microbiología
San José
Costa Rica

Gerardo R. Marchesini
Plasmore S.r.l.
Via G. Deledda 4
21020 Ranco (Varese)
Italy

Esther Marie Martin
University of Antwerpen
Department of Chemistry
Biomolecular and Analytical
Mass Spectrometry
Groenenborgerlaan 171
2020 Antwerpen
Belgium

Dora Mehn
Fondazione Don Carlo Gnocchi
Onlus, Via Capecelatro 66
20148 Milano
Italy

Ruin Moaddel
National Institute on Aging
National Institutes of Health
Bioanalytical Chemistry and
Drug Discovery Section
Biomedical Research Center
251 Bayview Boulevard
Suite 100
Baltimore, MD 21224-6825
USA

and

Jinan University
Department of Pharmacy and
Guangdong Province Key
Laboratory of Pharmacodynamic
Constituents of Traditional
Chinese Medicine and New Drug
Research
Guangzhou 510632
China

Wilfried M.A. Niessen
hyphen MassSpec
de Wetstraat 8
2332 XT Leiden
The Netherlands

and

VU University Amsterdam
Faculty of Science
Amsterdam Institute for
Molecules Medicines and
Systems
De Boelelaan 1083
1081 HV Amsterdam
The Netherlands

Emily Pipan
Auburn University
Harrison School of Pharmacy
Department of Drug Discovery
and Development
4306 Walker Building
Auburn, AL 36849
USA

Davinia Pla
Institut de Biomedicina de
València-CSIC
C/Jaume Roig, 11
46010 València
Spain

Libia Sanz
Institut de Biomedicina de
València-CSIC
C/Jaume Roig, 11
46010 València
Spain

Michal Sharon
Weizmann Institute of Science
Department of Biological
Chemistry
234 Herzl Street
Rehovot 76100
Israel

Nagendra S. Singh
National Institute on Aging
National Institutes of Health
Bioanalytical Chemistry and
Drug Discovery Section
Biomedical Research Center
251 Bayview Boulevard
Suite 100
Baltimore, MD 21224-6825
USA

Andrea Sinz
Martin-Luther University
Halle-Wittenberg
Institute of Pharmacy
Wolfgang-Langenbeck-Straße 4
06120 Halle (Saale)
Germany

Frank Sobott
University of Antwerpen
Department of Chemistry
Biomolecular and Analytical
Mass Spectrometry
Groenenborgerlaan 171
2020 Antwerpen
Belgium

Govert W. Somsen
VU University Amsterdam
Faculty of Science
Amsterdam Institute for
Molecules Medicines and
Systems
Division of BioAnalytical
Chemistry/BioMolecular
Analysis
De Boelelaan 1083
1081 HV Amsterdam
The Netherlands

Klaus T. Wanner
Ludwig-Maximilians-Universität
Department für Pharmazie –
Zentrum für Pharmaforschung
Butenandtstraße 7
81377 München
Germany

Preface

The introduction, in 1988, of two new ionization methods for mass spectrometry (MS) has greatly changed the application areas of MS, especially in the biochemical and biological fields. Electrospray ionization (ESI) and matrix-assisted laser desorption ionization (MALDI) enabled the efficient analysis of highly polar biomolecules as well as complex biomacromolecules in an easy and user-friendly way and with excellent sensitivity. Multiple charging of proteins in ESI-MS enables the use of simple and relatively cheap mass analyzers in the analysis of peptides and proteins and even opened the way to study intact noncovalent complexes of proteins and drugs or other molecules, including protein–protein complexes. In addition, ESI provided an excellent means to perform online coupling of liquid chromatography (LC) to MS. MALDI-MS with its high level of user-friendliness and excellent sensitivity also boosted the applications of MS in studying biomacromolecules, being more recently even extended to the characterization of complete microorganisms. These developments encouraged further instrumental developments toward highly advanced (and more expensive) mass spectrometers, which provide additional possibilities in the study of biomolecules and their interactions. These new technologies opened a wide range of new application areas, of which perhaps proteomics and all derived strategies and applications belong to the most marked accomplishments. ESI-MS and MALDI-MS changed the way biochemists and biologists perform their research into molecular structures and (patho)physiological processes. Along similar lines, it also changed the ways drug discovery and development is being performed within the pharmaceutical industries. And in the slipstream of this, it changed analytical chemical research efforts in many other application areas.

The ability to study intact biomacromolecules and especially noncovalent complexes between biomolecules as well as other developments in the field, initiated by the introduction of ESI-MS and MALDI-MS, opened extensive research into the way MS can be used in the study of biomolecular interactions. Different distinct areas for analysis of bioaffinity interactions, and for analysis of biologically active molecules in general, can be recognized in this regard. These areas include precolumn-based ligand trapping followed by MS analysis, affinity chromatography following MS, and postcolumn online affinity profiling. Other methodologies are more indirect and relate to separately performed bioassays and (LC)-MS analysis, such as effect-directed analysis, metabolic profiling, and antivenomics approaches. Besides these, direct approaches without the use of chromatography are nowadays also used in several research areas. These include direct MS-based bioassays and native MS studies in which the latter looks at intact protein complexes in the gas phase. Affinity techniques for trapping proteins and protein complexes toward bottom-up proteomics analysis could also be mentioned in this regard although these techniques are actually specific sample preparation strategies for proteomics research.

With so many new approaches and technologies being introduced in this area in the past 10–15 years, it seems appropriate to compile a thorough review of the current state of the art in the analysis of biomolecular interactions by MS. That is what this book provides in 12 chapters. Apart from a tutorial chapter on MS in the beginning and a conclusive overview at the end of the book, the various chapters are grouped into four themes:

- Native MS, that is, the study of liquid-phase and gas-phase protein–protein interactions by MS and ion-mobility MS
- The use of LC–MS to study biomolecular interactions via indirect assays, as, for instance, applied in effect-directed analysis and related approaches, MS-based binding and activity assays, and other ways to study and identify bioactive molecules, for example, via metabolic profiling or antivenomics.
- Precolumn and on-column technologies to assess bioaffinity, involving frontal and zone affinity chromatography, ultrafiltration and size exclusion chromatography, affinity capillary electrophoresis, and biosensor affinity analysis coupled to MS.
- Online postcolumn continuous-flow bioassays to study bioactivity or bioaffinity of compounds after chromatographic separation.

The contributors to this book did a great job in writing very good reviews and providing beautiful artwork to illustrate the principles and applications of their specific areas within the analysis of biomolecular interactions by MS. For us, it was a pleasure to work with them in this project. We would like to thank them all for their work and for their patience with us in finalizing the final versions of the various chapters.

We hope the readers will benefit from this book, value the overview provided in the various chapters, and perhaps even get stimuli for new research areas or new approaches to perform their research, for instance, by combining ideas and approaches from various chapters of the book into new advanced technologies.

Enjoy reading and get a high affinity with MS!

August 2014

Jeroen Kool and Wilfried Niessen
VU University Amsterdam, Faculty of Science,
Amsterdam Institute for Molecules,
Medicines and Systems, Division of BioAnalytical,
Chemistry/BioMolecular Analysis
Amsterdam,
Netherlands

Abbreviations

μ	Electrophoretic mobility
2DE	Two-dimensional electrophoresis
5-HT	5-Hydroxytryptamine, serotonin
5-HT_{2A}	5-Hydroxytryptamine (serotonin) receptor subtype 2A
Ab	Antibody
ACE	Affinity capillary electrophoresis
ACE	Angiotensin converting enzyme
AChBP	Acetyl choline binding protein
Ag	Antigen
Ag–Ab	Antigen–antibody complex
AhR	Aryl hydrocarbon receptor
AMAC	Accelerated membrane assisted clean-up
APCI	Atmospheric pressure chemical ionization
API	Atmospheric pressure ionization
AR-CALUX	Androgen chemically activated luciferase expression
BGE	Background electrolyte
BGF	Bioassay guided fractionation
BGT1	Betaine-GABA transporter
BLAST	Basic local alignment search tool
BS^2G	Bis(sulfosuccinimidyl)suberate
CCT	Chaperonin containing Tcp1
CDER	Center for drug evaluation and research
CE	Capillary electrophoresis
CECs	Chemicals of emerging concern
CHCA	α-Cyano-4-hydroxy cinnamic acid
CI	Chemical ionization
CID	Collision-induced dissociation
CID-MS/MS	Collision-induced dissociation tandem mass spectrometry
CRISPR	Clustered regularly interspaced short palindromic repeat
CZE	Capillary zone electrophoresis
D_{1-5}	Dopamine receptor subtypes D1 to D5
DAD	Diode array detector
DCC	Dynamic combinatorial chemistry

DCL	Dynamic combinatorial library
DDA	Data dependent acquisition
DVB/CAR/PDMS	Divinyl-benzene/carboxen/polydimethylsiloxane
EC	Electrochemical conversion
ECD	Electron-capture dissociation
EDA	Effect-directed analysis
EI	Electron ionization
EIC	Extracted ion chromatograms
EICs	Extracted ion currents
ELSD	Evaporative light scattering detection
EOF	Electroosmotic flow
ER	Estrogen receptor
EROD	Ethoxyresorufin-*O*-deethylase
ESI	Electron spray ionization
ESI	Electrospray ionization
ESI-MS	Electrospray-ionization mass spectrometry
ETD	Electron-transfer dissociation
FA	Formic acid
FA	Frontal analysis
Fab	Fragment antigen-binding
FACCE	Frontal analysis continuous capillary electrophoresis
FDA	US Food and Drug Administration
FIA	Flow-injection analysis
FLD	Fluorescence detection
FRAP	Ferric reducing antioxidant power
FRET	Fluorescence resonance energy transfer
FWHM	Full width at half maximum
GABA	γ-Aminobutyric acid
GAT1–3	GABA transporter subtypes 1–3 (according to HUGO)
GC–MS	Gas chromatography mass spectrometry
GC-O	Gas chromatography olfactometry
GCxGC	Comprehensive two dimensional gas chromatography
GPC	Gel permeation chromatography
GPCR	G protein-coupled receptor
GSI	Global snakebite initiative
GST	Glutathione-*S*-transferase
HBH	Histidine–biotin–histidine
HDX	Hydrogen–deuterium exchange
HEK	Human embryonic kidney cells
HPLC	High performance liquid chromatography
HRS	High-resolution screening
HTLC	High-temperature liquid chromatography
HTS	High throughput screening
I.D.	Inner diameter
IA-CE	Immunoaffinity capillary electrophoresis

IC_{50}	Half maximal inhibitory concentration
ICP	Inductively coupled plasma
ICP-MS	Inductively coupled plasma MS
ID	Inner diameter
IMS	Ion mobility spectrometry
ISD	In-source decay
IT	Ion-trap MS
IT-TOF	Tandem ion-trap – time-of-flight MS
K_a	Association constant
K_d	Dissociation constant
K_d	Equilibrium dissociation constant
kDa	kilodalton (10^3 Da)
K_i	Affinity constant
k_{off}	Rate constant of complex dissociation
k_{on}	Rate constant of complex formation
L	Ligand
LC	Liquid chromatography
LC–MS	Liquid chromatography mass spectrometry
LC–MS^E	Liquid chromatography mass spectrometry in an alternating energy mode
LIF	Laser induced fluorescence
LLE	Liquid liquid extraction
LLOQ	Lower limit of quantification
MALDI	Matrix assisted laser desorption ionization
MS	Mass spectrometry/mass spectrometer
MS/MS	Tandem mass spectrometry
MTS	3-(4,5-Dimethylthiazol-2-yl)-5-(3-carboxymethoxyphenyl)-2-(4-sulfophenyl)-2H-tetrazolium
MTT	3-(4,5-Dimethyldiazol-2-yl)-2,5 diphenyl tetrazolium bromid
nAChR	Nicotinic acetylcholine receptor
NECEEM	Non-equilibrium capillary electrophoresis of equilibrium mixtures
NHS	*N*-Hydroxysuccinimide
NMR	Nuclear magnetic resonance
NMR	Nuclear magnetic resonance spectrometry
np-HPLC	Normal phase high performance liquid chromatography
p38	p38 mitogen-activated protein kinase
PAHs	Poly aromatic hydrocarbons
PDE	Phosphodiesterase
PEEK	Polyether ether ketone
PEG	Polyethylene glycol
PLE	Pressurized liquid extraction
POCIS	Polar organic chemical integrative sampler
PTFE	Polytetrafluoroethylene

QSAR	Quantitative structure–activity relationships
QTAX	Quantitative analysis of tandem affinity purified *in vivo* cross-linked protein complexes
Q-TOF	Quadrupole time-of-flight
q-TOF	Tandem quadropule – time-of-flight MS
R	Receptor
rhSHBG	Recombinant human sex hormone binding globulin
RL	Receptor–ligand complex
RP	Reverse-phase
RP-HPLC	Reverse-phase high-performance liquid chromatography
RP-LC	Reversed phase LC
rTTR	Recombinant transthyretin
SAFE	Solvent assisted flavor extraction
SAXS	Small-angle X-ray scattering
SBSE	Stir bar sorptive extraction
SD	Standard deviation
SDS-PAGE	Sodium dodecyl sulfate polyacrylamide gel electrophoresis
SEC	Size exclusion chromatography
sEH	Soluble epoxide hydrolase
SERT	Serotonin transporter
SID	Surface-induced dissociation
SILAC	Stable isotope labeling of amino acids in cell culture
SLC6	Solute carrier family 6
SPE	Solid phase extraction
SPMD	Semi permeable membrane device
SPME	Solid phase microrxtraction
SRM	Selected reaction monitoring mode
T_4	Thyroxin
T_4^*	Radiolabeled thyroxin
TAP	Tandem affinity purification
TCA	Tricyclic antidepressants
TFA	Trifluoroacetic acid
TIC	Total ion chromatograms
TIE	Toxicity identity evaluation
TLC	Thin layer chromatography
TOF	Time-of-flight
TP	Transformation product
TTR	Transthyretin
UPLC	Ultra performance liquid chromatography
UV	Ultraviolet
UV/vis	Ultra violet/visible spectroscopy
WHO	World Health Organization
YAS	Yeast androgen screen
YES	Yeast estrogen screen

1
Introduction to Mass Spectrometry, a Tutorial

Wilfried M.A. Niessen and David Falck

1.1
Introduction

In the past 30 years, mass spectrometry (MS) has undergone a spectacular development, in terms of both its technological innovation and its extent of application. On-line liquid chromatography–mass spectrometry (LC–MS) has become a routine analytical tool, important in many application areas. The introduction of electrospray ionization (ESI) and matrix-assisted laser desorption/ionization (MALDI) has enabled the MS analysis of highly polar and large molecules, including biomacromolecules. MS is based on the generation of gas-phase analyte ions, the separation of these ions according to their mass-to-charge ratio (m/z), and the detection of these ions. A wide variety of ionization techniques are available to generate analyte ions (Section 1.3). Mass analysis can be performed by six types of mass analyzers (Section 1.4), although quite frequently tandem mass spectrometers, featuring the combination of two mass analyzers, are used (Section 1.5). The data acquired by MS allow quantitative analysis of target analytes, determination of the molecular mass/weight, and/or structure elucidation or sequence determination of (unknown) analytes (Section 1.6).

This chapter provides a general introduction to MS, mainly from a functional point of view. Next to basic understanding of operating principles of ionization techniques and mass analyzers, the focus is on data interpretation and analytical strategies required in the study of biomolecular interactions using MS.

1.2
Figures of Merit

1.2.1
Introduction

An MS experiment typically consists of five steps: (i) sample introduction, (ii) analyte ionization, (iii) mass analysis, (iv) ion detection, and (v) data processing and

Analyzing Biomolecular Interactions by Mass Spectrometry, First Edition.
Edited by Jeroen Kool and Wilfried M.A. Niessen.

interpretation of the results. Sample introduction may involve individual samples or may follow (on-line) chromatographic separation. Mass analysis and ion detection require a high vacuum (pressure $\leq 10^{-5}$ mbar). Analyte ionization may take place either in high vacuum or at atmospheric pressure. In the latter case, a vacuum interface is required to transfer ions from the atmospheric-pressure ionization (API) source into the high-vacuum mass analyzer region.

In its basic operation with on-line chromatography or other forms of continuous sample introduction, the mass spectrometer continuously acquires mass spectra, that is, the instrument is operated in the full-spectrum (or full-scan) mode. This means that a three-dimensional data array is acquired, defined by three axes: time, *m*/*z*, and ion intensity (counts). This data array can be visualized in different ways (Figure 1.1). In the *total-ion chromatogram* (TIC), the sum of the ion counts in the individual mass spectra are plotted as a function of time. A *mass spectrum* represents a slice of the data array of the ion counts as a function of *m*/*z* at a particular time point. Summed, averaged, and/or background subtracted mass spectra can be generated. Mass spectra may be searched against libraries, when available, to assist in compound identification. In an *extracted-ion chromatogram* (XIC), the counts for the ion with a selected *m*/*z* are plotted as a function of time. The *m*/*z* selection window may be adapted to the resolution of the mass spectrometer. In instruments providing unit-mass resolution, the selection window in most cases is ±0.5 *m*/*z* units (u), whereas with high-resolution mass spectrometry (HRMS, see below) selection windows as small as ±10 mu can be used (narrow-window XIC). In a *base-peak chromatogram* (BPC), the ion count recorded for the most abundant ion in each spectrum is plotted as a function of time. BPCs are especially useful for peak searching in chromatograms with relatively high chemical background. More advanced tools of data processing are discussed in Section 1.6.1.

Three figures of merit are relevant: mass spectrometric resolution, mass accuracy, and the acquisition speed, that is, the time needed to acquire one spectrum (or one data point in a chromatogram).

1.2.2
Resolution

Despite the fact that mass spectrometrists readily discuss (and boast) on the resolution of their instruments, it seems that there is no unambiguous definition available. The IUPAC (International Union of Pure and Applied Chemistry) recommendations [1] and ASMS (American Society for Mass Spectrometry) guidelines [2] are different in that respect [3, 4]. Most people in the MS community define *resolution* as $m/\Delta m$, where *m* is the mass of the ion (and obviously should be read as *m*/*z*) and Δm is either the peak width (mostly measured at full-width half-maximum, FWHM) or the spacing between two equal-intensity peaks with a valley of, for instance, 10% [1]. The FWHM definition is generally used with all instruments, except sector instruments where the valley definition is used. The *resolving power* is defined as the ability to distinguish two ions with a small difference in *m*/*z* However, resolving power has also been defined as $m/\Delta m$ and the

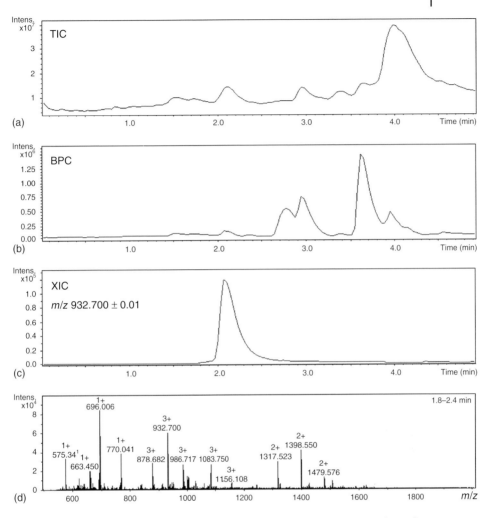

Figure 1.1 Visualization of the three-dimensional data array acquired in a full-spectrum MS experiment. (a) Total-ion chromatogram (TIC), (b) base-peak chromatogram (BPC), (c) extracted-ion chromatogram (XIC), and (d) mass spectrum. Data for an N-glycopeptide from the LC–MS analysis of a tryptic digest of a commercial immunoglobulin G (IgG) standard, analyzed using a Dionex Ultimate 3000 nano-LC coupled via ESI to a Bruker Maxis Impact Q-TOF MS in the laboratory of one of the authors (D. Falck).

resolution as the inverse of resolving power [3]. The IUPAC definition is used throughout this text.

In a simple and straightforward way, mass analyzers can be classified as either unit-mass-resolution or high-resolution instruments (see Table 1.1). For unit-mass-resolution instruments such as quadrupoles and ion traps, calculation of the resolution as $m/\Delta m$ is not very useful, as the FWHM is virtually constant over the entire mass range.

Table 1.1 Characteristics and features of different mass analyzers.

Analyzer	Resolution[a]	Mass accuracy	Full-spectrum performance[b]	Selected-ion performance[b]	Pressure (mbar)
Quadrupole	Unit-mass	±0.1	+	++	$<10^{-5}$
Ion-trap	Unit-mass	±0.1	++	+	10^{-5}
Time-of-flight	≤70 000	<3 ppm	++	−	$<10^{-7}$
Orbitrap	≤140 000	<1 ppm	++	−	$<10^{-9}$
FT-ICR	≤400 000	<1 ppm	++	−	$<10^{-9}$
Sector	≤60 000	<3 ppm	+	++	$<10^{-7}$

a) Resolution based on FWHM definition, except for sector (5% valley definition).
b) ++, instrument highly suitable for this operation; +, instrument less suitable for this operation; and −, instrument not suitable for this operation (post-acquisition XIC possible).

1.2.3
Mass Accuracy

In MS, the mass of a molecule or the *m/z* of an ion is generally expressed as a monoisotopic mass (molecular mass) or *m/z*, referring to the masses of the most abundant natural isotopes of the elements present in the ion or molecule. In chemistry, the average mass or molecular weight is used, based on the average atomic masses of the elements present in the molecule. The *exact mass* (or better *m/z*) of an ion is its calculated mass, that is, its theoretical mass. In this respect, the charge state of the ion is relevant, because the electron mass (0.55 mDa) may not be negligible. The *accurate mass* (or better *m/z*) of an ion is its experimentally determined mass, measured with an appropriate degree of accuracy and precision. The accurate mass is the experimental approximation of the exact mass. The *nominal mass* (or better *m/z*) is the mass of a molecule or an ion calculated using integer values for the masses of the most abundant isotopes of the elements present in the molecule or ion. The *mass defect* is the difference between the exact mass and the nominal mass of ion or molecule [1, 5].

The achievable mass accuracy in practice depends on the resolution of the mass analyzer and the quality and stability of the calibration of the *m/z* axis. An instrument providing unit-mass resolution generally allows *m/z* determination for single-charge ions with an accuracy of ±0.1 u (nominal mass determination). In HRMS, the *mass accuracy* is generally expressed either as an absolute mass error (accurate mass − exact mass, in mu) or as a relative error (in ppm), calculated from

$$\frac{(\text{accurate mass} - \text{exact mass})}{(\text{exact mass})} \times 10^6$$

In HRMS of small molecules, the error in *m/z* determination will typically be in the third decimal place (accurate mass determination).

From the accurate *m/z* of an ion, one can use software tools to calculate its possible elemental compositions. The number of hits from such a calculation obviously depends on the m/z value, the number of elements considered, and the mass accuracy achieved [6]. The number of hits may also be reduced by taking an accurately measured isotope pattern of the ion into consideration [7, 8]. For a given ion with m/z M, the relative abundances of the ions with m/z M+1, M+2, and M+3 reveal the presence (or absence) and even the number of specific elements, for example, Cl, Br, and S from the M+2 ion. For small molecules (<1 kDa), the maximum number of carbon atoms in the molecule can be estimated by dividing the relative abundance (in percent) of the M+1 peak by 1.1. Ultra-HRMS instruments have additional possibilities to derive elemental composition, as they can even separate the contributions of different atoms to the M+2 isotope peak. This is illustrated for an unknown compound with $C_{13}H_{24}N_3O_6S_2$ in an onion bulb in Figure 1.2 (see also [9]).

As discussed in Section 1.6.6, mass accuracy also has a distinct influence on the ease and quality of protein identification from peptide-mass fingerprints or peptide-sequence analysis approaches.

1.2.4
General Data Acquisition in MS

The general mode of data acquisition of a mass spectrometer is the full-spectrum (or full-scan) mode. In this mode, mass spectra are continuously acquired between

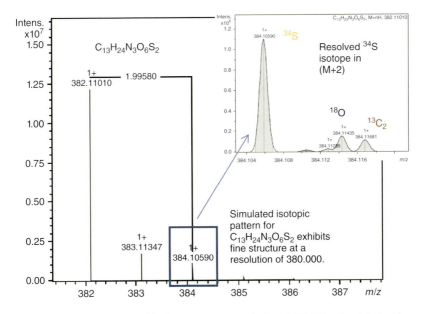

Figure 1.2 Demonstration of high-resolution mass spectrometry. Simulated isotopic pattern for an unknown compound with $C_{13}H_{24}N_3O_6S_2$ in an onion bulb with isotopic fine structure exhibited at a resolution of 380 000. (Reprinted with permission from Prof. Kazuki Saito (RIKEN Plant Science Center, Yokohama, Japan) and Bruker Daltonics Application Note # LC-MS 85, ©2013, Bruker Daltonics, 1822187.)

a low m/z and a high m/z within a preset period of time (mostly ≤1 s). Obviously, the information content of the spectrum depends on (i) the selected ionization technique, (ii) the resolution of the instrument, and (iii) data system parameters. The mass spectra are acquired in continuous or profile mode, that is, with a number of data points per m/z value. For unit-mass-resolution instruments, ~10 data points per m/z suffice, whereas in HRMS far more data points per m/z are required to provide the appropriate resolution and mass accuracy. Either the profile data are saved by the data system, eventually after some data reduction such as apodization to reduce the data file size (see, e.g., [10]), or centroiding is performed, where only a weighted average of the mass peak is saved [4]. The latter greatly reduces the data file size. Post-acquisition data processing tools may require either profile or centroid data.

Some mass analyzers (see Table 1.1) can also acquire data in the selected-ion mode, which means that the mass analyzer is programmed to select a particular m/z for transmission to the detector during a preset period (the so-called dwell time, typically 5–200 ms) and to subsequently jump to other preselected m/z values; after monitoring all selected m/z values, the same function is repeated for some time, for example, during (part of) the chromatographic run time. Thus, compared to the full-spectrum mode, the MS has a longer measurement time of the selected ion, and thus provides enhanced signal-to-noise ratio (S/N). The data can be displayed in terms of XICs. This acquisition mode is especially applied in targeted quantitative analysis. With HRMS instruments not capable of a selected-ion mode, improved S/N and targeted quantitative analysis can be achieved post-acquisition in narrow-window XICs (see Section 1.2.1).

For a proper understanding of the possibilities and limitations of MS, one should be aware of the fact that a mass spectrometer can generally perform only one experiment at a time. However, various experiments can be performed consecutively. Functions may be defined to perform various experiments repeatedly. As outlined in Section 1.6.1, decisions for the next experiment may be based on the data acquired in the previous experiment (data-dependent acquisition, DDA). The time required for individual MS experiments very much depends on the type of instrument used (and its purchase date). Because of the huge progress in faster electronics, modern instruments can perform faster than older instruments.

1.3
Analyte Ionization

1.3.1
Introduction

More than 50 analyte ionization techniques are available for MS. An ionization technique has to generate gas-phase analyte ions, either in (high) vacuum or transferable from atmospheric pressure into high vacuum, to enable subsequent mass analysis. The various ionization techniques can be classified in different ways.

Analyte ionization techniques can be classified based on the physical state of the analyte molecules: (i) gas or vapor, (ii) liquid or in solution, or (iii) solid or dry on a target. Traditional ionization techniques such as electron ionization (EI) and chemical ionization (CI) are examples of gas-phase ionization techniques, and are thus frequently used in on-line gas chromatography–mass spectrometry (GC–MS). ESI, which is extensively used in on-line LC–MS and peptide and protein analysis, is a liquid-phase ionization technique, whereas MALDI and desorption electrospray ionization (DESI) are examples of solid-phase or surface ionization techniques. Gas-phase ionization requires either gas-phase samples or evaporation of the analytes before ionization. Surface ionization techniques are frequently so-called energy-sudden techniques [11], in which intense localized energy is applied to the sample, for example, by means of a laser pulse, to simultaneously ionize and transfer the ion from the solid phase to the gas phase. In liquid-phase ionization, the sample solution, for example, the LC (liquid chromatography) mobile phase, is nebulized into small droplets, from which gas-phase analyte ions are generated, for example, in ESI.

A second classification is based on the amount of internal energy that is put into the molecule on generation of the ion. In a hard ionization technique such as EI, typically a few electron volts internal energy is transferred to the molecular ion, $M^{+\bullet}$. This internal energy results in rapid in-source compound-specific fragmentation. The mixture of intact molecular ions and fragment ions is subsequently mass-analyzed. In a soft-ionization technique, hardly any internal energy is transferred to the ion during the ionization process. Often, a protonated or deprotonated molecule, $[M+H]^+$ or $[M-H]^-$, is generated and no in-source fragmentation occurs. Some ionization techniques allow some control over the amount of energy deposited in the ion on its formation.

A third classification is based on the type of primary ions generated in the ionization process. The molecular ion, $M^{+\bullet}$, generated in EI, is an odd-electron ion ($OE^{+\bullet}$), whereas the protonated or deprotonated molecule, $[M+H]^+$ or $[M-H]^-$, generated in ESI or MALDI, are either positive- or negative-charge even-electron ions (EE^+ or EE^-). In this context, the nitrogen rule is important, which states that a molecule, a $M^{+\bullet}$, or any other $OE^{+\bullet}$ with an odd mass or m/z should contain an odd number of N atoms, whereas a (single-charge) EE^+ with an odd m/z contains an even number of N atoms. For a known elemental composition, the nitrogen rule allows discriminating between $OE^{+\bullet}$ and EE^+ ions.

Another useful tool is the double-bond equivalent (DBE), "degree of unsaturation," or "ring double bond" (RDB) parameter. The DBE can be calculated from the elemental composition of the molecule of ion, using the equation:

$$DBE = 1 + C - \frac{1}{2}(H + F + Cl + Br + I) + \frac{1}{2}(N + P)$$

The DBE is a measure of the number of unsaturations in the molecule, that is, the number of rings and/or double bonds. DBE is an integer number for a molecule or an $OE^{+\bullet}$ and a number ending at 0.5 for an EE^+ ion. For molecules with P and S atoms, the DBE does not consider the double bonds in, for instance, a phosphate $((RO)_3P=O)$, a phosphorothioate $((RO)_3P=S)$, a sulfoxide $(S=O)$, or a sulfone (SO_2) as a double bond.

1.3.2
Electrospray Ionization

In ESI, a solution, for example, the mobile phase from an LC column, is nebulized into an API source as a result of a strong electric field, eventually assisted by N_2 as a nebulizing gas and heating. Small, highly charged droplets $(1-10\,\mu m)$ are generated. Gas-phase ions are generated in the process of droplet evaporation and field-induced electrohydrodynamic disintegration of the droplets [12]. The gas-vapor mixture (N_2 and mobile-phase solvents) with analyte ions is sampled from the ion source into the vacuum interface. Desolvation and collisional cooling of the ions occur when they move through the vacuum interface toward the high-vacuum mass analyzer [13]. In most cases, either $[M+H]^+$ or $[M-H]^-$ is generated, depending on the operating polarity, but other adduct ions such as $[M+Na]^+$ or $[M+CH_3COO]^-$ may be generated as well (or instead).

In terms of instrumentation, significant improvements in the performance of ESI interfaces have been achieved. In the ion source itself, the orthogonal rather than axial positioning of the electrospray needle is important to reduce contamination of the ion-sampling orifice. In the vacuum interface, ion transmission has been improved by the use of continuously more advanced RF-only (radiofrequency) ion focusing and transport devices, that is, next to RF-only quadrupoles (see also Section 1.4.2), hexapoles, and octapoles and also the implementation of ion funnels [14] and traveling-wave stacked-ring ion guides [15]. Whereas in most analytical applications of ESI-MS low pressure is pursued in the vacuum interface, the preservation of protein complexes in native-MS is best achieved at somewhat higher pressures in this region [16]; valves have been implemented to allow pressure adjustments in native-MS experiments.

The ionization mechanism of ESI is not fully understood [12, 17, 18]. The two prevailing models are the *charge-residue model* of Dole [19] and the *ion-evaporation model* of Iribarne and Thomson [20, 21]. Both models assume that analyte molecules are present in solution as preformed ions, which can, for instance, be achieved by choosing an appropriate *pH* of the solution or the mobile phase. According to the *charge-residue model*, the sequence of solvent evaporation and electrohydrodynamic droplet disintegration proceeds until the microdroplets contain only one preformed analyte ion per droplet. By evaporation of the solvent, the preformed analyte ion is released to the gas phase. According to the *ion-evaporation model*, gas-phase ions are generated from the highly charged microdroplets, because the local field strength is high enough for preformed ions to be emitted into the gas phase. Although the two models are to some extent complementary, the relative importance of either mechanism in the actual ion production of a particular analyte is difficult to decide. Smith and Light-Wahl [18] discussed whether the preservation of noncovalent associates of proteins and drugs in ESI is reasonable within the context of proposed ionization mechanisms for ESI. They argue that the stripping of noncovalently associated solvent molecules from the highly desolvated multiprotonated molecules can be achieved without influencing other noncovalent drug–protein associations. They suggested that the initial highly charged droplets in the 1-µm-I.D. range repeatedly disintegrate to generate nanodroplets in the 10-nm-I.D. range. The latter further shrink by evaporation to yield the ions detected in MS.

Given the importance of droplet evaporation during ESI, the generation of smaller droplets is more favorable in term of sensitivity and the ability to preserve noncovalent molecular associates. This can be achieved by using nanoelectrospray ionization (nESI), where the analyte is sprayed, for example, from a gold-coated fused-silica capillary with a tip diameter of $1-5\,\mu m$ rather than from the 100 to 150-µm tips that are used in conventional (pneumatically assisted) ESI. Flow rates as low as $20\,nl\,min^{-1}$ can be nebulized [22]. Thus, gentler operating conditions (temperature, gas flows, needle voltage) can be achieved. nESI is extensively used in the analysis of biomacromolecules, that is, in native MS, where intact protein complexes are studied [16], as well as in combination with nano-LC for proteomics studies [23, 24]. Integrated chip-based nano-LC-nESI devices have also been developed [25].

ESI enables the soft ionization of highly labile and nonvolatile compounds such as (oligo)nucleotides, (oligo)saccharides, peptides, and proteins without significant fragmentation. In the analysis of biomacromolecules, an ion envelope of multiple-charge ions, $[M+nH]^{n+}$ or $[M-nH]^{n-}$, is generated (see Figure 1.3), from which the molecular weight of the molecule can be accurately calculated (better than 0.01%) using software procedures (see Section 1.6.5).

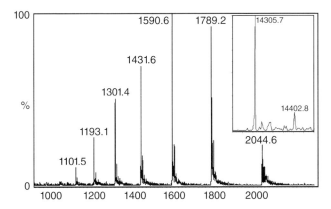

Figure 1.3 Electrospray ionization mass spectrum of hen egg lysozyme, showing an ion envelope of multiple-charge ions $[M+nH]^{n+}$. Inset: Deconvoluted or transformed spectrum. (Reprinted after modification with permission of Prof. Alison Ashcroft (Biomolecular Mass Spectrometry Group at University of Leeds, UK) from *http://www.personal.leeds.ac.uk/~fbsmaspe/ tutorial.html.*)

1.3.3
Matrix-Assisted Laser Desorption Ionization

Next to ESI, MALDI is the most important ionization technique in the analysis of biomacromolecules. MALDI was introduced in 1988 [26, 27]. In a typical MALDI experiment, 0.3–1 µl of an aqueous analyte solution is mixed with 0.5–1 µl of a ~5 mM solution of an appropriate matrix, for example, sinapinic acid, 2,5-dihydroxybenzoic acid, or α-cyano-4-hydroxycinnamic acid, in 50% aqueous acetonitrile containing 0.1% trifluoroacetic acid, and then deposited onto a metal target. On drying, cocrystallization of matrix and analyte molecules takes place. When these crystals are laser-bombarded with photons that fit the absorption maximum of the matrix, for example, with 337 nm from a N_2 laser for the matrices mentioned, gas-phase analyte ions are generated in the selvedge, which can be mass-analyzed by a time-of-flight mass spectrometer (TOF-MS) [28, 29]. Unlike ESI, where ion envelopes of multiple-charge ions are generated, mostly single-charge protonated molecules $[M+H]^+$ are generated in MALDI-MS, together with less abundant $[M+2H]^{2+}$ and $[2M+H]^+$ ions.

MALDI-TOF-MS is widely applied especially in the analysis of proteins, peptides, and oligosaccharides [28–30]. It is used in bottom-up protein identification using peptide-map fingerprinting approaches (Section 1.6.6). Currently, MALDI-MS plays an important role in two emerging application areas of MS: imaging MS [31] and identification of bacteria and microbial fingerprinting [32].

Surface-enhanced laser desorption ionization (SELDI-TOF-MS) is a variation of MALDI. Instead of an inert metal target, a modified target is applied to achieve biochemical affinity with the analyte molecules [33]. In SELDI, the protein mixture is spotted onto the surface with a specific (bio)chemical functionality, for example, cation- or anion-exchange materials, hydrophobic materials, or materials with immobilized metal affinity, lectin, or even protein or antibody affinity. Specific proteins in the mixture bind to the surface, while others can be removed by washing. Thus, on-target biomolecular interactions are used as part of the measurement strategy. The specific binding to the SELDI target acts as a sample pretreatment and/or analyte isolation step. After washing, a matrix is applied and the experiments proceeds as in MALDI-TOF-MS [33]. SELDI-TOF-MS is currently extensively used for clinical diagnostics and in clinical biomarker discovery studies.

1.3.4
Other Ionization Methods

Next to EI, ESI, and MALDI, already discussed, there is a wide variety of other analyte ionization techniques. In coupling of LC and MS, atmospheric-pressure chemical ionization (APCI) is an important alternative to ESI. In APCI, the LC mobile phase is nebulized by pneumatic nebulization. The aerosol is evaporated in a heated zone of the interface probe, enabling soft desolvation of analyte molecules. The gas-phase CI is initiated by a corona discharge needle, which ionizes the mobile-phase constituents, which in turn ionize the analyte molecules by gas-phase ion–molecule reactions. Again, mainly $[M+H]^+$ or $[M-H]^-$ ions are generated with little internal energy. APCI is generally limited to molecules with masses below 1 kDa, but can ionize somewhat less polar analytes than ESI.

In the past decade, a large number of, sometimes closely related, atmospheric-pressure surface ionization techniques have been introduced [34]. DESI may serve as an example. If a high-velocity spray of charged microdroplets from a (pneumatically assisted) electrospray needle is directed at a surface, mounted in front of the ion-sampling orifice of an API source, gas-phase ions from the surface material or surface constituents can be observed by MS [35]. In this way, analytes may be studied at surfaces without extensive sample pretreatment, for example, to analyze drugs of abuse in tablets. The introduction of such surface ionization techniques also opens possibilities to perform chemical imaging of surfaces such as thin-layer chromatography plates and tissue sections [36].

In paperspray ionization (PSI), biomolecules are ionized from a paper tip emerged with nonpolar solvents such as hexane or toluene, placed in the electric field close to the ion-sampling orifice of an API source [37]. Although the polar biomolecules are largely insoluble in these solvents, $[M+H]^+$, $[M+Na]^+$, or $[M-H]^-$ ions are observed. Meanwhile, the technique has found several applications, including the analysis of noncovalent protein complexes [38].

1.3.5
Solvent and Sample Compatibility Issues

In both ESI and MALDI, the ionization efficiency of a particular analyte may be influenced by the solvent composition, but also by (coeluting, when LC separation is applied) sample constituents. Either ionization enhancement or ionization suppression may take place. If the effect is due to sample constituents, it is mostly called a *matrix effect* [39]. If not dealt with properly, matrix effects can have highly detrimental effects on accuracy and precision in routine quantitative analysis. However, matrix effects, especially severe ionization suppression, may also have distinct influence on the analyte response of constituents of complex mixtures during qualitative analysis. High salt concentrations, nonvolatile salts, surface-active components, for example, phospholipids in blood-related samples, and proteins are known for their matrix effects in small-molecule quantitative bioanalysis. Nonvolatile compounds, be it sample constituents or solvent additives, may result in ionization suppression, in reducing response or S/N by extensive formation of background ions at every m/z, formation of alkali metal ion adducts or H^+/Na^+-exchange products, and/or ion source contamination.

Optimization of the sample composition and/or mobile-phase composition, if LC separation is involved, is important to achieve the best response and selectivity in MS. Systematic studies on the influence of additives on ESI performance have been reported [40, 41] and give an initial idea on which additives can be applied and which ones should be avoided. With respect to buffers, preferably only volatile buffer constituents are permitted, that is, ammonium salts of formic, acetic, or carbonic acid. Borate, citrate, and phosphate buffers, frequently applied in LC, but also many of the buffer systems such as TRIS and HEPES, frequently used in biochemistry, are not compatible with ESI-MS. The same holds for components frequently applied to help solubilizing proteins, that is, detergents such as TWEEN and sodium dodecylsulfate (SDS) and chaotropic agents such as urea and guanidine·HCl. This limits the applicability of ESI-MS in the analysis of proteins that are difficult to solubilize, for example, membrane proteins. Although MALDI-MS is considered less impacted by the presence of such additives on the target, various reports indicate that severe ionization suppression may occur in MALDI-MS as well [42, 43]. This is especially important as MALDI-MS approaches are less likely to be preceded by a separation step.

High concentrations of proteins should not be introduced in ESI-MS. This means that on analysis of undigested blood-related samples (whole blood, plasma, serum), care must be taken. In most applications of intact protein analysis by MS, either the concentrations applied are relatively low or the amount introduced is low due to the use of nESI flow rates.

1.4
Mass Analyzer Building Blocks

1.4.1
Introduction

In this section, the different mass analyzers are briefly discussed, with special attention on their possibilities and limitations. A summarizing overview on the available mass analyzers is provided in Table 1.1. Most of them may be used as stand-alone systems, when equipped with an ion source of choice, a detection system, and a data system. Alternatively, they may be combined into tandem MS (MS–MS) systems (see Section 1.5).

Any discussion on mass analyzers should perhaps start with introducing the sector instrument, which is historically at the basis of all MS developments. Ions with mass m and z elementary charges e are accelerated with a voltage V into a magnetic field B with a path with a radius of curvature r. One can derive the equation $m/z = B^2 r^2 e / 2V$, which indicates that the separation of ions with different m/z can be achieved in three different ways: by variation of the radius of curvature ions with different m/z are separated in space, while by variation of either B or V ions of different m/z are separated in time, that is, they can be detected one after another by a single-point detector at a fixed position behind a slit [44]. Better performance of the sector instrument in terms of mass resolution is achieved by combining the magnetic sector with an electrostatic analyzer, resulting in a double-focusing instrument, which provides high resolution and accurate mass determination. Until the mid-1990s, the sector instrument was the instrument of choice for HRMS and accurate mass determination. Since then, it has been rapidly replaced by alternatives that are less expensive and much easier to operate, especially in combination with API sources.

1.4.2
Quadrupole Mass Analyzer

A quadrupole mass analyzer or mass filter consists of four hyperbolic- or circular-shaped rods that are accurately positioned parallel in a radial array (Figure 1.4a). Opposite rods are charged by either a positive or a negative direct-current (DC) potential, at which an alternating-current (AC) potential in the radiofrequency region (megahertz, thus indicated as RF) is superimposed. At a given DC/RF combination, only the ions of a particular *m/z* show a stable trajectory and are transmitted to the detector, while ions with unstable trajectories do not pass the mass filter, because the amplitude of their oscillation becomes infinite. Thus, the quadrupole acts as a variable band pass filter [46]. By changing DC and RF in time, usually at a fixed ratio, ions with different *m/z* values can be transmitted to the detector one after another.

The quadrupole mass filter can be operated in four modes. (i) It can be applied in full-spectrum mode by scanning DC and RF in a fixed ratio, providing generally unit-mass resolution and nominal monoisotopic mass determination with great ease of operation, versatility, fast scanning, and limited costs. Scan speeds as high as $10\,000\,\mathrm{u\,s^{-1}}$ can be achieved. (ii) Alternatively, it can be applied in selected-ion monitoring (SIM) mode, dwelling on selected *m/z* values, and capable of rapidly switching (within ≤5 ms) between different *m/z* values. In SIM mode, significantly improved S/N can be achieved, making the SIM mode of a quadrupole ideal for routine targeted quantitative analysis. (iii) In RF-only mode, the quadrupole can be used as an ion-transport and focusing device. As such, RF-only quadrupole and related hexapole or octapole devices have been used in vacuum interfaces of API-MS systems and as collision cells and/or ion-transport devices in MS–MS instruments (Section 1.5.2). (iv) Finally, a quadrupole mass analyzer can be applied as a linear ion trap (LIT), providing similar features as the conventional three-dimensional ion traps (see Section 1.4.3) [47–49]. The majority of mass spectrometers installed and used are based on quadrupole mass analyzer technology.

1.4.3
Ion-Trap Mass Analyzer

A typical (three-dimensional quadrupole) ion trap consists of a cylindrical ring electrode and two end-cap electrodes (Figure 1.4b) [50, 51]. The end-cap electrodes contain holes for the introduction of ions from an external ion source and for the ejection of ions toward an external detector. A He bath gas (~1 mbar) is

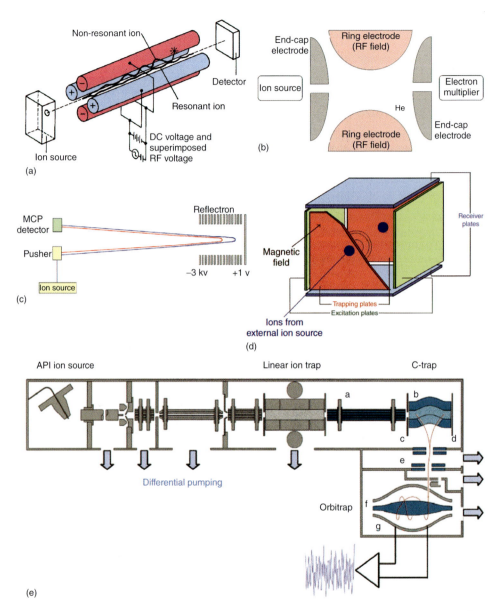

Figure 1.4 Schematic diagrams of various mass analyzers, with (a) quadrupole mass analyzer, (b) ion-trap mass analyzer, (c) orthogonal-acceleration reflectron time-of-flight mass analyzer, (d) cell of a Fourier-transform ion-cyclotron resonance mass spectrometer, and (e) linear-ion-trap–orbitrap hybrid mass spectrometer. (All diagrams from the authors' collection, except (e) which is reprinted with permission from Ref. [45]. © 2006, American Chemical Society.)

used to stabilize the ion trajectories in the trap. The basic mass analysis process consists of two steps, performed consecutively in time: (i) injection of ions by means of an ion injection pulse of variable duration and storage of the ions in the trap by application of an appropriate low RF voltage to the ring electrode and (ii) ramping the RF voltage at the ring electrode to consecutively eject ions with different m/z values from the trap toward the external detector (resonant ion ejection) [50, 51]. As too high numbers of ions in the ion trap adversely influences mass resolution and accuracy due to space-charge effects, the number of ions that can be stored in the trap is limited. Software control of the duration of the ion injection pulse from the external ion source with the ion current at the time has been developed [51]. Ion ejection and subsequent detection can be achieved with unit-mass resolution, or at enhanced resolution by slowing down the scan rate.

A typical feature of an ion-trap MS is its excellent full-spectrum sensitivity, resulting from the accumulation of ions in the first step. The system can be operated in SIM mode, but the gain achieved is far less than in a quadrupole. The resolution achieved depends on the scan speed and the age of the instrument. Older ion traps provide peak widths (FWHM) of 0.2 u at a scan speed of ~300 u s^{-1}, unit-mass resolution (FWHM ~0.7 u) at 5500 u s^{-1}, and degraded resolution (FWHM of 3.0 u at 55 000 u s^{-1}), whereas more recently introduced systems provide better resolution at higher scan speeds, for example, FWHM of 0.1 u at 4600 u s^{-1} and of 0.58 u at 52 000 u s^{-1}. A FWHM of 0.1 u enables almost baseline resolution for [4+] ions.

More recently, linear two-dimensional ion traps (LIT) were introduced as an alternative to three-dimensional ion traps [47–49]. Because a LIT is less prone to space-charging effects, a higher number of ions can be accumulated, and enhanced sensitivity (~60×) can be achieved. LITs are extensively used in hybrid MS–MS instruments (see Section 1.5.6), but stand-alone versions of a LIT have been introduced as well, thus competing with the three-dimensional ion traps. A dual-pressure two-stage LIT has been reported as well: the first high-pressure ion trap serves to capture, select, and fragment ions, whereas the second low-pressure ion trap is used to perform fast scanning of product ions, eventually at enhanced resolution [52].

1.4.4
Time-of-Flight Mass Analyzer

A time-of-flight (TOF) instrument consists of a pulsed ion source, an accelerating grid, a field-free flight tube, and a detector [53, 54]. The flight time t needed by the ions with a particular m/z, accelerated by a potential V, to reach the detector placed at a distance d, can be calculated from $t = d \times \sqrt{\{m/(2zeV)\}}$. Pulsing of the ion source is required to avoid the simultaneous arrival of ions of different m/z at the detector. In MALDI-TOF-MS experiments, the pulse rate of the source is directly related to the pulse rate of the laser (typically <1 kHz). One of the benefits is, in principle, the unlimited mass range of the TOF-MS. When combined with continuous ion sources, such as ESI, much higher pulse frequencies are applied (20–50 kHz). The spectra from multiple pulses are accumulated, providing enhanced spectrum quality by averaging random noise. In LC–MS, spectra acquisition rates up to 50 spectra per second have been reported.

The resolution of a TOF-MS is limited by the speed of the detection and acquisition electronics, which nowadays is hardly a limiting factor, and by the initial ion kinetic energy spread of the ions. Delayed extraction [55], orthogonal acceleration [53], and especially the use of a reflectron [53, 56] are powerful tools to reduce the deteriorating effect of the ion kinetic energy spread on the resolution. The reflectron, consisting of a series of equally spaced grid or ring electrodes connected to a resistive network, creates a homogenous or curved retarding field that acts like an ion mirror. Assume that two ions with the same *m/z* but slightly different kinetic energy enter the reflectron. The ion with the higher kinetic energy penetrates deeper into the field and thus has a slightly longer flight path, and in effect reaches the detector more simultaneously with the ions with the lower kinetic energy (Figure 1.4c). Since its introduction, more advanced dual-stage reflectrons have been produced. With a reflectron-TOF in combination with either delayed extraction (in MALDI) or orthogonal acceleration (with continuous ion sources, such as ESI), a mass resolution in excess of 15 000 (FWHM) can be readily achieved, enabling accurate mass determination (<3 ppm) [57]. Currently, commercial TOF-MS systems are available with a mass resolution in excess of 70 000 (FWHM).

1.4.5
Fourier Transform Ion Cyclotron Resonance Mass Spectrometer

A Fourier-transform ion cyclotron resonance mass spectrometer (FT-ICR-MS) can be considered an ion-trap system, where the ions are trapped in a magnetic rather than in a quadrupole electric field. The ICR (ion-cyclotron resonance) cell is positioned in a strong magnetic field *B* (up to 15 T). The cylindrical or cubic ICR cell consists of two opposite trapping plates, two opposite excitation plates, and two opposite receiver plates (Figure 1.4d). The ions with *m/z* describe circles with radius *r* perpendicular to the magnetic field lines. This results in a cyclotron frequency $\omega_c = 2\pi f = v/r = Bez/m$, where *f* is the frequency in Hertz. The cyclotron frequency is thus inversely proportional to the *m/z* value. The ions, trapped in their cyclotron motion in the cell, are excited by means of an RF pulse at the excitation plates. As a result, the radius of the cyclotron motion increases and ions with the same *m/z* values start moving in phase. The coherent movement of the ions generates an image current at the receiver plates. The image current signal decays in time, because the coherency of the ion movement is disturbed in time. The time-domain signal from the receiver plates contains all frequency information of the ions present in the cell. By Fourier transformation, the time-domain signal is transformed into a frequency-domain signal, which can then be transformed into a mass spectrum [10, 58]. As ion trapping is involved in FT-ICR, concerns with space-charging effect are valid as well.

FT-ICR-MS instruments provide extremely high (mass-dependent) resolution, typically in excess of 10^5 (FWHM), accurate mass determination (≤ 1 ppm), and a dynamic range of five orders of magnitude. The resolution increases with measurement time, and longer measurement times are only possible if extreme high vacuum ($\sim 10^{-9}$ mbar) is achieved in the ICR cell. Higher spectrum acquisition rates can be achieved in instruments with a higher magnetic field strength, that is, the same high-resolution spectrum can be achieved in a shorter time. At present, commercial FT-ICR-MS systems are available that provide a resolution of $\sim 650\,000$ (at m/z 400, FWHM) at 1 spectra per second. For many years, FT-ICR-MS was primarily used in fundamental studies of gas-phase ion-molecule reactions. However, because of its high-resolution capabilities, FT-ICR-MS in combination with ESI is an ideal tool for the characterization of large biomolecules [59]. At present, FT-ICR-MS plays an important role in top-down strategies to characterize proteins [60, 61] (see Section 1.6.6).

1.4.6
Orbitrap Mass Analyzer

Similar to FT-ICR-MS, the acquisition of mass spectra in an orbitrap MS is based on the Fourier transformation of image currents of trapped ions. However, in the orbitrap, no magnetic field is involved. The ions perform an axial oscillation while rotating around a cylindrical inner electrode. The image currents are detected by the two outer electrodes (Figure 1.4e). The mass spectra are generated by Fourier transformation of the time-domain signals into the frequency-domain signals. The m/z value is inversely proportional to the square of the frequency. Orbitrap instruments, which have been introduced only recently [10, 45, 62], allow ultra-high-resolution measurements (in excess of 10^5, FWHM) at relatively high speed. Although a stand-alone version of the orbitrap has been produced [63], in most cases hybrid systems are applied (see Section 1.5.8).

An important practical aspect to orbitrap MS performance is the adequate delivery of ions into the orbitrap. In commercial systems, this is done in a pulsed way by means of a C-trap, which is essentially a curved high-pressure quadrupole capable of trapping ions and sending them as a concise ion package into the orbitrap [45, 62]. A schematic diagram of the instrumental setup is shown in Figure 1.4e.

Similar to FT-ICR-MS, the mass resolution of an orbitrap improves with longer measurement times. However, significant progress has been made recently in improving the acquisition rates. Whereas initial commercially available orbitrap systems needed ~ 1.6 s to achieve a resolution of $100\,000$ (at m/z 200, FWHM), more recent systems can achieve a resolution of $140\,000$ (at m/z 200, FWHM) in 1 s, and an orbitrap-based system enabling a resolution of $450\,000$ (at m/z 200,

FWHM) has been described as well. This enables accurate mass determination with an accuracy within 1 ppm.

1.4.7
Ion Detection

Different types of ion detection devices are in use. All ion detection systems must be backed by sufficiently fast electronics, including analog-to-digital converters (ADCs), to enable the high-speed data acquisition required in MS [44].

The most widely applied detection system is an electron multiplier, based on the repeated emissions of secondary electrons, resulting from the repeated collisions of energetic particles at a suitable surface. The electron multiplier may be either of the discrete dynode type or of the continuous dynode type [64]. The typical gain of an electron multiplier is 10^6. The electron multiplier is used in combination with quadrupole, ion-trap, and sector instruments. They have limited lifetime ($\sim 1-2$ years). A conversion dynode, held at a high potential ($5-20$ kV), is positioned in front of the multiplier, to enable the detection of negative ions and to increase the signal intensity of ions, especially in the high-mass region.

With TOF instruments, microchannel plate (MCP) detectors are applied, as they are more suitable for ion detection when the ion beam is more spread in space. An MCP is an array of miniature electron multipliers oriented parallel to one another, often with a small angle to the surface [65]. In order to generate a spectrum from ion arrival events in TOF instruments, either a time-to-digital converter (TDC) or an ADC is applied. TDCs provide excellent time resolution and low random noise, but do not discriminate in the intensity of the pulse. High ion densities may lead to saturation effects. In an ADC, the integrated circuit chip receives a time-dependent signal and generates a typically 10-bit digital output: both arrival time and the number of colliding ions are recorded. ADCs can provide $1-4$ GHz time resolution and discriminate 1024 different ion intensity levels.

In FT-ICR-MS and orbitrap MS systems, ion detection is based on the detection of high-frequency image currents if the ions move coherently, as described in Sections 1.4.5 and 1.4.6 [58]. The signals of all ions, that is, with different m/z values, are detected simultaneously.

1.5
Tandem Mass Spectrometry

1.5.1
Introduction: "Tandem-in-Time" and "Tandem-in-Space"

In tandem mass spectrometry (MS−MS), two mass analyzers are combined in series (in time or space, see below) with a reaction chamber in between. The m/z values of ions are measured before and after the reaction within the reaction chamber. In most cases, a change in mass and thus in m/z is involved, although

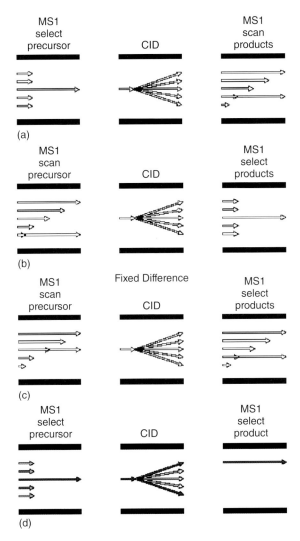

Figure 1.5 Analysis modes of an MS–MS instrument ("tandem-in-space"), with (a) product-ion analysis, (b) precursor-ion analysis, (c) neutral-loss analysis, and (d) selected-reaction monitoring mode.

a change in charge, for example, charge stripping of multiple-charge ions, is also possible. For positive ions, the precursor or parent ion m_p^+ is converted into the product or daughter ion m_d^+ via the loss of a neutral fragment m_n. Whereas the neutral fragment m_n is not detected, its mass can be deduced from the m/z difference of m_p^+ and m_d^+. In the product ion analysis mode (see Figure 1.5a), which is the most basic MS–MS experiment, the precursor ion m_p^+ is selected in the first stage of mass analysis (MS1), while the product ions m_d^+ are mass-analyzed and detected in the second stage (MS2) [66].

The historical starting point of MS–MS is the observation and explanation in the 1940s of the occurrence of metastable ions in a magnetic-sector instrument [67]. Subsequently, it was discovered that the abundance of the metastable ions can be increased by energetic collisions in a collision cell. From the mid-1970s onward, instruments were especially designed for MS–MS, for example, triple-quadrupole (TQ) instruments [68], multistage MS–MS in ion-trap instruments [69], hybrid quadrupole–time-of-flight instruments (Q–TOF) [70], hybrid quadrupole-linear ion-trap instruments (Q–LIT) [47], TOF–TOF instruments [71], and hybrid LIT–orbitrap instruments [45, 62].

The MS–MS instrument includes a combination of two mass analyzers. The first and second stage of mass analysis may be performed by the same type of mass analyzer, as in a TQ or an ion-trap instrument. In TQ instruments, the three steps of the MS–MS process, that is, the precursor ion selection, collision-induced dissociation (CID; see Section 1.5.2), and mass analysis of the product ions, are performed in spatially separated devices ("tandem-in-space"), whereas in an ion-trap instrument, the three steps are performed one after another in the same device ("tandem-in-time") [72]. In a hybrid instrument, the first and second stage of mass analysis are performed in two different types of mass analyzers, for example, a first-stage quadrupole and second-stage linear ion trap in a Q–LIT instrument, or in a first-stage quadrupole and second-stage TOF in a Q–TOF instrument.

An MS–MS instrument allows studying the fragment ions of selected precursor ions and is therefore an indispensable tool in fundamental studies on ion generation, ion–molecule reactions, unimolecular fragmentation reactions, and identity of ions. It plays an important role in qualitative analytical applications of MS involving the on-line coupling of MS to GC (gas chromatography) or LC, for example, in the identification of drug metabolites (Section 1.6.4) or in peptide sequencing (Section 1.6.6) and protein identification (Section 1.6.7).

1.5.2
Ion Dissociation Techniques

MS–MS is based on the gas-phase dissociation or fragmentation of selected ions. It may involve either metastable ions or activated ions. Metastable ions are ions with sufficient internal energy that survive long enough to be extracted from the ion source before they fragment, but may then fragment in the mass analyzer region before detection. The charged fragments of metastable ions may be detected, for example, using various linked-scan procedures in double-focusing sector instruments [44, 66]. However, fragmentation may also be induced by activation of selected ions, that is, by increasing their internal energy. The most widely applied method of ion activation is collisional activation. On acceleration and collision of the selected ions with a target gas (He, N_2, or Ar) in a collision cell, the ion translational energy can partially be converted into ion internal energy. If subsequent dissociation of the ion occurs in the collision cell, the process is called *collision-induced dissociation*. CID is a two-step process: after converting ion translational energy into ion internal energy in an ultrafast collision event

($\sim 10^{-15}$ s), unimolecular decomposition of the excited ions may yield various product ions by competing reaction pathways. CID can be performed in two different energy regimes [73]. Low-energy CID is performed in most instruments. During the residence time in the collision cell, the selected precursor ions undergo multiple collisions with a target gas ($\sim 10^{-3}$ mbar He, N_2, or Ar). In sector and TOF–TOF instruments, high-energy CID can be performed, involving single kilo electron-volts collisions with He as target gas. High-energy collisions may result in more informative and more complex MS–MS spectra, because a wider range of fragmentation reaction pathways is opened. In the low-energy CID regime, one may further discriminate between collision-cell CID and ion-trap CID. In collision-cell CID, applicable to TQ and Q–TOF instruments, collisions are performed with N_2 or Ar after acceleration of the precursor ions with 10–60 V. In ion-trap CID, collisions are performed with a smaller target (He instead of Ar), ion excitation is achieved by an RF waveform pulse, and the interaction time is milliseconds in ion-trap CID rather than microseconds [50].

Various other ion activation methods have been used, mostly in specific applications and/or instruments [73, 74]. Some of these methods are primarily developed to induce fragmentation in FT-ICR-MS instruments, for example, infrared multiphoton photodissociation (IRMPD), sustained off-resonance irradiation (SORI), and black-body infrared radiative dissociation (BIRD) [73, 74]. Others such as surface-induced dissociation and laser photodissociation have been used on various instruments, but mostly by a limited number of research groups. Currently, the most widely applied alternative ion-activation methods are electron-capture dissociation (ECD) and electron-transfer dissociation (ETD) [75, 76]. A nice comparison of some of these ion dissociation techniques has been reported for glycopeptide analysis [77].

ECD is based on the interaction of multiple-charge protein ions with low-energy free electrons (~ 1 eV) in an FT-ICR cell, resulting in the formation of multiple-charge odd-electron ions, $[M+nH]^{(n-1)+\bullet}$ [78]. In CID, peptides predominantly show backbone fragmentation at the peptide bond, resulting in the so-called b and y'' ions [79] (see also Section 1.6.6). In ECD and ETD, cleavage of the N–C_α bond is observed, thus resulting in even-electron c' and odd-electron $z^\bullet$ ions. Moreover, in ECD, labile bonds to posttranslational modifications such as glycans and phosphates are mostly preserved [75]. The low-energy electrons used in ECD are generally generated from an (indirectly heated) electron-emitting surface. In ETD, the peptide fragmentation is induced by transfer of electrons from radical anions of compounds such as fluoranthene to the multiple-charge ions [76, 80]. ETD is more readily implemented on other type of mass analyzers such as ion-trap, Q-TOF, and orbitrap instruments [75].

1.5.3
Tandem Quadrupole MS–MS Instruments

Probably the most widely used MS–MS configuration is the TQ instrument, where mass analysis is performed in the first and third quadrupoles, while the second quadrupole is used as a collision cell in the RF-only mode, that is, in a $Q-q_{coll}-Q$ configuration [68]. Although usually called a *triple-quadrupole instrument*, because of the initial lineup of two analyzing quadrupoles and a quadrupole collision cell, the term tandem quadrupole (TQ) would nowadays be more appropriate to describe most of the commercially available instruments. The gas-filled RF-only collision cell, which provides refocusing of ions scattered by collisions, results in significant transmission losses. In attempts to reduce these losses, alternative RF-only collision cells have been developed, for example, RF-only hexapoles or octapoles. In a linear-acceleration high-pressure collision cell (LINAC), an axial voltage and tilted rods are used to reduce the residence time of the ions in the collision cell and to reduce crosstalk [81]. Crosstalk may occur on rapid switching between two selected-reaction monitoring (SRM, see below) transitions with the same product ion from two different precursor ions. Product ions of the first precursor ion are erroneously attributed to the second precursor ion, because these still reside in the collision cell while MS1 is already switched to the second precursor ion. A stacked-ring collision cell, featuring an axial traveling-wave or transient DC voltage to propel the ions and to reduce the transit times, has been reported as well [15].

The introduction and wide application of soft-ionization techniques such as ESI has in fact greatly stimulated the use of MS–MS. The most frequently applied MS–MS data acquisition mode is SRM, applied in routine targeted quantitative analysis [82]. In the SRM mode (see Figure 1.5d), both stages of mass analysis perform the selection of ions with a particular *m/z* value. In MS1, a precursor ion is selected, mostly the protonated or deprotonated molecule of the target analyte. The selected ion is subjected to dissociation in the collision cell. In MS2, a preferably structure-specific product ion of the selected precursor is selected and detected. Thus, an SRM transition is a combination of a precursor ion *m/z*, a product ion *m/z*, and all MS parameters, for example, collision energy, required to measure this transition with the best sensitivity in a particular TQ instrument. Because of the high selectivity involved in SRM, excellent sensitivity may be achieved in target quantitative analysis (see Section 1.6.2). The SRM mode is the method of choice in quantitative bioanalysis using LC–MS, for example, in (pre)clinical studies for drug development within the pharmaceutical industry [83–85], and is also widely used for targeted quantitative analysis in many other application areas, including environmental, food safety, and clinical analysis. It has also been implemented in quantitative analytical strategies using GC–MS as well [86]. The SRM mode is frequently (but erroneously) called "*multiple-reaction monitoring*" (*MRM*) to indicate that multiple product ions of one precursor ion are monitored, even when only one product ion is monitored. In addition to SRM, various structure-specific screening procedures for the TQ instruments were introduced, for example, the precursor ion analysis (PIA) and neutral-loss analysis (NLA) mode [87, 88] (see Section 1.6.1 and Figure 1.5b,c).

1.5.4
Ion-Trap MS^n Instruments

An ion-trap instrument provides three features, which makes it ideally suited for MS–MS. It has the possibility to m/z-selectively eject ions from the trap, it is operated with a constant He pressure in the trap that may serve as collision gas, and it is possible to apply an m/z-selective RF waveform to the end-cap electrodes to excite ions of the selected m/z [50]. In ion-trap MS–MS, one starts with a population of ions, from which the precursor ion is selected, excited, and fragmented, resulting in a new population of (product or daughter) ions. The latter population can be scanned out to be detected, or can serve in a new series of subsequent steps of the process: selection of a product ion as precursor ion in a new MS–MS experiment, which is to be excited and fragmented, leading to a new population of (granddaughter) ions. Excitation means that the selected ions move with wider amplitude, and thus at greater speed, through the ion trap. This leads to more energetic collisions with the He bath gas, which in turn leads to a gain in ion internal energy and subsequent fragmentation of the excited precursor ion. The various steps of the process are performed one after another in the same space (the ion trap), and can thus be considered to be "tandem-in-time" [72]. The process described can be repeated to achieve multiple stages of MS–MS or MS^n (up to 10 stages in most instruments).

As indicated in Section 1.5.2, the ion-trap CID process differs from collision-cell CID: smaller target, RF ion excitation, and longer interaction time. As a result, different fragmentation pathways may be accessible in ion-trap CID compared to collision-cell CID, especially in fragmenting fragment ions in MS^n experiments. For some compounds, for example, glycosylated saponins [89], stepwise fragmentation can be achieved, for example, subsequent losses of sugar monomers in subsequent MS–MS steps (Figure 1.6). The lower energy involved in ion-trap CID facilitates the acquisition of a wealth of structural information, for example, by stepwise fragmentation and the generation of fragmentation trees [90]. A fragmentation tree is generated by further fragmenting selected fragment ions of a particular stage of MS^n into the next stage, that is, MS^{n+1}. This provides a wealth of information in structure elucidation and identification of unknowns, as is clearly demonstrated for polyphenols [91].

1.5.5
Tandem TOF (TOF–TOF) Instruments

MALDI is considered to be a soft-ionization technique. However, considerable fragmentation of MALDI-generated ions occurs after acceleration; that is, metastable ions are generated. This process is sometimes indicated as laser-induced dissociation [92]. Obviously, the fragmentation of metastable ions can be induced by additional (high-energy) collisions in a collision cell, which may also result in CID. In MALDI-TOF-MS terms, the process of fragmentation of metastable ions is frequently called post-source decay (PSD) [93]. The kinetic

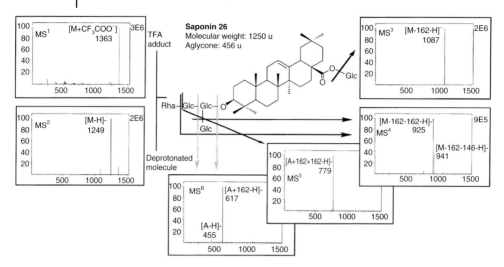

Figure 1.6 LC–MS^n in an ion-trap instrument: stepwise fragmentation of glucosylated Saponin 26. (Reprinted with permission from Ref. [89]. ©1998, Elsevier Science.)

energy of these fragment ions m_d in PSD significantly differs from that of the parent ions m_p, whereas their velocities are the same. The difference in kinetic energy is proportional to m_d/m_p. They also suffer from a wide ion kinetic energy distribution. Therefore, these fragment ions are not observed in the TOF-MS process, unless specific actions are taken, including the use of a gridless curved-field reflectron or a reflectron with a time-dependent field. A linear-TOF experiment can be used to measure the precursor ions, whereas a number of reflectron-TOF experiments are needed to acquire the wide-range product ion spectrum. Finally, the product ion mass spectra have to be concatenated by the data system [93].

The problems with the different kinetic energies of precursor and products ions, described above, can also be solved by decelerating the precursor ions before dissociation, for example, from 20 keV down to 1–2 keV, and then reaccelerating the product ions generated in a collision cell. This is generally done, in one way or another, in a TOF–TOF system [94]. MALDI-TOF–TOF-MS systems can currently be routinely applied to acquire MS–MS information from MALDI-generated ions, for example, in proteomics and glycomics [30, 95, 96].

1.5.6
Hybrid Instruments (Q–TOF, Q–LIT, IT–TOF)

In this section, three types of hybrid MS–MS systems are discussed, two with TOF mass analyzers as MS2 and one with a LIT in MS2.

The Q–TOF instrument can be considered as a modified TQ instrument, where the MS2 quadrupole has been replaced by an orthogonal-acceleration reflectron-TOF mass analyzer [70]. That means that in MS mode, the quadrupole (MS1) is operated in RF-only mode and the RF-only collision cell with low collision energy, whereas in MS–MS mode the quadrupole performs the selection of the precursor ion with unit-mass resolution and fragmentation of the precursor ion is achieved in the collision cell. In both modes, the ions are orthogonally accelerated into the flight tube and high-resolution mass analysis is performed in a reflectron-TOF analyzer (MS2). Q–TOF instruments are now available from various instrument manufacturers [57]. They are widely used in structure elucidation and biomacromolecule sequencing (see Section 1.6). Because collision-cell CID is applied, the fragmentation characteristics are the same as in TQs. In structure elucidation, a significant advantage of Q–TOF is the ability to perform accurate mass determination (<5 ppm) for both precursor and product ions. Principles and applications of Q–TOF hybrid instruments have been reviewed [97, 98].

Hybrid MS–MS instruments have been developed, featuring ion traps in either the first (MS1) or the second stage of mass analysis (MS2). If the ion trap is implemented as MS1, it generally acts as a "filter" with respect to the number of ions that is transferred to MS2, based on the duration of the ion accumulation in the ion trap. In addition, MS^n experiments may be performed before transferring a package of ions to MS2. Thus, in such hybrid systems, no collision cell has to be present between the first and second stage of mass analysis. This is true for ion-trap hybrids with FT-ICR-MS (see Section 1.5.7) and orbitrap (see Section 1.5.8) instruments, but also for the hybrid ion-trap-time-of-flight (IT–TOF) system. IT–TOF systems have been pioneered by the group of Lubman [99, 100]. It has subsequently become commercially available for both MALDI and LC–MS applications [101]. It readily provides high-resolution MS and MS^n data, and has been applied for structure elucidation, for instance, in metabolite identification [101] or otherwise [102]. Unlike CID in other IT devices, where He is used to stabilize the ion trajectories and as collision gas, pulses of Ar are used to prevent precursor ions from being lost from the trap and to perform MS^n in an IT–TOF instrument.

In the Q–LIT hybrid instruments [47, 103], a (linear) IT is implemented as the second stage of mass analysis (MS2), for accumulation of ions to achieve improved sensitivity after collision-cell CID, and/or to perform MS^3. The Q–LIT instrument can be operated either as a conventional TQ instrument, where it is capable of all TQ acquisition modes including SRM, or as the hybrid instrument. In the hybrid mode, full-spectrum data can be acquired in the enhanced product ion (EPI) analysis mode with up to 60-fold enhanced sensitivity compared to TQ instruments, while still acquiring collision-cell CID spectra. The system also allows the acquisition of MS^3 spectra, with the second dissociation step to be performed in the LIT [103, 104]. Hardware, electronics, and software control of the Q–LIT instrument have been optimized to allow very rapid switching between various MS and MS–MS experiments. Q–LIT instruments have found wide application in LC–MS.

1.5.7

MS–MS and MSn in FT-ICR-MS

As targeted ions can be selectively trapped in the ICR cell, while unwanted ions can be eliminated by the application of RF pulses, the MS^n procedures in an FT-ICR-MS instrument greatly resemble those in an ion-trap instrument. However, successful operation of an FT-ICR-MS instrument requires extreme low pressures in the cell ($\sim 10^{-9}$ mbar). Thus, the vacuum constraints hamper the possibilities of performing CID in the FT-ICR cell [58]. This problem can be elegantly solved either by the use of hybrid systems where fragmentation is performed before transfer of ions to the ICR cell or by the use of alternative ion dissociation techniques. With respect to hybrid systems, both quadrupole–FT-ICR-MS systems [105] and LIT–FT-ICR-MS systems [106, 107] have been described and widely applied. These hybrid systems provide great versatility, user-friendliness, and excellent performance characteristics. With respect to alternative ion-activation methods, IRMPD, SORI, and, more recently, ECD and ETD have been frequently applied (cf. Section 1.5.2).

The potential of FT-ICR-MS in studying biomolecular interactions can be illustrated by some early examples [108, 109]. ESI-MS was used to generate gas-phase ions of noncovalent complexes of 16 benzenesulfonamide inhibitors

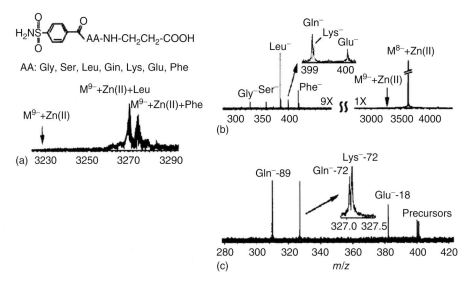

Figure 1.7 Negative-ion ESI spectra (using FT-ICR-MS) of 7.0 µM bovine carbonic anhydrase II with benzenesulfonamide inhibitors (7.0 µM each) in 10 mM NH₄OAc (pH 7.0). With (a) the mass spectrum of the [9–]-ion prior to ion isolation, (b) isolation and subsequent CID of the [9–]-ion, where the intensity of dissociated inhibitors correlate with their binding affinity, and (c) SWIFT isolation of three of the inhibitor ions (from B), Glu-, Lys-, and Gln- at m/z 399–401 and subsequent CID to obtain fragment ions for the inhibitors. (Reprinted with permission from Ref. [108]. ©1995, American Chemical Society.)

with bovine carbonic anhydrase II (BCA-II), which are introduced into the FT-ICR cell (see Figure 1.7). The tightly folded complexes generally formed only two charge states. Complexes of BCA-II were observed with all 16 benzenesulfonamides with relative abundances consistent with the binding constants of the inhibitors in solution. Competition experiments could be performed as well by adding an excess of a high-affinity inhibitor to the electrosprayed solution. For some inhibitors of closely related masses, gas-phase dissociation of the (isolated) complexes was achieved to release the inhibitors, measure their accurate mass, and subsequently perform a second dissociation step to study the fragmentation of the inhibitors (see Figure 1.7) [108]. This approach was coined "bioaffinity characterization mass spectrometry" [109]. It basically comprises three steps: (i) selective accumulation of a noncovalent protein–ligand complex, (ii) gas-phase dissociation of the complex to release the ligand, and (iii) fragmentation of the ligand for further structure elucidation [109].

1.5.8
Orbitrap-Based Hybrid Systems

Although stand-alone orbitrap systems have been produced, the full power of the orbitrap mass analyzer is achieved by its use in an MS–MS setting. The initial instrumental setup of the orbitrap consisted of a hybrid LIT–orbitrap configuration, featuring an LIT to control the number of ions transferred to the orbitrap and to perform MS^n, a C-trap to direct the package ions into the orbitrap, and the orbitrap itself (see Figure 1.4e) [45]. As the ion-trap system in this commercial LIT–orbitrap instrument is equipped with separate off-axis detectors, simultaneous acquisition of high-resolution precursor ion and unit-mass-resolution product ion mass spectra can be achieved. If both precursor ions and product ions are detected using the orbitrap, high resolution (~100 000, FWHM) is used for the precursor ions, whereas medium resolution (~15 000–30 000, FWHM) is generally sufficient for the product ions of a well-characterized precursor ion.

Later on, it was demonstrated that CID could be achieved in the C-trap, which is a gas-filled quadrupole type of device [110]. The C-trap fragmentation resembles the collision-cell CID more than it resembles the ion-trap–CID. Subsequently, separate higher-energy RF-only collision octapoles (higher-energy collision-induced dissociation, HCD) were mounted on LIT–orbitrap hybrid systems to make optimum use of this feature. Such a system can be considered a gas-phase ion-chemistry laboratory on its own, featuring different ways to perform fragmentation, that is, ion-trap CID, HCD, and eventually ETD, as well as different ways to measure the *m/z* values of the resulting ions, that is, by unit-mass resolution with the ion trap or by ultra-high resolution with the orbitrap. In fact, HCD also led to the stand-alone orbitrap [63], mentioned earlier, thus providing the possibility to efficiently fragment ions without precursor ion selection, as well as to hybrid quadrupole–orbitrap systems [111].

Most recently, a tribrid orbitrap-based system was introduced, featuring parallel quadrupole and LIT systems, next to an orbitrap mass analyzer [112].

In this system, precursor ion selection is performed either in the quadrupole or in the LIT. In the former case, fragmentation can take place either in the HCD cell, followed by orbitrap mass analysis, or in the ion trap, followed by mass analysis and detection in either the ion trap or the orbitrap. In the latter case, (eventually multistage) fragmentation takes place in the ion trap, again with mass analysis and detection in either the ion trap or the orbitrap. Fragmentation by HCD is performed in the HCD cell, while ion-trap–CID or ETD can be performed in the ion trap. The high-field orbitrap analyzer implemented provides at a resolution of $\sim$500 per ms acquisition time, thus $\sim$120 000 in 0.26 s (at m/z 200, FWHM), with a maximum of $\sim$500 000 in $\sim$1.1 s (at m/z 200, FWHM).

1.5.9
Ion-Mobility Spectrometry–Mass Spectrometry

Ion-mobility spectrometry (IMS) is a powerful tool in the study of gas-phase ions [113, 114]. An ion-mobility spectrometer consists of (i) an ion-generation region, mostly a radioactive ^{63}Ni foil, (ii) a charge-separation region, (iii) an ion shutter, (iv) the actual drift-reaction region, and (v) an ion detector, for example, a mass spectrometer. The ion shutter provides pulse-wise introduction of ions into the drift tube, in which a uniform axial electric field gradient (typically $1–1000\,V\,cm^{-1}$) is maintained with a series of guard rings, separated by electrically insulating spacers and connected with an appropriate precision resistor network. The IMS measures how fast a given ion moves through the buffer gas (He, N_2, or Ar) in a uniform electric field, and can thus be used to determine (reduced) ion mobility from the drift time. IMS may be considered being gas-phase electrophoresis. Larger ions undergo more collisions with the buffer gas and thus will have longer drift times than smaller ions. Higher charge states of an ion experience a greater effective drift force, and thus show higher mobility than the lower charge states. From the measured reduced mobility, the experimental collision cross section can be determined, after appropriate calibration with reference compounds [115]. Theoretical prediction of the collision cross section, for example, for peptides, is also possible, generally showing good correlation to experimental values [116, 117].

The on-line combination of IMS with MS results in a very attractive tool to combine analysis of conformation and shape, as performed in IMS, with the analysis of m/z and structural features, as performed in MS. IMS–MS has been pioneered by the groups of Bowers [118] and Clemmer [119, 120], developing IMS interfaced to quadrupole or TOF instruments. IMS–MS provides a rapid gas-phase separation step before MS analysis, enabling the identification of ions with different drift times, thus with different collisional cross sections. Instrumental developments in IMS–MS have been reviewed [121]. IMS–MS is extensively used to study gas-phase protein conformations, for instance, in relation to neurodegenerative and neuropathic diseases such as Parkinson's and Alzheimer's disease [122]. Currently, there are three ways to implement IMS in IMS–MS.

The groups of Bowers and Clemmer [118, 119, 123] use the type of drift tubes also applied in stand-alone IMS. Whereas the conventional drift-tube approach has been the first way to perform ion mobility in combination with MS, drift-tube IMS–MS systems have become commercially available only very recently (2014). Nevertheless, successful application of this type of IMS–MS instruments has been demonstrated from various other laboratories as well for both small molecules and biomacromolecules [121–124].

A successful alternative approach to IMS–MS is based on the use of traveling-wave stacked-ring ion guides, which were initially developed to replace RF-only hexapole ion guides in vacuum interfaces for API or as collision cells [15, 125]. To this end, a Q–TOF instrument featuring traveling-wave ion guides in both the vacuum-interface region and the collision-cell region was constructed. Initially, ion-mobility spectra were acquired by storing ions in the ion-source ion guide and gating them periodically to the collision-cell ion guide, operated at a 0.2-mbar pressure of Ar. The mobility-separated ions were subsequently analyzed using the TOF-MS system [15]. The initial setup was developed into a hybrid Q–IMS–TOF instrument [125]. The collision-cell region of this instrument features three traveling-wave stacked-ring ion guides, of which the middle one is used as a 185-mm-long ion-mobility drift tube, operated at pressures up to 1 mbar and up to 200 ml Ar gas, and the other two may be used as 100-mm-long collision cell, operated at 10^{-2} mbar when applicable. The system can be applied for a wide variety of ion-mobility studies, such as protein conformation studies [122] and differentiation of heterogeneities in glycoproteins [126].

A third way to perform IMS–MS is high-field asymmetric waveform ion-mobility spectrometry (FAIMS) [127–129]. In FAIMS, the gas-phase mobility separation of ions in an electric field is achieved at atmospheric pressure. In its simplest design, the FAIMS device comprises two parallel rectangular electrodes at a uniform distance. One of the electrodes is grounded, while at the other an asymmetrical waveform is applied, characterized by a significant difference in voltage in the positive and negative polarities of the waveform. Ions drift through the gas between the electrodes and are separated depending on their mobility. Whereas at low field the ion mobility is proportional to the field strength, at high field the ion mobility becomes dependent on the applied electric field. Because of the applied asymmetric waveform, the ions show a net displacement toward the grounded plate, which, however, is compensated for by a DC voltage (compensation voltage). This ensures that the ions remain between the plates. Scanning of the compensation voltage allows ions with different mobilities to be monitored. Next to the ion-mobility separation of the ions, focusing of the beam is achieved, thus improving the sensitivity in an FAIMS–MS system. Currently, a variety of (often cylindrical) electrode configurations are applied [127–129]. Commercially available FAIMS devices are primarily applied to improve sensitivity and to reduce matrix effects in quantitative analysis using LC–ESI-MS [129, 130].

In its various forms, IMS–MS plays an important role in many forefront application areas of biological MS, including structural proteomics [131, 132], characterization of protein assemblies [133], and chiral and structural analysis of biomolecules [134].

1.6
Data Interpretation and Analytical Strategies

1.6.1
Data Acquisition in MS Revisited

As indicated in Section 1.2.4, the two basic acquisition modes of MS are the full-spectrum mode and the selected-ion mode. The full-spectrum acquisition mode is applicable to all mass analyzers, in MS, MS−MS, and MS^n modes. The full-spectrum mode in MS−MS and MS^n is the product ion analysis mode (Section 1.5.1). The selected-ion acquisition mode, that is, SIM in single-MS instruments and SRM in MS−MS instruments, is a powerful tool in especially the beam instruments, that is, quadrupole and sector instruments, to improve S/N in targeted analysis by elongating the dwell time at a particular m/z. The resulting data are somewhat comparable to XICs (Figure 1.1c), but often with better S/N than when acquired in full-spectrum mode. Although ion-trapping devices, both ion-traps and FT-ICR-MS systems, can perform a selected-ion acquisition mode as well, the gain in S/N will generally be less than in beam instruments under similar conditions. In TQ instruments, SRM is a powerful tool to greatly enhance selectivity, and thereby achieve excellent lower limits of quantification in targeted quantitative analysis (see Section 1.6.2). In HRMS, especially demonstrated for TOF and orbitrap instruments, post-acquisition narrow-window XIC is a powerful tool to achieve excellent S/N in targeted quantitative analysis. Recently, a high-resolution variant of SRM was proposed, involving the post-acquisition use of accurate mass product ions in MS2 (see Section 1.6.2).

Another powerful data acquisition mode is the DDA mode (also called *information-dependent acquisition*) [135, 136]. In DDA, the instrument performs a rule-based automatic switching between a survey and a dependent mode. The most widely applied DDA experiments switches between full-spectrum MS survey mode and a full-spectrum product ion (MS−MS) analysis dependent mode, but switching between SRM as a survey mode and product ion analysis as a dependent mode, for instance, is also demonstrated (see Section 1.6.4 for an example). The switching is controlled by the intensity of a possible precursor ion observed and eventually by additional criteria such as isotope pattern, charge state, specific m/z values on an inclusion list or an exclusion list. In this way, highly efficient data acquisition is possible: MS and MS−MS data of unknown compounds in a mixture are acquired simultaneously in one chromatographic run. An alternative to DDA is data-independent acquisition (MS^E) [137], where scanwise switching between MS and MS−MS is performed to obtain fragments for all precursor ions present. The MS^M approach, described for LIT−orbitrap instruments, is a bit similar [138]. Both data-dependent and data-independent acquisition procedures are frequently used in qualitative analysis of mixtures with unknown constituents, for instance, in metabolite identification (see Section 1.6.4) and proteomics (see Section 1.6.7) workflows.

In "tandem-in-space" instruments, two other data acquisition modes can be performed, that is, the PIA and the NLA modes (see Figure 1.5b,c), which are especially powerful for the screening for related structures in complex mixtures. It allows screening for compounds that on CID show either a characteristic fragment ion (in PIA mode) or a characteristic neutral loss (in NLA mode). In the PIA mode, MS1 is continuously scanned, while in MS2 a characteristic fragment ion is selected and detected. In PIA mode, a signal is detected if an ion transmitted in MS1 on CID generates the common product ion selected in MS2. In the PIA mass spectrum, peaks are shown with their precursor ion m/z value. An early example of PIA involves the screening for phthalate plasticizers in environmental samples by means of the common fragment ion with m/z 149, due to protonated phthalic anhydride [87]. In the NLA mode, both MS1 and MS2 are scanned, but with a fixed m/z difference. In NLA mode, a signal is detected if an ion transmitted in MS1 on CID loses a neutral molecule with a mass fitting the fixed m/z difference. An early example of NLA is the monitoring of CO_2 losses from deprotonated aromatic carboxylic acids [87]. NLA and PIA are frequently applied in drug metabolism studies (see Section 1.6.4) [139] and in phosphoproteomics [140].

1.6.2
Quantitative Bioanalysis and Residue Analysis

LC–MS on a TQ instrument operated in SRM mode is the gold standard in routine quantitative analysis, as, for instance, performed in pharmacokinetics/pharmacodynamics (PKPD) and absorption, distribution, metabolism, excretion (ADME) studies during drug discovery and drug development by the pharmaceutical industry and related contract-research organizations [83–85]. Whereas in such pharmaceutical applications just one SRM transition per compound is applied, in many areas of quantitative analysis at least two SRM transitions per compound are applied. The latter can be attributed to protocols issued by regulatory bodies, defining procedures in residues analysis [141, 142], which demand not only quantitative analysis of a target compound, but also conformation of its identity, based on retention time, precursor and product ion, and intensity ratio between two specific ions, for example, two compound-specific SRM transitions. This protocol, initially made for the analysis of veterinary drug residues in food of animal origin, has been widely adapted in other areas of residue analysis, for example, for pesticide analysis in fruits and vegetables and in drinking water, in sports doping, and in forensic/toxicological analysis. Strengths and weaknesses of such an approach have been critically assessed [143, 144]. Perhaps the most obvious limitation of this approach in residue analysis is that a targeted method is applied to search for "in principle" unknown contaminants.

More recently, the use of HRMS has been advocated as an alternative to the targeted approach based on SRM transitions and TQ instruments [57, 145]. The approach is sometimes called a *qual/quant strategy*. HRMS provides the possibility to acquire full-spectrum data at up to 50 spectra per second and yields accurate mass data. The analysis is untargeted and does not need prior optimization of (multiple) SRM transitions. It also provides the possibility for post-acquisition data mining and retrospective analysis, for example, to search in already acquired data for the presence of a later discovered metabolite or previously not anticipated contaminant. Initial limitations of HRMS in quantitative analysis in terms of lower limit of quantitation and linear dynamic range have been greatly removed by recent technological developments.

1.6.3
Identification of Small-Molecule "Known Unknowns"

Identification of unknowns may be directed to either "known unknowns" or "unknown unknowns." The term *known unknown* was introduced to indicate compounds that are unknown to the researcher, but actually described somewhere in the scientific literature and/or available in compound databases [146]. The "known unknowns" differ from the target compounds searched for in targeted residue analysis, described in Section 1.6.2, which can be considered as "known knowns." The identification of "known unknowns" is a highly challenging task, even more so for "unknown unknowns." This task can generally not be performed by using MS technologies alone, especially because MS often is not a very powerful tool in clearing stereoisomerism issues. Thus, in this, MS analysis should be combined with other techniques, especially nuclear magnetic resonance (NMR) spectroscopy. In practice, this is not straightforward, if one keeps in mind that NMR needs ~100–1000 times more (pure) compound to get an interpretable spectrum. Besides, MS and MS^n are readily performed within the timescale of high-resolution chromatography, whereas NMR requires far longer data acquisition times, typically 8–16 h when only low concentrations are available. Thus, either fraction collection or time-consuming stop-flow operations have to be performed, when multiple unknowns within one LC run are to be identified by NMR.

    The general procedure of the identification of unknowns consists of a number of steps, which primarily are described for "known unknowns" [7, 146, 147]. (i) One needs to collect as much information on the unknowns as possible. Parameters such as origin of the sample, solubility, thermal stability, and possibly underlying chemistry may provide valuable pieces of information. (ii) One needs to establish whether the sample is actually amenable to MS analysis, by GC–MS in EI mode, LC–MS in either positive-ion or negative-ion mode (or preferably both), MALDI-TOF-MS, or by any of the other available MS techniques. (iii) If the first MS data are acquired by HRMS, the calculation of the elemental composition of the unknown is possible, especially when a soft-ionization technique is applied. (iv) On the basis of the elemental composition and the general information on the unknown, compound databases may be searched for known structures, which will be successful for the "known unknowns," but not for the "unknown unknowns."

(v) Subsequently acquired MS – MS or MS^n data allow filtering the known structures from the database search by checking the observed fragmentation behavior against predicted fragmentation of the database-retrieved structures. In favorable cases, this leads to a (number of) potential structure proposal(s) for the unknown. (vi) In the end, standards should be purchased or synthesized and analyzed to check retention time, fragmentation behavior, and possibly other properties. (vii) For the "unknown unknowns", the database search did not provide results, thus requiring more complicated de novo data interpretation, possibly incorporating substructure searches. Additional tools such NMR, IR, and others will certainly be necessary to solve the puzzle.

At this stage, as a result it may be reported that a structure proposal for the unknown is available, for which the calculated elemental composition is in agreement with the measured accurate mass of the precursor ion, the main fragments in the product ion spectra could be assigned and seem to agree with the proposed structure, and chromatographic and MS characteristics seem to be in agreement with that of a synthetic standard (or an "known unknowns"). Further experiments may need to be performed, for example, preparative LC in order to collect sufficient material for NMR analysis, to further confirm the structure and rule out isomerism issues.

1.6.4
Identification of Drug Metabolites

The structure elucidation of related substances, be it synthesis by-products or degradation products of active pharmaceutical ingredients drug metabolites, or natural products within a particular compound class, can be performed by more or less similar strategies [148 – 151]. Suitable sample pretreatment and LC methods must be developed for the isolation and separation of the related substances. The acquisition of MS, MS – MS, and/or MS^n spectra of the parent drug and the thorough interpretation of these spectra is of utmost importance to the success of the study. After the analysis of relevant samples in LC – MS mode and data processing to search for potential related substances, MS – MS or MS^n have to be acquired. Nowadays, this is mostly done by DDA or data-independent MS^E procedures, using automatic switching between survey MS and (dependent) MS – MS or MS^n experiments, preferably using HRMS [57, 150]. Finally, interpretation of the data has to be performed, often followed by additional LC – MS^n experiments, isolation of particular compounds, synthesis of standards, and NMR analysis.

Drug metabolism is directed at increasing the polarity of the drug to enhance its excretion via the kidneys into the urine. It generally occurs in two steps, that is, first by, among others, oxygenation and dealkylation (Phase I metabolism) and second by conjugation with, for instance, sulfate, glucuronic acid, or glutathione (Phase II metabolism). From an MS point of view, the biotransformation of drugs results in metabolic m/z shifts, for example, of −14.052 Da due to demethylation and +15.995 Da due to hydroxylation, or N-oxide or sulfoxide formation. From m/z shifts in the metabolites, relative to the parent drug, one can often conclude which metabolic change occurred. Understanding of the MS–MS fragmentation of the parent drug allows to keep track of these metabolic shifts in the structure of the compound. To this end, the parent drug can be subdivided into a number of so-called profile groups, to keep track of metabolic m/z shifts of the precursor ions and to pinpoint them to particular structural elements [152, 153].

The experimental and data interpretation procedures in the identification of in vitro drug metabolites may be illustrated for the antidepressant drug nefazodone, which in the past few years has frequently been used to demonstrate advances in LC–MS and MS technology [154–156]. The structure of nefazodone, its MS–MS spectrum, the identity of a number of its fragments, and relevant profile groups are given in Figure 1.8a. Initially, a Q-LIT system was applied to profile and identify as many as 22 nefazodone metabolites, 7 of which were not previously reported [154]. DDA was applied to acquire EPI spectra (see Section 1.5.6). An iterative data processing strategy was applied, based on (i) the recognition of characteristic product ions, for example, the ions with m/z 274 and 140; (ii) the generation of XICs of samples and controls to find possible metabolites; and (iii) the inspection of the product ion spectra of the newly identified metabolites in order both (a) to recognize other characteristic ions that may be used in the generation of additional XICs, for example, the ions with m/z 290, being the oxidized form of the ion with m/z 274, and (b) to identify the metabolites, based on the profile group approach [154]. Subsequently, the identification of the metabolites was confirmed by accurate mass determination using an LIT–orbitrap system. Additional metabolites resulting from N-dealkylation or hydrolysis reactions of large groups were detected as well [155]. An overview of the metabolites found is given in Figure 1.8b. Fourteen out of the 26 identified metabolites involved oxidation (hydroxylation or N-oxidation) of the parent drug. In many cases, the exact position could not be established. It must be pointed out that the described post-acquisition iterative data processing strategy closely resembles an experimental data acquisition strategy based on PIA and/or NLA. Screening for nefazodone metabolites in PIA using the product ions with m/z 246, 274, and 290 (cf., Figure 1.8a) and NLA using constant neutral losses of 196, 224, and 212 Da has also been demonstrated [156]. Another powerful tool in the screening and profiling of drug metabolites is the use of mass-defect filtering (MDF) in HRMS [157]. The metabolic m/z shifts are accompanied by a shift in the mass defect of the metabolite, for example, of -0.0157 Da due to demethylation and -0.0051 Da due to hydroxylation. If the biotransformation does not lead to major structural changes in the parent drug, for example, N- or O-dealkylation or hydrolysis of large side groups, the mass-defect shifts are limited to ±40 mDa for Phase I metabolites and ±70 mDa for Phase II metabolites. Software tools have been developed to process HRMS data and to perform MDF within a

Figure 1.8 Metabolite identification of nefazodone. With (A) MS–MS spectrum, interpretation and profile groups of the parent compound nefazodone. (B) Overview of Phase-I metabolites of nefazodone. MS–MS spectrum of nefazodone redrawn from Ref. [155]. (Reprinted with permission from Ref. [155], ©2007, John Wiley & Sons Ltd.)

(A)

(B)

±70 mDa wide window around the exact m/z of the parent drug [157]. This has been successfully demonstrated for nefazodone [156]. However, as in the biotransformation of nefazodone, major structural changes do occur (see above), and additional mass-defect filters must be defined and used [156].

In many cases, multiple oxygenated metabolites may be observed. H/D-exchange experiments may be performed to discriminate between C-hydroxylation and N- or S-oxidation. All three types of metabolites show a +15.9949 Da mass shift. The hydroxylated metabolites show H/D exchange, whereas the N- or S-oxidated products do not [158]. Differentiation between various isomeric C-hydroxylated metabolites is often difficult. In this respect, IMS–MS may be of help. For isomeric hydroxylated metabolites, the differences in drift time are generally too small, compared to the resolution of the IMS separation [159]. A selective derivatization of aromatic hydroxyl groups of drug metabolites into N-methyl pyridinium derivatives has been demonstrated to sufficiently increase the differences in collision cross sections, and thus in drift times, to allow differentiation between isomeric forms [160]. In an earlier study, the same group differentiated between isomeric glucuronic acid conjugates using IMS–MS [161]. By correlating theoretically predicted collision cross sections to measured drift times of a parent drug and its fragments, a calibration plot was generated, which could subsequently be used to differentiate between chromatographically separated isomeric glucuronic acid conjugates or N-methyl pyridinium derivatives.

In total, the described data acquisition and processing strategies form an elaborate toolbox that may be used in profiling and identification of drug metabolites (see Figure 1.9). The acquired full-scan high-resolution MS data sets may be interrogated using XIC, MDF, isotope-pattern filtering (if specific isotopic features, e.g., Cl or Br atoms, are present in the parent drug), and background subtraction, whereas data-dependent and/or data-independent MS–MS and/or MS^n data sets may be interrogated using product ion filtering or PIA and neutral-loss filtering or NLA [157]. NLA is especially powerful in the screening for Phase II drug metabolites, using constant neutral losses of 80 and 176 Da for sulfate and glucuronic acid conjugates, respectively [139]. The potential of IMS–MS in this area has just started to be explored.

Similar strategies are applied in the screening for reactive metabolites by glutathione trapping [162]. In positive-ion mode, either NLA using losses of 129 Da (pyroglutamic acid) or 307 Da (glutathione) or PIA with the characteristic product ion with m/z 130 (protonated pyroglutamic acid) can be applied to screen for glutathione conjugates, whereas in negative-ion mode PIA using the common fragment ion with m/z 272 can be used [163]. Enhanced selectivity in NLA is achieved in MS^E with a post-acquisition selection of the precursor ions that show the loss of 129.043 Da, the accurate mass of pyroglutamic acid [164]. Alternatively, SRM-triggered DDA on a Q–LIT instrument can be applied using a wide range of predicted SRM transitions for possible glutathione conjugates [165]. NLA in combination with reactive-metabolite trapping by stable-isotope-labeled or chemically modified glutathione has been another screening strategy [166, 167].

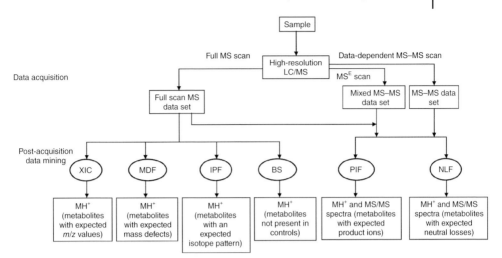

Figure 1.9 Toolbox for profiling and identification of drug metabolites. The acquired full-scan high-resolution MS data sets may be interrogated using extracted-ion chromatograms (XIC), mass-defect filtering (MDF), isotope-pattern filtering (IPF), and background subtraction (BS), whereas data-dependent and/or data-independent MS–MS and/or MS^n data sets may be interrogated using product-ion filtering (PIF) or precursor-ion analysis (PIA) and neutral-loss filtering (NLF). (Reprinted with permission from Ref. [157], ©2009, John Wiley & Sons Ltd.)

Obviously, MDF can also be applied to screen for glutathione conjugates of reactive metabolites [168].

1.6.5
Protein Molecular Weight Determination

One of the great benefits of ESI-MS is the ability to generate multiple-charge ions for biomacromolecules such as peptides and proteins, oligonucleotides, and oligosaccharides [169]. In proteins, the multiple charging is due to protonation at the basic amino acids, that is, Lys, His, and Arg, as well as at the N-terminal [170]. On the basis of the m/z values of the ions in the ion envelope of multiple-charge ions, the molecular weight can be calculated using an averaging algorithm [169, 171]. The algorithm assigns the number of charges to the peaks in the ion envelope and then averages the calculated values for the molecular weight. This simple and straightforward approach can be applied manually to relatively simple spectra with good S/N. Computer algorithms for the deconvolution or transformation of ESI-MS mass spectra of proteins have been developed [171] (see Figure 1.3). An even more powerful software tool is based on maximum-entropy algorithms for transformation of ESI-MS mass spectra [172]. The latter provides far batter resolution in the transformed spectrum, enabling detection of small mass differences between proteins in mixtures. These software tools are commercially available from various instrument manufacturers. The maximum-entropy-based algorithms are especially important if ESI spectra of heterogeneous proteins, for example, glycosylated proteins, are to be transformed.

The deconvolution of ESI spectra of oligonucleotides follows the same lines, except that in the negative-ion ESI mode, deprotonation has to be considered. The deconvolution may be hampered by the presence of H^+/Na^+-exchange ions in the ion envelope. Therefore, attention has to be paid to adequate desalting of the samples before molecular weight determination [173].

1.6.6
Peptide Fragmentation and Sequencing

MS using either ESI or MALDI as an ionization technique plays an important role in the characterization of proteins. Next to molecular weight determination of the intact protein (see Section 1.6.5), MS readily enables amino acid sequencing of peptides and proteins (see also Section 1.6.7). On fragmentation in MS–MS, using CID or any of the other ion dissociation techniques, a peptide shows primarily backbone fragmentation. Nomenclature rules have been proposed to readily annotate the spectra [79]. The N-terminal fragments are annotated with *a*, *b*, or *c*, depending on the position of the backbone cleavage (see Figure 1.10), whereas the corresponding C-terminal fragments are annotated with *x*, *y*, and *z*. Under high-energy CID conditions, amino acid side-chain cleavages may occur, leading to *d*, *v*, and *w* ions. Although initially accents were used to indicate protonation and hydrogen rearrangements, for example, the *c* and *y* ions in positive-ion mode were annotated as *c″* and *y″* to indicate that the *y*-ion was generated after rearrangement of a hydrogen to a shorter peptide and subsequent protonation; these accents are nowadays left out in most cases. In low-energy CID, but also in, for instance, SORI and IRMPD, fragmentation occurs at the peptide bond. Thus, predominantly *b* and *y* ions are observed, together with some *a* ions due to CO losses from *b* ions. In peptide sequencing by MS–MS, mostly double-charge ions are

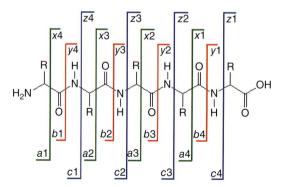

Figure 1.10 Nomenclature for backbone cleavages of peptides, according to Roepstorff [79]. The N-terminal fragments are annotated with *a*, *b*, or *c*, depending on the position of the backbone cleavage, whereas the corresponding C-terminal fragments are annotated with *x*, *y*, and *z*.

selected as precursor ions. In that case, next to *b* ions both single-charge and double-charge *y* ions may be observed. This leads to a series of sequence ions. In ECD and ETD, cleavage of the $N–C_\alpha$ bond is observed, thus resulting in even-electron *c′* ions and odd-electron *z•* ions [75].

For a known amino acid sequence, the *m/z* values of the fragment or sequence ions can be easily predicted. To this end, residue masses of amino acids are used, defined as the mass of the amino acid minus water (H_2O, 18.011 Da). The *m/z* of the b_1 ion equals 1.007 u (H^+) plus the residue mass of the N-terminal amino acid; the *m/z* of the b_2 ion is that of the b_1 ion plus the residue mass of the second N-terminal amino acid, and so on. The *m/z* of the y_1 ion equals 19.018 u (H_3O^+) plus the residue mass of the C-terminal amino acid; the *m/z* of the y_2 ion is that of the y_1 ion plus the residue mass of the second C-terminal amino acid, and so on. Interpretation of the product ion spectrum of a peptide thus involves identifying the series of *b* ions and *y* ions and fitting the amino acid residue masses in between adjacent peaks. As an example, the annotated product ion spectrum of [Glu1]-fibrinopeptide B is shown in Figure 1.11.

1.6.7
General Proteomics Strategies: Top-Down, Middle-Down, Bottom-Up

Proteomics involves the large-scale study of structure and functions of the proteins in the complete proteome of an organism or system. MS plays an important role in proteomics studies. In this respect, a proteomics study may be considered as a combination of two workflows, that is, an experimental workflow based on MS and MS–MS analysis of the proteome, and a bioinformatics workflow to efficiently interpret the data acquired. For clarity, these two workflows are best explained for a single protein and its identification first.

In the MS-based experimental workflow, there are three approaches to identify a protein in a proteome, that is, top-down, bottom-up, and middle-down [60, 61].

In the top-down approach, the intact protein, mostly a selected multiple-charge ion generated by ESI as the precursor ion, is subjected to an ion dissociation technique in an MS–MS experiment [60, 61]. High-sensitivity HRMS is needed, preferably using FT-ICR-MS or orbitrap MS systems, as interpretation of the data requires differentiation between the many different charge states that might be present in the product ion spectrum. Significant progress has been made in top-down proteomics [174, 175]. The isolation of the relevant proteins from a proteome is one of the challenges in top-down proteomics as, due to the low diffusion coefficient of proteins, the chromatographic separation of proteins is not very efficient; mostly C_4 rather than C_{18} reversed-phase LC is used. Electrophoretic techniques, especially solution-phase isoelectric focusing, can be a powerful alternative. Isolation of individual proteins from complex mixtures by immunoaffinity techniques may be applied. Although CID may be used, the most prominent ion dissociation techniques for top-down proteomics are ECD (with FT-ICR-MS) and ETD (with orbitrap and Q–TOF instruments). In some cases, the internal energy of the selected precursor ion is first increased by CID or infrared laser

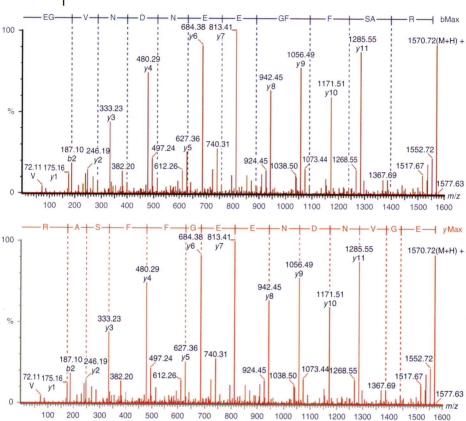

Figure 1.11 Annotated MS–MS spectrum of the [Glu1]-fibrinopeptide (amino-acid sequence: EGVNDNEEGFFSAR), acquired on a Waters Q–TOF 2 instrument and processed using Waters BioLynx® software, part of the Mass-Lynx suite. In the top spectrum the *b*-series are annotated, the bottom one the *y*-series.

light before ECD or ETD. Advanced software tools have been developed for the interpretation of the data [174, 175]. Along these lines, top-down proteomics has been applied for the characterization of intact proteins, especially in mapping posttranslational modifications such as phosphorylation and glycosylation [176], of the study of protein conformations and of noncovalent protein–ligand and protein–protein complexes, where among other techniques H/D-exchange experiments play an important role [177]. Top-down proteomics is expected to play an increasingly important role in the characterization of therapeutic proteins (biopharmaceuticals). Besides ESI-MS approaches, MALDI-TOF-MS has also been applied in top-down proteomics, either by in-source decay [178] or by TOF–TOF approaches [179]. MALDI-FT-ICR-MS may be considered as well.

In the bottom-up approach to protein identification, the intact protein is first enzymatically digested into smaller peptides, after reduction of the Cys–Cys bridges by dithiothreitol and alkylation of the thiol groups of Cys by iodoac-etamide [61, 180]. Peptides can be separated more efficiently using reverse-phase liquid chromatography (RPLC) and can be efficiently fragmented by CID in a wide range of MS–MS instruments. Trypsin is the most widely applied protease for protein digestion [181]. Trypsin cleaves the protein at the C-terminal side of the basic amino acids Arg or Lys, except when there is Pro. This results in peptides (the peptide map or peptide-map fingerprint) that readily form double-charge ions in ESI-MS, which can be selected as precursor ions for efficient CID in MS–MS. The tryptic peptides of an isolated protein or a complete proteome can be separated by RPLC before ESI-MS analysis with DDA to obtain both MS and MS–MS (or MS^n) data within the same chromatographic run. In the analysis of complex proteomes, on-line two-dimensional (nano-)LC can be used [23, 24]. In this case, the (complex) peptide mixture is fractionated by cation-exchange chromatography by means of a step salt gradient. Each fraction is desalted and preconcentrated on a short RPLC trapping column. The peptide mixture in each fraction is eluted from the trapping column by a solvent gradient (typically consisting of water, acetonitrile, and an acid) onto a (nano-)LC column for peptide separation and on-line ESI-MS analysis with DDA. Powerful software tools have been developed for data processing and interpretation [180].

Data interpretation by the (parallel) bioinformatics workflow can be readily explained for the bottom-up approach [182] (cf. Table 1.2). The starting point of the bioinformatics workflow is the availability of protein databases, eventually derived from DNA or genomic databases. Given the known protease selectivity of trypsin, as well as that of many other proteases, the *in silico* digestion of all proteins in the protein database results in a huge collection of peptides. For each peptide from the *in silico* digestion of the protein database, its mass or *m/z* value, amino acid sequence, and protein ID are stored. MALDI-TOF-MS or LC–MS

Table 1.2 Bioinformatics workflow complements the parallel biochemical workflow to achieve protein identification based on peptide map.

Experimental workflow	Bioinformatics tools
Isolate the proteome of relevant biological system	(cDNA-derived) Protein sequence database for relevant species
Perform tryptic digestion into complex peptide mixture	Perform *in silico* digestion of protein database
	List calculated mass of peptides with three pieces of information
	• Peptide mass • Amino acid sequence of peptide • Protein ID origin of peptide
Obtain peptide-mass fingerprint using MALDI-MS or LC–MS	Search experimental masses against calculated masses
List of peptide-masses	Scoring algorithm for possible hits (protein IDs)

analysis of the mixture of tryptic peptides results in a list of m/z values, the so-called peptide-mass fingerprint or peptide map. This list of m/z values can be scored against the m/z values of the peptides from the *in silico* digestion of the protein database. Bioinformatics software tools have been developed to perform this process. In most cases, the scoring algorithm converges to a particular protein from the database, which means that a number of peptides in the peptide map correspond to partial sequences of a particular protein in the database. The software actually provides scoring values on the likeliness that a particular experimental peptide map is derived from a particular protein. In this way, the peptide map may lead to protein identification [183, 184]. When HRMS data can be used, a surprisingly small number of peptides is needed to achieve protein identification.

An alternative workflow involves separation of proteins by one- or two-dimensional gel electrophoresis and staining to visualize the proteins. Spots of interest, for example, specific protein spots or up- or downregulated proteins in a comparative study, can be cut out of the gel. After destaining, in-gel trypsin digestion can be performed. The formed peptides can be readily extracted from the gel and analyzed by either MALDI-TOF-MS or LC–MS to obtain the peptide map or peptide-mass fingerprint, which can be interpreted by means of a similar bioinformatics workflow [185, 186].

Table 1.3 Bioinformatics workflow complements the parallel biochemical workflow to achieve protein identification based on peptide map and MS^n information.

Experimental workflow	Bioinformatics tools
Isolate the proteome	Protein sequence database
Perform tryptic digestion	Perform *in silico* digestion
Obtain peptide-mass fingerprint	Search against calculated masses
List of peptide-masses	Short list of peptides
Acquire MS–MS spectra of peptides in 2D-LC–MS^n	Predict sequence ions for selected peptides
	Each entry contains:
	• Peptide mass (Precursor m/z in MS–MS) • Amino acid sequence of peptide • Predicted sequence ions (b and y ions) • Protein ID origin of peptide
Data-dependent acquisition in (2D-)-LC–MS^n results in:	Predict sequence ions for peptides on short list
• List of peptide-masses • MS–MS spectra	
	Search experimental MS–MS data against predicted sequence ions Scoring algorithm for possible hits (protein IDs)

This process can even be made more powerful if additionally experimental MS–MS data are available for the peptides in the peptide map. As outlined in Section 1.6.6, the m/z values of (possible) sequence ions of a peptide with a known sequence can be readily predicted. Thus, the above-mentioned stored information for each peptide from *in silico* digestion (mass or m/z, amino acid sequence, and protein ID) can be easily complemented by predicted m/z values of the sequence ions of each of these peptides (cf. Table 1.3). Thus, the m/z values for the fragment ions in the experimental product ion spectra of all fragmented peptides can be cross-correlated to the information in the database in order to achieve an even more reliable protein identification [187, 188]. Powerful bioinformatics tools have been developed that are based on these principles, for example, MASCOT [189], SEQUEST [190], and advanced modifications thereof [191].

Finally, the middle-down approach shows resemblances to both bottom-up and top-down strategies. The larger protein is first digested in larger fragments, for instance, using the OmpT protease [192], which selectively cleaves between two basic amino acids. The resulting protein fragments are subjected to common top-down proteomics strategies for identification and/or characterization. The application of all three approaches in the characterization of histone variants has recently been reported [193].

1.7
Conclusion and Perspectives

MS is a powerful tool in qualitative and quantitative analysis of compounds with biological relevance. Currently available analyte ionization techniques, especially ESI and MALDI, provide ionization for a wide variety of biomolecules, from small cellular metabolites up to large biomolecular aggregates. The ions generated in this way may be studied with a wide range of mass analyzers, providing molecular mass or weight information on intact molecules. A plethora of tandem MS instruments, featuring a range of different ion activation and dissociation techniques, enables structure elucidation and sequencing. As such, MS is a powerful tool to study biomolecular interactions at different levels, as readily demonstrated in subsequent chapters of the book.

References

1. Murray, K.K., Boyd, R.K., Eberlin, M.N., Langley, G.J., Li, L., and Naito, Y. (2013) Definitions of terms relating to mass spectrometry (IUPAC Recommendations 2013). *Pure Appl. Chem.*, **85**, 1515–1609.
2. Price, P. (1991) Standard definitions of terms relating to mass spectrometry. A report from the committee on measurements and standards of the American Society for Mass Spectrometry. *J. Am. Soc. Mass Spectrom.*, **2**, 336–348.
3. Sparkman, O.D. (2000) *Mass Spec Desk Reference*, Global View Publishing, Pittsburgh, PA.
4. Urban, J., Afseth, N.K., and Štys, D. (2014) Fundamental definitions and confusions in mass spectrometry about

mass assignment, centroiding and resolution. *Trends Anal. Chem.*, **52**, 126–136.

5. Brenton, A.G. and Godfrey, A.R. (2010) Accurate mass measurement: terminology and treatment of data. *J. Am. Soc. Mass Spectrom.*, **21**, 1821–1835.

6. Kind, T. and Fiehn, O. (2006) Metabolomic database annotations via query of elemental compositions: mass accuracy is insufficient even at less than 1 ppm. *BMC Bioinf.*, **7**, 234.

7. Kind, T. and Fiehn, O. (2007) Seven golden rules for heuristic filtering of molecular formulas obtained by accurate mass spectrometry. *BMC Bioinf.*, **8**, 105.

8. Pelander, A., Tyrkkö, E., and Ojanperä, I. (2009) In silico methods for predicting metabolism and mass fragmentation applied to quetiapine in liquid chromatography/time-of-flight mass spectrometry urine drug screening. *Rapid Commun. Mass Spectrom.*, **23**, 506–514.

9. Nakabayashi, R., Sawada, Y., Yamada, Y., Suzuki, M., Hirai, M.Y., Sakurai, T., and Saito, K. (2013) Combination of liquid chromatography-Fourier transform ion cyclotron resonance-mass spectrometry with ^{13}C-labeling for chemical assignment of sulfur-containing metabolites in onion bulbs. *Anal. Chem.*, **85**, 1310–1315.

10. Scigelova, M., Hornshaw, M., Giannokopulos, A., and Makarov, A. (2011) Fourier transform mass spectrometry. *Mol. Cell. Proteomics*, **10**, M111.009431.1–M111.009431.19.

11. Vestal, M.L. (1983) Ionization techniques for nonvolatile molecules. *Mass Spectrom. Rev.*, **2**, 447–480.

12. Cech, N.B. and Enke, C.G. (2001) Practical implications of some recent studies in electrospray ionization fundamentals. *Mass Spectrom. Rev.*, **20**, 362–387.

13. Gabelica, V. and De Pauw, E. (2005) Internal energy and fragmentation of ions produced in electrospray sources. *Mass Spectrom. Rev.*, **24**, 566–587.

14. Kelly, R.T., Tolmachev, A.V., Page, J.S., Tang, K., and Smith, R.D. (2010) The ion funnel: theory, implementations, and applications. *Mass Spectrom. Rev.*, **29**, 294–312.

15. Giles, K., Pringle, S.D., Worthington, K.R., Little, D., Wildgoose, J.L., and Bateman, R.H. (2004) Applications of a travelling wave-based radio-frequency-only stacked ring ion guide. *Rapid Commun. Mass Spectrom.*, **18**, 2401–2414.

16. Lorenzen, K. and van Duijn, E. (2010) Native mass spectrometry as a tool in structural biology. *Curr. Protoc. Protein Sci.*, **62**, 17.12.1–17.12.17.

17. Kebarle, P. and Peschke, M. (2000) On the mechanisms by which the charged droplets produced by electrospray lead to gas phase ions. *Anal. Chim. Acta*, **406**, 11–35.

18. Smith, R.D. and Light-Wahl, K.J. (1993) The observation of non-covalent interactions in solution by electrospray ionization mass spectrometry: promise, pitfalls and prognosis. *Biol. Mass Spectrom.*, **22**, 493–501.

19. Dole, M., Hines, R.L., Mack, L.L., Mobley, R.C., Ferguson, L.D., and Alice, M.B. (1968) Molecular beams of macroions. *J. Chem. Phys.*, **49**, 2240–2249.

20. Iribarne, J.V. and Thomson, B.A. (1976) On the evaporation of small ions from charged droplets. *J. Chem. Phys.*, **64**, 2287–2294.

21. Fenn, J.B., Mann, M., Meng, C.K., Wong, S.F., and Whitehouse, C.M. (1990) Electrospray ionization – principles and practice. *Mass Spectrom. Rev.*, **9**, 37–70.

22. Wilm, M.S. and Mann, M. (1996) Analytical properties of the nanoelectrospray ion source. *Anal. Chem.*, **68**, 1–8.

23. Wolters, D.A., Washburn, M.P., and Yates, J.R. III, (2001) An automated multidimensional protein identification technology for shotgun proteomics. *Anal. Chem.*, **73**, 5683–5690.

24. Nägele, E., Vollmer, M., and Hörth, P. (2003) Two-dimensional nano-liquid chromatography-mass spectrometry system for applications in proteomics. *J. Chromatogr. A*, **1009**, 197–205.

25. Yin, H., Killeen, K., Brennen, R., Sobek, D., Werlich, M., and van de Goor, T.

(2005) Microfluidic chip for peptide analysis with an integrated HPLC column, sample enrichment column, and nanoelectrospray tip. *Anal. Chem.*, **77**, 527–533.

26. Tanaka, K., Waki, H., Ido, Y., Akita, S., Yoshida, Y., and Yoshida, T. (1988) Protein and polymer analyses up to m/z 100 000 by laser ionization time-of-flight mass spectrometry. *Rapid Commun. Mass Spectrom.*, **2**, 151–153.

27. Karas, M. and Hillenkamp, F. (1988) Laser desorption ionization of proteins with molecular masses exceeding 10 000 Daltons. *Anal. Chem.*, **60**, 2299–2301.

28. Karas, M. (1996) Matrix-assisted laser desorption ionization MS: a progress report. *Biochem. Soc. Trans.*, **24**, 897–900.

29. Marvin, L.F., Roberts, M.A., and Fay, L.B. (2003) Matrix-assisted laser desorption/ionization time-of-flight mass spectrometry in clinical chemistry. *Clin. Chim. Acta*, **337**, 11–21.

30. Harvey, D.J. (2012) Analysis of carbohydrates and glycoconjugates by matrix-assisted laser desorption/ionization mass spectrometry: an update for 2007-2008. *Mass Spectrom. Rev.*, **31**, 183–311.

31. Angel, P.M. and Caprioli, R.M. (2013) Matrix-assisted laser desorption ionization imaging mass spectrometry: in situ molecular mapping. *Biochemistry*, **52**, 3818–3828.

32. Clark, A.E., Kaleta, E.J., Arora, A., and Wolk, D.M. (2013) Matrix-assisted laser desorption ionization-time of flight mass spectrometry: a fundamental shift in the routine practice of clinical microbiology. *Clin. Microbiol. Rev.*, **26**, 547–603.

33. Tang, N., Tornatore, P., and Weinberger, S.R. (2004) Current developments in SELDI affinity technology. *Mass Spectrom. Rev.*, **23**, 34–44.

34. Van Berkel, G.J., Pasilis, S.P., and Ovchinnikova, O. (2008) Established and emerging atmospheric pressure surface sampling/ionization techniques for mass spectrometry. *J. Mass Spectrom.*, **43**, 1161–1180.

35. Takáts, Z., Wiseman, J.M., and Cooks, R.G. (2005) Ambient mass spectrometry using desorption electrospray ionization (DESI): instrumentation, mechanisms and applications in forensics, chemistry, and biology. *J. Mass Spectrom.*, **40**, 1261–1275.

36. Wu, C., Dill, A.L., Eberlin, L.S., Cooks, R.G., and Ifa, D.R. (2013) Mass spectrometry imaging under ambient conditions. *Mass Spectrom. Rev.*, **32**, 218–243.

37. Li, A., Wang, H., Ouyang, Z., and Cooks, R.G. (2011) Paper spray ionization of polar analytes using non-polar solvents. *Chem. Commun.*, **47**, 2811–2813.

38. Zhang, Y., Ju, Y., Huang, C., and Wysocki, V.H. (2014) Paper spray ionization of noncovalent protein complexes. *Anal. Chem.*, **86**, 1342–1346.

39. Niessen, W.M.A., Manini, P., and Andreoli, R. (2006) Matrix effects in quantitative pesticide analysis using liquid chromatography-mass spectrometry. *Mass Spectrom. Rev.*, **25**, 881–899.

40. García, M.C., Hogenboom, A.C., Zappey, H., and Irth, H. (2002) Effect of the mobile phase composition on the separation and detection of intact proteins by reversed-phase liquid chromatography-electrospray mass spectrometry. *J. Chromatogr. A*, **957**, 187–199.

41. Kostiainen, R. and Kauppila, T.J. (2009) Effect of eluent on the ionization process in liquid chromatography-mass spectrometry. *J. Chromatogr. A*, **1216**, 685–699.

42. Börnsen, K.O., Gass, M.A., Bruin, G.J., von Adrichem, J.H., Biro, M.C., Kresbach, G.M., and Ehrat, M. (1997) Influence of solvents and detergents on matrix-assisted laser desorption/ionization mass spectrometry measurements of proteins and oligonucleotides. *Rapid Commun. Mass Spectrom.*, **11**, 603–609.

43. Amini, A., Dormady, S.J., Riggs, L., and Regnier, F.E. (2000) The impact of buffers and surfactants from micellar electrokinetic chromatography on

matrix-assisted laser desorption ionization (MALDI) mass spectrometry of peptides. Effect of buffer type and concentration on mass determination by MALDI-time-of-flight mass spectrometry. *J. Chromatogr. A*, **894**, 345–355.

44. De Hoffmann, E. and Stroobant, V. (2007) *Mass Spectrometry. Principles and Applications*, 3rd edn, John Wiley & Sons, Ltd, Chichester.

45. Makarov, A., Denisov, E., Kholomeev, A., Balschun, W., Lange, O., Strupat, K., and Horning, S. (2006) Performance evaluation of a hybrid linear ion trap/orbitrap mass spectrometer. *Anal. Chem.*, **78**, 2113–2120.

46. Miller, P.E. and Denton, M.B. (1986) The quadrupole mass filter: basic operating concepts. *J. Chem. Educ.*, **63**, 617–622.

47. Hager, J.W. (2002) A new linear ion trap mass spectrometer. *Rapid Commun. Mass Spectrom.*, **16**, 512–526.

48. Schwartz, J.C., Senko, M.W., and Syka, J.E.P. (2002) A two-dimensional quadrupole ion trap mass spectrometer. *J. Am. Soc. Mass Spectrom.*, **13**, 659–669.

49. Douglas, D.J., Frank, A.J., and Mao, D. (2005) Linear ion traps in mass spectrometry. *Mass Spectrom. Rev.*, **24**, 1–29.

50. Jonscher, K.R. and Yates, J.R. III, (1997) The quadrupole ion trap mass spectrometer – a small solution to a big challenge. *Anal. Biochem.*, **244**, 1–15.

51. March, R.E. (1997) An introduction to quadrupole ion trap mass spectrometry. *J. Mass Spectrom.*, **32**, 351–369.

52. Olsen, J.V., Schwartz, J.C., Griep-Raming, J., Nielsen, M.L., Damoc, E., Denisov, E., Lange, O., Remes, P., Taylor, D., Splendore, M., Wouters, E.R., Senko, M., Makarov, A., Mann, M., and Horning, S. (2009) A dual pressure linear ion trap Orbitrap instrument with very high sequencing speed. *Mol. Cell. Proteomics*, **8**, 2759–2769.

53. Guilhaus, M., Selby, D., and Mlynski, V. (2000) Orthogonal acceleration time-of-flight mass spectrometry. *Mass Spectrom. Rev.*, **19**, 65–107.

54. Lacorte, S. and Fernandez-Alba, A.R. (2006) Time of flight mass spectrometry applied to the liquid chromatographic analysis of pesticides in water and food. *Mass Spectrom. Rev.*, **25**, 866–880.

55. Vestal, M.L., Juhasz, P., and Martin, S.A. (1995) Delayed extraction matrix-assisted laser desorption time-of-flight mass spectrometry. *Rapid Commun. Mass Spectrom.*, **9**, 1044–1050.

56. Doroshenko, V.M. and Cotter, R.J. (1989) Ideal velocity focusing in a reflectron time-of-flight mass spectrometer. *J. Am. Soc. Mass Spectrom.*, **10**, 992–999.

57. Xie, C., Zhong, D., Yu, K., and Chen, X. (2012) Recent advances in metabolite identification and quantitative bioanalysis by LC-Q-TOF MS. *Bioanalysis*, **4**, 937–959.

58. Marshall, A., Hendrickson, C.L., and Jackson, G.S. (1998) Fourier transform ion cyclotron resonance mass spectrometry: a primer. *Mass Spectrom. Rev.*, **17**, 1–35.

59. Römpp, A., Taban, I.M., Mihalca, R., Duursma, M.C., Mize, T.H., McDonnel, L.A., and Heeren, R.M. (2005) Examples of Fourier transform ion cyclotron resonance mass spectrometry developments: from ion physics to remote access biochemical mass spectrometry. *Eur. J. Mass Spectrom.*, **11**, 443–456.

60. Kelleher, N.L., Lin, H.Y., Valaskovic, G.A., Aserud, D.J., Fridriksson, E.K., and McLafferty, F.W. (1999) Top-down versus bottom-up protein characterization by tandem high-resolution mass spectrometry. *J. Am. Chem. Soc.*, **121**, 806–812.

61. Bogdanov, B. and Smith, R.D. (2005) Proteomics by FTICR mass spectrometry: top down and bottom up. *Mass Spectrom. Rev.*, **24**, 168–200.

62. Hu, Q., Noll, R.J., Li, H., Makarov, A., Hardman, M., and Cooks, G.R. (2005) The Orbitrap: a new mass spectrometer. *J. Mass Spectrom.*, **40**, 430–443.

63. Geiger, T., Cox, J., and Mann, M. (2010) Proteomics on an Orbitrap benchtop mass spectrometer using

all-ion fragmentation. *Mol. Cell. Proteomics*, **9**, 2252–2261.

64. Allen, J.S. (1947) An improved electron multiplier particle counter. *Rev. Sci. Instrum.*, **18**, 739–749.

65. Wiza, J.L. (1979) Microchannel plate detectors. *Nucl. Instrum. Methods*, **162**, 587–601.

66. Busch, K.L., Glish, G.L., and McLuckey, S.A. (1988) *Mass Spectrometry–Mass Spectrometry. Techniques and Applications of Tandem Mass Spectrometry*, VCH Publishers, Inc., New York.

67. Hipple, J.A., Fox, R.E., and Condon, E.U. (1946) Metastable ions formed by electron impact in hydrocarbon gases. *Phys. Rev.*, **69**, 347–356.

68. Yost, R.A. and Enke, C.G. (1978) Selected ion fragmentation with a tandem quadrupole mass spectrometer. *J. Am. Chem. Soc.*, **100**, 2274–2275.

69. Louris, J.N., Cooks, R.G., Syka, J.E.P., Kelley, P.E., Stafford, G.C. Jr., and Todd, J.F.J. (1987) Instrumentation, applications, and energy deposition in quadrupole ion-trap tandem mass spectrometry. *Anal. Chem.*, **59**, 1677–1685.

70. Morris, H.R., Paxton, T., Dell, A., Langhorne, J., Berg, M., Bordoli, R.S., Hoyes, J., and Bateman, R.H. (1996) High-sensitivity collisionally-activated decomposition tandem mass spectrometry on a novel quadrupole–orthogonal acceleration time-of-flight mass spectrometer. *Rapid Commun. Mass Spectrom.*, **10**, 889–896.

71. Bienvenut, W.V., Déon, C., Pasquarello, C., Campbell, J.M., Sanchez, J.C., Vestal, M.L., and Hochstrasser, D.F. (2002) Matrix-assisted laser desorption/ionization-tandem mass spectrometry with high resolution and sensitivity for identification and characterization of proteins. *Proteomics*, **2**, 868–876.

72. Johnson, J.V., Yost, R.A., Kelley, P.E., and Bradford, D.C. (1990) Tandem-in-space and tandem-in-time mass spectrometry: triple quadrupoles and quadrupole ion traps. *Anal. Chem.*, **62**, 2162–2172.

73. Sleno, L. and Volmer, D.A. (2004) Ion activation methods for tandem mass spectrometry. *J. Mass Spectrom.*, **39**, 1091–1112.

74. Laskin, J. and Futrell, J.H. (2005) Activation of large ions in FT-ICR mass spectrometry. *Mass Spectrom. Rev.*, **24**, 135–167.

75. Zhurov, K.O., Fornelli, L., Wodrich, M.D., Laskay, Ü.A., and Tsybin, Y.O. (2013) Principles of electron capture and transfer dissociation mass spectrometry applied to peptide and protein structure analysis. *Chem. Soc. Rev.*, **42**, 5014–5030.

76. Kim, M.-S. and Pandey, A. (2012) Electron transfer dissociation mass spectrometry in proteomics. *Proteomics*, **12**, 530–542.

77. Wuhrer, M., Catalina, M.I., Deelder, A.M., and Hokke, C.H. (2007) Glycoproteomics based on tandem mass spectrometry of glycopeptides. *J. Chromatogr. B*, **849**, 115–128.

78. Zubarev, R.A., Kelleher, N.L., and McLafferty, F.W. (1998) Electron capture dissociation of multiply charged protein cations. A nonergodic process. *J. Am. Chem. Soc.*, **120**, 3265–3266.

79. Roepstorff, P. and Fohlman, J. (1984) Proposal for a common nomenclature for sequence ions in mass spectra of peptides. *Biomed. Mass Spectrom.*, **11**, 601.

80. Syka, J.E., Coon, J.J., Schroeder, M.J., Shabanowitz, J., and Hunt, D.F. (2004) Peptide and protein sequence analysis by electron transfer dissociation mass spectrometry. *Proc. Natl. Acad. Sci. U.S.A.*, **101**, 9528–9533.

81. Mansoori, B.A., Dyer, E.W., Lock, C.M., Bateman, K., Boyd, R.K., and Thomson, B.A. (1998) Analytical performance of a high-pressure radiofrequency-only quadrupole collision cell with a axial field applied by using conical rods. *J. Am. Soc. Mass Spectrom.*, **9**, 775–788.

82. van Dongen, W.D. and Niessen, W.M.A. (2012) LC-MS systems for quantitative bioanalysis. *Bioanalysis*, **4**, 2391–2399.

83. Hopfgartner, G. and Bourgogne, E. (2003) Quantitative high-throughput analysis of drugs in biological matrices by mass spectrometry. *Mass Spectrom. Rev.*, **22**, 195–214.

84. Xu, R.N., Fan, L., Rieser, M.J., and El-Shourbagy, T.A. (2007) Recent advances in high-throughput quantitative bioanalysis by LC-MS/MS. *J. Pharm. Biomed. Anal.*, **44**, 342–355.

85. Jemal, M., Ouyang, Z., and Xia, Y.Q. (2010) Systematic LC-MS/MS bioanalytical method development that incorporates plasma phospholipids risk avoidance, usage of incurred sample and well thought-out chromatography. *Biomed. Chromatogr.*, **24**, 2–19.

86. Cherta, L., Portolés, T., Beltran, J., Pitarch, E., Mol, J.G., and Hernández, F. (2013) Application of gas chromatography-(triple quadrupole) mass spectrometry with atmospheric pressure chemical ionization for the determination of multiclass pesticides in fruits and vegetables. *J. Chromatogr. A*, **1314**, 224–240.

87. Hunt, D.F., Shabanowitz, J., Harvey, T.M., and Coates, M.L. (1983) Analysis of organics in the environment by functional group using a triple quadrupole mass spectrometer. *J. Chromatogr.*, **271**, 93–105.

88. Johnson, J.V. and Yost, R.A. (1985) Tandem mass spectrometry for trace analysis. *Anal. Chem.*, **57**, 758A–768A.

89. Wolfender, J.-L., Rodriguez, S., and Hostettmann, K. (1998) Liquid chromatography coupled to mass spectrometry and nuclear magnetic resonance spectroscopy for the screening of plant constituents. *J. Chromatogr. A*, **794**, 299–316.

90. Kind, T. and Fiehn, O. (2010) Advances in structure elucidation of small molecules using mass spectrometry. *Bioanal. Rev.*, **2**, 23–60.

91. van der Hooft, J.J., Vervoort, J., Bino, R.J., Beekwilder, J., and de Vos, R.C. (2011) Polyphenol identification based on systematic and robust high-resolution accurate mass spectrometry fragmentation. *Anal. Chem.*, **83**, 409–416.

92. Macht, M., Asperger, A., and Deininger, S.O. (2004) Comparison of laser-induced dissociation and high-energy collision-induced dissociation using matrix-assisted laser desorption/ionization tandem time-of-flight (MALDI-TOF/TOF) for peptide and protein identification. *Rapid Commun. Mass Spectrom.*, **18**, 2093–2105.

93. Spengler, B. (1997) Post-source decay analysis in matrix-assisted laser desorption/ionization mass spectrometry of biomolecules. *J. Mass Spectrom.*, **32**, 1019–1036.

94. Cotter, R.J., Griffith, W., and Jelinek, C. (2007) Tandem time-of-flight (TOF/TOF) mass spectrometry and the curved-field reflectron. *J. Chromatogr. B*, **855**, 2–13.

95. Mechref, Y., Novotny, M.V., and Krishnan, C. (2003) Structural characterization of oligosaccharides using MALDI-TOF/TOF tandem mass spectrometry. *Anal. Chem.*, **75**, 4895–4903.

96. Franc, V., Řehulka, P., Raus, M., Stulík, J., Novak, J., Renfrow, M.B., and Šebela, M. (2013) Elucidating heterogeneity of IgA1 hinge-region O-glycosylation by use of MALDI-TOF/TOF mass spectrometry: role of cysteine alkylation during sample processing. *J. Proteomics*, **92**, 299–312.

97. Chernushevich, I.V., Loboda, A.V., and Thomson, B.A. (2001) An introduction to quadrupole–time-of-flight mass spectrometry. *J. Mass Spectrom.*, **36**, 849–865.

98. Campbell, J.L. and Le Blanc, J.C.Y. (2012) Using high-resolution quadrupole TOF technology in DMPK analyses. *Bioanalysis*, **4**, 487–500.

99. Michael, S.M., Chien, B.M., and Lubman, D.M. (1992) An ion trap storage/time-of-flight mass spectrometer. *Rev. Sci. Instrum.*, **63**, 4277–4284.

100. Michael, S.M., Chien, B.M., and Lubman, D.M. (1993) Detection of electrospray ionization using a quadrupole ion trap storage/reflectron time-of-flight mass spectrometer. *Anal. Chem.*, **65**, 2614–2620.

101. Liu, Z.Y. (2012) An introduction to hybrid ion trap/time-of-flight mass spectrometry coupled with liquid chromatography applied to drug metabolism studies. *J. Mass Spectrom.*, **47**, 1627–1642.

102. Giera, M., de Vlieger, J.S.B., Niessen, W.M.A., Lingeman, H., and Irth, H.

(2010) Structure elucidation of biologically active neomycin N-octyl derivatives in a regioisomeric mixture by means of liquid chromatography – ion trap time-of-flight mass spectrometry. *Rapid Commun. Mass Spectrom.*, **24**, 1439–1446.

103. Hopfgartner, G., Varesio, E., Tschäppät, V., Grivet, C., Bourgogne, E., and Leuthold, L.A. (2004) Triple quadrupole linear ion trap mass spectrometer for the analysis of small molecules and macromolecules. *J. Mass Spectrom.*, **39**, 845–855.

104. Xia, Y.Q., Miller, J.D., Bakhtiar, R., Franklin, R.B., and Liu, D.Q. (2003) Use of a quadrupole linear ion trap mass spectrometer in metabolite identification and bioanalysis. *Rapid Commun. Mass Spectrom.*, **17**, 1137–1145.

105. Patrie, S.M., Charlebois, J.P., Whipple, D., Kelleher, N.L., Hendrickson, C.L., Quinn, J.P., Marshall, A.G., and Mukhopadhyay, B. (2004) Construction of a hybrid quadrupole–Fourier transform ion cyclotron resonance mass spectrometer for versatile MS–MS above 10 kDa. *J. Am. Soc. Mass Spectrom.*, **15**, 1099–1108.

106. Wu, S.L., Jardine, I., Hancock, W.S., and Karger, B.L. (2004) A new and sensitive on-line liquid chromatography/mass spectrometric approach for top-down protein analysis: the comprehensive analysis of human growth hormone in an E. coli lysate using a hybrid linear ion trap/Fourier transform ion cyclotron resonance mass spectrometer. *Rapid Commun. Mass Spectrom.*, **18**, 2201–2207.

107. Syka, J.E., Marto, J.A., Bai, D.L., Horning, S., Senko, M.W., Schwartz, J.C., Ueberheide, B., Garcia, B., Busby, S., Muratore, T., Shabanowitz, J., and Hunt, D.F. (2004) Novel linear quadrupole ion trap/FT mass spectrometer: performance characterization and use in the comparative analysis of histone H3 post-translational modifications. *J. Proteome Res.*, **3**, 621–626.

108. Cheng, X.H., Chen, R.D., Bruce, J.E., Schwartz, B.L., Anderson, G.A., Hofstadler, S.A., Gale, D.C., Smith, R.D., Gao, J.M., Sigal, G.B., Mammen, M., and Whitesides, G.M. (1995) Using ESI-FT-ICR-MS to study competitive binding of inhibitors to carbonic anhydrase. *J. Am. Chem. Soc.*, **117**, 8859–8860.

109. Bruce, J.E., Anderson, G.A., Chen, R., Cheng, X., Gale, D.C., Hofstadler, S.A., Schwartz, B.L., and Smith, R.D. (1995) Bio-affinity characterization mass spectrometry. *Rapid Commun. Mass Spectrom.*, **9**, 644–650.

110. Olsen, J.V., Macek, B., Lange, O., Makarov, A., Horning, S., and Mann, M. (2007) Higher-energy C-trap dissociation for peptide modification analysis. *Nat. Methods*, **4**, 709–712.

111. Michalski, A., Damoc, E., Hauschild, J.P., Lange, O., Wieghaus, A., Makarov, A., Nagaraj, N., Cox, J., Mann, M., and Horning, S. (2011) Mass spectrometry-based proteomics using Q Exactive, a high-performance benchtop quadrupole orbitrap mass spectrometer. *Mol. Cell. Proteomics*, **10**, M111.011015(1-12).

112. Senko, M.W., Remes, P.M., Canterbury, J.D., Mathur, R., Song, Q., Eliuk, S.M., Mullen, C., Earley, L., Hardman, M., Blethrow, J.D., Bui, H., Specht, A., Lange, O., Denisov, E., Makarov, A., Horning, S., and Zabrouskov, V. (2013) Novel parallelized quadrupole/linear ion trap/orbitrap tribrid mass spectrometer improving proteome coverage and Peptide identification rates. *Anal. Chem.*, **85**, 11710–11714.

113. Hill, H.H. Jr., Siems, W.F., St. Louis, R.H., and McMinn, D.G. (1990) Ion mobility spectrometry. *Anal. Chem.*, **62**, 1201A–1209A.

114. Creaser, C.S., Griffiths, J.R., Bramwell, C.J., Noreen, S., Hill, C.A., and Thomas, C.L.P. (2004) Ion mobility spectrometry: a review. Part 1. Structural analysis by mobility measurement. *Analyst*, **129**, 984–994.

115. Huang, Y. and Dodds, E.D. (2013) Ion mobility studies of carbohydrates as group I adducts: isomer specific collisional cross section dependence on metal ion radius. *Anal. Chem.*, **85**, 9728–9735.

116. Valentine, S.J., Counterman, A.E., and Clemmer, D.E. (1999) A database of

660 peptide ion cross sections: use of intrinsic size parameters for bona fide predictions of cross sections. *J. Am. Soc. Mass Spectrom.*, **10**, 1188–1211.

117. Wang, B., Valentine, S., Plasencia, M., Raghuraman, S., and Zhang, X. (2010) Artificial neural networks for the prediction of peptide drift time in ion mobility mass spectrometry. *BMC Bioinf.*, **11**, 182.

118. Wyttenbach, T., von Helden, G., and Bowers, M.T. (1996) Gas-phase conformation of biological molecules: bradykinin. *J. Am. Chem. Soc.*, **118**, 8355–8364.

119. Clemmer, D.E. and Jarrold, M.F. (1997) Ion mobility measurements and their applications to clusters and biomolecules. *J. Mass Spectrom.*, **32**, 577–592.

120. Srebalus, C.A., Li, J., Marshall, W.S., and Clemmer, D.E. (1999) Gas-phase separations of electrosprayed peptide libraries. *Anal. Chem.*, **71**, 3918–3927.

121. Kanu, A.B., Dwivedi, P., Tam, M., Matz, L., and Hill, H.H. Jr., (2008) Ion mobility-mass spectrometry. *J. Mass Spectrom.*, **43**, 1–22.

122. Williams, D.M. and Pukala, T.L. (2013) Novel insights into protein misfolding diseases revealed by ion mobility-mass spectrometry. *Mass Spectrom. Rev.*, **32**, 169–187.

123. Wyttenbach, T., Pierson, N.A., Clemmer, D.E., and Bowers, M.T. (2014) Ion mobility analysis of molecular dynamics. *Annu. Rev. Phys. Chem.*, **65**, 175–196.

124. Lapthorn, C., Pullen, F., and Chowdhry, B.Z. (2013) Ion mobility spectrometry-mass spectrometry (IMS–MS) of small molecules: separating and assigning structures to ions. *Mass Spectrom. Rev.*, **32**, 43–71.

125. Pringle, S.D., Giles, K., Wildgoose, J.L., Williams, J.P., Slade, S.E., Thalassinos, K., Bateman, R.H., Bowers, M.T., and Scrivens, J.H. (2007) An investigation of the mobility separation of some peptide and protein ions using a new hybrid quadrupole/travelling wave IMS/oa-ToF instrument. *Int. J. Mass Spectrom.*, **261**, 1–12.

126. Olivova, P., Chen, W., Chakraborty, A.B., and Gebler, J.C. (2008) Determination of *N*-glycosylation sites and site heterogeneity in a monoclonal antibody by electrospray quadrupole ion-mobility time-of-flight mass spectrometry. *Rapid Commun. Mass Spectrom.*, **22**, 29–40.

127. Guevremont, R. (2004) High-field asymmetric waveform ion mobility spectrometry: a new tool for mass spectrometry. *J. Chromatogr. A*, **1058**, 3–19.

128. Kolakowski, B.M. and Mester, Z. (2007) Review of applications of high-field asymmetric waveform ion mobility spectrometry (FAIMS) and differential mobility spectrometry (DMS). *Analyst*, **132**, 842–864.

129. Tsai, C.W., Yost, R.A., and Garrett, T.J. (2012) High-field asymmetric waveform ion mobility spectrometry with solvent vapor addition: a potential greener bioanalytical technique. *Bioanalysis*, **4**, 1363–1375.

130. Xia, Y.Q., Wu, S.T., and Jemal, M. (2008) LC-FAIMS–MS/MS for quantification of a peptide in plasma and evaluation of FAIMS global selectivity from plasma components. *Anal. Chem.*, **80**, 7137–7143.

131. Zhong, Y., Hyung, S.J., and Ruotolo, B.T. (2012) Ion mobility-mass spectrometry for structural proteomics. *Expert Rev. Proteomics*, **9**, 47–58.

132. Jurneczko, E. and Barran, P.E. (2011) How useful is ion mobility mass spectrometry for structural biology? The relationship between protein crystal structures and their collision cross sections in the gas phase. *Analyst*, **136**, 20–28.

133. Uetrecht, C., Rose, R.J., van Duijn, E., Lorenzen, K., and Heck, A.J. (2010) Ion mobility mass spectrometry of proteins and protein assemblies. *Chem. Soc. Rev.*, **39**, 1633–1655.

134. Enders, J.R. and McLean, J.A. (2009) Chiral and structural analysis of biomolecules using mass spectrometry and ion mobility-mass spectrometry. *Chirality*, **21**, E253–E264.

135. Stahl, D.C., Swiderek, K.M., Davis, M.T., and Lee, T.D. (1996) Data-controlled automation of liquid

chromatography/tandem mass spectrometry analysis of peptide mixtures. *J. Am. Soc. Mass Spectrom.*, **7**, 532–540.

136. Wenner, B.R. and Lynn, B.C. (2004) Factors that affect ion trap data-dependent MS/MS in proteomics. *J. Am. Soc. Mass Spectrom.*, **15**, 150–157.

137. Plumb, R.S., Johnson, K.A., Rainville, P., Smith, B.W., Wilson, I.D., Castro-Perez, J.M., and Nicholson, J.K. (2006) UPLC/MS^E; a new approach for generating molecular fragment information for biomarker structure elucidation. *Rapid Commun. Mass Spectrom.*, **20**, 1989–1994.

138. Cho, R., Huang, Y., Schwartz, J.C., Chen, Y., Carlson, T.J., and Ma, J. (2012) MS^M, an efficient workflow for metabolite identification using hybrid linear ion trap Orbitrap mass spectrometer. *J. Am. Soc. Mass Spectrom.*, **23**, 880–888.

139. Kostiainen, R., Kotiaho, T., Kuuranne, T., and Auriola, S. (2003) Liquid chromatography/atmospheric pressure ionization-mass spectrometry in drug metabolism studies. *J. Mass Spectrom.*, **38**, 357–372.

140. Boersema, P.J., Mohammed, S., and Heck, A.J. (2009) Phosphopeptide fragmentation and analysis by mass spectrometry. *J. Mass Spectrom.*, **44**, 861–878.

141. Commission Decision 2002/657/EC implementing Council Directive 96/23/EC concerning the performance of analytical methods and the interpretation of results. *Off. J. Eur. Commun.*, **L221**, 8–36.

142. Stolker, A.A.M., Stephany, R.W., and van Ginkel, L.A. (2000) Identification of residues by LC-MS. The application of new EU guidelines. *Analusis*, **28**, 947–951.

143. Sauvage, F.-L., Gaulier, J.M., Lachâtre, G., and Marquet, P. (2008) Pitfalls and prevention strategies for liquid chromatography-tandem mass spectrometry in the selected reaction-monitoring mode for drug analysis. *Clin. Chem.*, **54**, 1519–1527.

144. Berendsen, B.J., Stolker, L.A., and Nielen, M.W. (2013) The (un)certainty of selectivity in liquid chromatography

tandem mass spectrometry. *J. Am. Soc. Mass Spectrom.*, **24**, 154–163.

145. Ramanathan, R., Jemal, M., Ramagiri, S., Xia, Y.Q., Humpreys, W.G., Olah, T., and Korfmacher, W.A. (2011) It is time for a paradigm shift in drug discovery bioanalysis: from SRM to HRMS. *J. Mass Spectrom.*, **46**, 595–601.

146. Little, J.L., Cleven, C.D., and Brown, S.D. (2011) Identification of "known unknowns" utilizing accurate mass data and chemical abstracts service databases. *J. Am. Soc. Mass Spectrom.*, **22**, 348–359.

147. Thurman, E.M., Ferrer, I., and Fernández-Alba, A.R. (2005) Matching unknown empirical formulas to chemical structure using LC/MS TOF accurate mass and database searching: example of unknown pesticides on tomato skins. *J. Chromatogr. A*, **1067**, 127–134.

148. Staack, R.F. and Hopfgartner, G. (2007) New analytical strategies in studying drug metabolism. *Anal. Bioanal. Chem.*, **388**, 1365–1380.

149. Holčapek, M., Kolářová, L., and Nobilis, M. (2008) High-performance liquid chromatography–tandem mass spectrometry in the identification and determination of phase I and phase II drug metabolites. *Anal. Bioanal. Chem.*, **391**, 59–78.

150. Zhu, M., Zhang, H., and Humphreys, W.G. (2011) Drug metabolite profiling and identification by high-resolution mass spectrometry. *J. Biol. Chem.*, **286**, 25419–25425.

151. Singh, S., Handa, T., Narayanam, M., Sahu, A., Junwal, M., and Shah, R.P. (2012) A critical review on the use of modern sophisticated hyphenated tools in the characterization of impurities and degradation products. *J. Pharm. Biomed. Anal.*, **69**, 148–173.

152. Kerns, E.H., Volk, K.J., Hill, S.E., and Lee, M.S. (1995) Profiling new taxanes using LC/MS and LC/MS/MS substructural analysis techniques. *Rapid Commun. Mass Spectrom.*, **9**, 1539–1545.

153. Kerns, E.H., Rourick, R.A., Volk, K.J., and Lee, M.S. (1997) Buspirone metabolite structure profile using a

standard liquid chromatographic–mass spectrometric protocol. *J. Chromatogr. B*, **698**, 133–145.

154. Li, A.C., Gohdes, M.A., and Shou, W.Z. (2007) "N-in-one" strategy for metabolite identification using a liquid chromatography/hybrid triple quadrupole linear ion trap instrument using multiple dependent product ion scans triggered with full mass scan. *Rapid Commun. Mass Spectrom.*, **21**, 1421–1430.

155. Li, A.C., Shou, W.Z., Mai, T.T., and Jiang, X.Y. (2007) Complete profiling and characterization of in vitro nefazodone metabolites using two different tandem mass spectrometric platforms. *Rapid Commun. Mass Spectrom.*, **21**, 4001–4008.

156. Zhu, M., Ma, L., Zhang, D., Ray, K., Zhao, W., Humphreys, W.G., Skiles, G., Sanders, M., and Zhang, H. (2006) Detection and characterization of metabolites in biological matrices using mass defect filtering of liquid chromatography/high resolution mass spectrometry data. *Drug Metab. Dispos.*, **34**, 1722–1733.

157. Zhang, H., Zhang, D., Ray, K., and Zhu, M. (2009) Mass defect filter technique and its applications to drug metabolite identification by high-resolution mass spectrometry. *J. Mass Spectrom.*, **44**, 999–1016.

158. Liu, D.Q. and Hop, C.E.C.A. (2005) Strategies for characterization of drug metabolites using liquid chromatography–tandem mass spectrometry in conjunction with chemical derivatization and on-line H/D exchange approaches. *J. Pharm. Biomed. Anal.*, **37**, 1–18.

159. Dear, G.J., Munoz-Muriedas, J., Beaumont, C., Roberts, A., Kirk, J., Williams, J.P., and Campuzano, I. (2010) Sites of metabolic substitution: investigating metabolite structures utilising ion mobility and molecular modelling. *Rapid Commun. Mass Spectrom.*, **24**, 3157–3162.

160. Shimizu, A. and Chiba, M. (2013) Ion mobility spectrometry-mass spectrometry analysis for the site of aromatic

hydroxylation. *Drug Metab. Dispos.*, **41**, 1295–1299.

161. Shimizu, A., Ohe, T., and Chiba, M. (2012) A novel method for the determination of the site of glucuronidation by ion mobility spectrometry-mass spectrometry. *Drug Metab. Dispos.*, **40**, 1456–1459.

162. Ma, S. and Zhu, M. (2009) Recent advances in applications of liquid chromatography–tandem mass spectrometry to the analysis of reactive drug metabolites. *Chem. Biol. Interact.*, **179**, 25–37.

163. Dieckhaus, C.M., Fernández-Metzler, C.L., King, R., Krolikowski, P.H., and Baillie, T.A. (2005) Negative ion tandem mass spectrometry for the detection of glutathione conjugates. *Chem. Res. Toxicol.*, **18**, 630–638.

164. Castro-Perez, J., Plumb, R., Liang, L., and Yang, E. (2005) A high-throughput liquid chromatography/tandem mass spectrometry method for screening glutathione conjugates using exact mass neutral loss acquisition. *Rapid Commun. Mass Spectrom.*, **19**, 798–804.

165. Zheng, J., Ma, L., Xin, B., Olah, T., Humphreys, W.G., and Zhu, M. (2007) Screening and identification of GSH-trapped reactive metabolites using hybrid triple quadruple linear ion trap mass spectrometry. *Chem. Res. Toxicol.*, **20**, 757–766.

166. Yan, Z. and Caldwell, G.W. (2004) Stable-isotope trapping and high-throughput screenings of reactive metabolites using the isotope MS signature. *Anal. Chem.*, **76**, 6835–6847.

167. Rousu, T., Pelkonen, O., and Tolonen, A. (2009) Rapid detection and characterization of reactive drug metabolites in vitro using several isotope-labeled trapping agents and ultra-performance liquid chromatography/time-of-flight mass spectrometry. *Rapid Commun. Mass Spectrom.*, **23**, 843–855.

168. Zhu, M., Ma, L., Zhang, H., and Humphreys, W.G. (2007) Detection and structural characterization of glutathione-trapped reactive metabolites using liquid chromatography-high-resolution mass spectrometry and

mass defect filtering. *Anal. Chem.*, **79**, 8333–8341.

169. Mann, M., Meng, C.K., and Fenn, J.B. (1989) Interpreting mass spectra of multiply charged ions. *Anal. Chem.*, **61**, 1702–1708.

170. Smith, R.D., Loo, J.A., Edmonds, C.G., Barinaga, C.J., and Udseth, H.R. (1990) New developments in biochemical mass spectrometry: electrospray ionization. *Anal. Chem.*, **62**, 882–899.

171. Covey, T.R., Bonner, R.F., Shushan, B.I., and Henion, J.D. (1988) The determination of protein, oligonucleotide and peptide molecular weights by ESI-MS. *Rapid Commun. Mass Spectrom.*, **2**, 249–256.

172. Ferrige, A.G., Seddon, M.J., Green, B.N., Jarvis, S.A., and Skilling, J. (1992) Disentangling electrospray spectra with maximum entropy. *Rapid Commun. Mass Spectrom.*, **6**, 707–711.

173. Castleberry, C.M., Rodicio, L.P., and Limbach, P.A. (2008) Electrospray ionization mass spectrometry of oligonucleotides. *Curr. Protoc. Nucleic Acid Chem.*, **10**, 10.2.1–10.2.19.

174. Cui, W., Rohrs, H.W., and Gross, M.L. (2011) Top-down mass spectrometry: recent developments, applications and perspectives. *Analyst*, **136**, 3854–3864.

175. Zhou, H., Ning, Z., Starr, A.E., Abu-Farha, M., and Figeys, D. (2012) Advancements in top-down proteomics. *Anal. Chem.*, **84**, 720–734.

176. Lanucara, F. and Eyers, C.E. (2013) Top-down mass spectrometry for the analysis of combinatorial post-translational modifications. *Mass Spectrom. Rev.*, **32**, 27–42.

177. Konermann, L., Vahidi, S., and Sowole, M.A. (2014) Mass spectrometry methods for studying structure and dynamics of biological macromolecules. *Anal. Chem.*, **86**, 213–232.

178. Demeure, K., Quinton, L., Gabelica, V., and De Pauw, E. (2007) Rational selection of the optimum MALDI matrix for top-down proteomics by in-source decay. *Anal. Chem.*, **79**, 8678–8685.

179. Fagerquist, C.K. and Sultan, O. (2011) Induction and identification of disulfide-intact and disulfide-reduced β-subunit of Shiga toxin 2 from Escherichia coli O157:H7 using MALDI-TOF-TOF-MS/MS and top-down proteomics. *Analyst*, **136**, 1739–1746.

180. Zhang, Y., Fonslow, B.R., Shan, B., Baek, M.C., and Yates, J.R. III, (2013) Protein analysis by shotgun/bottom-up proteomics. *Chem. Rev.*, **113**, 2343–2394.

181. Switzar, L., Giera, M., and Niessen, W.M.A. (2013) Protein digestion: an overview of the available techniques and recent developments. *J. Proteome Res.*, **12**, 1067–1077.

182. Yates, J.R. III, (1998) Mass spectrometry and the age of the proteome. *J. Mass Spectrom.*, **33**, 1–19.

183. Henzel, W.J., Billeci, T.M., Stults, J.T., Wong, S.C., Grimley, C., and Watanabe, C. (1993) Identifying proteins from two-dimensional gels by molecular mass searching of peptide fragments in protein sequence databases. *Proc. Natl. Acad. Sci. U.S.A.*, **90**, 5011–5015.

184. Mann, M., Højrup, P., and Roepstorff, P. (1993) Use of mass spectrometric molecular weight information to identify proteins in sequence databases. *Biol. Mass Spectrom.*, **22**, 338–345.

185. Shevchenko, A., Jensen, O.N., Podtelejnikov, A.V., Sagliocco, F., Wilm, M., Vorm, O., Mortensen, P., Shevchenko, A., Boucherie, H., and Mann, M. (1996) Linking genome and proteome by mass spectrometry: large-scale identification of yeast proteins from two dimensional gels. *Proc. Natl. Acad. Sci. U.S.A.*, **93**, 14440–14445.

186. Gevaert, K. and Vandekerckhove, J. (2000) Protein identification methods in proteomics. *Electrophoresis*, **21**, 1145–1154.

187. Yates, J.R. III, Eng, J.K., and McCormack, A.L. (1995) Mining genomes: correlating tandem mass spectra of modified and unmodified peptides to sequences in nucleotide databases. *Anal. Chem.*, **67**, 3202–3210.

188. Liska, A.J. and Shevchenko, A. (2003) Combining mass spectrometry with database interrogation strategies in proteomics. *Trends Anal. Chem.*, **22**, 291–298.

189. Perkins, D.N., Pappin, D.J., Creasy, D.M., and Cottrell, J.S. (1999) Probability-based protein identification by searching sequence databases using mass spectrometry data. *Electrophoresis*, **20**, 3551–3567.

190. Ducret, A., Van Oostveen, I., Eng, J.K., Yates, J.R. III, and Aebersold, R. (1998) High throughput protein characterization by automated reverse-phase chromatography/electrospray tandem mass spectrometry. *Protein Sci.*, **7**, 706–719.

191. Ahrné, E., Müller, M., and Lisacek, F. (2010) Unrestricted identification of modified proteins using MS/MS. *Proteomics*, **10**, 671–686.

192. Wu, C., Tran, J.C., Zamdborg, L., Durbin, K.R., Li, M., Ahlf, D.R., Early, B.P., Thomas, P.M., Sweedler, J.V., and Kelleher, N.L. (2012) A protease for "middle-down" proteomics. *Nat. Methods*, **9**, 822–824.

193. Moradian, A., Kalli, A., Sweredoski, M.J., and Hess, S. (2014) The top-down, middle-down, and bottom-up mass spectrometry approaches for characterization of histone variants and their post-translational modifications. *Proteomics*, **14**, 489–497.

Part I
Direct MS Based Affinity Techniques

Analyzing Biomolecular Interactions by Mass Spectrometry, First Edition.
Edited by Jeroen Kool and Wilfried M.A. Niessen.
© 2015 Wiley-VCH Verlag GmbH & Co. KGaA. Published 2015 by Wiley-VCH Verlag GmbH & Co. KGaA.

2
Studying Protein–Protein Interactions by Combining Native Mass Spectrometry and Chemical Cross-Linking

Michal Sharon and Andrea Sinz

2.1
Introduction

Nearly every biological process is coordinated by protein–protein interactions, as reflected in the vast number of protein interactions (between 300 000 and 650 000) that are predicted to form between the ~21 000 human genes [1–3]. In fact, it is estimated that the majority of proteins exist as multisubunit assemblies, forming hundreds of different functional protein complexes within the cellular environment [4]. These protein assemblies, functioning in concert, enable cell viability and homeostasis [5]. Therefore, to understand the workings of a cell, it is crucial to reveal the structure of protein assemblies, and determine how interactions between subunits of complexes translate into function. Such knowledge will not only provide a basic, mechanistic understanding of cellular processes, but may also yield attractive targets for drugs that disrupt protein–protein binding interfaces [6, 7].

The gold standard in structural characterization of protein complexes is to elucidate the structures of protein assemblies at atomic resolution. This can be accomplished by means of X-ray crystallography and nuclear magnetic resonance (NMR) [8, 9]; however, many protein targets are not compatible with these techniques due to their low solubility, weak stability, large size, or inability to form crystals. For example, membrane protein complexes are still notably underrepresented in structural data sets, despite their critical biological importance [10]. This gap in structural information can be bridged by low-resolution methods that provide complementary structural data. Interacting protein partners, for instance, can be identified by yeast two-hybrid screens [11], fluorescence resonance energy transfer (FRET) [12], native mass spectrometry (MS) [13–15], and chemical cross-linking [16]. Cryoelectron microscopy [17], small-angle X-ray scattering (SAXS) [18], and ion-mobility MS [19] can reveal the overall architecture of a protein complex. Residues that mediate protein–protein interactions can be identified by hydrogen–deuterium exchange (HDX) [20], chemical cross-linking [16, 21], and chemical footprinting [22], while the relative

Analyzing Biomolecular Interactions by Mass Spectrometry, First Edition.
Edited by Jeroen Kool and Wilfried M.A. Niessen.
© 2015 Wiley-VCH Verlag GmbH & Co. KGaA. Published 2015 by Wiley-VCH Verlag GmbH & Co. KGaA.

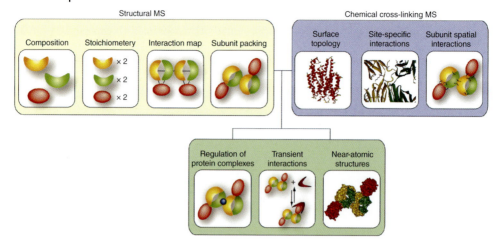

Figure 2.1 Advancing structural investigation of large protein assemblies by integrating native MS and chemical cross-linking MS. Analysis of the intact protein complexes by native MS can provide information on the composition, stoichiometry, subunit architecture, and topological organization of an assembly. Chemical cross-linking MS can probe protein surface topology, reveal specific protein–protein interaction sites, and map the intermolecular subunit contacts of protein complexes. Consequently, "marrying" these two complementary approaches and using the structural data provided by each method will facilitate investigations of challenging protein systems, for example, dynamic assemblies or proteins that are not amenable to high-resolution structural techniques such as X-ray crystallography or NMR. Here, this combination is demonstrated for studies of the interaction between a large multisubunit complex and a relatively small protein, for characterizing labile protein interactions, and for generating three-dimensional, near-atomic models, when combined with computational modeling.

orientations of subunits can be inferred from cryoelectron microscopy [17], SAXS [18], and ion-mobility MS [19].

In this chapter, we will outline the specific contributions of native MS [13–15] and chemical cross-linking MS [16, 21] to reveal the structural features of protein assemblies. While each method has its own advantages and limitations, we will focus on their enhanced value in combination: a whole that is more than the sum of its parts (Figure 2.1). We begin by describing the types of information that can be gained through the use of each individual technique, and then highlight the additional capabilities that emerge when the two methods are integrated. Lastly, we point out the challenges that lie ahead, and the future directions we foresee for this hybrid approach.

2.2
Protein Analysis by Mass Spectrometry

MS is a method that measures the mass-to-charge ratio (m/z) of an ion. In the course of analysis, the sample of interest is first ionized and transferred into the gas phase. One of the most prominent ways to ionize a protein sample involves electrospray ionization (ESI) [23], the method of choice for the applications we describe here; however, an alternative soft ionization technique, matrix-assisted laser desorption/ionization (MALDI), is complementarily used [24–27]. Once ions are generated, they are focused into an ion beam and separated within a mass analyzer, according to their m/z ratios. Finally, ions that emerge from the mass analyzer are sensed by the detector.

Instruments based on a combination of two or more mass analyzers enable the performance of tandem mass spectrometry experiments (MS/MS) [28]. In this type of analysis, a specific m/z ratio is isolated from the rest of the ions, and activated within a collision cell to induce their dissociation into smaller components. The ions thus produced are then analyzed in the second mass analyzer. Multiple types of mass analyzers (e.g., time-of-flight (TOF), quadrupole, orbitrap) are available to date, as well as various dissociation methods (e.g., collision-induced dissociation (CID), electron-transfer dissociation (ETD), electron-capture dissociation (ECD), and surface-induced dissociation (SID)); reviewed in [29–31]. As a result, a number of configurations of mass spectrometers, each of which vary in their capabilities and limitations, can be designed for different purposes.

Analysis of large protein assemblies by native MS and chemical cross-linking MS, the MS approaches we discuss here, require prior purification of the protein complex of interest. This can be achieved through reconstitution methods, that is, overexpressing and purifying recombinant versions of each component and then reconstituting the complex, either *in vitro* or, alternatively, *in vivo*, by coexpressing multiple subunits [32]. While reconstituted systems often benefit from large yields that facilitate the structural analysis, they are often employed using bacteria as the host system. As a consequence, posttranslational modifications, and associations with interacting protein partners, may be lost. Therefore, various biochemical approaches have been developed that enable the isolation of endogenous protein complexes directly from cells. For example, one of the subunits is labeled with a tag (e.g., FLAG, His, GST (glutathione-*S*-transferase), TAP (tandem affinity purification)), which is then used as a bait for isolating the entire protein complex by means of an appropriate affinity column, under conditions that maintain the integrity of the protein complex [33]. Notably, the degree to which the sample is purified will dictate the quality of the data. Elimination of protein contaminants and nonvolatile adducts such as salts, chemical compounds, and detergents will yield more clearly resolved peaks.

2.3
Native MS

The field of native MS is based on the ability to maintain protein complexes intact within the mass spectrometer, by preserving the weak noncovalent interactions between protein subunits and associated biomolecules such as DNA, cofactors, and ligands [13–15]. Although mass is a one-dimensional property, combining information from analysis of the intact complex with that of the smaller subcomplexes generated can yield three-dimensional data on the assembly being analyzed. Details on the composition, stoichiometry, subunit network, and overall topology of the protein complex can be obtained, in addition to insights regarding its heterogeneity and dynamic interactions [13–15, 20, 34].

2.3.1
Instrumentation for High-Mass Ion Detection

Native MS analysis is mostly performed using a nanoflow electrospray ionization source (nESI), and a quadrupole-time-of-flight (Q-TOF) mass analyzer, modified for high mass analysis [35, 36]. nESI, in essence, is a miniaturized version of ESI [37] and requires lower flow rates (nl/min) of the ionized solution in comparison to conventional ESI (μl/min). Consequently, the desolvation process is more efficient, requiring smaller amounts of sample. Typically, spectra are produced by using 1–3 μl of sample in the micromolar range, which is then loaded into gold-coated borosilicate or quartz capillaries with a tip diameter on the order of 1 μm [38]. To maintain the noncovalent interactions of large complexes within the Q-TOF mass spectrometer, higher pressures are often used in the initial vacuum stages of the instrument, which stabilize even relatively labile macromolecular protein complexes [36]. In addition, low-frequency quadrupoles are employed, in order to broaden the mass resolving capabilities to include species with m/z values larger than 6000, yet at the expense of resolution [35].

Recently, analyses of large protein complexes have also become feasible by using a modified orbitrap instrument [39]. While this instrument is still incapable of performing tandem MS analysis, its enhanced resolution ensures direct analysis of posttranslational modifications on intact proteins and assemblies, enabling resolution of mass differences down to 25 Da when analyzing, for example, intact immunoglobulin G (146 kDa) [40]. Moreover, its impressive sensitivity reduces the sample requirements to the femtomole range, a feature that will likely assist in broadening the scope of application of native MS to less abundant protein assemblies.

2.3.2
Defining the Exact Mass of the Composing Subunits

In general, investigation of protein assemblies by means of native MS is a multistep process. First, the unique mass of each of the protein components in the sample is defined. This is a crucial stage, as assigning a mass to subunits based on information contained in a protein sequence database is often impossible, as proteins are truncated and posttranslationally modified to such an extent that there is no correlation between the final product and the theoretical mass. Especially, this scenario takes place when studying uncharacterized endogenous protein complexes isolated directly from cells.

One way to define the exact mass of each of the individual subunits is through MS analysis of a denatured solution of the complex [41]. While this method enables accurate mass measurements of the individual components in a single step, overlapping charge states may often complicate the analysis. In such situations, a liquid chromatography (LC) separation method introduced before MS analysis may be utilized. Such an approach was recently demonstrated using a monolithic column [42]. The constituent subunits of a protein complex were separated on the column, based on their sizes and chemical properties. The eluted flow was then split into two parts: One fraction was sprayed directly into the mass spectrometer to accurately determine the mass of the individual subunits, whereas the second fraction was collected into a 96-well plate, for subsequent identification by proteomic analysis. On the basis of the elution profile, the identities of the subunits were accurately correlated with their mass [42].

2.3.3
Analyzing the Intact Complex

Once the unique mass of each subunit has been defined, the structural properties of the intact complex can be investigated. To this end, conditions within the mass spectrometer are optimized to preserve noncovalent interactions, and the spectra of the intact complex are then recorded. By measuring the mass of the assembly, taking into consideration the masses of the individual subunits, the composition and stoichiometry of subunits are tentatively assigned. In cases of multicomponent assemblies, in which multiple subunit combinations are possible, it is often beneficial to use algorithms that, given the molecular masses of subunits, are capable of calculating all possible compositions within a given error range [43]. Tandem MS analysis can then be used to validate the assignment (as described below).

One powerful advantage of MS is that all coexisting subpopulations of a protein complex can be detected within a single spectrum. As a consequence, a spectrum of the intact complex may reveal its heterogeneity, as well as its dynamic nature. Nevertheless, the investigation of heterogeneous systems is not a trivial task, as overlapping peaks may preclude the assignment. Several software packages have been developed, to assist in the analysis of such complicated spectra [44–46]. Furthermore, different volatile buffers that are compatible with MS (e.g., ammonium acetate, ethylenediammonium diacetate, and triethylammonium acetate) or reagents that can modulate ion charge can be screened for their ability to enhance the separation between sample components, thereby enabling an unambiguous assignment of the data [47–50].

The ability to isolate a discrete m/z value and induce its dissociation within the mass spectrometer, as may be achieved in tandem MS experiments, is a critical step in the structural characterization process, as assignment of the individual subunits and stripped complexes generated in the course of MS/MS analysis enables confirmation of the composition of the intact complex or the subcomplexes that comprise it. Moreover, the pattern of MS/MS spectra allows distinguishing between core and peripheral subunits of the assembly. Peripheral subunits characterized by a relatively small contact interface with the surrounding subunits will be the first ones to be expelled from the complex following an increase in collision energies, while core subunits will be protected from such exclusion [41].

Tandem MS analysis can also be applied to quantify the relative distribution of the species comprising polydisperse ensembles or other heterogeneous complex systems that cannot be deconvoluted by MS alone [51]. Since the dissociation process that occurs during the MS/MS process is typically asymmetric in terms of charge, the removal of highly charged monomers leads to "stripped" complexes of relatively low charge. As a consequence, a greater separation between adjacent peaks of the stripped complexes is achieved, aiding the assignment of the spectra. Fortunately, data analysis is also facilitated by the fact that dissociation products must be complementary in charge and in mass to those of the precursor ions.

Mapping the network of subunit interactions within a complex can be carried out by generating an array of smaller subcomplexes with overlapping components [52, 53]. Such subcomplexes can be generated either in solution, by manipulating the pH, increasing the ionic strength, or adding organic solvents to the buffer, or by CID within the mass spectrometer [54] (Figure 2.2). MS and MS/MS data are then utilized, to identify the composition of each of the subcomplexes. The connectivity network between subunits is then built, based on the overlapping components within the subcomplexes [43].

How subunits are packed, and what is the nature of the complex's three-dimensional shape, are questions that can be addressed by means of ion-mobility MS [19, 55, 56]. This method measures the time it takes for an ion to travel through a dense environment of gas molecules of neutral charge, under the influence of a weak electric field. The drift time is dependent not only on the ion's mass, but also on the overall shape of the analyzed protein complex: An assembly with a large volume will experience more collisions with the background gas, and consequently travel more slowly than a complex with the same mass, but a more compact structure. The measured drift time can be converted to collision cross-section (Ω) values, which reflect the three-dimensional shape of the ion [19, 55, 56]. Collision cross-section values derived from ion-mobility MS data can be used to produce information on the overall size and three-dimensional organization of the protein assembly, especially when combined with subunit interaction maps and homology modeling approaches [57–59].

A comprehensive list of achievements, which reflects the diverse capabilities of native MS, would encompass assemblies as large as virus capsids [60, 61], multi-component and asymmetric assemblies as the CRISPR (clustered regularly interspaced short palindromic repeat) [62], the 19S and 20S proteasomes [41, 63, 64] and the ribosome [65], assemblies as complicated as polydisperse systems [66], and as challenging as membrane protein complexes [67, 68]. On the basis of MS data, atomic models of protein assemblies have been generated [43], and insights into the molecular mechanism of action of protein complexes were obtained [47]. Moreover, the speed of analysis and separation afforded by MS place the method in a central position to capture dynamic processes such as protein folding [47],

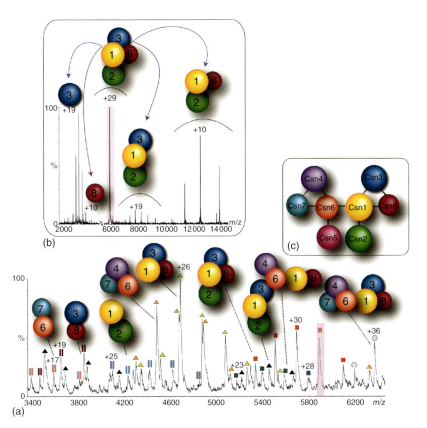

Figure 2.2 The subunit interaction map of a protein complex can be generated by defining the composition of multiple subcomplexes with overlapping components, as illustrated here for the COP9 signalosome (CSN), an eight-subunit (Csn1–8) complex. The intact CSN was decomposed into a set of smaller subcomplexes by manipulating the sample conditions both in solution (by addition of up to 10% methanol) and in the gas phase (by increasing the accelerating voltage). Assigning the composition of the generated subcomplexes, which are shown in the mass spectrum (a), was facilitated by analyses of corresponding MS/MS data. (b) An example of an MS/MS spectrum generated for an isolated ion at 5900 *m/z*. Peaks centered at 8000 *m/z* correspond to the loss of the Csn8 subunit (magenta arrows), whereas the series at 10 500–14 400 *m/z* correspond to the loss of the Csn3 subunit (blue arrows). At low (1600–3000) *m/z*, a series of peaks are assigned to the corresponding individual Csn8 and Csn3 subunits. By extrapolation, we could conclude that the subcomplex is composed of four subunits (Csn1/Csn2/Csn3/Csn8). Similar assignments were carried out for the other subcomplexes. Subsequently, the list of assigned subcomplexes was used to build the interaction map (c). (Adapted with permission from Sharon *et al.* [52].)

assembly [63, 69], and subunit exchange reactions [66]. Nevertheless, as in the case of any experimental technique, native MS has its own limitations, which are discussed below. Therefore, synergy with a complementary experimental technique, such as chemical cross-linking MS, will not only offer a more comprehensive structural description, but also enable examination of systems that until now were resistant to characterization by this method.

2.4
Chemical Cross-linking MS

During the past decade, chemical cross-linking, combined with MS analysis of the reaction products, has evolved into an alternative strategy to structurally resolve protein–protein interactions [16]. Usually chemical cross-linking MS is done in a bottom-up manner that includes digestion of the cross-linked proteins, but there have also been a few top-down studies [70, 71]. Cross-linking enables establishment of a set of structurally defined interactions, by covalently connecting pairs of functional groups within a protein. From the constraints obtained by the chemical cross-links, distance maps can be created within a protein or a protein complex, which can then serve as the basis for deducing low-resolution, three-dimensional structures [72–77]. The strengths of the cross-linking/MS strategy include the theoretically unlimited size of the protein or protein complex under investigation (for bottom-up analysis), and the minimal sample requirements in the femto- to attomole range, owing to the high sensitivity of MS analysis (see above) [72–84]. Recently, the chemical cross-linking MS approach has generated interest among systems biologists, and highly promising results indicate that the technique may be applied to map complex protein networks in cells [85].

2.4.1
Types of Cross-linkers

Nearly 40 years ago, N-hydroxysuccinimide (NHS) esters were introduced as homobifunctional, highly amine-reactive, cross-linking reagents [86, 87]. To date, NHS esters are the reagents most widely used to conduct protein–protein cross-linking, with bis(sulfosuccinimidyl)suberate (BS^{2}G) as one of the most prominent examples (Figure 2.3a). NHS esters react with nucleophiles to release the NHS or sulfo-NHS group and to create covalent amide and imide bonds with primary or secondary amines, such as the N-terminus and ε-amino groups in the lysine side chains of proteins. It has further been shown that NHS esters also react with hydroxyl groups in serines, threonines, and tryrosines, albeit to a lesser extent [88, 89]. The length of the carbon chain connecting both NHS groups determines the distance between the two reactive amino acid side chains that the cross-linker can bridge. In the case of BS^{2}G, Cα–Cα distances of up to 25 Å have been found to be connected by the linker [90, 91]. In order to facilitate the identification of cross-links in the mass spectra, isotope labels (i.e., deuterated

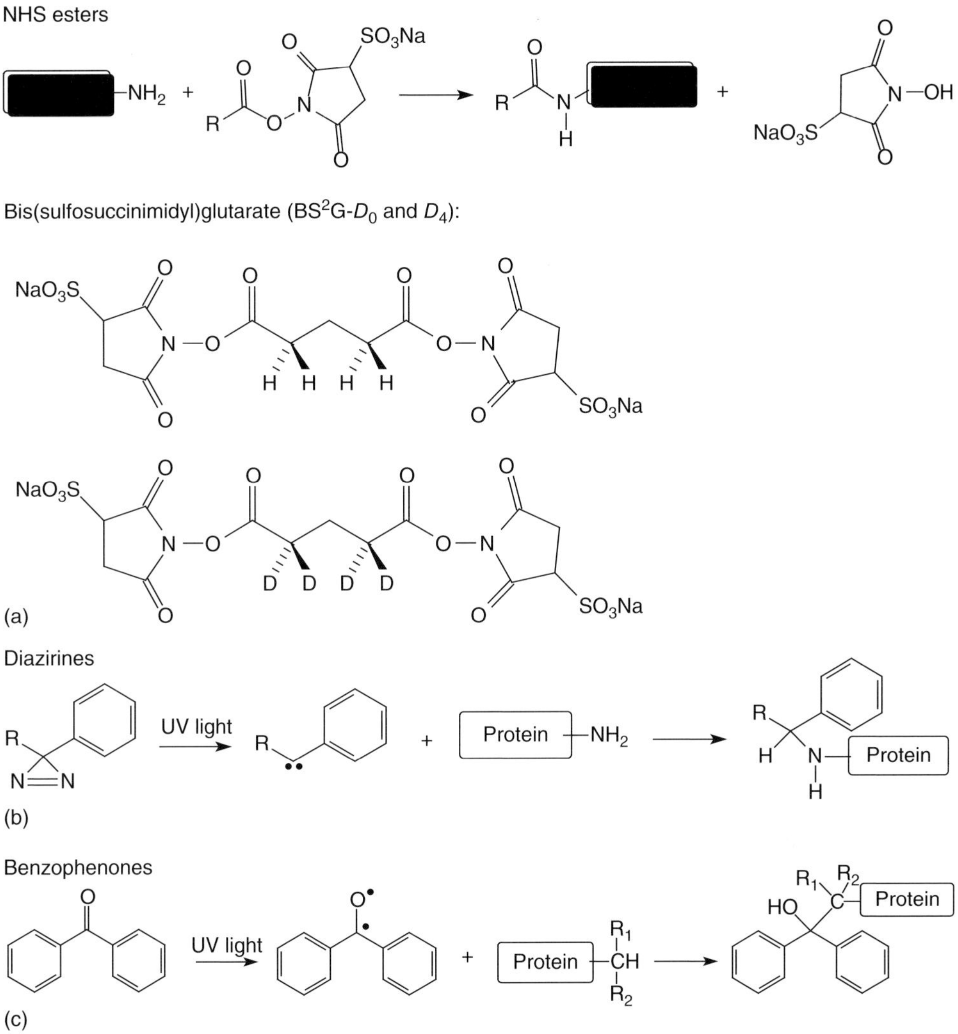

NHS esters

Bis(sulfosuccinimidyl)glutarate (BS^{2}G-D_0 and D_4):

(a)

Diazirines

(b)

Benzophenones

(c)

Figure 2.3 (a–c) Types of cross-linkers and their reactions with proteins.

cross-linkers) are employed. However, it is impossible to ignore the retention time shift of deuterated compounds in reverse-phase liquid chromatography (RP-LC), which prevents cross-linked species from coeluting [88].

An alternative class of compounds known as photoreactive cross-linkers holds great promise for conducting chemical cross-linking under *in vivo* conditions. The lower specificity of the cross-linking reaction, as well as the potential to conduct cross-linking reactions in a two-step manner, makes photoreactive cross-linkers an attractive choice for mapping protein–protein interactions. The photoreactive group is induced to react with the target molecules by exposure to long-wavelength UV light. The largest number of photoreactive reagents is based on nitrene or carbene chemistry; the photolabile precursors include azides, diazirines, diazo compounds, and benzophenones [16]. Most of the photoreactive cross-linkers are heterobifunctional reagents, which in addition, possess an amine-reactive group [89]. In our studies, we found diazirine- and benzophenone-based cross-linkers, in combination with MS, to be most useful for protein–protein interaction studies.

Diazirines are remarkably stable in a variety of chemical conditions, and are efficiently photolyzed at wavelengths of ~360 nm to generate a highly reactive carbene that reacts with an N–H or C–H bond (Figure 2.3b). A different photochemistry than that of aryl azides or diazirines is exhibited by benzophenones [16, 91], which create a biradical on irradiation. Subsequently, the oxygen radical abstracts a hydrogen radical from a bond of the reaction partner (Figure 2.3c). The alkyl radicals thus created react by forming a new C–C bond between the photophor and the reacting protein. Unlike diazirine compounds, activation of benzophenones does not proceed according to a photo-dissociative mechanism. The biradical created is susceptible to reactions with water, which result in an intact benzophenone molecule that can be reactivated by UV light. The reversibility of the photoreaction is one of the major advantages of benzophenone cross-linkers. Benzophenones react preferably with methionines [92], but reactions with other hydrophobic amino acids as well as with proline, arginine, and lysine, have been observed [93].

2.4.2
MS/MS Cleavable Cross-linkers

Cross-linkers that dissociate under CID conditions in the mass spectrometer may be utilized to facilitate identification of cross-linked products, based on the characteristic fragment ions and constant neutral losses in MS/MS spectra. In a series of different MS/MS cleavable linkers [94–96], an amine-reactive cross-linker containing a cleavable urea group was found to be most useful for creating characteristic marker ions capable of discriminating between inter- and intrapeptide cross-links, as well as between peptides that are modified by a partially hydrolyzed ("dead-end") cross-linker (Figure 2.4). The linker was found to be susceptible to fragmentation, both in the lower (ESI-CID-MS/MS) and in the higher (MALDI–in-source decay (ISD)–MS/MS) energy regime, which makes it of great interest for a wide number of protein structural applications. While

Figure 2.4 An MS/MS cleavable cross-linker enables to distinguish between an intra- and an interpeptide cross-link. (a) The structure of the cross-linker, as well as its fragmentation in an interpeptide cross-link, is presented. Two fragments are generated with a mass difference of two 26 u, due to the mass increases of 85 u (butyl, Bu) or 111 u (butyl + urea, BuUr) to the connected peptides. (b) An ISD-MALDI mass spectrum of an interpeptide cross-linked product, showing two 26 u doublets. (c) With the MS/MS cleavable linker, different cross-linked species are easily discriminated, based on their characteristic fragmentation patterns and neutral losses.

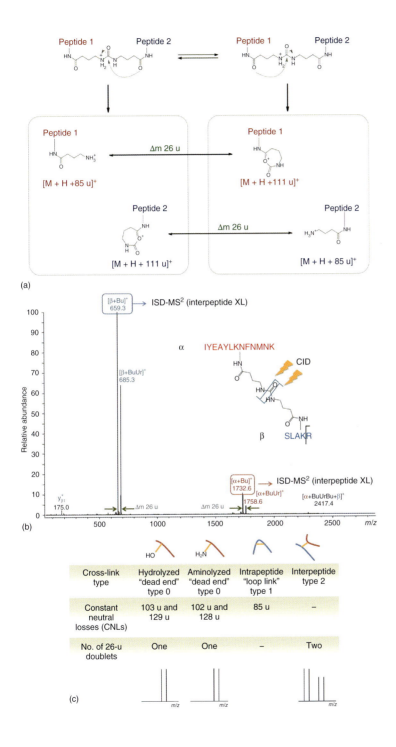

(a)

(b)

(c)

Cross-link type	Hydrolyzed "dead end" type 0	Aminolyzed "dead end" type 0	Intrapeptide "loop link" type 1	Interpeptide type 2
Constant neutral losses (CNLs)	103 u and 129 u	102 u and 128 u	85 u	–
No. of 26-u doublets	One	One	–	Two

ESI-MS is routinely used for the analysis of large protein complexes, chemical cross-linking before MS analysis also enables the use of MALDI-MS, which benefits from simpler spectra [97]. The use of MS/MS cleavable cross-linkers is especially advantageous in studies of large protein assemblies, where the mass spectra are highly complex, and one has to sift through large data sets. In these cases, one can rely on the presence of the two 26 amu doublets (Figure 2.4b) as indicators of an interpeptide cross-link – the most interesting cross-link type from a structural point of view – and use automatized data analysis to identify the cross-linked products. Intrapeptide cross-links or "dead-end" cross-links, on the other hand, will exhibit a constant neutral loss of 85 amu or one 26 amu doublet, respectively (Figure 2.3c) [94].

2.4.3
Data Analysis

Identifying cross-linked peptides poses additional difficulties for data analysis, as the number of potential cross-links increases quadratically, with increasing sample complexity. Thus, bioinformatics tools are required, in order to examine the large data sets generated during MS and MS/MS analyses of the peptide mixtures. Considerable efforts have been made to develop specific software tools that would be capable of analyzing these complex data sets; nevertheless, software that enables the fully automated analysis of MS and MS/MS data created from cross-linked product mixtures is still lacking. The current bottleneck in data analysis must be resolved, if chemical cross-linking is to evolve into a generally applicable and rapid method for global structural proteomics studies, underscoring the need to develop novel, powerful bioinformatics strategies. Among the currently available software packages used to analyze cross-linked products are GPMAW [98], xQuest [99], X-Link Identifier [100], xComb [101], MS-Bridge, which is part of *Protein Prospector* [102], and StavroX [103]. A summary of currently available cross-linking software is provided by Mayne and Patterton [104].

2.5
Value of Combining Native MS with Chemical Cross-linking MS

Each of the two MS-based methods described above has its specific advantages. Native MS is a very powerful tool for determining the stoichiometry and connectivity between subunits, as well as in defining the overall architecture of the assembly. Yet, this method cannot identify the direct binding region within the complex, and localize specific binding interfaces. These features, however, can be resolved by chemical cross-linking analysis – a method that may be challenged when forced to reveal the structural organization of multicomponent complexes, which lack any prior structural knowledge. Consequently, integrating these two complementary techniques would be a natural step toward a more complete characterization of protein assemblies.

In the next section, we discuss the added value that we believe is achieved when native MS and chemical cross-linking MS are combined. We specifically focus on challenging scenarios that can be resolved by integrating the two approaches, such as is the case when studying the interaction between a large multisubunit complex and a relatively small protein, or when probing transient interactions. We then describe a hybrid method approach, in which the two MS-based methods join forces with other structural biology approaches to provide a comprehensive analysis of difficult-to-study biological systems. Finally, challenges and future directions are presented.

2.6
Regulating the Giant

Large protein complexes that orchestrate fundamental cellular activities are often activated or deactivated through interactions with individual proteins that are much smaller in size. For example, the folding activity of the ~1 000 000 Da eukaryotic chaperonin containing Tcp1 (CCT) complex, responsible for maintaining protein folding homeostasis in cells, is modulated by interactions with phosducin-like protein 2 (PLP2, 38 000 Da) [105]. Similarly, protein degradation by the ~750 000 Da 20S proteasome complex is inhibited by interactions with the homodimeric enzyme NAD(P)H:quinone-oxidoreductase-1 (NQO1, 31 000 Da) [106]. Unraveling the mechanism of action of such regulators requires the structural characterization of their interaction with the complex they control. This task, however, is not simple, considering the relatively small size of the regulator, and the large dimensions of their associated protein assembly.

Native MS analysis of a sample containing a mixture of the target protein complex and its regulator can primarily provide information on the existence of such an interaction and the affinity of the composing parts [107]. A combination of MS and MS/MS experiments makes it possible to define the stoichiometry of the interaction, that is, the number of regulators that are bound to the protein complex. However, the specific sites of association between the regulator and the molecular machine, which would yield clues to the mechanism of its action, will remain elusive. The lack of information, however, can be filled by chemical cross-linking results.

To map the exact sites where the regulator interacts with the protein complex, cross-linking experiments can be performed using either a homobifunctional or a heterobifunctional cross-linker. Reactions with a homobifunctional cross-linker involve a single-step reaction, which yields both inter- and intramolecular cross-linked peptides. The advantage of this type of analysis is that if part or all of the protein components have known high-resolution structures, the quality of the results can be assessed by validating the applicability of the identified intrasubunit cross-linked peptides in the crystal structures [85, 108]. The drawback, on the other hand, lies in the generation of a complex mixture of peptides, comprising a high background of unmodified linear peptides. As a consequence, the signals for cross-linked peptides are low, and data interpolation is challenging. The use of a heterobifunctional cross-linker can ease these difficulties [79]. In this type of reaction, two steps are involved. First, one reactive site of the cross-linker is reacted with the small regulator. After removal of excess cross-linker, a different type of chemical reaction, preferably a photoreaction, is induced, using the second functional group connecting the regulator and its associated protein complex. This two-step approach reduces the formation of high molecular weight aggregates, and facilitates data analysis.

Functional regulation of a protein complex can also be achieved by posttranslational modifications, especially by phosphorylation [109]. The addition of multiple phosphate groups, which add both negative charge and bulk, can reversibly alter the conformation and stability of the assembly, resulting in modulation of intrinsic biological activity. This was recently demonstrated for the membrane motor F_1F_O–ATPase complex, in a study that effectively combined native MS and chemical cross-linking MS [110]. Mass spectra of the intact ATPase complex clarified the number of nucleotides bound to the complex. A dephosphorylated enzyme, however, displayed reduced nucleotide occupancy and decreased stability. Chemical cross-linking experiments of untreated and dephosphorylated ATPase enabled monitoring of conformational changes at the catalytic interface and provided a rationale for reduced nucleotide occupancy. Thus, the synergy of approaches uncovered the effect of phosphorylation on the dynamic subunit interactions within a membrane-embedded protein complex.

2.7
Capturing Transient Interactions

Transient protein–protein interactions, which form and break easily, are important to many aspects of cellular function, among them protein folding, modification, transport, signaling, and cell cycling [111]. Characterizing these interactions is therefore critical for understanding various biological processes; yet due to their labile nature, they are technically difficult to study and harder to detect than more stable interactions. Chemical cross-linking, which introduces a covalent bond between interacting partners, can be used as a tool for freezing such transiently formed complexes. For example, the QTAXs (quantitative analysis of tandem affinity purified *in vivo* cross-linked protein complexes) approach is a powerful cross-linking technique for comprehensive profiling of protein interactions network, including its weak and/or transient components [80]. The method freezes stable and transient interactions *in vivo*, using formaldehyde. The captured protein complexes and their interacting proteins are isolated by two-step affinity purification under fully denaturing conditions, using a histidine–biotin–histidine (HBH) tag. The purified proteins are then identified by LC/MS/MS, and the protein–protein interactions can be differentiated from background proteins based on their SILAC (stable isotope labeling of amino acids in cell culture) ratios. If a protein is a background protein, it is purified in equal amounts from both the tagged and control strains, and all peptides representing that protein will elute as a pair with a SILAC ratio of ≈1. In contrast, protein interaction partners will be enriched in the tagged sample. This strategy was used to identify the 26S proteasome interaction network, which involves hundreds of proteins intertwined in several critical cellular processes [80].

We envision an alternative approach, combining *in vivo* chemical cross-linking with native MS (Figure 2.5). To identify interactions within the cellular environment, diazirine-containing photo-activatable amino acids such as photo-methionine and photo-leucine may be used [112]. These photo-activated amino acids are known to be endogenously incorporated into the sequence of proteins during translation, substituting for methionine and leucine. Cross-links are

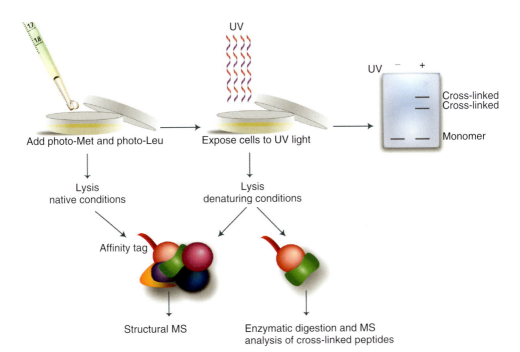

Figure 2.5 Schematic representation of an *in vivo* strategy for identifying protein–protein interactions. Cells are grown in media that are devoid of leucine and methionine. Photo-activatable leucine and methionine derivatives substitute the naturally occurring amino acids. Chemical cross-linking is then induced by UV illumination, forming a covalent bond with nearby protein side chains and backbones. Protein–protein interactions are identified by combining denaturing and native lysis conditions with pull-down experiments, using an affinity tag.

Pull-down experiments under native conditions, before the formation of cross-links, will enable the isolation and characterization of all interacting proteins within the complex, using native MS. Illuminating the sample with UV light will produce cross-links capable of capturing even transient and weak interactions. These will be identified following pull-down experiments under denaturing conditions and analysis by both native MS and peptide MS profiling. The former provide direct identification of the protein–protein interaction sites.

formed with spatially close protein side chains and backbones, on activation with UV irradiation of the cell culture. This approach enables efficient control of covalent bond formation, in comparison to the formaldehyde cross-linking used in the QTAX approach [80]. Cross-linked protein complexes can be detected by incorporating an affinity tag into the protein of interest, which is used as a bait for pulling down interacting proteins. Isolation of the tagged protein can then be performed under either denaturing or native conditions. Pull-down experiments under native conditions enable characterization of the complex stoichiometry, heterogeneity, and architecture, by means of native MS. This step makes it possible to directly determine stoichiometries of the protein complex components, without the need for SILAC labeling or digestion of the protein. In addition, validating the integrity of the isolated protein assembly yields a reduction in false positives, that is, copurified components, which are a major source of error in high-throughput affinity purifications, followed by MS. Pull-down experiments of the bait protein under denaturing conditions will maintain interactions with neighboring proteins, fixing even weakly associated components, which will be identified by native MS and LC/MS/MS analysis of the enzymatic peptide mixtures. The latter will provide interaction contacts at the individual peptide level. The advantage of this strategy is the coupling of two complementary MS approaches in a single biological setup.

2.8
An Integrative Approach for Obtaining Low-Resolution Structures of Native Protein Complexes

Native MS and cross-linking MS are low-resolution structural techniques. However, by integrating the data thus obtained with information produced by complementary structural biology techniques and computational molecular modeling approaches, atomic resolution structures can be generated [43, 84, 85, 108]. This is exemplified by the atomic model that was created for the 10-subunit yeast RNA exosome complex, by combining native MS data with homology modeling [43]. Similarly, high-resolution structural information was obtained by joining cross-linked-derived distance restraints with X-ray crystallography data and computational approaches, defining the subunit arrangement of the CCT complex [84] and its interaction with the human protein phosphatase 2A (PP2A) regulatory subunit 2ABG [85]. The model of the 26S proteasome complex is yet another example of a structure that benefited from a hybrid approach, combining chemical cross-linking, cryoelectron microscopy, X-ray crystallography, and proteomic approaches [108]. The separation capabilities of MS were also harnessed to isolate a specific protein complex from a heterogeneous ensemble for imaging by transmission electron microscopy and atomic force microscopy as demonstrated for the GroEL chaperonin [113]. Thus, the synergy between MS-based techniques and structural methods such as X-ray crystallography, NMR, and electron microscopy gives rise to a powerful approach for structural investigations of protein assemblies, especially for the examination of systems resistant to characterization by any single technique.

In view of the impact that integrative approaches have had on structural biology, we trust that coupling native MS and chemical cross-linking MS will yield mutual benefits. As highlighted previously, both techniques are complementary on many levels [110, 114]. The two methods can handle large, heterogeneous, and dynamic protein complexes, and tolerate protein impurities. They can both identify protein interactions, with chemical cross-linking MS providing the specific interaction sites at a peptide resolution, and native MS defining the interaction network and stoichiometry on the subunit level. Similarly, the spatial arrangement of subunits can be determined at the local and global levels by chemical cross-linking and ion-mobility MS, respectively. Thus, the attributes obtained from each method can be merged into a comprehensive structural representation that, if supplemented with computational approaches, may even yield near-atomic level models. Although the number of studies combining these two MS-based approaches is thus far exceedingly low, we expect that the urge to tackle more challenging biological systems will trigger the ever-increasing use of this integrative mode of analysis. In this respect, one hurdle that must yet be overcome is that each MS method demands its own specialized mass spectrometer. With the advent of the modern orbitrap mass spectrometers [39], capable of analyzing small peptides as well as very large protein complexes at high resolution, the instrumental gap between the native MS and the cross-linking MS methods will be bridged.

2.9
Future Directions

Progress toward a mechanistic understanding of cellular processes will involve the characterization of protein complex plasticity. The goal is to reveal the variability of a single protein complex within different cell types, tissues, and cellular compartments, and determine its dynamics at different time points during the cell cycle, or in response to various stimuli. Identifying protein–protein interactions directly, within the cell, will constitute an important step toward this goal. The cross-linking approach is starting to move in this direction, by mapping protein networks [85], but it is still far from being routinely used in global functional proteomics studies. The most promising strategy thus far from *in vivo* cross-linking relies on the use of photoactivated amino acids that are incorporated into proteins by mammalian cells, yielding specific cross-linking following photoactivation [112]. It is anticipated that by combining *in vivo* cross-linking, native purification, and native MS of intact assemblies, it will be possible to characterize protein networks, and even capture fleeting interactions. In conjunction with technological developments, such as an increase in instrument sensitivity, resolving power, and mass accuracy [36], the remarkable impact of MS in structural biology and in analysis of complex cellular networks, will only be enhanced.

References

1. Zhang, Q.C., Petrey, D., Deng, L., Qiang, L., Shi, Y. *et al.* (2012) Structure-based prediction of protein-protein interactions on a genome-wide scale. *Nature*, **490**, 556–560.

2. Stumpf, M.P., Thorne, T., de Silva, E., Stewart, R., An, H.J. *et al.* (2008) Estimating the size of the human interactome. *Proc. Natl. Acad. Sci. U.S.A.*, **105**, 6959–6964.

3. Pennisi, E. (2012) Genomics. ENCODE project writes eulogy for junk DNA. *Science*, **337**, 1159–1161.

4. von Mering, C., Krause, R., Snel, B., Cornell, M., Oliver, S.G. *et al.* (2002) Comparative assessment of large-scale data sets of protein-protein interactions. *Nature*, **417**, 399–403.

5. Alberts, B. (1998) The cell as a collection of protein machines: preparing the next generation of molecular biologists. *Cell*, **92**, 291–294.

6. Mullard, A. (2012) Protein-protein interaction inhibitors get into the groove. *Nat. Rev. Drug Discovery*, **11**, 173–175.

7. Wells, J.A. and McClendon, C.L. (2007) Reaching for high-hanging fruit in drug discovery at protein-protein interfaces. *Nature*, **450**, 1001–1009.

8. Ilari, A. and Savino, C. (2008) Protein structure determination by x-ray crystallography. *Methods Mol. Biol.*, **452**, 63–87.

9. Tugarinov, V., Hwang, P.M., and Kay, L.E. (2004) Nuclear magnetic resonance spectroscopy of high-molecular-weight proteins. *Annu. Rev. Biochem.*, **73**, 107–146.

10. Walian, P., Cross, T.A., and Jap, B.K. (2004) Structural genomics of membrane proteins. *Genome Biol.*, **5**, 215.

11. Parrish, J.R., Gulyas, K.D., and Finley, R.L. Jr., (2006) Yeast two-hybrid contributions to interactome mapping. *Curr. Opin. Biotechnol.*, **17**, 387–393.

12. Sun, Y., Hays, N.M., Periasamy, A., Davidson, M.W., and Day, R.N. (2012) Monitoring protein interactions in living cells with fluorescence lifetime imaging microscopy. *Methods Enzymol.*, **504**, 371–391.

13. Sharon, M. (2010) How far can we go with structural mass spectrometry of protein complexes? *J. Am. Soc. Mass Spectrom.*, **21**, 487–500.

14. Benesch, J.L., Ruotolo, B.T., Simmons, D.A., and Robinson, C.V. (2007) Protein complexes in the gas phase: technology for structural genomics and proteomics. *Chem. Rev.*, **107**, 3544–3567.

15. Heck, A.J. (2008) Native mass spectrometry: a bridge between interactomics and structural biology. *Nat. Methods*, **5**, 927–933.

16. Sinz, A. (2006) Chemical cross-linking and mass spectrometry to map three-dimensional protein structures and protein-protein interactions. *Mass Spectrom. Rev.*, **25**, 663–682.

17. Milne, J.L., Borgnia, M.J., Bartesaghi, A., Tran, E.E., Earl, L.A. *et al.* (2013) Cryo-electron microscopy–a primer for the non-microscopist. *FEBS J.*, **280**, 28–45.

18. Hura, G.L., Menon, A.L., Hammel, M., Rambo, R.P., Poole, F.L. II, *et al.* (2009) Robust, high-throughput solution structural analyses by small angle X-ray scattering (SAXS). *Nat. Methods*, **6**, 606–612.

19. Uetrecht, C., Rose, R.J., van Duijn, E., Lorenzen, K., and Heck, A.J. (2010) Ion mobility mass spectrometry of proteins and protein assemblies. *Chem. Soc. Rev.*, **39**, 1633–1655.

20. Konermann, L., Pan, J., and Liu, Y.H. (2011) Hydrogen exchange mass spectrometry for studying protein structure and dynamics. *Chem. Soc. Rev.*, **40**, 1224–1234.

21. Leitner, A., Walzthoeni, T., Kahraman, A., Herzog, F., Rinner, O. *et al.* (2010) Probing native protein structures by chemical cross-linking, mass spectrometry, and bioinformatics. *Mol. Cell. Proteomics*, **9**, 1634–1649.

22. Gau, B.C., Sharp, J.S., Rempel, D.L., and Gross, M.L. (2009) Fast photochemical oxidation of protein footprints faster than protein unfolding. *Anal. Chem.*, **81**, 6563–6571.

23. Fenn, J.B., Mann, M., Meng, C.K., Wong, S.F., and Whitehouse, C.M.

(1989) Electrospray ionization for mass spectrometry of large biomolecules. *Science*, **246**, 64–71.

24. Karas, M. and Hillenkamp, F. (1988) Laser desorption ionization of proteins with molecular masses exceeding 10 000 daltons. *Anal. Chem.*, **60**, 2299–2301.

25. Tanaka, K., Waki, H., Ido, Y., Akita, S., Yoshida, Y. *et al.* (1988) Protein and polymer analyses up to m/z 100,000 by laser ionization time-of-flight mass spectrometry. *Rapid Commun. Mass Spectrom.*, **2**, 151–153.

26. Charvat, A. and Abel, B. (2007) How to make big molecules fly out of liquid water: applications, features and physics of laser assisted liquid phase dispersion mass spectrometry. *Phys. Chem. Chem. Phys.*, **9**, 3335–3360.

27. Bich, C. and Zenobi, R. (2009) Mass spectrometry of large complexes. *Curr. Opin. Struct. Biol.*, **19**, 632–639.

28. Glish, G.L. and Vachet, R.W. (2003) The basics of mass spectrometry in the twenty-first century. *Nat. Rev. Drug Discovery*, **2**, 140–150.

29. Jones, A.W. and Cooper, H.J. (2011) Dissociation techniques in mass spectrometry-based proteomics. *Analyst*, **136**, 3419–3429.

30. Wysocki, V.H., Joyce, K.E., Jones, C.M., and Beardsley, R.L. (2008) Surface-induced dissociation of small molecules, peptides, and non-covalent protein complexes. *J. Am. Soc. Mass Spectrom.*, **19**, 190–208.

31. Yates, J.R., Ruse, C.I., and Nakorchevsky, A. (2009) Proteomics by mass spectrometry: approaches, advances, and applications. *Annu. Rev. Biomed. Eng.*, **11**, 49–79.

32. Demartino, G.N. (2012) Reconstitution of PA700, the 19S regulatory particle, from purified precursor complexes. *Methods Mol. Biol.*, **832**, 443–452.

33. Oeffinger, M. (2012) Two steps forward–one step back: advances in affinity purification mass spectrometry of macromolecular complexes. *Proteomics*, **12**, 1591–1608.

34. Ben-Nissan, G. and Sharon, M. (2011) Capturing protein structural kinetics by mass spectrometry. *Chem. Soc. Rev.*, **40**, 3627–3637.

35. Sobott, F., Hernandez, H., McCammon, M.G., Tito, M.A., and Robinson, C.V. (2002) A tandem mass spectrometer for improved transmission and analysis of large macromolecular assemblies. *Anal. Chem.*, **74**, 1402–1407.

36. Chernushevich, I.V. and Thomson, B.A. (2004) Collisional cooling of large ions in electrospray mass spectrometry. *Anal. Chem.*, **76**, 1754–1760.

37. Wilm, M. and Mann, M. (1996) Analytical properties of the nanoelectrospray ion source. *Anal. Chem.*, **68**, 1–8.

38. Kirshenbaum, N., Michaelevski, I., and Sharon, M. (2010) Analyzing large protein complexes by structural mass spectrometry. *J. Vis. Exp.*, **40**, 1954

39. Rose, R.J., Damoc, E., Denisov, E., Makarov, A., and Heck, A.J. (2012) High-sensitivity Orbitrap mass analysis of intact macromolecular assemblies. *Nat. Methods*, **9**, 1084–1086.

40. Rosati, S., Rose, R.J., Thompson, N.J., van Duijn, E., Damoc, E. *et al.* (2012) Exploring an Orbitrap analyzer for the characterization of intact antibodies by native mass spectrometry. *Angew. Chem. Int. Ed.*, **51**, 12992–12996.

41. Sharon, M., Taverner, T., Ambroggio, X.I., Deshaies, R.J., and Robinson, C.V. (2006) Structural organization of the 19S proteasome lid: insights from MS of intact complexes. *PLoS Biol.*, **4**, e267.

42. Rozen, S., Tieri, A., Ridner, G., Stark, A.K., Schmaler, T. *et al.* (2013) Exposing the subunit diversity within protein complexes: a mass spectrometry approach. *Methods*, **59**, 270–277.

43. Taverner, T., Hernandez, H., Sharon, M., Ruotolo, B.T., Matak-Vinkovic, D. *et al.* (2008) Subunit architecture of intact protein complexes from mass spectrometry and homology modeling. *Acc. Chem. Res.*, **41**, 617–627.

44. van Breukelen, B., Barendregt, A., Heck, A.J., and van den Heuvel, R.H. (2006) Resolving stoichiometries and oligomeric states of glutamate synthase protein complexes with curve fitting and simulation of electrospray mass spectra. *Rapid Commun. Mass Spectrom.*, **20**, 2490–2496.

45. Stengel, F., Baldwin, A.J., Bush, M.F., Hilton, G.R., Lioe, H. *et al.* (2012) Dissecting heterogeneous molecular chaperone complexes using a mass spectrum deconvolution approach. *Chem. Biol.*, **19**, 599–607.

46. Morgner, N. and Robinson, C.V. (2012) Massign: an assignment strategy for maximizing information from the mass spectra of heterogeneous protein assemblies. *Anal. Chem.*, **84**, 2939–2948.

47. Dyachenko, A., Gruber, R., Shimon, L., Horovitz, A., and Sharon, M. (2013) Allosteric mechanisms can be distinguished using structural mass spectrometry. *Proc. Natl. Acad. Sci. U.S.A.*, **110**, 7235–7239.

48. Lomeli, S.H., Peng, I.X., Yin, S., Loo, R.R., and Loo, J.A. (2010) New reagents for increasing ESI multiple charging of proteins and protein complexes. *J. Am. Soc. Mass Spectrom.*, **21**, 127–131.

49. Sterling, H.J., Daly, M.P., Feld, G.K., Thoren, K.L., Kintzer, A.F. *et al.* (2010) Effects of supercharging reagents on noncovalent complex structure in electrospray ionization from aqueous solutions. *J. Am. Soc. Mass Spectrom.*, **21**, 1762–1774.

50. Valeja, S.G., Tipton, J.D., Emmett, M.R., and Marshall, A.G. (2010) New reagents for enhanced liquid chromatographic separation and charging of intact protein ions for electrospray ionization mass spectrometry. *Anal. Chem.*, **82**, 7515–7519.

51. Aquilina, J.A., Benesch, J.L., Bateman, O.A., Slingsby, C., and Robinson, C.V. (2003) Polydispersity of a mammalian chaperone: mass spectrometry reveals the population of oligomers in alphaB-crystallin. *Proc. Natl. Acad. Sci. U.S.A.*, **100**, 10611–10616.

52. Sharon, M., Mao, H., Boeri Erba, E., Stephens, E., Zheng, N. *et al.* (2009) Symmetrical modularity of the COP9 signalosome complex suggests its multifunctionality. *Structure*, **17**, 31–40.

53. Hernandez, H., Dziembowski, A., Taverner, T., Seraphin, B., and Robinson, C.V. (2006) Subunit architecture of multimeric complexes isolated

directly from cells. *EMBO Rep.*, **7**, 605–610.

54. Hernandez, H. and Robinson, C.V. (2007) Determining the stoichiometry and interactions of macromolecular assemblies from mass spectrometry. *Nat. Protoc.*, **2**, 715–726.

55. Zhong, Y., Hyung, S.J., and Ruotolo, B.T. (2012) Ion mobility-mass spectrometry for structural proteomics. *Expert Rev. Proteomics*, **9**, 47–58.

56. Jurneczko, E. and Barran, P.E. (2011) How useful is ion mobility mass spectrometry for structural biology? The relationship between protein crystal structures and their collision cross sections in the gas phase. *Analyst*, **136**, 20–28.

57. Teplow, D.B., Lazo, N.D., Bitan, G., Bernstein, S., Wyttenbach, T. *et al.* (2006) Elucidating amyloid beta-protein folding and assembly: a multidisciplinary approach. *Acc. Chem. Res.*, **39**, 635–645.

58. van Duijn, E., Barendregt, A., Synowsky, S., Versluis, C., and Heck, A.J. (2009) Chaperonin complexes monitored by ion mobility mass spectrometry. *J. Am. Chem. Soc.*, **131**, 1452–1459.

59. Politis, A., Park, A.Y., Hyung, S.J., Barsky, D., Ruotolo, B.T. *et al.* (2010) Integrating ion mobility mass spectrometry with molecular modelling to determine the architecture of multiprotein complexes. *PLoS One*, **5**, e12080.

60. Uetrecht, C., Versluis, C., Watts, N.R., Roos, W.H., Wuite, G.J. *et al.* (2008) High-resolution mass spectrometry of viral assemblies: molecular composition and stability of dimorphic hepatitis B virus capsids. *Proc. Natl. Acad. Sci. U.S.A.*, **105**, 9216–9220.

61. Morton, V.L., Stockley, P.G., Stonehouse, N.J., and Ashcroft, A.E. (2008) Insights into virus capsid assembly from non-covalent mass spectrometry. *Mass Spectrom. Rev.*, **27**, 575–595.

62. van Duijn, E., Barbu, I.M., Barendregt, A., Jore, M.M., Wiedenheft, B. *et al.* (2012) Native tandem and ion mobility mass spectrometry highlight structural and modular similarities in clustered-regularly-interspaced shot-palindromic-repeats (CRISPR)-associated protein complexes from *Escherichia coli* and *Pseudomonas aeruginosa*. *Mol. Cell. Proteomics*, **11**, 1430–1441.

63. Sharon, M., Witt, S., Glasmacher, E., Baumeister, W., and Robinson, C.V. (2007) Mass spectrometry reveals the missing links in the assembly pathway of the bacterial 20 S proteasome. *J. Biol. Chem.*, **282**, 18448–18457.

64. Sharon, M., Witt, S., Felderer, K., Rockel, B., Baumeister, W. *et al.* (2006) 20S proteasomes have the potential to keep substrates in store for continual degradation. *J. Biol. Chem.*, **281**, 9569–9575.

65. Gordiyenko, Y., Videler, H., Zhou, M., McKay, A.R., Fucini, P. *et al.* (2010) Mass spectrometry defines the stoichiometry of ribosomal stalk complexes across the phylogenetic tree. *Mol. Cell. Proteomics*, **9**, 1774–1783.

66. Aquilina, J.A., Shrestha, S., Morris, A.M., and Ecroyd, H. (2013) Structural and functional aspects of hetero-oligomers formed by the small heat shock proteins alphaB-crystallin and HSP27. *J. Biol. Chem.*, **288**, 13602–13609.

67. Barrera, N.P., Di Bartolo, N., Booth, P.J., and Robinson, C.V. (2008) Micelles protect membrane complexes from solution to vacuum. *Science*, **321**, 243–246.

68. Zhou, M., Morgner, N., Barrera, N.P., Politis, A., Isaacson, S.C. *et al.* (2011) Mass spectrometry of intact V-type ATPases reveals bound lipids and the effects of nucleotide binding. *Science*, **334**, 380–385.

69. Liu, J. and Konermann, L. (2013) Assembly of hemoglobin from denatured monomeric subunits: heme ligation effects and off-pathway intermediates studied by electrospray mass spectrometry. *Biochemistry*, **52**, 1717–1724.

70. Novak, P., Young, M.M., Schoeniger, J.S., and Kruppa, G.H. (2003) A top-down approach to protein structure studies using chemical cross-linking and Fourier transform mass spectrometry. *Eur. J. Mass Spectrom. (Chichester)*, **9**, 623–631.

71. Novak, P., Kruppa, G.H., Young, M.M., and Schoeniger, J. (2004) A top-down method for the determination of residue-specific solvent accessibility in proteins. *J. Mass Spectrom.*, **39**, 322–328.

72. Young, M.M., Tang, N., Hempel, J.C., Oshiro, C.M., Taylor, E.W. *et al.* (2000) High throughput protein fold identification by using experimental constraints derived from intramolecular cross-links and mass spectrometry. *Proc. Natl. Acad. Sci. U.S.A.*, **97**, 5802–5806.

73. Back, J.W., de Jong, L., Muijsers, A.O., and de Koster, C.G. (2003) Chemical cross-linking and mass spectrometry for protein structural modeling. *J. Mol. Biol.*, **331**, 303–313.

74. Muller, D.R., Schindler, P., Towbin, H., Wirth, U., Voshol, H. *et al.* (2001) Isotope-tagged cross-linking reagents. A new tool in mass spectrometric protein interaction analysis. *Anal. Chem.*, **73**, 1927–1934.

75. Dihazi, G.H. and Sinz, A. (2003) Mapping low-resolution three-dimensional protein structures using chemical cross-linking and Fourier transform ion-cyclotron resonance mass spectrometry. *Rapid Commun. Mass Spectrom.*, **17**, 2005–2014.

76. Kalkhof, S., Ihling, C., Mechtler, K., and Sinz, A. (2005) Chemical cross-linking and high-performance Fourier transform ion cyclotron resonance mass spectrometry for protein interaction analysis: application to a calmodulin/target peptide complex. *Anal. Chem.*, **77**, 495–503.

77. Sinz, A. (2010) Investigation of protein-protein interactions in living cells by chemical crosslinking and mass spectrometry. *Anal. Bioanal. Chem.*, **397**, 3433–3440.

78. Zhang, H., Tang, X., Munske, G.R., Tolic, N., Anderson, G.A. *et al.* (2009) Identification of protein-protein interactions and topologies in living cells with chemical cross-linking and mass spectrometry. *Mol. Cell. Proteomics*, **8**, 409–420.

79. Krauth, F., Ihling, C.H., Ruttinger, H.H., and Sinz, A. (2009) Heterobifunctional isotope-labeled amine-reactive photo-cross-linker for structural investigation of proteins by matrix-assisted laser desorption/ionization tandem time-of-flight and electrospray ionization LTQ-Orbitrap mass spectrometry. *Rapid Commun. Mass Spectrom.*, **23**, 2811–2818.

80. Guerrero, C., Milenkovic, T., Przulj, N., Kaiser, P., and Huang, L. (2008) Characterization of the protea-some interaction network using a QTAX-based tag-team strategy and protein interaction network analysis. *Proc. Natl. Acad. Sci. U.S.A.*, **105**, 13333–13338.

81. Petrotchenko, E.V. and Borchers, C.H. (2010) Crosslinking combined with mass spectrometry for structural pro-teomics. *Mass Spectrom. Rev.*, **29**, 862–876.

82. Rappsilber, J. (2011) The beginning of a beautiful friendship: cross-linking/mass spectrometry and modelling of proteins and multi-protein complexes. *J. Struct. Biol.*, **173**, 530–540.

83. Jin Lee, Y. (2008) Mass spectromet-ric analysis of cross-linking sites for the structure of proteins and protein complexes. *Mol. Biosyst.*, **4**, 816–823.

84. Kalisman, N., Adams, C.M., and Levitt, M. (2012) Subunit order of eukaryotic TRiC/CCT chaperonin by cross-linking, mass spectrometry, and combinatorial homology modeling. *Proc. Natl. Acad. Sci. U.S.A.*, **109**, 2884–2889.

85. Herzog, F., Kahraman, A., Boehringer, D., Mak, R., Bracher, A. *et al.* (2012) Structural probing of a protein phos-phatase 2A network by chemical cross-linking and mass spectrometry. *Science*, **337**, 1348–1352.

86. Bragg, P.D. and Hou, C. (1975) Sub-unit composition, function, and spatial arrangement in the Ca2+- and Mg2+-activated adenosine triphosphatases of Escherichia coli and Salmonella typhimurium. *Arch. Biochem. Biophys.*, **167**, 311–321.

87. Lomant, A.J. and Fairbanks, G. (1976) Chemical probes of extended biological structures: synthesis and properties of the cleavable protein cross-linking reagent [35S]dithiobis(succinimidyl propionate). *J. Mol. Biol.*, **104**, 243–261.

88. Kalkhof, S. and Sinz, A. (2008) Chances and pitfalls of chemical cross-linking with amine-reactive N-hydroxysuccinimide esters. *Anal. Bioanal. Chem.*, **392**, 305–312.

89. Madler, S., Bich, C., Touboul, D., and Zenobi, R. (2009) Chemical cross-linking with NHS esters: a systematic study on amino acid reactivities. *J. Mass Spectrom.*, **44**, 694–706.

90. Chen, Z.A., Jawhari, A., Fischer, L., Buchen, C., Tahir, S. *et al.* (2010) Architecture of the RNA polymerase II-TFIIF complex revealed by cross-linking and mass spectrometry. *EMBO J.*, **29**, 717–726.

91. Dimova, K., Kalkhof, S., Pottratz, I., Ihling, C., Rodriguez-Castaneda, F. *et al.* (2009) Structural insights into the calmodulin-Munc13 inter-action obtained by cross-linking and mass spectrometry. *Biochemistry*, **48**, 5908–5921.

92. Wittelsberger, A., Thomas, B.E., Mierke, D.F., and Rosenblatt, M. (2006) Methio-nine acts as a "magnet" in photoaffinity crosslinking experiments. *FEBS Lett.*, **580**, 1872–1876.

93. Schwarz, R., Tanzler, D., Ihling, C.H., Muller, M.Q., Kolbel, K. *et al.* (2013) Monitoring conformational changes in peroxisome proliferator-activated receptor alpha by a genetically encoded photoamino acid, cross-linking, and mass spectrometry. *J. Med. Chem.*, **56**, 4252–4263.

94. Muller, M.Q., Dreiocker, F., Ihling, C.H., Schafer, M., and Sinz, A. (2010) Cleavable cross-linker for protein struc-ture analysis: reliable identification of cross-linking products by tandem MS. *Anal. Chem.*, **82**, 6958–6968.

95. Muller, M.Q., Dreiocker, F., Ihling, C.H., Schafer, M., and Sinz, A. (2010) Fragmentation behavior of a thiourea-based reagent for protein structure analysis by collision-induced disso-ciative chemical cross-linking. *J. Mass Spectrom.*, **45**, 880–891.

96. Dreiocker, F., Muller, M.Q., Sinz, A., and Schafer, M. (2010) Collision-induced dissociative chemical cross-linking reagent for protein structure characterization: applied Edman chemistry in the gas phase. *J. Mass Spectrom.*, **45**, 178–189.

97. Chen, F., Gerber, S., Heuser, K., Korkhov, V.M., Lizak, C. *et al.* (2013) High-mass matrix-assisted laser desorption ionization-mass spectrometry of integral membrane proteins and their complexes. *Anal. Chem.*, **85**, 3483–3488.

98. Peri, S., Steen, H., and Pandey, A. (2001) GPMAW – a software tool for analyzing proteins and peptides. *Trends Biochem. Sci.*, **26**, 687–689.

99. Rinner, O., Seebacher, J., Walzthoeni, T., Mueller, L.N., Beck, M. *et al.* (2008) Identification of cross-linked peptides from large sequence databases. *Nat. Methods*, **5**, 315–318.

100. Du, X., Chowdhury, S.M., Manes, N.P., Wu, S., Mayer, M.U. *et al.* (2011) Xlink-identifier: an automated data analysis platform for confident identifications of chemically cross-linked peptides using tandem mass spectrometry. *J. Proteome Res.*, **10**, 923–931.

101. Panchaud, A., Singh, P., Shaffer, S.A., and Goodlett, D.R. (2010) xComb: a cross-linked peptide database approach to protein-protein interaction analysis. *J. Proteome Res.*, **9**, 2508–2515.

102. Chu, F., Baker, P.R., Burlingame, A.L., and Chalkley, R.J. (2010) Finding chimeras: a bioinformatics strategy for identification of cross-linked peptides. *Mol. Cell. Proteomics*, **9**, 25–31.

103. Gotze, M., Pettelkau, J., Schaks, S., Bosse, K., Ihling, C.H. *et al.* (2012) StavroX – a software for analyzing crosslinked products in protein interaction studies. *J. Am. Soc. Mass Spectrom.*, **23**, 76–87.

104. Mayne, S.L. and Patterton, H.G. (2011) Bioinformatics tools for the structural elucidation of multi-subunit protein complexes by mass spectrometric analysis of protein-protein cross-links. *Brief. Bioinform.*, **12**, 660–671.

105. Stirling, P.C., Srayko, M., Takhar, K.S., Pozniakovsky, A., Hyman, A.A. *et al.*

(2007) Functional interaction between phosducin-like protein 2 and cytosolic chaperonin is essential for cytoskeletal protein function and cell cycle progression. *Mol. Biol. Cell*, **18**, 2336–2345.

106. Asher, G., Tsvetkov, P., Kahana, C., and Shaul, Y. (2005) A mechanism of ubiquitin-independent proteasomal degradation of the tumor suppressors p53 and p73. *Genes Dev.*, **19**, 316–321.

107. Moscovitz, O., Tsvetkov, P., Hazan, N., Michaelevski, I., Keisar, H. *et al.* (2012) A mutually inhibitory feedback loop between the 20S proteasome and its regulator, NQO1. *Mol. Cell*, **47**, 76–86.

108. Lasker, K., Forster, F., Bohn, S., Walzthoeni, T., Villa, E. *et al.* (2012) Molecular architecture of the 26S proteasome holocomplex determined by an integrative approach. *Proc. Natl. Acad. Sci. U.S.A.*, **109**, 1380–1387.

109. Cohen, P. (2000) The regulation of protein function by multisite phosphorylation-a 25 year update. *Trends Biochem. Sci.*, **25**, 596–601.

110. Schmidt, C., Zhou, M., Marriott, H., Morgner, N., Politis, A. *et al.* (2013) Comparative cross-linking and mass spectrometry of an intact F-type ATPase suggest a role for phosphorylation. *Nat. Commun.*, **4**, 1985.

111. Perkins, J.R., Diboun, I., Dessailly, B.H., Lees, J.G., and Orengo, C. (2010) Transient protein-protein interactions: structural, functional, and network properties. *Structure*, **18**, 1233–1243.

112. Suchanek, M., Radzikowska, A., and Thiele, C. (2005) Photo-leucine and photo-methionine allow identification of protein-protein interactions in living cells. *Nat. Methods*, **2**, 261–267.

113. Benesch, J.L., Ruotolo, B.T., Simmons, D.A., Barrera, N.P., Morgner, N. *et al.* (2010) Separating and visualising protein assemblies by means of preparative mass spectrometry and microscopy. *J. Struct. Biol.*, **172**, 161–168.

114. Stengel, F., Aebersold, R., and Robinson, C.V. (2012) Joining forces: integrating proteomics and cross-linking with the mass spectrometry of intact complexes. *Mol. Cell. Proteomics*, **11**, R111.014027.

3
Native Mass Spectrometry Approaches Using Ion Mobility-Mass Spectrometry

Frederik Lermyte, Esther Marie Martin, Albert Konijnenberg, Filip Lemière, and Frank Sobott

3.1
Introduction

Over the last two decades, mass spectrometry (MS) has become a powerful tool for investigating biomolecules and their interactions. Traditional top–down MS approaches have included the study of intact proteins under denaturing conditions (acidic solution and/or high concentration of organic solvent) to determine accurate protein masses or to map posttranslational modifications (PTMs) using MS/MS (tandem mass spectrometry) fragmentation (Figure 3.1). Proteomics scientists are becoming increasingly interested in the use of "native" approaches, where conformations of individual proteins and noncovalent complexes are preserved throughout the ionization process and analysis inside the mass spectrometer ("native" MS; Figure 3.1).

The formation of biologically active protein structures is mediated by numerous noncovalent contacts. It is therefore very desirable to study these systems under conditions that preserve these interactions, in order to gain information about their function. Native ion mobility-mass spectrometry (IM-MS) is one route that is being increasingly used in the field of structural biology as a complementary tool to nuclear magnetic resonance (NMR) spectroscopy and X-ray crystallography, as well as electron microscopy and other biophysical and spectroscopic approaches. This new MS technology has opened up numerous possibilities for the characterization of protein complexes, and information on protein–protein [1–4], protein–DNA [5], protein–RNA [6, 7], protein–metal ion [8], protein–carbohydrate [9], protein–lipid [10], protein–drug interactions [11], and even membrane-bound proteins [12, 13] is now readily accessible.

Here we describe the necessary steps to be taken in order to study protein complexes successfully by MS (illustrated in Figure 3.2). We first discuss the way sample preparation is typically performed in native MS, then we describe the mass analyzers used in these studies and discuss possibilities for tandem MS experiments. Finally we discuss the added value of IM (ion mobility) for native MS. These points are highlighted using a selection of novel examples from the rapidly expanding body of literature on native MS.

Analyzing Biomolecular Interactions by Mass Spectrometry, First Edition.
Edited by Jeroen Kool and Wilfried M.A. Niessen.
© 2015 Wiley-VCH Verlag GmbH & Co. KGaA. Published 2015 by Wiley-VCH Verlag GmbH & Co. KGaA.

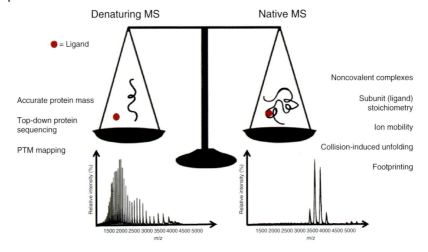

Figure 3.1 A comparison of the information obtained from denaturing and native MS of proteins. While denaturing MS yields accurate sequence masses, native MS is able to preserve higher-order structure allowing ion mobility (IM) and footprinting experiments. Typical charge state distributions are indicated as follows: a broad distribution around *m/z* 1500–2000 irrespective of mass for denatured proteins (as well as some partially folded states at higher *m/z*), and a narrow distribution for native proteins at high *m/z* depending on their mass.

3.2
Sample Preparation

Solution conditions are especially important when performing native MS experiments. The analyte obviously has to be in a native-like environment in solution, which requires a near-physiological pH and prevents the use of organic solvents typically employed in liquid chromatography–mass spectrometry (LC–MS). It is best to keep the percentage of organic co-solvents (e.g., DMSO) below 5–10% in order to preserve the native structure in electrospray ionization (ESI). Analyte concentrations used are typically in the 1–10 μM range, although 0.1–100 μM is possible. Only a few microliters are required per analysis at flow rates of 10–100 nl min^{-1}, so sample consumption is on the order of a few micrograms of protein.

In order to provide sufficient charge carriers and stabilize the pH of the solution, volatile buffers are used in native electrospray ionization-mass spectrometry (ESI-MS) which evaporate in the gas phase. In practice, the most frequently used buffer systems are 10–500 mM ammonium acetate solutions with pH close to 7 [14]. However, it should be emphasized that the search for an optimal buffer for use in native MS experiments is still very much an active area of research. For instance, it was recently shown that the use of ethylene diammonium diacetate (EDDA), a buffer that had not previously been used in MS experiments, can lead to significantly better resolved peaks for the intact 800 kDa GroEL tetradecamer, presumably by aiding desolvation [1] (Case Study 1).

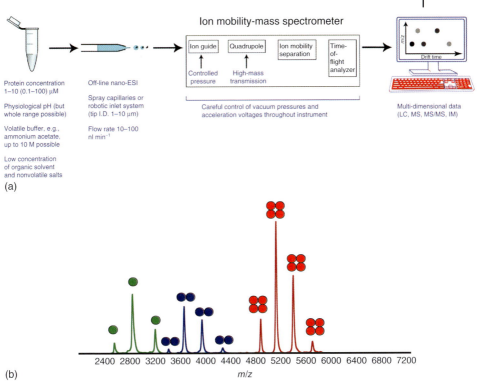

(a)

(b)

Figure 3.2 (a) Typical workflow used in native IM-MS experiments. Sample preparation, ionization, and conditions inside the mass spectrometer are all optimized for preservation of noncovalent interactions. (b) Example of native MS of concanavalin A, showing a mixture of monomer, dimer, and (native) tetramer.

Desalting is a key step in sample preparation for native MS. Solutions of biomolecules are routinely purified in buffers that contain nonvolatile salts (e.g., NaCl), detergents, or co-solvents (e.g., glycerol), which either cluster to the analyte ions or suppress ionization. For example, Na^+ ions can displace protons on the analyte, leading to peak broadening by formation of satellite peaks with varying numbers of these cations, which reduces sensitivity as the total ion intensity is distributed over several species [14]. NH_4^+ ions, on the other hand, can displace these alkali metal ions and dissociate into NH_3 (which is released) and H^+ (which stays attached) during the ESI process [14]. Desalting and buffer exchange can be performed immediately prior to analysis by using gel filtration or size-exclusion micro-columns. Alternatively, dialysis or centrifugal filters can be used to remove impurities with a low molecular weight, with the latter enabling simultaneous increase of the concentration of the analyte. A useful workflow in the analysis of protein–ligand complexes is to add an excess of ligand to the

---

Case Study 1: GroEL

An exciting application of native MS is the structural study of large protein complexes such as chaperonins, which are a class of homo-oligomeric protein complexes. Tetradecameric GroEL (800 kDa) from *Escherichia coli* works together with GroES to promote the correct folding of newly synthesized proteins [15]. GroEL has been used as a model system in tandem MS approaches, which show that upon collision-induced dissociation (CID), highly charged monomers were ejected from the complex, resulting in the observation of a 13mer [16]. Experimental collisional cross sections (CCSs) from IM-MS are also consistent with those calculated by computational methods, and confirm that the ring cavity does not undergo structural collapse in the gas phase [17]. Native MS conditions have been optimized to resolve adenosine triphosphate (ATP) binding to GroEL as shown in Figure 3.3a.

MS allows the different ligand-bound states to be distinguished and studied side by side, for example, how conformational change links to ligand binding, unlike methods that provide information averaged across the different species [1]. The inset of Figure 3.3a shows well-resolved zero to four ATP molecules bound to GroEL. Stepwise binding of ATP to GroEL in a cooperative manner was observed when the concentration of ATP added to the protein was gradually increased (Figure 3.3b). Using this distribution the binding constants of the interaction could be determined. This could facilitate the analysis of other complexes that involve cooperative ligand binding. An earlier important study that highlights this approach is by Rogniaux and colleagues, who directly measured the cooperative binding of nicotinamide adenine dinucleotide (NAD^+) to a multimeric enzyme [18].

---

protein and then desalt the resulting solution, as the excess low–molecular weight ligand will be removed in the desalting step.

3.3
Electrospray Ionization

Native MS relies on transfer of the ionized analyte from solution into the gas phase of the mass spectrometer. This requires desolvation (removal of water and buffer) while maintaining the weak, noncovalent interactions that stabilize tertiary and quaternary structure. We owe our ability to perform these experiments to the development of soft ionization techniques such as ESI by John Fenn in 1989 (2002 Nobel Prize in Chemistry) [19]. In ESI, a potential difference of a few kilovolts (typically $1.2 - 1.8$ kV for nano-ESI) is applied between the sprayer and the vacuum entrance of the mass spectrometer. The resulting electric field causes the surface of the liquid to distort into a so-called Taylor cone from which droplets are released that carry an excess of charge carriers (e.g., H^+ or NH_4^+).

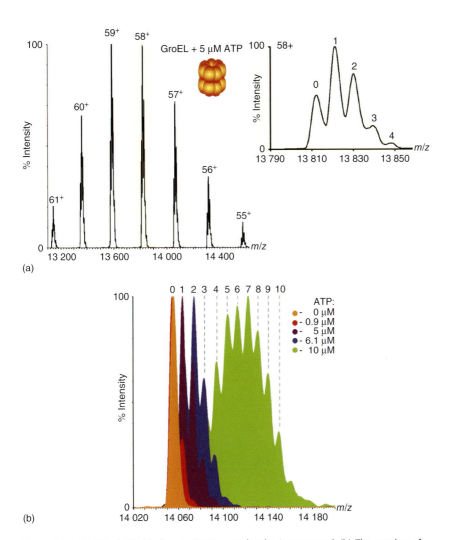

Figure 3.3 ESI-MS of ATP binding to GroEL. (a) Charge state distribution of GroEL. The inset shows a close-up view of the 58+ charge states showing the fine structure of the peak. The number of bound ATP molecules is annotated. (b) The number of bound ATP molecules increases as a function of the concentration added to the protein, highlighting co-operativity of binding. (Adapted from Ref. [1].)

As solvent molecules evaporate, the density of these charges increases until the resulting electrostatic repulsion overcomes the surface tension of the liquid. At this point, known as the *Rayleigh limit*, fission of the droplet into smaller droplets occurs [14].

The final step of the process by which free, gaseous ions are formed is believed to occur via a mechanism described by the charged residue model (CRM) [20, 21]. This proposes that the cycle of droplet fission and solvent evaporation continues until extremely small droplets are formed. A (small) portion of these droplets contain a single copy of the analyte ion, while the majority of them are empty. As the final droplet "dries up," the remaining charges are then deposited on the analyte. Strong experimental evidence for the CRM comes from the observed charge states of proteins in native MS experiments (see later). Unlike denatured proteins, native proteins do not proportionally pick up more charges with increasing mass which results in charge states higher up the *m/z* scale. (see Figure 3.1).

Like the ESI mechanism, the charge state distribution of native proteins is also debated [22]. Several possible explanations for the observed degree of charging have been proposed, including the availability of basic sites [23], intramolecular interactions within the protein [24], and the surface tension of the solvents used (which relates to the Rayleigh limit) [25]. While all of these factors probably contribute to some extent, it is striking that the observed charge states of globular proteins and protein complexes systematically approach, but do not exceed, the Rayleigh limit [26]. This is consistent with the formation of free ions from a droplet of essentially the same size as the ion, that is, the charge states of the analyte are defined in the very last stages of ion release when the droplet is drying up, according to the CRM [27–29]. In this way, the charge states of native proteins in ESI are thought to strongly correlate with the exposed surface area of the folded protein structure.

If very high analyte concentrations are used in ESI ($>20\,\mu$M), in principle more than one copy of the analyte can be present in the final ESI droplet, potentially leading to formation of clusters, that is, artificial interactions during solvent removal [30, 31]. Usually, though, it is possible to tell these apart from biologically relevant interactions, as in the case of artificial interactions, a statistical distribution of dimer, trimer, and so on, will be observed, while in native MS, one would expect to see only the specific complex being studied. In order to avoid these artifacts, it is best to dilute the solution stepwise to monitor this behavior, and use concentrations in the low micromolar range. At these concentrations, nonspecific dimers typically do not occur. The risk of observing nonspecific, artifactual interactions is further mitigated by the fact that mostly highly purified, well-known samples are used to study interactions that are already known to occur *in vivo*.

For native MS, the key requirements are the use of purely aqueous solutions and a minimized consumption of the often precious sample. Both these demands are met by a miniaturized version of the ESI source developed by Wilm and Mann, known as nano-ESI [32, 33]. Crucially, the much smaller spray orifices used here also lead to improved dispersion of the liquid into nano-droplets with a favorable surface-to-volume ratio, enabling the formation of "naked" ions without requiring harsh desolvation conditions or the use of organic (co-)solvents. Normally, a spray orifice with an inner diameter of $1-2\,\mu m$ is achieved by pulling a glass or quartz capillary to a very fine tip and coating this with conductive material such as gold, or inserting a metal wire [33] (Figure 3.2). This can be done in-house, but ready-to-use capillaries are also available from commercial sources (New Objective, Woburn/MA, USA or Proxeon Biosystems, Odense, Denmark – now Thermo Fisher Scientific). Typically, these capillaries are then loaded with $2-3\,\mu l$ of analyte solution, which is infused directly (i.e., without an additional online separation step) into the source of the mass spectrometer. An alternative, automated method for performing nano-ESI also exists, in which a few microliters of the analyte solution are aspirated into a sampling tip, which is then positioned against a chip, containing an array of about 390 nano-ESI nozzles. This system is produced by Advion (Ithaca, NY/USA) and has been used to analyze noncovalent multiprotein and protein–ligand complexes [34], among others.

A central, important aspect of native MS is the question of whether it is justified to talk about native structures in a solvent-free environment. Indeed, the high vacuum found in a mass analyzer is far removed from biological conditions, whereas the hydrophobic effect (which is absent *in vacuo*) plays an important role in protein folding in solution. However, from the early days of ESI-MS, studies have been performed on noncovalent protein–ligand [35–38] and multiprotein complexes [39], as well as nucleic acid duplexes [40] and protein–DNA/RNA complexes [39]. This type of work was later expanded to increasingly larger systems [41, 42], and, recently, mass spectra of an 18 MDa virus assembly were reported [2].

The evolution of protein conformations in native ESI-MS experiments has been studied in some detail [43], and it is believed that significant changes to the global fold of a protein occur on a longer timescale than that of most MS experiments. This can be explained by kinetic trapping: it is known that the conformational flexibility of a protein is greatly reduced in the absence of water [44], due to an increased kinetic barrier for refolding (mainly due to a lack of H-bonding partners). Even though the native structure is not necessarily the most stable one in the gas phase, it is pre-formed in solution and large-scale structural rearrangement is rather slow under normal MS conditions [45]. While collapse of charged side chains can occur [46], it is usually only observed in structures that contain disordered, extended regions or that are "hollow" (e.g., ring-shaped protein complexes). This is further confirmed by indications that this compaction can be reduced by microsolvation of these side chains using cyclic polyethers (e.g., crown ethers) [47]. IM measurements have provided strong evidence that the native structure of protein complexes is largely maintained in the gas phase [3, 48, 49], and the (rotationally averaged) size of most folded proteins matches the expected value from X-ray structures, at least for the lowest observed charge states (see later).

In the context of ligand binding studies, the question of gas-phase versus solution affinity arises. How strong does binding have to be so that the complex can be detected in native MS, and can dissociation constants (K_d) be estimated from the relative abundance of bound and unbound species in the spectra? As a rule of thumb, complexes with a K_d below 50 μM are detected routinely if the analyte concentration is sufficiently high, while for weaker complexes ($K_d > 50$ μM) the mix of electrostatic versus hydrophobic contacts becomes important. Hydrophobic interactions are lost in the gas phase and there is usually no energetic barrier for dissociation, due to the absence of H bonds. Electrostatic interactions, on the other hand, are strengthened in vacuum since the shielding effect of surrounding water is lost. In almost all cases of ligand binding, or more generally protein interactions, the mix of dispersive van-der-Waals forces, H-bonds, and electrostatic and hydrophobic interactions guarantees a gas-phase stability of the complex, which frequently correlates in a semi-quantitative way with the solution K_d (Case Study 2). Rankings of related ligands (or sequence variants of proteins) are thus often possible.

Case Study 2: K_d Determination

As an example for semi-quantitative K_d determination by native MS, Figure 3.4 shows a titration of increasing amounts of nicotinamide adenine diphosphate (NADP) to the enzyme keto-pantoate reductase in aqueous ammonium acetate buffer. Three example spectra are shown (Figure 3.4a); they were recorded on a Waters Q-TOF (Quadrupole/time-of-flight) II instrument with nano-ESI and careful control of voltages to preserve noncovalent interactions. The K_d determined by this approach is on the same order of magnitude as measured by isothermal calorimetry (ITC), demonstrating that this protein–ligand complex which is formed in solution survives transfer into the gas phase of the mass spectrometer intact (F. Sobott, A. Ciulli, C. Abell, C.V. Robinson; unpublished data).

3.4
Mass Analyzers and Tandem MS Approaches

Native MS experiments impose a number of requirements on the mass analyzer [4, 50–53], due to the high mass and the relatively low charge state (see earlier) of native proteins. Q-TOF instruments are widely used for these experiments. Due to the high m/z of the analytes – for protein complexes up to m/z 50 000 – it is necessary to reduce the frequency of the quadrupole to allow transmission and selection of high m/z species [4, 50–53]. Collisions

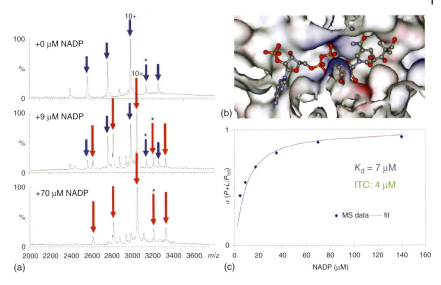

Figure 3.4 (a) Titration of NADP ligand (3.5–140 µM) to keto-pantoate reductase (5 µM; blue arrows in top panel; the asterisk denotes dimer) shows increasing numbers of protein–ligand complexes formed (red arrows in middle and bottom panel).

(b) Graphical representation of NADP in the binding pocket of the enzyme. (c) The K_d value of about 7 µM derived from ligand titration and native MS is on the same order of magnitude as the value obtained by isothermal calorimetry (ITC).

with background gas molecules must be carefully controlled, particularly in the early vacuum stages of the instrument. Upon transition into vacuum, the ion cloud undergoes fast expansion and the ions are accelerated. Through gentle collisions with background gas, they can be slowed down and focused into a tight beam, which improves transmission of high m/z ions significantly. Also, efficient removal of residual solvent molecules can be achieved without disrupting protein interactions. This process is known as *collisional cooling* and is especially important for supramolecular assemblies. In general, this requires the pressure in the initial vacuum stages to be increased (or a heavier collision gas to be used), compared to the analysis of small molecules.

In contrast, collisional heating is the process in which collisions increase the internal energy of the analyte ion, causing unfolding and fragmentation. The applied pressures and voltages determine which of the two regimes the analytes experience. In order to avoid collision-induced unfolding (CIU) and CID, potential differences are deliberately kept low in native MS experiments. The detector also needs to be more sensitive for analyses in the high m/z range. In the past, this required custom modification of instruments; however, currently, high-mass-modified instruments are commercially available. The Q-TOF instruments manufactured by Waters (Manchester, UK) and AB SCIEX

(Concord, Ontario/Canada) should be specifically mentioned, although Orbitrap instruments (Thermo, Bremen, Germany) modified for native work are also currently available.

While CIU and CID destroy a protein's native conformation, they can be very beneficial for top–down analysis of protein complexes. CIU can be used to study the stability of noncovalent interactions, as illustrated in Case Study 6. As the ionized complex is collisionally heated, one monomer unfolds before dissociating from the complex. As it unfolds, its surface increases and a disproportionately large portion of the mobile charges on the complex is ejected along with the monomer [54]. This process is known as *asymmetric dissociation*. In this way, accurate masses of the individual building blocks of a protein complex can be determined and the subunits or ligand(s) identified.

In CID-based MS/MS of protein complexes, the native structure is destroyed because noncovalent interactions are broken before dissociation of covalent bonds occurs. In contrast, electron capture dissociation (ECD) [55, 56] and electron transfer dissociation (ETD) [57] are fast, nonergodic dissociation techniques that selectively break the $C\alpha-N$ bond in the protein backbone while maintaining the higher-order structure. In a particularly interesting example, ECD was used on a 1 : 1 complex of α-synuclein with spermine [58]. Most of the detected fragments still carried the spermine ligand, allowing the researchers to locate the binding site of the ligand. These results were consistent with earlier NMR experiments [59]. There have also been attempts to use ECD and ETD as conformational footprinting techniques [60–62]. In these studies, ECD/ETD was performed on a complex in its native conformation. In this way, dissociation occurs selectively near regions of the protein sequence that are exposed at the surface, yielding information similar to that obtained from techniques such as hydrogen–deuterium exchange (HDX) or covalent labeling [26]. The main difference is that in HDX or covalent labeling, the surface-modifying reaction (either exchange of amide hydrogens for deuterium or labeling of reactive side chains) occurs in the (native) solution phase, after which (tandem) MS experiments are carried out under denaturing conditions. One downside in the use of ECD/ETD in this way is that, so far, these results have been limited to probing regions of the protein near the C or N terminus.

3.5
Ion Mobility

IM can be used purely as an additional separation dimension, in which case differential mobility spectrometry (DMS) or high-field asymmetric waveforms (field asymmetric ion mobility spectrometry (FAIMS)) are typically used [63–65], with resolving power up to 500 [66]. However, other variants of this technique also offer low-resolution structural information on a timescale of milliseconds, in addition

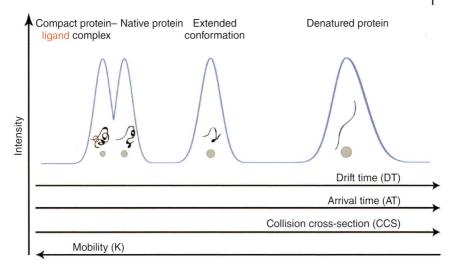

Figure 3.5 Ion-mobility spectrometry (IMS) allows the separation of analytes based on shape. Assuming (nearly) identical mass and charge, analytes in a more compact, native conformation will have a shorter drift time (equivalent to a smaller CCS or an increased mobility constant).

to separating analytes. In IM, the ion of interest is propelled forward by an electric field in an inert background gas, typically He or N_2 (Figure 3.5). On a macroscopic scale, the ions will quickly reach a constant velocity due to friction caused by collisions with the mobility gas. This process is somewhat similar to electrophoresis (albeit in the gas phase), and IM has historically also been referred to as *plasma chromatography* [67]. It is important to note that the pressure and voltage gradients need to be carefully controlled, so that the collisions with background gas do not cause unwanted CIU or CID (see earlier).

The ratio of an ion's velocity to the applied electric field is expressed as the mobility constant (K), and depends primarily on the reduced mass (μ) of both collision partners (which for protein–gas collisions becomes almost identical to the gas mass), the ion charge (z), and the CCS (Ω). In the case of a constant, homogeneous electric field, this is expressed mathematically as follows [67]:

$$K = \frac{3}{16} \frac{ze}{N} \sqrt{\frac{2\pi}{\mu k_b T} \frac{1}{\Omega}}$$

where N is the number density of the drift gas, T is the temperature, and k_b is the Boltzmann constant. As the mass of the drift gas is known, and the mass and charge state of the analyte can be easily obtained through MS, coupling IM with MS can reveal an analyte's CCS. Since the analyte ions tumble fast and freely while moving through the mobility cell (as a result of collisions with molecules of the background gas), the CCS is determined by the rotationally averaged 3D structure of the analyte. It follows that IM-MS can distinguish ions with the same m/z, but with a different shape. The most obvious application of this is the analysis of different conformations of a large biomolecule, but IM-MS can also separate oligomers with a different aggregation number and (nearly) identical m/z (e.g., a monomer with charge state n and a dimer with charge state $2n$). In this case, the relative mobility of the oligomers should be noted. As the observed CCS depends on the rotationally averaged structure, it will not increase as quickly as the mass with the addition of more subunits – unless the oligomers grow rod-like rather than in a more spherical fashion (see Figure 3.8b).

Several "subfamilies" of ion mobility spectrometry (IMS) can be distinguished. Drift-tube-ion mobility spectrometry (DT-IMS) and differential mobility analysis (DMA) [75] utilize electric fields that are constant over time. Traveling-wave ion mobility spectrometry (TWIMS) [76] utilizes a series of traveling electrostatic waves instead of a uniform field to propel the ions forward.

Case Study 3: Native MS Applied to Metalloproteins

Many enzymes require metal ions to be biologically active, and it is estimated that a third of proteins require a metal cofactor [68]. Changes to protein conformation upon metal ion adduction can be small and IM-MS is a sensitive analytical tool to detect these. Flick *et al.* [69] used IM-MS to compare the CCS of specific and nonspecific metal ion interactions. Upon binding of nonspecific La^{3+} to α-lactalbumin, a compaction in the protein conformation was observed as indicated by a shorter drift time. In contrast, the specific interaction between α-lactalbumin and one Ca^{2+} showed evidence for an enlargement of the structure. Highlighted in Figure 3.6 are the drift time distributions for apo- and holo-α-lactalbumin, and it is clear that an additional conformation of the protein (two peaks in the drift time distribution) is seen compared to the single one observed without Ca^{2+}. Using these data it was calculated that the CCS for the protein increased upon Ca^{2+} binding. Therefore IM could potentially be used to assess the specificity of metal ion binding to proteins.

With regard to the Zn^{2+} ion, MS has become even more invaluable as it is spectroscopically silent and therefore few spectroscopic techniques can be used to probe protein–Zn interactions. IM-MS was employed to analyze the interactions between a Cys_2His_2-type zinc finger peptide, vCP1 [70]. In this case the CCS values for apo- and holo-vCP1 were similar, and these values correlated well with molecular dynamics simulations of the structures. Another study has confirmed that MS can be utilized not only to study shape changes in the biomolecules but also to extract quantitative information. Deng *et al.* [71] achieved this by analyzing a reference protein that bound nonspecifically to Ca^{2+} which they could use to correct their data on β-lactoglobulin for the nonspecific effect.

Case Study 4: Drug Binding to Proteins

The fast screening of protein–ligand interactions by MS is an important development in the field of drug discovery as the majority of drug targets are proteins [72]. IM-MS has been used to probe the possible conformational changes upon covalent binding of cisplatin to ubiquitin as any structural changes detected may have biological implications *in vivo* [11]. Alterations in the ubiquitin system could have an impact upon tumor progression [73].

The overlaid IM drift times for various platinated ubiquitin species in Figure 3.7b demonstrate how changes in conformation can be identified. Firstly, unmodified ubiquitin (black line) already shows evidence for more than one conformation observed for the 7+ charge state. Upon binding of cisplatin, it is clear that the number of peaks in the drift time distribution increases: the mono-dentate cisplatin–ubiquitin adduct is shown in blue and the bi-dentate species in red. Overall, the cisplatin-bound forms have a higher mobility, and therefore are more compact, than the free protein. Calculating CCS revealed that there was a reduction in size of 2–5% compared to the free ubiquitin. CID experiments were able to establish that the platination site was the N-terminal methionine, and therefore the authors proposed that the decrease in size upon cisplatin binding could result from the reduced flexibility in the N-terminus. New conformers that have a lower propensity to unfold in the gas phase have also been observed in the case of a palladium(II) complex binding to ubiquitin [74].

When an ion is retarded sufficiently due to collisions with the drift gas, it "rolls" over the wave and stays in place until it is picked up by the next one. Ion transmission is higher than in DT-IMS, but the simple mathematical relation between mobility and CCS is lost, and a calibration procedure is necessary in order to derive structural information [77–79]. Resolving power is currently limited to around 50, which allows the distinction of analytes with CCS differing by 2–3%. A TWIMS cell is incorporated in the first commercially available IM-MS instrument, the Synapt high definition mass spectrometry (HDMS) [80] (Waters, Manchester, UK), which was launched in 2006, and its successors, the Synapt G2 and G2-S. In 2013, Agilent Technologies (Santa Clara/CA, USA) launched the 6560 IM Q-TOF with a DT-IMS cell. Less common, recently developed IMS variants such as overtone mobility spectrometry (OMS) [81–85], transversal modulation ion mobility spectrometry (TM-IMS) [86], trapped ion mobility spectrometry (TIMS) [87], and ion cyclotron mobility spectrometry [88, 89] will not be discussed here.

Native MS and IM are increasingly being used in combination with other techniques in structural biology. When comparing IM-MS to more traditional methods for the elucidation of structural details of large biomolecules and their binding partners, such as X-ray diffraction (XRD) and NMR spectroscopy, the

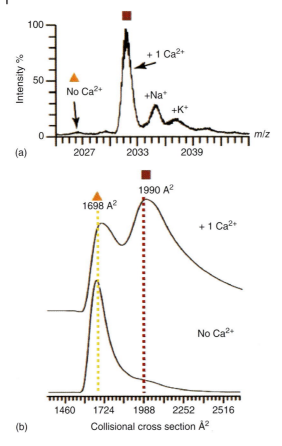

(a)

(b)

Figure 3.6 Comparison of the cross sections of α-lactalbumin in the presence and absence of Ca^{2+}. (a) Partial ESI positive ion spectrum from an aqueous solution containing 10 μM α-lactalbumin and 100 μM $CaCl_2$. (b) Drift profiles of the 7+ charge state with no metal ion adducts (bottom) and one Ca^{2+} adduct (top). (Adapted with permission from Ref. [69].)

significantly higher resolution of XRD and NMR must be noted. However, these techniques do have their limitations: For XRD, it is necessary that a protein crystallizes, which is not trivial to achieve, especially in the case of proteins containing flexible regions, or when different conformational or oligomeric states coexist. NMR spectroscopy can deal with heterogeneous structures, but is limited to small proteins, due to relaxation rates. Both these techniques require considerable instrument time and fairly large quantities of highly purified sample. Using IM-MS, analyses only require roughly 1/1000 of the sample compared to XRD and NMR and are sped up considerably, allowing the study of dynamic conformational or association phenomena, for example, in response to variation of solution-phase conditions.

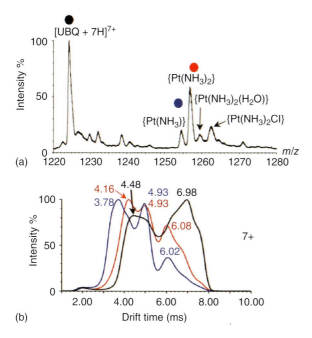

Figure 3.7 Effect of cisplatin binding on ubiquitin. (a) Partial ESI-MS spectrum showing the 7+ charge state of ubiquitin and the peaks that result from cisplatin binding. (b) Drift time distributions for the 7+ charge state of various ubiquitin species. The black line represents unmodified UBQ, the blue line [ubiquitin + Pt(NH$_3$)], and the red line [ubiquitin + Pt(NH$_3$)$_2$]. Note that unmodified ubiquitin (UBQ) already shows more than one conformation for the 7+ charge state. (Adapted with permission from Ref. [11].)

3.6
Data Processing

IM-MS data are usually represented as a 2D plot (heat map) of m/z value versus drift time. The most popular software for this is DriftScope (Waters, Manchester, UK), as it is also the most convenient way of processing Synapt IM-MS data. In DriftScope it is possible to select a specific drift time or range of drift times, and extract the associated (partial) mass spectrum for that range. Conversely, it is also possible to extract the intensity versus drift time plot associated with a specific m/z window. In the latter case, and especially when working with large molecules or complexes, one must be careful to select only one analyte species. If, for instance, a ligand or metal cation induces a conformational change, inadvertent simultaneous selection of the protein or complex with and without the ligand would lead one to incorrectly conclude that the analyte itself has two distinct conformations.

Case Study 5: Amyloidogenic Proteins

IM-MS has been applied to study amyloidogenic peptides and proteins whose aggregation *in vivo* has been associated with a number of diseases including Alzheimer's [90], Parkinson's [91], and dialysis-related amyloidosis [92]. Although the mass spectra arising from such insoluble aggregates can be complex, using IM-MS offers another dimension of separation by size, which is extremely beneficial for heterogeneous systems. It has been suggested that the toxicity of these misfolded proteins is related to the early formation of oligomers, rather than the fibers. Therefore it is pertinent that the formation of these oligomers be addressed and IM-MS is one such technique where snapshots of the intermediates can be obtained throughout oligomerization.

ESI-MS can be used to show that β_2-microglobulin assembles into dimers, trimers, and tetramers, which lead to the formation of fibers [93]. As highlighted in Figure 3.8a, IM-MS has the capability to separate these species even if they have the same *m/z*. Application of this technique was also able to clarify the structural progression of these oligomers and compare their cross sections with those of globular proteins [94]. Figure 3.8b shows the measured CCS for all charge states of monomeric β_2-microglobulin and its oligomeric forms compared to those calculated for other proteins at pH 6.5. Firstly, monomeric β_2-microglobulin can adopt different conformational states in low–ionic strength buffer: β_2m-red = reduced; β_2m-A = acid-unfolded; β_2m-P = partially unfolded; β_2m-N = native. Secondly, these results show that the β_2-microglobulin assemblies are elongated in size and grow as rod-like rather than spherical structures. The CCS for oligomers of increasing size do not fit the typical curve observed for globular proteins (solid line in Figure 3.8b). For example, the β_2-microglobulin tetramer (47 kDa) shows a CCS of 3721 Å^2, which is considerably larger than the value of 2900 Å^2 determined for the transthyretin (TTR) tetramer, a globular protein of similar size (55 kDa).

To map the dynamics of oligomer assembly by ESI-MS, ^{14}N/^{15}N labeled β_2-microglobulin monomers were combined together. This revealed that the dimers undergo rapid exchange and a model was proposed whereby a dimer associates with a monomer to form an unstable trimer. The dynamic complexes then reorganize into more compact, stable species.

IM measurements can inform on subunit connectivity within protein complexes by ruling out possible architectures that would lead to CCS values significantly different from experimental observations [49]. Several computational methods have been developed in order to relate the measured CCS to structures determined by using XRD or NMR [96–98], or to other candidate structures. These include the projection approximation (PA), the exact hard-sphere scattering (EHSS) method, and the trajectory method (TM), all of which calculate CCS based on Cartesian coordinates of the atoms in the model. All three of these algorithms are incorporated in the open-source program MOBCAL,

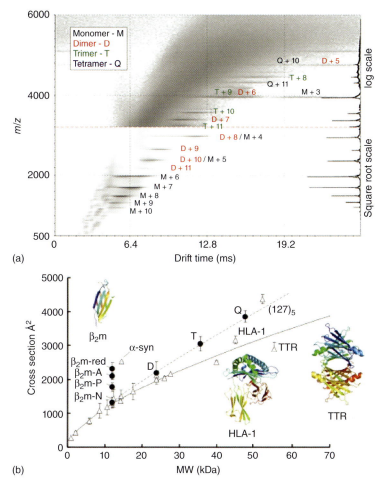

Figure 3.8 Cross sections of β_2-microglobulin oligomers compared to other, globular protein standards. (a) ESI-IMS/MS data for β_2m measured at 1 min into the lag time of fibril assembly. Labels refer to monomer (M), dimer (D), trimer (T), and tetramer (Q). The scaling on the right-hand axis is adjusted to "square root" or "log" to display the weaker ions clearly. (b) Measured CCS of β_2-microglobulin oligomers (black circles) compared to other proteins (open triangles). Two protein structures highlighted are HLA-1 and TTR as these have a similar size to the β_2-microglobulin tetramer but are shown to be more compact. (Adapted with permission from Ref. [94].)

developed by Shvartsburg *et al.* [97]. Further discussion of these methods would be beyond the scope of this chapter; however, in general, IM data are in good agreement with other biophysical techniques, with gas-phase structures typically being slightly more compact than crystal structures [99–101], possibly due to the aforementioned side-chain collapse during the later stages of the ESI process. One should also note that it has been suggested that the ultimate limitation to resolving power in IMS is not due to instrumentation, but conformational heterogeneity inherent to the large biomolecules being studied close to room temperature [100, 102].

Case Study 6: Binding Modes of Kinase Inhibitors

IM-MS is not limited to the determination of static protein conformations, but can also monitor unfolding events within a protein complex. This experiment is termed *collision induced unfolding* (see earlier) and involves activation of ions via collisions with gas molecules, but not to the point that the ligand of interest dissociates from the protein. When applied to the protein kinase domain of Abl, this approach can allow differentiation between type I and type II inhibitors depending on how the complex unfolds in the gas phase [95]. A single charge state of the Abl–inhibitor complex is selected in the quadrupole and then subjected to collisional activation. Analyzing the complex over a range of collision energies and plotting this against the drift time of the ions observed creates a "CIU fingerprint" as shown in Figure 3.9a. To validate this assay the 11+ charge state of Abl interacting with a range of seven known inhibitors was subjected to CIU and the fingerprints compared (Figure 3.9a). The small region from 40 to 44 V (highlighted in Figure 3.9a) was found to be effective in differentiating between type I (blue) and type II inhibitors (red). Integrating the boxed regions shown resulted in an average drift time for both type I and type II inhibitors. Interestingly, type II inhibitors show approximately eight distinct drift times (shapes) whereas type I inhibitors only exhibit two as illustrated in Figure 3.9b. Consequently, there appears to be a distinct correlation between the unfolded conformations observed for the kinase–inhibitor complex, and where the inhibitor binds to the protein. This novel method could be used to rapidly screen inhibitors and assess their binding modes.

3.7
Challenges and Future Perspectives

While there have been significant advances in native MS in recent years, challenges still remain in sample preparation, instrumentation, and data analysis. In this final section, we will discuss some recent, exciting developments that allow analysis of ever-larger complexes, and have opened up the field for the study of integral membrane proteins and their interactions.

A first example is the investigation of lipid binding by soluble proteins with so-called catch-and-release MS. Here nanodiscs – membrane bilayers restricted in size by a scaffolding protein wrapped around them – are used as a platform to simulate protein–lipid interactions for membrane-associated proteins. Many

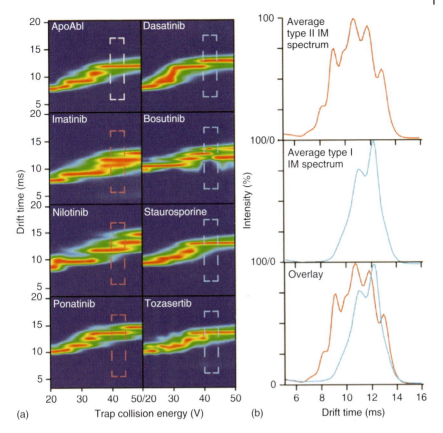

(a) Trap collision energy (V)

(b) Drift time (ms)

Figure 3.9 The 11+ charge state of the protein kinase Abl bound to type I (blue) or type II inhibitors (red) was subjected to CIU. (a) Ion intensity plot as a function of the collision energy applied and the drift time observed. (b) Average drift times extracted from the boxed regions (40–44 V collision energy) show a marked difference in the number of conformational states observed for type I and type II inhibitors. (Reproduced with permission from Ref. [95].)

proteins, especially when involved in signal transduction or antigen/microbial recognition, bind to specific lipids or receptors on the cell surface. By preparing the lipid bilayer in the nanodiscs from lipids extracted from specific ranges of cells, it has been shown that it is possible to probe selectively for lipid affinity of membrane-associated proteins [103]. To demonstrate their approach, the authors investigate the binding of the glycosphingolipid (GSL) GM1 to the cholera toxin. The toxins were mixed with nanodiscs with a dimyristoyl phosphatidylcholine (DMPC)/GM1 bilayer and introduced into the mass spectrometer by nano-ESI. By applying mild CID in the source region, the lipid–lipid interactions were broken, allowing for the release of the toxin–GM1 complex from the nanodisc–protein complex. Not only does this simplify the spectrum, but higher-energy CID also allows the identification of the type of lipid attached, together with IM, which can further confirm the assignment. This approach seems to hold promise for investigating protein–lipid interactions and might enable high-throughput screening of nanodiscs loaded with a broad variety of lipids.

Case Study 7: Native MS of Membrane-Associated Protein–Lipid Interactions

One of the most recent developments in native MS is the analysis of intact membrane proteins. The advent of this research began with the realization that micelles could be transferred into the vacuum of the mass spectrometer [104], and that they can be used to protect membrane proteins during this transfer [13, 105–107]. Current methodology relies on the use of CID inside the mass spectrometer to remove detergent and facilitate analysis of the released membrane–protein complex. In a benchmark paper, Robinson and colleagues show one of the very exciting possibilities this technique has to offer, by investigating two rotary ATPases [12]. Not only did their methodology allow for elucidation of the subunit stoichiometry of these protein structures, but for the first time it was demonstrated that this approach enabled the study of the lipids and nucleotides that were attached.

The membrane-spanning subcomplex VO of the TtATPase was analyzed in the presence of both ATP and adenosine diphosphate (ADP). Figure 3.10a shows the drift time distributions of the proton channel domain of the TtATPase under ATP-rich and ATP-depleted (ADP-rich) conditions. Under ADP-rich conditions, the authors observed the loss of subunit I. However, under ATP-rich conditions, a broadening of the peak series was observed, which was consistent with the binding of six lipids and two nucleotides. To investigate if subunit I can sense intracellular levels of nucleotides, as was indicated by previous research, the authors applied an IM-MS approach. Figure 3.10b,c show the drift time distributions of different charge states under high levels of either ATP or ADP, respectively. They observed a narrow distribution in drift times for the CL12 complex under ADP-rich conditions, whereas the broader drift time distribution for the ICL12 complex under ATP-rich conditions suggests that a heterogeneous distribution of structures is present. Based on these observations, a mechanism was proposed wherein the removal of subunit I due to low concentrations of either ATP or protons leads to inactivation of the proton channel. As depicted in Figure 3.10d, removal of subunit I would enable lipids to block the proton channel.

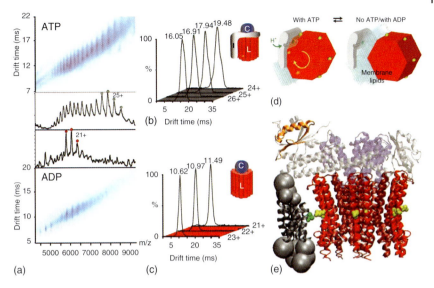

Figure 3.10 IM-MS of the V-type ATPase reveals the closing mechanism for the H$^+$ channel based on nucleotide binding. (a) IM-MS spectra of ICL12/CL12 under ATP-rich and ATP-depleted (ADP-rich) conditions. (b,c) The broader drift time distribution of the ICL12 subcomplex (top) compared to CL12 (bottom) indicates heterogeneity in the subunit I. (d) Proposed mechanism for nucleotide-dependent closing of the proton channel. (Reproduced with permission from Ref. [12].)

Surface-induced dissociation (SID) [108] is a technique that has proven superior to CID for a more efficient and controlled tandem MS of protein complexes. In SID, ions collide with a solid target that carries a fluorinated monolayer and energy transfer is believed to require only a few picoseconds [109, 110]. As a result, there is insufficient time for proteins to unfold and for charge carriers to relocate along the surface of the dissociating complex. It has been shown that this method allows extensive dissociation and can provide a great deal of information about the structure of large molecular systems [111].

A current challenge for large complexes is the time-consuming analysis of complicated spectra, for example, from heterogeneous, polydisperse samples. An unambiguous assignment of all observed peak series is a prerequisite for structural studies, and this problem is being addressed by a number of research groups. Massign [112] and Amphitrite [112] were developed in order to extract *m/z* and mobility data from spectra obtained from large biological systems. To deconvolute the data generated from heterogeneous small heat-shock proteins, the CHAMP algorithm (Calculating the Heterogeneous Assembly and Mass Spectra of Proteins) was recently developed together with the CHAMPION extension for IM data [113].

In summary, native MS is now increasingly used to characterize protein conformations and study the effect of ligands and other binding partners on the stoichiometry and stability of complexes. In combination with IM, the shape and size of molecular assemblies can be determined as well in this way. Native approaches together with computational methods are thus beginning to pull their weight in the fields of biomolecular analysis and structural proteomics, and address questions of biochemical and biomedical importance.

References

1. Dyachenko, A., Gruber, R., Shimon, L., Horovitz, A., and Sharon, M. (2013) Allosteric mechanisms can be distinguished using structural mass spectrometry. *Proc. Natl. Acad. Sci. U.S.A.*, **110**, 7235–7239.

2. Snijder, J., Rose, R.J., Veesler, D., Johnson, J.E., and Heck, A.J. (2013) Studying 18 MDa virus assemblies with native mass spectrometry. *Angew. Chem., Int. Ed. Engl.*, **52**, 4020–4023.

3. Ruotolo, B.T., Giles, K., Campuzano, I., Sandercock, A.M., Bateman, R.H., and Robinson, C.V. (2005) Evidence for macromolecular protein rings in the absence of bulk water. *Science*, **310**, 1658–1661.

4. Rose, R.J., Damoc, E., Denisov, E., Makarov, A., and Heck, A.J. (2012) High-sensitivity Orbitrap mass analysis of intact macromolecular assemblies. *Nat. Methods*, **9**, 1084–1086.

5. Ma, X., Shah, S., Zhou, M.Q., Park, C.K., Wysocki, V.H., and Horton, N.C. (2013) Structural analysis of activated SgrAI-DNA oligomers using ion mobility mass spectrometry. *Biochemistry*, **52**, 4373–4381.

6. Vincent, H.A., Henderson, C.A., Stone, C.M., Cary, P.D., Gowers, D.M., Sobott, F., Taylor, J.E., and Callaghan, A.J. (2012) The low-resolution solution structure of vibrio cholerae Hfq in complex with Qrr1 sRNA. *Nucleic Acids Res.*, **40**, 8698–8710.

7. Mei, H.Y., Mack, D.P., Galan, A.A., Halim, N.S., Heldsinger, A., Loo, J.A., Moreland, D.W., Sannes-Lowery, K.A., Sharmeen, L., Truong, H.N., and Czarnik, A.W. (1997) Discovery of selective, small-molecule inhibitors of RNA complexes--I. The Tat protein/TAR RNA complexes required for HIV-1 transcription. *Bioorg. Med. Chem.*, **5**, 1173–1184.

8. Wyttenbach, T., Grabenauer, M., Thalassinos, K., Scrivens, J.H., and Bowers, M.T. (2010) The effect of calcium ions and peptide ligands on the relative stabilities of the calmodulin dumbbell and compact structures. *J. Phys. Chem. B*, **114**, 437–447.

9. El-Hawiet, A., Kitova, E.N., Liu, L., and Klassen, J.S. (2010) Quantifying labile protein-ligand interactions using electrospray ionization mass spectrometry. *J. Am. Soc. Mass Spectrom.*, **21**, 1893–1899.

10. Liu, L., Kitova, E.N., and Klassen, J.S. (2011) Quantifying protein-fatty acid interactions using electrospray ionization mass spectrometry. *J. Am. Soc. Mass Spectrom.*, **22**, 310–318.

11. Williams, J.P., Phillips, H.I.A., Campuzano, I., and Sadler, P.J. (2010) Shape changes induced by N-terminal platination of ubiquitin by cisplatin. *J. Am. Soc. Mass Spectrom.*, **21**, 1097–1106.

12. Zhou, M., Morgner, N., Barrera, N.P., Politis, A., Isaacson, S.C., Matak-Vinkovic, D., Murata, T., Bernal, R.A., Stock, D., and Robinson, C.V. (2011) Mass spectrometry of intact V-type ATPases reveals bound lipids and the effects of nucleotide binding. *Science*, **334**, 380–385.

13. Barrera, N.P., Di Bartolo, N., Booth, P.J., and Robinson, C.V. (2008) Micelles protect membrane complexes from solution to vacuum. *Science*, **321**, 243–246.

14. Kebarle, P. and Verkerk, U.H. (2009) Electrospray: from ions in solution to

ions in the gas phase, what we know now. *Mass Spectrom. Rev.*, **28**, 898–917.

15. Hartl, F.U. and Hayer-Hartl, M. (2002) Molecular chaperones in the cytosol: from nascent chain to folded protein. *Science*, **295**, 1852–1858.

16. Sobott, F. and Robinson, C.V. (2004) Characterising electrosprayed biomolecules using tandem-MS – the noncovalent GroEL chaperonin assembly. *Int. J. Mass Spectrom.*, **236**, 25–32.

17. van Duijn, E., Barendregt, A., Synowsky, S., Versluis, C., and Heck, A.J. (2009) Chaperonin complexes monitored by ion mobility mass spectrometry. *J. Am. Chem. Soc.*, **131**, 1452–1459.

18. Rogniaux, H., Sanglier, S., Strupat, K., Azza, S., Roitel, O., Ball, V., Tritsch, D., Branlant, G., and Van Dorsselaer, A. (2001) Mass spectrometry as a novel approach to probe cooperativity in multimeric enzymatic systems. *Anal. Biochem.*, **291**, 48–61.

19. Fenn, J.B., Mann, M., Meng, C.K., Wong, S.F., and Whitehouse, C.M. (1989) Electrospray ionization for mass spectrometry of large biomolecules. *Science*, **246**, 64–71.

20. Dole, M., Mack, L.L., and Hines, R.L. (1968) Molecular beams of macroions. *J. Chem. Phys.*, **49**, 2240–2249.

21. Mack, L.L., Kralik, P., Rheude, A., and Dole, M. (1970) Molecular beams of macroions. *J. Chem. Phys.*, **52**, 4977–4986.

22. Hall, Z. and Robinson, C.V. (2012) Do charge state signatures guarantee protein conformations? *J. Am. Soc. Mass Spectrom.*, **23**, 1161–1168.

23. Chowdhury, S.K., Katta, V., and Chait, B.T. (1990) Probing conformational-changes in proteins by mass-spectrometry. *J. Am. Chem. Soc.*, **112**, 9012–9013.

24. Grandori, R. (2003) Origin of the conformation dependence of protein charge-state distributions in electrospray ionization mass spectrometry. *J. Mass Spectrom.*, **38**, 11–15.

25. Samalikova, M. and Grandori, R. (2005) Testing the role of solvent surface tension in protein ionization by electro-spray. *J. Mass Spectrom.*, **40**, 503–510.

26. Konijnenberg, A., Butterer, A., and Sobott, F. (2013) Native ion mobility-mass spectrometry and related methods in structural biology. *Biochim. Biophys. Acta*, **1834**, 1239–1256.

27. Winger, B.E., Lightwahl, K.J., Loo, R.R.O., Udseth, H.R., and Smith, R.D. (1993) Observation and implications of high mass-to-charge ratio ions from electrospray-ionization mass-spectrometry. *J. Am. Soc. Mass Spectrom.*, **4**, 536–545.

28. Kebarle, P. and Peschke, M. (2000) On the mechanisms by which the charged droplets produced by electrospray lead to gas phase ions. *Anal. Chim. Acta*, **406**, 11–35.

29. de la Mora, J.F. (2000) Electrospray ionization of large multiply charged species proceeds via Dole's charged residue mechanism. *Anal. Chim. Acta*, **406**, 93–104.

30. Smith, R.D., Lightwahl, K.J., Winger, B.E., and Loo, J.A. (1992) Preservation of noncovalent associations in electrospray ionization mass-spectrometry – multiply charged polypeptide and protein dimers. *Org. Mass Spectrom.*, **27**, 811–821.

31. Smith, R.D. and Lightwahl, K.J. (1993) The observation of non-covalent interactions in solution by electrospray-ionization mass-spectrometry – promise, pitfalls and prognosis. *Biol. Mass. Spectrom.*, **22**, 493–501.

32. Wilm, M.S. and Mann, M. (1994) Electrospray and Taylor-Cone theory, Doles beam of macromolecules at last. *Int. J. Mass Spectrom.*, **136**, 167–180.

33. Wilm, M. and Mann, M. (1996) Analytical properties of the nanoelectrospray ion source. *Anal. Chem.*, **68**, 1–8.

34. Keetch, C.A., Hernanndez, H., Sterling, A., Baumert, M., Allen, M.H., and Robinson, C.V. (2003) Use of a microchip device coupled with mass spectrometry for ligand screening of a multi-protein target. *Anal. Chem.*, **75**, 4937–4941.

35. Ganem, B., Li, Y.T., and Henion, J.D. (1991) Detection of noncovalent

receptor ligand complexes by mass-spectrometry. *J. Am. Chem. Soc.*, **113**, 6294–6296.

36. Ganem, B., Li, Y.T., and Henion, J.D. (1991) Observation of noncovalent enzyme substrate and enzyme product complexes by ion-spray mass-spectrometry. *J. Am. Chem. Soc.*, **113**, 7818–7819.

37. Katta, V. and Chait, B.T. (1991) Observation of the heme globin complex in native myoglobin by electrospray-ionization mass-spectrometry. *J. Am. Chem. Soc.*, **113**, 8534–8535.

38. Nguyen, A. and Moini, M. (2008) Analysis of major protein-protein and protein-metal complexes of erythrocytes directly from cell lysate utilizing capillary electrophoresis mass spectrometry. *Anal. Chem.*, **80**, 7169–7173.

39. Loo, J.A. (1997) Studying noncovalent protein complexes by electrospray ionization mass spectrometry. *Mass Spectrom. Rev.*, **16**, 1–23.

40. Hofstadler, S.A. and Griffey, R.H. (2001) Analysis of noncovalent complexes of DNA and RNA by mass spectrometry. *Chem. Rev.*, **101**, 377–390.

41. Han, X., Jin, M., Breuker, K., and McLafferty, F.W. (2006) Extending top-down mass spectrometry to proteins with masses greater than 200 kilodaltons. *Science*, **314**, 109–112.

42. van Duijn, E., Bakkes, P.J., Heeren, R.M., van den Heuvel, R.H., van Heerikhuizen, H., van der Vies, S.M., and Heck, A.J. (2005) Monitoring macromolecular complexes involved in the chaperonin-assisted protein folding cycle by mass spectrometry. *Nat. Methods*, **2**, 371–376.

43. Breuker, K. and McLafferty, F.W. (2008) Stepwise evolution of protein native structure with electrospray into the gas phase, 10(-12) to 10(2) S. *Proc. Natl. Acad. Sci. U.S.A.*, **105**, 18145–18152.

44. Doukyu, N. and Ogino, H. (2010) Organic solvent-tolerant enzymes. *Biochem. Eng. J.*, **48**, 270–282.

45. Hamdy, O.M. and Julian, R.R. (2012) Reflections on charge state distributions, protein structure, and the mystical mechanism of electrospray ionization. *J. Am. Soc. Mass Spectrom.*, **23**, 1–6.

46. Steinberg, M.Z., Elber, R., McLafferty, F.W., Gerber, R.B., and Breuker, K. (2008) Early structural evolution of native cytochrome c after solvent removal. *ChemBioChem*, **9**, 2417–2423.

47. Warnke, S., von Helden, G., and Pagel, K. (2013) Protein structure in the gas phase: the influence of side-chain microsolvation. *J. Am. Chem. Soc.*, **135**, 1177–1180.

48. Ruotolo, B.T. and Robinson, C.V. (2006) Aspects of native proteins are retained in vacuum. *Curr. Opin. Chem. Biol.*, **10**, 402–408.

49. Baldwin, A.J., Lioe, H., Hilton, G.R., Baker, L.A., Rubinstein, J.L., Kay, L.E., and Benesch, J.L.P. (2011) The polydispersity of alpha B-crystallin is rationalized by an interconverting polyhedral architecture. *Structure*, **19**, 1855–1863.

50. Tahallah, N., Pinkse, M., Maier, C.S., and Heck, A.J. (2001) The effect of the source pressure on the abundance of ions of noncovalent protein assemblies in an electrospray ionization orthogonal time-of-flight instrument. *Rapid Commun. Mass Spectrom.*, **15**, 596–601.

51. Sobott, F., McCammon, M.G., Hernandez, H., and Robinson, C.V. (2005) The flight of macromolecular complexes in a mass spectrometer. *Philos. Trans. A Math. Phys. Eng. Sci.*, **363**, 379–389; discussion 389–391.

52. Chernushevich, I.V. and Thomson, B.A. (2004) Collisional cooling of large ions in electrospray mass spectrometry. *Anal. Chem.*, **76**, 1754–1760.

53. Sobott, F., Hernandez, H., McCammon, M.G., Tito, M.A., and Robinson, C.V. (2002) A tandem mass spectrometer for improved transmission and analysis of large macromolecular assemblies. *Anal. Chem.*, **74**, 1402–1407.

54. Schwartz, B.L., Bruce, J.E., Anderson, G.A., Hofstadler, S.A., Rockwood, A.L., Smith, R.D., Chilkoti, A., and Stayton, P.S. (1995) Dissociation of tetrameric ions of noncovalent streptavidin complexes formed by electrospray-ionization. *J. Am. Soc. Mass Spectrom.*, **6**, 459–465.

55. Zubarev, R.A., Kelleher, N.L., and McLafferty, F. (1998) Electron capture

dissociation of multiply charged protein cations. A nonergodic process. *J. Am. Chem. Soc.*, **120**, 2.

56. Horn, D.M., Zubarev, R.A., and McLafferty, F.W. (2000) Automated de novo sequencing of proteins by tandem high-resolution mass spectrometry. *Proc. Natl. Acad. Sci. U.S.A.*, **97**, 10313–10317.

57. Syka, J.E., Coon, J.J., Schroeder, M.J., Shabanowitz, J., and Hunt, D.F. (2004) Peptide and protein sequence analysis by electron transfer dissociation mass spectrometry. *Proc. Natl. Acad. Sci. U.S.A.*, **101**, 9528–9533.

58. Xie, Y., Zhang, J., Yin, S., and Loo, J.A. (2006) Top-down ESI-ECD-FT-ICR mass spectrometry localizes noncovalent protein-ligand binding sites. *J. Am. Chem. Soc.*, **128**, 14432–14433.

59. Fernandez, C.O., Hoyer, W., Zweckstetter, M., Jares-Erijman, E.A., Subramaniam, V., Griesinger, C., and Jovin, T.M. (2004) NMR of alpha-synuclein-polyamine complexes elucidates the mechanism and kinetics of induced aggregation. *EMBO J.*, **23**, 2039–2046.

60. Zhang, H., Cui, W., Wen, J., Blankenship, R.E., and Gross, M.L. (2010) Native electrospray and electron-capture dissociation in FTICR mass spectrometry provide top-down sequencing of a protein component in an intact protein assembly. *J. Am. Soc. Mass Spectrom.*, **21**, 1966–1968.

61. Zhang, H., Cui, W., Wen, J., Blankenship, R.E., and Gross, M.L. (2011) Native electrospray and electron-capture dissociation FTICR mass spectrometry for top-down studies of protein assemblies. *Anal. Chem.*, **83**, 5598–5606.

62. Lermyte, F., Konijnenberg, A., Williams, J.P., Brown, J.M., Valkenborg, D., and Sobott, F. (2014) ETD allows for native surface mapping of a 150 kDa non-covalent complex on a commercial Q-TWIMS-TOF instrument. *J. Am. Soc. Mass Spectrom.*, **25**, 343–350.

63. Buryakov, I.A., Krylov, E.V., Nazarov, E.G., and Rasulev, U.K. (1993) A new method of separation of multi-atomic ions by mobility at atmospheric-pressure using a high-frequency amplitude-asymmetric strong electric-field. *Int. J. Mass Spectrom.*, **128**, 143–148.

64. Purves, R.W. and Guevremont, R. (1999) Electrospray ionization high-field asymmetric waveform ion mobility spectrometry-mass spectrometry. *Anal. Chem.*, **71**, 2346–2357.

65. Shvartsburg, A.A., Li, F., Tang, K., and Smith, R.D. (2006) High-resolution field asymmetric waveform ion mobility spectrometry using new planar geometry analyzers. *Anal. Chem.*, **78**, 3706–3714.

66. Shvartsburg, A.A., Seim, T.A., Danielson, W.F., Norheim, R., Moore, R.J., Anderson, G.A., and Smith, R.D. (2013) High-definition differential ion mobility spectrometry with resolving power up to 500. *J. Am. Soc. Mass Spectrom.*, **24**, 109–114.

67. Revercomb, H.E. and Mason, E.A. (1975) Theory of plasma chromatography gaseous electrophoresis – review. *Anal. Chem.*, **47**, 970–983.

68. Waldron, K.J. and Robinson, N.J. (2009) How do bacterial cells ensure that metalloproteins get the correct metal? *Nat. Rev. Microbiol.*, **7**, 25–35.

69. Flick, T.G., Merenbloom, S.I., and Williams, E.R. (2013) Effects of metal ion adduction on the gas-phase conformations of protein ions. *J. Am. Soc. Mass Spectrom.*, **24**, 1654–1662.

70. Berezovskaya, Y., Armstrong, C.T., Boyle, A.L., Porrini, M., Woolfson, D.N., and Barran, P.E. (2011) Metal binding to a zinc-finger peptide: a comparison between solution and the gas phase. *Chem. Commun. (Camb.)*, **47**, 412–414.

71. Deng, L., Sun, N., Kitova, E.N., and Klassen, J.S. (2010) Direct quantification of protein-metal ion affinities by electrospray ionization mass spectrometry. *Anal. Chem.*, **82**, 2170–2174.

72. Pacholarz, K.J., Garlish, R.A., Taylor, R.J., and Barran, P.E. (2012) Mass spectrometry based tools to investigate protein-ligand interactions for drug discovery. *Chem. Soc. Rev.*, **41**, 4335–4355.

73. Hoeller, D. and Dikic, I. (2009) Targeting the ubiquitin system in cancer therapy. *Nature*, **458**, 438–444.

74. Giganti, V.G., Kundoor, S., Best, W.A., and Angel, L.A. (2011) Ion mobility-mass spectrometry study of folded ubiquitin conformers induced by treatment with cis-[Pd(en)(H(2)O)(2)](2+). *J. Am. Soc. Mass Spectrom.*, **22**, 300–309.

75. Labowsky, M. and de la Mora, J.F. (2006) Novel ion mobility analyzers and filters. *J. Aerosol Sci.*, **37**, 340–362.

76. Giles, K., Pringle, S.D., Worthington, K.R., Little, D., Wildgoose, J.L., and Bateman, R.H. (2004) Applications of a travelling wave-based radio-frequencyonly stacked ring ion guide. *Rapid Commun. Mass Spectrom.*, **18**, 2401–2414.

77. Ruotolo, B.T., Benesch, J.L.P., Sandercock, A.M., Hyung, S.J., and Robinson, C.V. (2008) Ion mobility-mass spectrometry analysis of large protein complexes. *Nat. Protoc.*, **3**, 1139–1152.

78. Bush, M.F., Hall, Z., Giles, K., Hoyes, J., Robinson, C.V., and Ruotolo, B.T. (2010) Collision cross sections of proteins and their complexes: a calibration framework and database for gas-phase structural biology. *Anal. Chem.*, **82**, 9557–9565.

79. Hamilton, J.V., Renaud, J.B., and Mayer, P.M. (2012) Experiment and theory combine to produce a practical negative ion calibration set for collision cross-section determinations by travelling-wave ion-mobility mass spectrometry. *Rapid Commun. Mass Spectrom.*, **26**, 1591–1595.

80. Pringle, S.D., Giles, K., Wildgoose, J.L., Williams, J.P., Slade, S.E., Thalassinos, K., Bateman, R.H., Bowers, M.T., and Scrivens, J.H. (2007) An investigation of the mobility separation of some peptide and protein ions using a new hybrid quadrupole/travelling wave IMS/oa-ToF instrument. *Int. J. Mass Spectrom.*, **261**, 1–12.

81. Kurulugama, R.T., Nachtigall, F.M., Lee, S., Valentine, S.J., and Clemmer, D.E. (2009) Overtone mobility spectrometry: part 1. Experimental observations. *J. Am. Soc. Mass Spectrom.*, **20**, 729–737.

82. Valentine, S.J., Stokes, S.T., Kurulugama, R.T., Nachtigall, F.M., and Clemmer, D.E. (2009) Overtone mobility spectrometry: part 2. Theoretical considerations of resolving power. *J. Am. Soc. Mass Spectrom.*, **20**, 738–750.

83. Valentine, S.J., Kurulugama, R.T., and Clemmer, D.E. (2011) Overtone mobility spectrometry: part 3. On the origin of peaks. *J. Am. Soc. Mass Spectrom.*, **22**, 804–816.

84. Kurulugama, R.T., Nachtigall, F.M., Valentine, S.J., and Clemmer, D.E. (2011) Overtone mobility spectrometry: part 4. OMS-OMS analyses of complex mixtures. *J. Am. Soc. Mass Spectrom.*, **22**, 2049–2060.

85. Ewing, M.A., Zucker, S.M., Valentine, S.J., and Clemmer, D.E. (2013) Overtone mobility spectrometry: part 5. Simulations and analytical expressions describing overtone limits. *J. Am. Soc. Mass Spectrom.*, **24**, 615–621.

86. Vidal-de-Miguel, G., Macia, M., and Cuevas, J. (2012) Transversal Modulation Ion Mobility Spectrometry (TM-IMS), a new mobility filter overcoming turbulence related limitations. *Anal. Chem.*, **84**, 7831–7837.

87. Fernandez-Lima, F.A., Kaplan, D.A., and Park, M.A. (2011) Note: integration of trapped ion mobility spectrometry with mass spectrometry. *Rev. Sci. Instrum.*, **82**, 126106.

88. Merenbloom, S.I., Glaskin, R.S., Henson, Z.B., and Clemmer, D.E. (2009) High-resolution ion cyclotron mobility spectrometry. *Anal. Chem.*, **81**, 1482–1487.

89. Glaskin, R.S., Valentine, S.J., and Clemmer, D.E. (2010) A scanning frequency mode for ion cyclotron mobility spectrometry. *Anal. Chem.*, **82**, 8266–8271.

90. Murphy, M.P. and LeVine, H. 3rd, (2010) Alzheimer's disease and the amyloid-beta peptide. *J. Alzheimers Dis.*, **19**, 311–323.

91. Maries, E., Dass, B., Collier, T.J., Kordower, J.H., and Steece-Collier, K. (2003) The role of alpha-synuclein in Parkinson's disease: insights from animal models. *Nat. Rev. Neurosci.*, **4**, 727–738.

92. Williams, D.M. and Pukala, T.L. (2013) Novel insights into protein misfolding diseases revealed by ion mobility-mass spectrometry. *Mass Spectrom. Rev.*, **32**, 169–187.

93. Smith, A.M., Jahn, T.R., Ashcroft, A.E., and Radford, S.E. (2006) Direct observation of oligomeric species formed in the early stages of amyloid fibril formation using electrospray ionisation mass spectrometry. *J. Mol. Biol.*, **364**, 9–19.

94. Smith, D.P., Radford, S.E., and Ashcroft, A.E. (2010) Elongated oligomers in beta(2)-microglobulin amyloid assembly revealed by ion mobility spectrometry-mass spectrometry. *Proc. Natl. Acad. Sci. U.S.A.*, **107**, 6794–6798.

95. Rabuck, J.N., Hyung, S.J., Ko, K.S., Fox, C.C., Soellner, M.B., and Ruotolo, B.T. (2013) Activation state-selective kinase inhibitor assay based on ion mobility-mass spectrometry. *Anal. Chem.*, **85**, 6995–7002.

96. Wyttenbach, T., von Helden, G., Batka, J.J., Carlat, D., and Bowers, M.T. (1997) Effect of the long-range potential on ion mobility measurements. *J. Am. Soc. Mass Spectrom.*, **8**, 275–282.

97. Shvartsburg, A.A. and Jarrold, M.F. (1996) An exact hard-spheres scattering model for the mobilities of polyatomic ions. *Chem. Phys. Lett.*, **261**, 86–91.

98. Shvartsburg, A.A., Schatz, G.C., and Jarrold, M.F. (1998) Mobilities of carbon cluster ions: critical importance of the molecular attractive potential. *J. Chem. Phys.*, **108**, 2416–2423.

99. Jurneczko, E. and Barran, P.E. (2011) How useful is ion mobility mass spectrometry for structural biology? The relationship between protein crystal structures and their collision cross sections in the gas phase. *Analyst*, **136**, 20–28.

100. Benesch, J.L. and Ruotolo, B.T. (2011) Mass spectrometry: come of age for structural and dynamical biology. *Curr. Opin. Struct. Biol.*, **21**, 641–649.

101. Scarff, C.A., Thalassinos, K., Hilton, G.R., and Scrivens, J.H. (2008) Travelling wave ion mobility mass spectrometry studies of protein structure: biological significance and comparison with X-ray crystallography and nuclear magnetic resonance spectroscopy measurements. *Rapid Commun. Mass Spectrom.*, **22**, 3297–3304.

102. Zhong, Y., Hyung, S.J., and Ruotolo, B.T. (2011) Characterizing the resolution and accuracy of a second-generation traveling-wave ion mobility separator for biomolecular ions. *Analyst*, **136**, 3534–3541.

103. Zhang, Y., Liu, L., Daneshfar, R., Kitova, E.N., Li, C., Jia, F., Cairo, C.W., and Klassen, J.S. (2012) Protein-glycosphingolipid interactions revealed using catch-and-release mass spectrometry. *Anal. Chem.*, **84**, 7618–7621.

104. Sharon, M., Ilag, L.L., and Robinson, C.V. (2007) Evidence for micellar structure in the gas phase. *J. Am. Chem. Soc.*, **129**, 8740–8746.

105. Wang, S.C., Politis, A., Di Bartolo, N., Bavro, V.N., Tucker, S.J., Booth, P.J., Barrera, N.P., and Robinson, C.V. (2010) Ion mobility mass spectrometry of two tetrameric membrane protein complexes reveals compact structures and differences in stability and packing. *J. Am. Chem. Soc.*, **132**, 15468–15470.

106. Borysik, A.J. and Robinson, C.V. (2012) Formation and dissociation processes of gas-phase detergent micelles. *Langmuir*, **28**, 7160–7167.

107. Borysik, A.J., Hewitt, D.J., and Robinson, C.V. (2013) Detergent release prolongs the lifetime of native-like membrane protein conformations in the gas-phase. *J. Am. Chem. Soc.*, **135**, 6078–6083.

108. Jones, C.M., Beardsley, R.L., Galhena, A.S., Dagan, S., Cheng, G.L., and Wysocki, V.H. (2006) Symmetrical gas-phase dissociation of noncovalent protein complexes via surface collisions. *J. Am. Chem. Soc.*, **128**, 15044–15045.

109. Meroueh, O. and Hase, W.L. (2002) Dynamics of energy transfer in peptide-surface collisions. *J. Am. Chem. Soc.*, **124**, 1524–1531.

110. Christen, W., Even, U., Raz, T., and Levine, R.D. (1998) Collisional energy loss in cluster surface impact: experimental, model, and simulation studies of some relevant factors. *J. Chem. Phys.*, **108**, 10262–10273.

111. Wysocki, V.H., Jones, C.M., Galhena, A.S., and Blackwell, A.E. (2008) Surface-induced dissociation shows potential to be more informative than collision-induced dissociation for structural studies of large systems. *J. Am. Soc. Mass Spectrom.*, **19**, 903–913.

112. Morgner, N. and Robinson, C.V. (2012) Massign: an assignment strategy for maximizing information from the mass spectra of heterogeneous protein assemblies. *Anal. Chem.*, **84**, 2939–2948.

113. Stengel, F., Baldwin, A.J., Bush, M.F., Hilton, G.R., Lioe, H., Basha, E., Jaya, N., Vierling, E., and Benesch, J.L. (2012) Dissecting heterogeneous molecular chaperone complexes using a mass spectrum deconvolution approach. *Chem. Biol.*, **19**, 599–607.

Part II
LC−MS Based with Indirect Assays

Analyzing Biomolecular Interactions by Mass Spectrometry, First Edition.
Edited by Jeroen Kool and Wilfried M.A. Niessen.
© 2015 Wiley-VCH Verlag GmbH & Co. KGaA. Published 2015 by Wiley-VCH Verlag GmbH & Co. KGaA.

4

Methodologies for Effect-Directed Analysis: Environmental Applications, Food Analysis, and Drug Discovery

Willem Jonker, Marja H. Lamoree, Corine J. Houtman, and Jeroen Kool

4.1
Introduction

In many fields of research, complex samples undergo chemical analysis in combination with bioassays for the identification of compounds that induce a biological effect. In environmental applications this is often named effect-directed analysis (EDA). Besides using traditional techniques for analyzing environmental samples for well-known contaminants, EDA studies are carried out in order to screen for newly arising environmental contaminants or so-called chemicals of emerging concerns (CECs). The pharmaceutical industry frequently uses the term bioassay-guided fractionation (BGF) for the combination of chemical analysis with bioassays. In this field, bioactivity screening studies have been performed, from the beginning of drug discovery until lead development. Also, the identification of flavors and fragrances in food and perfume industries requires chemical analysis combined with biodetection. In the latter case, olfactometry is often used with the human nose functioning as a biological detector. Furthermore, the combination of chemical analysis and bioassays is used in functional foods and in addition might have a great potential for forensics and doping control. As performing a bioassay on a mixture does not return information about the identity of the specific compounds that are responsible for the effect, it is necessary to separate the compounds via chemical techniques before bioassay testing and, in addition, to identify candidates with a suitable analytical detector. Unfortunately, due to instrumental limitations in the past it was often not possible to identify all bioactive compounds and a lot of these compounds remained unknown. However, with the technical improvements achieved in separation techniques and detection systems over the last two decades, it became possible to chemically resolve and detect numerous biologically active compounds in a single extract. As a result, nowadays there are better analytical tools to identify which candidates are truly responsible for the biological effect, with mass spectrometry (MS) being

Analyzing Biomolecular Interactions by Mass Spectrometry, First Edition.
Edited by Jeroen Kool and Wilfried M.A. Niessen.
© 2015 Wiley-VCH Verlag GmbH & Co. KGaA. Published 2015 by Wiley-VCH Verlag GmbH & Co. KGaA.

the preferred option due to its high sensitivity and capacity of elucidation of molecular structures.

The nomenclature for the different approaches and applications of the combined use of chemical analytical techniques and bioassay testing is clearly described in a recent review published by Weller [1]. This review was the first that focused on the techniques used in different research fields although it did not cover the principle of high-resolution screening [2, 3]. Previous reviews mainly focused on a specific research field. For environmental samples, Brack [4] discussed sample preparation, fractionation techniques, the use of bioassays, and chemical identification [4]. In 2008, another review was published that discussed the difficulties in confirming a potential toxicant in EDA as even with high-resolution MS, multiple candidates remain [5]. Furthermore, in 2011 the book *Effect-Directed Analysis of Complex Environmental Contamination* was published in which bioassays, chemical analysis, and computer tools were discussed [6]. In addition, recently a review was published comparing EDA with toxicity identity evaluation (TIE). A distinct difference between both approaches is that EDA mainly uses *in vitro* assays whereas in TIE whole organism tests are applied. In this review the differences but also the similarities between both the approaches have been carefully discussed [7]. In contrast to the discussion on whole EDA procedures, another review provides an overview of commercially available bioassays for assessing chemical toxicity in aqueous samples [8]. In addition, with respect to food analysis, d'Acampora Zellner *et al.* [9] focused on sample preparation and detection of flavors from different food matrices. In that same year, the application of gas chromatography-olfactometry (GC-O) in flavor analysis from alcoholic beverages [10] was also reviewed. Kool *et al.* [3] reviewed MS-based postcolumn bioaffinity profiling of mixtures. This review was mainly focused on pharmacology; however, the potential of implementing similar approaches in other fields was mentioned as well. Another EDA-related review published in 2011 mainly focused on the use of *in vitro* bioassays for the study of endocrine-disrupting food additives and contaminants [11]. Finally, the use of both analytical techniques and computer tools for structure elucidation in EDA-like studies was also discussed [12]. The use of databases such as NIST (National Institute of Science and Technology) for gas chromatography–mass spectrometry (GC–MS) or PubChem as well as the use of high-resolution and accurate mass data for the generation of chemical formulas and mass accurate database searching was reviewed [12].

For EDA-like studies, a variety of analytical and biological techniques are used, and different strategies are applied in the different research fields. This chapter will provide a global and critical overview of these approaches and techniques and include and discuss the most recent advances including new developments in the application of chemical analysis with bioassays. The chapter is divided into subchapters that, for each field, discuss the following topics: sample preparation, fractionation and biotesting, and identification and confirmation.

4.2
Principle of Traditional Effect-Directed Analysis

In general, an EDA study starts with the extraction of compounds from a complex matrix. The extraction methods either are focused on a specific class of compounds or cover a broad range of compounds. It is usually not possible to have a single extraction method suitable for a high number of compounds with very different physicochemical properties; focus on certain compound classes is also necessary. After extraction, the extracts are fractionated before bioassay testing. Fractionation is necessary to separate compounds into different fractions to obtain enough resolution for the identification of individual compounds inducing a biological effect in the bioassay and to prevent mixture effects such as additive or synergistic toxicity. After fractionation, each fraction is analyzed by bioassays. There are a huge variety of bioassays available ranging from enzyme bioaffinity assays to *in vivo* tests using, for example, mice or zebra fish. It often turns out that despite fractionation, the collected fractions still contain multiple compounds and further fractionation is necessary. When, after final fractionation, all compounds are resolved and only a few or ideally a single compound is left to induce the biological effect, that compound is identified using analytical techniques and data analysis software. Once the bioactive compound is identified, it is necessary to confirm its bioactivity by testing a pure sample. Figure 4.1 schematically illustrates the EDA process.

4.3
Sample Preparation

Because it is impossible to measure compounds directly without any sample pretreatment, different approaches have been developed for different matrices. In some cases, sample pretreatment is necessary to concentrate compounds and complex matrices may require extensive cleanup to remove interferences or compounds that negatively affect instrument or bioassay performance. In the different research fields, a wide variety of approaches are used, some of them being fairly straightforward, with others being elaborate and requiring multiple steps. In this section, an overview is given of the techniques used in different research fields related to identification of biologically active compounds.

4.3.1
Environmental Analysis

4.3.1.1 Aqueous Samples

There are significant differences between environmental sample matrices, each demanding a different sample preparation. For aqueous samples, such as surface or groundwater and sewage treatment plant effluents, it is mainly important to concentrate compounds, whereas for landfill leachates both concentrating and

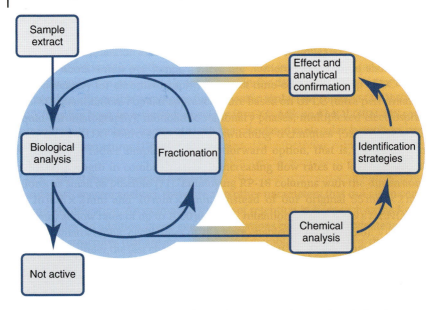

Figure 4.1 Schematic representation of the EDA approach. After sample preparation the next step is to assess which extracts are biologically active. Subsequently, the active fractions are fractionated in order to reduce their complexity. It is often necessary to perform multiple fractionations in order to reduce the complexity to a single or only a few compounds. In each cycle, fractions are biologically analyzed to identify the active fractions and chemically analyzed for assessing the complexity. Identification of the compounds is done via identification strategies where chemical analysis and data processing tools are used.

cleaning of the sample are required. There are various methods available for the extraction of contaminants from aqueous samples; however, not all of them are suitable for EDA studies. Previously, both liquid–liquid extraction (LLE) and extraction by means of XAD resins were popular extraction techniques [13, 14]; however, nowadays solid-phase extraction (SPE) is the most widely used technique for the concentration of environmental contaminants and the removal of interfering compounds from aqueous samples [15–17]. Currently, popular sorbents are StrataX (Phenomenex) [16, 17] and HLB (Waters) [15, 18, 19], which, in contrast to conventional C18 sorbents, are also able to extract the more polar compounds [15]. Nevertheless, in order to cover compounds ranging from nonpolar to polar and/or negatively or positively charged compounds at certain pH ranges, carefully choosing the right SPE procedure is pivotal. To extend the polarity range of extracted compounds as far as possible to both sides, sequential SPE can be very helpful. For example, a combination of a nonpolar sorbent and an ion exchange sorbent (such as MCX) can be used for the extraction of both nonpolar and cationic compounds. While previously nonpolar compounds were the main focus, polar compounds, such as drug metabolites, are nowadays gaining attention because polar fractions in EDA studies often demonstrate biological responses [19, 20]. The high activity of polar fractions was confirmed in fractions

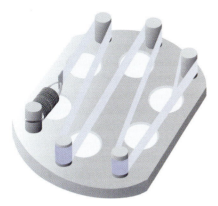

Figure 4.2 Example of an SPMD. A triolein-filled semipermeable membrane is placed on a rack, which in turn is placed in a protective cage (not shown). The latter prevents damage to the membrane from objects floating in the aqueous surroundings.

from sediment samples [21–23]. Besides SPE, other techniques for the extraction of contaminants from aqueous samples are solid-phase microextraction (SPME) [24] and stir-bar sorptive extraction (SBSE) [25]. These techniques are faster and simpler than LLE or SPE; however, they have, as far as we know, not been applied in EDA studies, most likely because of their low absorption capacity and their reliance on thermal desorption. A technique that has been used in a few cases for EDA studies is extraction via semipermeable membrane devices (SPMDs) [26, 27], also called a passive sampler device [28–30]. Figure 4.2 schematically illustrates an SPMD.

In contrast to SPE, this technique enables bioaccumulation of compounds without the losses associated with metabolism. Furthermore, sampling is done over a time range, instead of at certain points, and is thus less subjective to peak discharges and other events that might affect the compound concentration. Also, large volumes are extracted over time compared to SPE. Recently two papers have been published in which other passive samplers such as silicone rubber sheets and polar organic chemical integrative samplers (POCISs) have been discussed [26, 27]. Also, SPE was used and it was shown to be complementary to passive sampling with respect to extraction efficiency of different compound classes. In general, it can be concluded that the use of a variety of extraction techniques will result in the extraction of a broader range of compounds than the use of only a single technique. Furthermore, the selection of the extraction material is guided by the aims of the study (i.e., at a temporal or spatial scale). Nevertheless, thus far the majority of EDA studies that focused on water samples applied mainly SPE. This is most likely because most laboratories are more experienced with SPE than passive sampling and in addition, with passive sampling it is more difficult to relate the amount of trapped compound back to the water concentration. For the latter, partition coefficients have to be confirmed and these are often not available for emerging compounds. However, for EDA studies this is not very important as the main goal is identification of unknown toxicants with a certain bioactivity. For hazard assessment at a later stage, quantitative extraction and analysis methodologies should be applied.

It can be concluded from the different extraction methods that are available for water samples that the most popular are based on SPE and that passive samplers are also used. In 2007 Grung *et al.* used SPE for the extraction of wastewater effluents. Multiple compounds were identified that exhibited activity in the ethoxyresorufin-*O*-deethylase (EROD) assay, which is based on the aryl hydrocarbon receptor (AhR) [31]. In the same year, Urbatzka *et al.* [32] published an EDA study on water from the river Lambro (Italy). After SPE, the biological analysis with the yeast (anti)androgen assay (YAS) allowed identification of (anti)androgenic fractions besides several known compounds. The main contributors, unfortunately, remain unknown. Schulze *et al.* [18] undertook EDA for the identification of transformation products (TPs) of diclofenac that demonstrated reproduction inhibition toward the green algae *Scenedesmus vacuolatus*. Via EDA, 2-[2-(chlorophenyl)amino]benzaldehyde was identified as a toxic TP. In 2011, Smital *et al.* [19] analyzed municipal wastewaters with SPE, chemical analysis, and biological assays. In brief, a crude separation via column chromatography enabled biotesting of different fractions for cytotoxicity by means of the MTT (3-(4,5-dimethyldiazol-2-yl)-2,5 diphenyl tetrazolium bromide) reduction assay, chronic toxicity via an algae growth inhibition test, and EROD activity. Besides biologically active nonpolar fractions, polar fractions exhibited enhanced activity in multiple bioassays. The study was, however, not a complete EDA study as identification of bioactive compounds was not possible. Booij *et al.* [27] used POCIS and silicone rubber passive samplers for the extraction of estuarine and coastal waters. Atrazine, diuron, Irgarol, isoproturon, terbutryn, and terbutylazine were identified as the main contributors that negatively affected the effective photosystem II efficiency in marine microalgae. There are many similar studies that combine toxicity tests with chemical analysis. As isolation and identification of pure compounds is a difficult task, in many cases biological activity cannot be assigned to a specific compound but rather to a certain fraction that still contains several compounds some of which some are unknown. Even so, improvements in EDA procedures by collecting smaller fractions to provide higher separation resolution and the use of more specific bioassays may simplify the complex and labor-intensive EDA processes in the future. This would be especially interesting for water monitoring laboratories that would be able to monitor both known contaminants and emerging or unknown compounds that may induce a biological effect.

4.3.1.2 Biological Samples

Owing to the complexity of the matrix, biological samples need more extensive sample preparation than aqueous samples. In biota samples, interfering compounds such as lipids and endogenous hormones have to be removed before fractionation and further analysis. Lipids in particular interfere with both chemical and biological analysis and different methods are available for sample cleanup [33]. Suzuki *et al.* [34] employed PLE (pressurized liquid extraction)

extraction on blubber and liver samples of higher trophic level animals. Each extract underwent sulfuric acid treatment to remove lipids, and several dioxin and androgen-like compounds were successfully identified. Generally, the use of strong acids or purification via column chromatography using silica or alumina are popular cleanup techniques. However, these methods are developed for specific compound classes and are not suitable for analyzing a broad range of compounds because treatment with strong acids can lead to degradation. Additionally, purification by column chromatography is generally optimized for certain compound classes because of which other compounds may be missed. As a result, when a broad range of compounds with different physicochemical properties have to be analyzed, one needs a nondestructive methodology that is able to cover as many compounds as possible.

Because the development of such a method for solid biota samples is complex, thus far the focus has been on the development of suitable sample preparation methods rather than performing complete EDA studies. Both Streck *et al.* and Simon *et al.* [35, 36] published on the development of sample preparation techniques for biological tissues. Sample preparation in both approaches started with homogenization and drying with hydromatrix. In the next step the dried homogenate underwent PLE. The latter is schematically illustrated in Figure 4.3.

After this step the two author groups continued the protocol in different ways, as is illustrated in Figure 4.4. Lipid removal was done either via dialysis accelerated membrane-assisted cleanup (AMAC), which is also illustrated in Figure 4.4, or via manual dialysis followed by gel permeation chromatography (GPC); in addition, preparative normal-phase high-performance liquid chromatography (NP-HPLC) was undertaken for the removal of natural hormones and remaining lipids.

The main advantage of dialysis in both methods is the efficiency in matrix removal. Purification solely via GPC is another common cleanup technique; however, its low capacity makes it less suitable for biological samples [35]. Simon *et al.*'s method was capable of removing 98% of the matrix lipids, which is comparable to Streck's method (97–99%). The method followed by Streck *et al.* is, however, less labor-intensive and achieves similar matrix removal without an additional GPC and NP-HPLC step. Furthermore, the dialysis could proceed via automation by placing the PLE extracts in dialysis membranes and subsequently extract the filled membranes via the PLE system. On the other hand, natural hormone removal was not investigated by Streck *et al.* and could be favorable with respect to interferences in the bioassay. This, however, could also be achieved in subsequent fractionation steps reducing the risk of missing newly emerging contaminants. Finally, the remaining 1–3% of the lipids may still interfere in the bioassays, and this was investigated by Simon *et al.* for androgen chemically activated luciferase expression (AR-CALUX) and radiolabeled thyroxin–transthyretin (T4*-TTR) binding assay. No effect was observed in AR-CALUX; however, the radiolabeled T4*-TTR binding assay did show a decrease in TTR binding. Dilution resolved this problem without loss of analyte response, and normal-shaped dose–response curves were obtained.

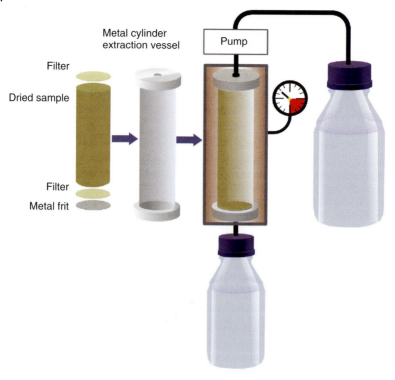

Figure 4.3 Schematic illustration of pressurized liquid extraction. During extraction the PLE cell is located in a chamber in which the temperature can be controlled. The solvent is brought into the PLE cell via an opening in the metal lid that closes the cell. This solvent is brought under pressure and in combination with increased temperature the extraction efficiency can be enhanced. When the extraction process is completed, the extraction solvent is released via an opening in the bottom lid and is collected in collection bottles. In addition, it is possible that after the first extraction a second extraction is carried out with a different solvent allowing the capture of compounds with a broad range of physicochemical properties.

A method for extraction and analysis of thyroid hormone-disrupting compounds in blood plasma has been published that is suitable for EDA purposes [37]. The method starts with protein denaturation and in the next step SPE extraction is carried out with an OASIS MCX (Waters) cartridge, which is followed by LLE in order to change to a more suitable solvent. Water removal was done over a Pasteur pipette filled with sodium (II) sulfate, and finally natural hormones were separated from contaminants via preparative NP-HPLC [37]. The method was used for the analysis of polar bear plasma for thyroid hormone-disrupting activity and several hydroxylated polychlorinated biphenyls were identified as the major contributors to show potency toward TTR. The process is illustrated in Figure 4.5. In addition, SPE has also been used as a sample preparation technique for the analysis of fish bile from fish exposed to wastewater effluents [38, 39]. Both estrogenic and antiandrogenic compounds were found in these studies. Furthermore,

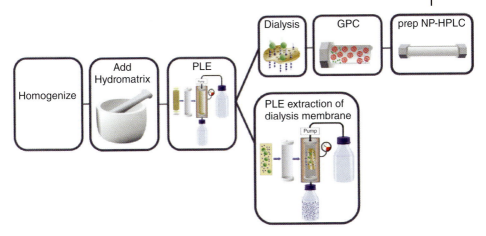

Figure 4.4 Schematic illustration of the sample preparation procedures for solid biological samples. Simon *et al.* followed the upper route and continued with dialysis followed by gel permeation chromatography (GPC) in which both steps aimed for the removal of lipids. Additionally, preparative normal-phase high-performance liquid chromatography NP-HPLC was applied for the removal of remaining small lipids and natural hormones. Streck *et al.* continued in a different way by transferring the PLE extract into dialysis bags and re-extracting these bags via PLE; this is called *accelerated membrane assisted cleanup* (*AMAC*). Both author groups finalized sample pretreatment with evaporation and changed to a solvent suitable for chemical and biological analysis.

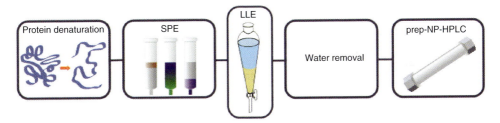

Figure 4.5 Illustrates the different steps in the sample preparation process for plasma samples.

in another study fish bile was extracted with LLE [40]. This study focused on estrogens and the endogenous hormones estrone; 17β-estradiol and estriol were identified as the major contributors to the measured estrogenic activity. Also, the birth control drug ethynylestradiol was found at effective concentrations. Finally, several other xenobiotics were detected in high concentrations, but their influence on the measured estrogenic activity was minimal.

The above-discussed methodologies are promising tools for future research because they overcome problems in sample preparation such as nondestructive matrix removal and, in contrast to many other methods, are able to cover a broad range of compounds. Hopefully, this will encourage scientists to study these matrices and future research will provide information about uptake, bioaccumulation, and metabolism of contaminants, together with toxicological information. In addition, it may also give an idea about the bioavailability of the contaminants. In this respect, it should be taken into account that differences in samples can be expected as different organisms may accumulate contaminants in different organs or tissues, and that the uptake and metabolism between different compounds can differ significantly.

4.3.1.3 Sediment, Soil, Suspended Matter

Many EDA studies focused on sediments, and, to a lesser extent, on suspended particulate matter and soil, for example, [21–23, 41–48]. In multiple cases, compounds that can induce an adverse biological effect were successfully identified [22, 23, 41, 43, 45, 48]. There are various techniques available for the extraction of organic pollutants, namely, Soxhlet and supercritical fluid extraction [49] and microwave-assisted [50] and PLE extraction [51]. However, not all available techniques are suitable for EDA purposes and currently the main extraction technique applied for EDA studies is extraction via PLE [23, 41–43, 46, 52]. A general starting point is drying and homogenization of the sample to prevent losses in the extraction efficiency due to the presence of water. After PLE extraction, two main techniques are used for the sample cleanup, sample cleanup either by means of GPC [23, 41, 53] or via AMAC [21, 43, 44], both aiming to remove interfering compounds such as humic acids. There are, however, other options. Reifferscheid *et al.* [42], for example, used LLE for the removal of elemental sulfur. The efficiency of sulfur removal was monitored via GC–MS; if the removal was insufficient LLE was repeated. This technique is specific for sulfur removal and hydrophilic interferences that are separated from the nonpolar compounds in the LLE, while other interferences might not be removed.

Despite the successes achieved in EDA studies on sediment samples, there is a disadvantage associated with the use of PLE as an extraction method. For hazard assessment it is necessary to take into account properties such as bioavailability. PLE extraction is an exhaustive extraction technique and does not take into account desorption kinetics. As a result, contaminants could be classified as important while in reality these compounds hardly desorb from their sediment and are consequently not accessible for any organism. For this reason, in addition to exhaustive extraction, there are also a number of publications that make use of soft extraction techniques or a combination of exhaustive extraction with a soft release technique [45, 54, 55]. One of the approaches is to perform an exhaustive PLE extraction followed by fractionation and subsequently fractions are spiked to artificial sediments. After solvent evaporation, the following day water was added and allowed to equilibrate. Finally the test organisms (*Potamopyrgus antipodarum*) were added [55]. In this way, the release of contaminants as would occur in a real sediment is mimicked with the difference that the number of compounds is reduced in the fractionation step. However, this approach to mimic the bioavailability of contaminants in EDA studies is time-consuming; it depends on living organisms and cannot be applied to other kinds of biological tests. Another technique to mimic bioavailability relies on the absorption of contaminants on silicone stir bars and is probably more generic [45, 54]. Also here, PLE extraction and subsequent fractionation on sediment is performed but instead of spiking fractions to sediments, a precleaned silicone rubber is added to the fractions. Within 5–50 min absorption of the contaminants is complete and the silicone rubber is placed in the growth medium. Here the silicone rubber gradually releases contaminants and is comparable to desorption of contaminants from the sediment. Finally, the growth medium was used for the green algae cell multiplication inhibition assay [54]. Besides these combinations of an exhaustive extraction technique with slowly desorbing release methods there is also the possibility of soft extraction methods with Tenax or similar sorbents. For example, there is an approach in which the sediment is shaken together with water and Tenax the Tenax sorbent is collected after shaking and the absorbed compounds are eluted with organic solvents [56]. In this way potentially hazardous compounds that are more readily bioavailable are further analyzed while the compounds that are slowly released are left out. Furthermore, as was already discussed for aqueous samples, SPMDs can be used in order to obtain information about bioaccumulation. Despite the fact that these approaches take into account factors such as bioavailability, they have thus far not been used frequently for EDA purposes. A possible explanation is that with soft release or soft extraction techniques, the risk of missing potential hazardous compounds is larger than with a more exhaustive technique. From this perspective it is perhaps a safer option to identify potentially hazardous compounds and to study their bioavailability at a later stage.

4.3.2
Food Analysis

In food analysis, BGF is used for various purposes. Numerous studies focus on flavors and fragrances to investigate which compounds are responsible for a specific kind of taste or odor. Typically, these studies are performed by means of GC-O, and a range of different matrices such as wine Chin, S.-T., Eyres, G.T., and Marriott, P.J. (2011) Identification of potent odourants in wine and brewed coffee using gas chromatography-olfactometry and comprehensive two-dimensional gas chromatography. J. Chromatogr. A, 1218 (42), 7487–7498, cheese [66–68], chocolate [69, 70], meat [71], truffle [72, 73], coffee Chin, S.-T., Eyres, G.T., and Marriott, P.J. (2011) Identification of potent odourants in wine and brewed coffee using gas chromatography-olfactometry and comprehensive two-dimensional gas chromatography. J. Chromatogr. A, 1218 (42), 7487–7498, and nuts [75] have been studied. Besides flavor analysis, other kinds of studies utilize GC-O to investigate malodors from food package materials [76] and agricultural activities [77]. Furthermore, several studies have investigated the presence of natural compounds present in food that promote health or prevent diseases. This trend is called *functional foods* and several articles on the use of BGF for the identification of health-improving substances can be found [78–84]. Techniques such as LLE, liquid chromatography (LC), and a broad range of bioassays, such as radical scavenging assays including antimicrobial and enzyme or receptor inhibition assays, are used for fractionation and identification of biologically active substances. Another field in food analysis is that of food safety where the abbreviation EDA is used. In this field, EDA-like approaches have thus far not received much attention. However, it might be interesting to use EDA-like approaches to screen foods for the presence of new kinds of plasticizers or growth hormones. The latter was done by Nielen *et al.* [85] when they analyzed cow urine for the presence of estrogen growth promoters. A study related to this area investigated the presence of anabolic steroids and derivatives in sport supplements and herbal mixtures with an EDA-like approach [86]. Finally, with respect to food preservation, an article was recently published that deals with the discovery of antioxidant compounds from barley husks by means of EDA and the potential of this compound to act as an antioxidant film in food packaging [87]. As can be seen from the previously discussed subareas within food analysis, EDA-like approaches are used in completely different ways and as one can imagine these different subareas all have different extraction methodologies.

4.3.2.1 Chromatography-Olfactometry

GC-O is the most widely used approach for food flavor analysis, and headspace analysis is the preferred identification technique. For the analysis of wine, various techniques such as LLE are often used with dichloromethane [60, 62] headspace extraction [57, 59, 63, 65], SBSE [88], and SPE [58, 64]. Another approach is the combination of LLE with solvent-assisted flavor extraction (SAFE). This approach is, however, used more frequently for fat-containing and/or more viscous matrices [61]. With respect to headspace analysis of wine, passive or dynamic sampling techniques are used. Passive sampling is mainly used for SPME approaches [65], and dynamic sampling is done via a purge and trap system [57, 59] either bubbling through the dissolved sample or blowing on the surface. This is schematically illustrated in Figure 4.6.

For wine analysis via dynamic headspace, LiChrolut EN polymeric sorbents are frequently used as an extraction sorbent. Compound elution is generally done

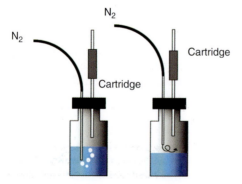

Figure 4.6 Schematic illustration of headspace analysis. Compounds released in the headspace by bubbling nitrogen through the sample solution or by blowing nitrogen on top of the sample solution are trapped in a sorbent-containing cartridge. Trapped compounds are released either via solvent elution or via thermal elution.

with dichloromethane and methanol. Other matrices such as that of cheese and chocolate require a slightly different approach. However, here too the main extraction technique is headspace extraction. Cheese samples are cut into small pieces, mixed with a salt, brought in a tube with a glass wool bottom, and purged with helium. The volatile compounds are trapped on a Tenax© trap and eluted via thermal desorption. Eluted compounds are focused on the GC (gas chromatography) column by cryofocusing [66, 67]. A variant of this approach is the use of SPME instead of Tenax, as was published by Guneser and Yuceer [68]. Analysis of chocolate is done in a similar way making use of either Tenax [69] or SPME [70]. Truffle, coffee, and pistachio nuts too have been analyzed by these techniques [72–75]. Finally, a fairly alternative approach is the use of multichannel open tubular traps for headspace sampling of milk volatiles as was done by Naudé *et al.* [89]. In this approach, small silicone rubber tubes are placed in a glass tube according to the method followed by Ortner and Rohwer [90] to form a multichannel open tubular trap. This results in a larger sorbent volume than is used with other headspace extraction techniques. In addition, the design allows recapture of aroma compounds at the outlet of the GC column. As can be concluded, various extraction methodologies are used in food flavor analysis, the main extraction approaches being headspace or LLE and SPE.

For flavors and fragrance analysis, volatile compounds are often the main compounds of interest and headspace extraction is the preferred analytical technique for sampling. Furthermore, considering practical issues, headspace approaches (especially SPME) are relatively easy to apply compared with LLE and SPE. With LLE and SPE careful separation of interfering compounds is necessary, this is not the case with headspace techniques as interfering compounds most often are of a nonvolatile nature and remain in solution [38]. The release of volatiles, however, is dependent not only on the volatility of these compounds but also on other factors such as the matrix in which they are present. Nonvolatile compounds that do not induce any perception of flavor themselves might enhance or suppress the release of flavor compounds but, unfortunately, remain unknown with headspace analysis. For extraction of these compounds more exhaustive extraction methods such as LLE and SPE are preferable. In addition, for SPME, only a small number of sorbents are available and the most commonly applied fiber for GC-O studies is the divinyl-benzene/carboxen/polydimethylsiloxane (DVB/CAR/PDMS) fiber. In principle, to obtain the most complete picture of the sample the best approach is a combination of headspace sampling as well as a more exhaustive approach in order to capture as many compounds as possible from the matrix.

4.3.2.2 Functional Foods
Besides food flavor analysis, currently functional foods are gaining increasing attention. Compared to olfactometry studies [81], completely different extraction, fractionation, and chemical as well as biological detection methods are used, and much more elaborate methodologies are applied. Different kinds of foods have been the subject of various studies aiming to identify health-improving compounds [78–84, 87, 91]. For example, in the onion *Allium cepa* L., a γ-glutamyl

peptide was found to inhibit bone resorption in rats [80]. Furthermore, Lagemann *et al.* [84] identified a compound in lettuce that inhibits the angiotensin converting enzyme (ACE), which is suspected to be involved in hypertension pathogenesis. In addition, grape seed extracts have been shown to exhibit antibacterial activity against the bacteria *Campylobacter* spp. [91]. Compounds identified in Chinese tea were shown to inhibit the growth of important food pathogens [92], and there are various other examples. Solid samples are in most of the cases cut, freeze-dried, ground, and in a subsequent step extracted via shaking, stirring, or reflux methods. Extraction of solid and liquid samples is most often done with ethanol or methanol and, in addition, sequential LLE is sometimes applied to selectively extract compound classes from the initial extract [78, 83] or the residue after extraction is reextracted with other organics [81]. In most cases this is the complete sample preparation and in a following step the extract undergoes rough fractionation by means of column chromatography often with silica [79, 81, 83]. Thin-layer chromatography (TLC) is used in these cases as a rapid identification tool. In a later stage more sophisticated fractionation techniques are applied. Nevertheless, in some cases slightly more elaborate sample cleanup and extraction might be necessary. In an article published by Catrin *et al.* lipid precipitation at $-78\,°C$ was performed for the removal of triglycerides. The resulting extract was enzymatically destarched and was removed from blue wheat UC66049 *Triticum aestivum* L. via saponification phytochemicals [78].

4.3.3
Drug Discovery

4.3.3.1 Combinatorial Libraries and Natural Extracts

Another important field that makes use of BGF-like approaches is that of drug discovery. Here, various strategies are applied in the search for new lead compounds. One of the most important areas in drug discovery that makes extensive use of BGF are drug discovery pipelines based on natural extract screening [93]. In addition, combinatorial libraries form another important source for drug discovery. Knowledge of the receptor pocket, physicochemical properties, structure of endogenous ligands, patents, drug likeness, and literature are important for *in silico* virtual screening and structure-based drug design.

With respect to combinatorial libraries, from the point of hit identification to the stage of lead development, high-throughput screening (HTS), and fragment-based lead discovery are important approaches for screening of compounds for different biological end points. These approaches are mainly focused on bioaffinity and microtiter *in vitro* bioassays are mostly used. In addition, in recent years microarrays have also gained increasing popularity for these purposes [79, 80]. Unfortunately, in HTS, with parallel synthesis, optimal reaction conditions are often not met and bioassays are performed on mixtures. Reaction conditions might have led to incomplete product formation, transformation of reactants, and various other side reactions. As a consequence, mixture analysis in bioassays can lead to false positives and/or negatives [94]. To circumvent this, one could apply a BGF-like approach. In addition, BGF is essential for drug discovery from natural extracts as these consist of numerous compounds. One can imagine that products from combinatorial libraries and other crude synthesis products demand a different sample preparation than natural sources. With respect to the former, dissolving the synthetic products in a suitable solvent for subsequent fractionation is the main sample preparation procedure. Sample preparation for natural extracts, however, is often much more elaborate due to the matrix and large dynamic range in which compounds are present. Medicinal plants are an example of interesting sources for potential drugs [95–98] and the elaborate extraction and cleanup of potential bioactive compounds from this matrix in general starts with drying and grinding of the sample. This is followed by extraction with an organic solvent via shaking, percolation, refluxing, or Soxhlet extraction. In addition, when necessary, filtration for particle removal is applied and finally, in a few cases, an additional LLE is carried out on the extract [95–100]. By using one extraction solvent, only a specific range of compounds is extracted and other compound classes remain out of scope of the study. In 2011, Yuliana *et al.* [101] published a comprehensive extraction method combined with nuclear magnetic resonance (NMR) metabolomics and multivariate statistics for bioactivity screening studies of plants. In this study the extraction was carried out with an in-house-developed extraction system with some similarities to a PLE-like approach as is illustrated in Figure 4.7. As with PLE, in this method also the sample is dried with hydromatrix and placed in a metal cylinder. Multiple solvents are sequentially applied to the dried sample enabling extraction of a broad range of compounds. This relatively easy sample preparation method has potential to be further improved possibly by making use of a PLE system. The

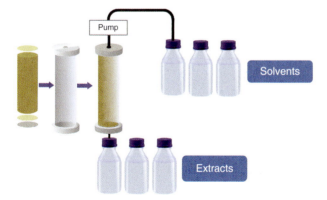

Figure 4.7 Scheme of the extraction method described by Yuliana *et al.* [101]. The powdered plant material is mixed with Kiezelguhr (hydromatrix) to dry the plant material and placed in a metal tube. The outer ends are closed with cotton and act as a filter. For the extraction, solvents with different polarities are pumped through the cylinder in order to extract a broad range of compounds. To reduce the complexity of the extract, the extracts obtained with the different solvents are collected separately instead of in the same vessel.

latter would enable automation and could possibly enhance extraction efficiencies because of its ability to regulate extraction temperatures and pressures.

In addition to plants, marine organisms such as algae, sponges, anemones, and marine bacteria have gained interest from the pharmaceutical industry, for example, for the presence of compounds with antioxidant properties [102], apoptosis-inducing properties [103], anticancer, antiparasite activity, and activity against microbial pathogens [104]. Biselyngbyasides identified from the marine cyanobacterium *Lyngbya* sp. was shown to induce apoptosis in HeLa S3 cells and HL60 cells and was extracted with methanol followed by LLE [105]. Algae are frozen in order to preserve them until sample preparation. After thawing, extraction is carried out with an organic solvent [102, 104]. A similar approach was carried out by Zimba *et al.* [103] for euglenoids; however, in this case solvents with different polarities were used to stepwise dissolve different cellular components [103]. Furthermore, venoms from, for example, scorpions and snakes are interesting for biopharmaceutical drug discovery against neurodegenerative diseases [91, 92]. Focusing on biopharmaceutical venom peptides and proteins, sample preparation is relatively straightforward. After collection, the venom is generally diluted in a slightly acidified aqueous solution and subsequently filtered or centrifuged. Finally, the sample is lyophilized and stored at −80 °C until analysis [106–109]. Besides drug discovery, natural extracts can also prove to be useful in other research fields; an example is the search for natural pesticides in the field of food and agriculture. In a recent example two compounds were identified from *Conyza canadensis* (horseweed) with potential for use as a natural pesticide [110].

4.4
Fractionation for Bioassay Testing

After sample preparation, the next step in EDA-like studies is the fractionation/separation of compounds present in the extract. As for sample preparation, different approaches are used in different research fields in addition to the mainstream approaches.

4.4.1
Environmental Analysis

In particular in environmental analysis, but also in other research areas, preparative LC-based fractionation methods are mainly used enabling large-volume injections and a crude separation of complex mixtures. Generally, a few relatively large time frame fractions are collected and multiple fractionation cycles are performed. Additionally, prior LLE and/or SPE steps may be included in the process. After initial fractionation, a part of each fraction is used for biological analysis and active fractions are further analyzed by GC–MS or liquid chromatography mass spectrometry (LC–MS). In case the fraction is still too complex, the fraction is further fractionated on preparative LC systems using orthogonal stationary

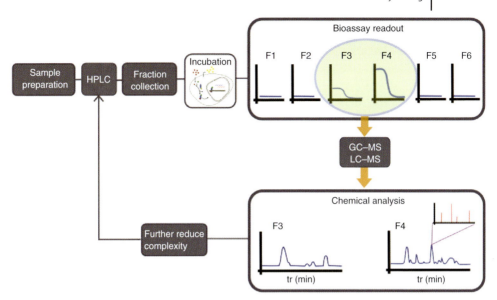

Figure 4.8 Workflow of an EDA study. The complexity of the sample extract is reduced by liquid chromatography and fractions of the column eluate are collected. Dilutions of each fraction are tested in a bioassay and dose–response curves are obtained. The active fractions are then analyzed by chemical analysis via GC–MS, LC–MS, or a combination of both. In case the fractions are still too complex these undergo another fractionation cycle.

phases. Figure 4.8 schematically illustrates the workflow of an EDA study. Once the complexity has been reduced sufficiently and a bioactive compound has been identified, the last step is to confirm this finding with a pure standard.

The advantage of utilizing GC–MS is the high-resolution chromatographic separation, stable retention times, and the ability to consult databases such as NIST, which is a valuable tool for the identification of unknowns via GC electron ionization (EI) MS. A popular GC column is the HP5 column, which has been used by most authors to separate the contaminants present in each fraction [42–44, 48]. To enable GC–MS analysis, it is necessary to change to a more volatile solvent. Because a high-resolution separation can be obtained by GC, it would be of interest to perform fraction collection of a GC separation. In 2003, Mandalakis and Gustafsson [111] published a method in which six fractions could be collected; this is, however, not sufficient to maintain high-resolution GC separation. In 2007, Meinert *et al.* [112] improved the method of Mandalakis *et al.* and was able to collect 12 fractions. Nevertheless, the problem is that, with the low number of fractions collected, resolved peaks end up in the same fraction. Recently, however, a new approach for GC fractionation was published. Instead of using cold or sorbent traps, in this approach, via a y-split, a carrier solvent is mixed with the separated compounds and collection is done in a 96-well plate [113]. The platform is illustrated in Figure 4.9 together with the bioassay that is performed at-line in the fraction collection plate. Its ability to

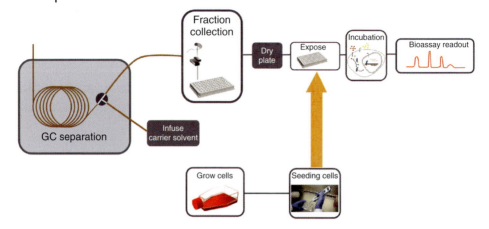

Figure 4.9 Illustration of a novel approach for GC fractionation. In contrast to other GC fractionation approaches where only a few fractions can be collected, with this platform up to 384 fractions can be collected. This ensures that chromatographic resolution is maintained, which significantly reduces the number of follow-up fractionation cycles needed for compound isolation.

collect a high number of fractions, with a collection time down to 6.5 s per well, allows maintenance of the chromatographic resolution and consequently aids in the throughput of an EDA study. Its disadvantage, however, is the amount of sample compound collected. To obtain sufficient material for bioanalysis, multiple injections have to be carried out and collected in the same fraction. With the platform, a mixture of polyaromatic hydrocarbons (PAHs) and a certified reference material was fractionated. Each fraction was tested in a cell-based gene reporter assay selective for the detection of dioxin-like activity, and several PAHs showed dioxin-like activity. Via parallel GC–MS measurements, the measured activity could successfully be applied to several PAHs.

Although GC–MS analysis is of great value in EDA, LC–MS has distinct advantages especially because more polar fractions are gaining interest due to the responses that are measured in bioassays [20]. As for LC–MS there are no databases available such as NIST for GC EI-MS, a high-resolution mass spectrometer is essential to narrow down the number of possible candidates for the unknowns. Suitable instruments are the time of flight (TOF) instrument and the orbitrap but even then, multiple candidates remain and a combination of software tools has to be used for successful identification of the unknowns. Weiss *et al.* have published the use of LC in combination with high-resolution MS and applied identification strategies in EDA. This was a follow-up study on [41] to further investigate bioactive fractions by means of a LC orbitrap MS [22]. Another study focused on thyroid hormone–disrupting compounds in polar bear plasma [114]. Target compounds analyzed by GC–MS with EI could only explain 40–47% of binding to TTR. The identity of the unidentified binders was further investigated by performing a GC–MS full-spectrum database search on the GC–MS data. In addition, LC–MS analysis with a TOF-MS (time-of-flight mass spectrometer) was performed to obtain accurate mass data. The data was analyzed with various software tools for compound identification. Mono- and dihydroxylated-octachlorinated biphenyls and branched nonylphenols were identified as the compounds responsible for the unexplained TTR binding. Despite the information that EDA studies can provide, unfortunately, EDA studies often fail in identifying bioactive compounds. The low-resolution preparative fractionation methodologies and in addition large time frame fraction collection procedures are often not able to sufficiently separate bioactive compounds. As such, the final bioactive fractions remain too complex. In addition, the many steps that have to be undertaken can lead to compound losses during the process.

Therefore, applying a higher-resolution separation and reduction of sample fraction size in an early EDA stage may reduce the number of consecutive fractionations that are necessary for identification of a bioactive compound, thereby also better exploiting the resolution of modern-day LC. A possible approach could be to collect fractions in a well plate as was done by Kool *et al.* [115] in the field of drug discovery where they made use of a nanospotter. After bioassay readout of the well plate, the response of each well can be plotted as a data point in time. The result is a bioassay chromatogram that can be compared directly with the chromatogram obtained by chemical analysis. The nanofractionation platform was applied for the identification of compounds with affinity toward the acetylcholine binding protein (AChBP). A mixture of ligands was successfully fractionated and correlated with the MS trace chromatogram. A schematic presentation of the process is given in Figure 4.10. Complexities can, however, arise in the miniaturization of the bioassays as, for example, difficulties can arise with growing a cell monolayer in a 1536-well plate.

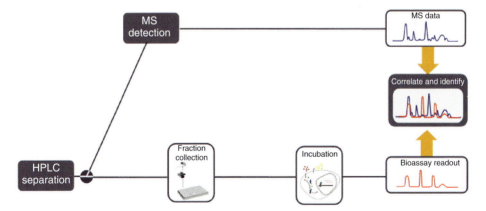

Figure 4.10 Schematic representation of a high-resolution nanofractionation platform for the at-line detection of bioactive compounds. The fraction collection allows bioanalysis with assays that require long incubation times. After HPLC separation, the column eluate is divided via a split. A small part is guided toward the MS whereas the major part goes to the fraction collector. After fraction collection, a bioassay is performed and after the bioassay readout, the data can be directly correlated with the MS data.

Besides HPLC (high-performance liquid chromatography)-based fractionation, the applicability of TLC has also been investigated. In 2004, Müller *et al.* [116] combined TLC successfully with the yeast estrogen assay (YES). Recently, this approach was further improved and applied on real samples [117]. Where the normal YES assay is performed in 96 wells, this p-yeast estrogen assay assay or planar-YES assay is performed directly on the TLC plate and lower limits of detection (LODs) are obtained than with the standard YES assay. In addition, further analysis by MS is possible by scratching off a small amount of silica and extracting it with MeOH. Although the separation obtained by TLC is not comparable with that of HPLC, the throughput of this approach is a great advantage. Where an average EDA study can take several weeks or even months, with this approach, results can be obtained within a day. Because of the speed of the method, this approach has the potential to be used as a complementary tool to HPLC-based fraction platforms.

4.4.2
Food Analysis

In the field of food analysis, GC-O is a special analytical technique in a way that, compared to other fields, it significantly differs in its fractionation and bioassay methodology. Fractionation/separation is done online with GC, which in most cases is equipped with a DB-WAX column, or comprehensive two-dimensional GC (GC × GC) is used [73, 98]. Olfactometric analysis is generally done via a GC-Sniff flame ionization detection (FID) platform in which, after separation, via a split, a part of the flow is guided toward an FID detector and the other part toward a sniffing port enabling flavor assessment by certified sniffing specialists. When the sniffing specialist recognizes a certain smell, this can be correlated to the FID signal. Later, for identification by GC–MS, the FID signal is compared with the MS signal and the compound can be identified. In this approach, it is possible to correlate the smell of a certain compound with a peak in the mass spectrum. Thus far, little has been published on fraction collection in GC-O studies. A recent article focused on a platform that was developed for GC or GC × GC analysis for extensive compound separation and the ability to concentrate low-level volatiles on a single trap [88]. However, this means that it is not possible to collect different compounds on different separate traps. For GC-O studies this is not necessarily problematic as long as the number of compounds to be collected is limited as biological detection via sniffing is performed online. Unfortunately, as the system depends on thermal desorption it is not possible to analyze the trapped compounds multiple times. In contrast to a single trap, previously a manual method for fraction collection was used to trap milk volatiles for GC-O analysis [89]. In this approach multichannel tubular traps used for headspace sampling were placed at the end of the GC column allowing the recapture of a specific peak of interest for subsequent identification with, for example, offline MS, repeated sniffing, or confirmation of the flavor compound. The results of this study indicate that synergism of 2-heptanone and 2-nonanone may be responsible for a cheese/sour milk aroma.

In contrast to GC-O approaches, the methodology in functional foods is completely different. Here, for the fractionation of extracts, column chromatography is one of the first applied fractionation techniques [79–81, 83, 84] and in various studies TLC is used as a rapid identification tool to investigate if certain fractions can be pooled. Further fractionation is done by repeating column chromatography with different solvents and/or column material, or a more high-resolution fractionation with HPLC is chosen. Final fractionation is done with HPLC, and for detection, NMR [79, 80, 83, 118] and MS [80, 84] are popular techniques but optical rotation [79] and X-ray crystallography [79] have also been used. An example is the study of Nitteranon *et al.* [81] in which the freeze-dried powder of the fruits of *Morinda citrifolia* were extracted with ethylacetate and the extracts were fractionated by silica gel flash chromatography. Several fractions were collected and a selection was further fractionated via TLC or HPLC. Bioassay testing was done with the nitric oxide assay and quinone reductase inhibition assay. The compounds scopoletin and quercetin were identified as potential anti-inflammatory and anti-cancer components. In environmental analysis, due to trace level concentrations, NMR cannot be applied; in this field, however, it is an extensively used technique [83]. In addition, MS is also popular and high-resolution instruments such as Fourier transform ion cyclotron resonance (FTICR) [79] and TOF [80, 81, 84] as well as triple quad MS instruments [84] are used. This combination of NMR and MS is powerful for compound identification as the data obtained by both instruments make a more complete picture. MS can, for example, serve to help with identification and confirmation by accurate mass measurements or by compound fragmentation and subsequent interpretation of fragmentation spectra.

4.4.3
Drug Discovery

In this field various strategies are used for the discovery of new drugs. In general, bioassay testing in BGF studies on natural extracts is performed offline and separation is done via HPLC [89, 90, 104, 119]. However, gel chromatography is also a popular technique, which is later followed by HPLC separation [97, 108, 120, 121]. Occasionally preparative TLC is also used for separation [97]. Chemical analysis is done via MS and NMR. In contrast to the above-described approach, in recent years developments have been made in the hyphenation of biological assays with HPLC and MS as was thoroughly reviewed by Kool *et al.* [3].

In brief, there are three main approaches for online bioaffinity analysis, which are illustrated in Figure 4.11. The first is that, after separation and online incubation for bioaffinity analysis, the bioassay readout is done via MS. A second approach is that instead of MS readout, bioassay readout is done via fluorescent measurement and chemical detection is done in parallel via MS. Finally, a third approach can be used for at-line bioanalysis in case the bioassay cannot be used for any of the other two approaches. This is, for example, the case when incubation times are too long for online bioanalysis. Via a splitter a part of the eluent is guided toward an MS for chemical detection and the remaining

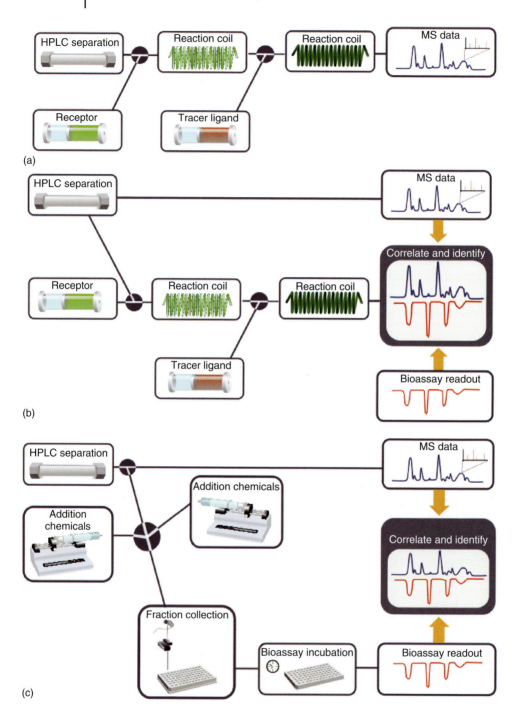

(a)

(b)

(c)

part is directed toward a mixing device and fractionated in a well plate. This allows long incubation times and subsequent bioassay readout is most often done via a plate reader as many bioassay responses are spectrophotometrically detectable. These automated platforms have been applied on metabolic mixtures [122], combinatorial libraries [94], and natural extracts [123, 124]. Nevertheless, especially for natural product screening, more classical approaches in which separation and chemical detection is done offline are still applied [97, 100, 125].

4.5
Miscellaneous Approaches

Besides the mainstream methodologies described for EDA studies in this chapter there are also some less common analytical approaches to identify biologically active compounds in mixtures. This section deals with some alternative approaches that are less widespread and usually applied only to a very specific research niche. An extensive review of most of these alternative approaches is provided by Jonker *et al.* [126].

In the first typical example based on ligand binding, different groups have used size exclusion chromatographic (SEC) approaches coupled to MS that allow the binding of multiple ligands to a receptor, such as the muscarinic acetylcholine receptor (AChR) M2 [127] or the histamine H2 receptor [128]. This step was then followed by SEC-based purification of the receptor–ligand complexes and finalized by a disruption step and analyses of the released ligands by MS. This way,

Figure 4.11 (a) Schematic view of an on-line bioaffinity analysis platform with MS based bioassay detection. Separation of the sample extract is done via an LC system after which the column eluate is mixed online via a superloop filled with a solution that contains the receptor. The eluate and receptor mixture pass through a reaction coil that allows binding of the eluate ligands. After the first reaction coil, tracer ligand is added via another superloop and passes to the next reaction coil where the tracer ligands can bind to the remaining free receptors. Readout of the amount of bound tracer ligand is done via MS. (b) Platform for on-line bioaffinity analysis with fluorescence bioassay detection. The mixture is separated via LC and the column eluate is split after which a part is guided towards the MS. In parallel, online bioaffinity analysis is performed. For this, receptor is added to the eluate and passes the first reaction coil. Subsequently, tracer ligand is added and after passing a second reaction coil bioaffinity can be detected via fluorescence detection. The parallel acquired MS chromatogram and fluorescence bioaffinity chromatogram can be directly correlated for bioactive identification. (c) A final approach is at-line bioaffinity analysis. After separation, via gradient LC, a part of the flow is guided towards the MS. To the remaining part, necessary biochemicals are added and the mixture is fractionated in microtiter plates ranging from 96 to 1536 wells per plate. This allows to use bioassays that require long incubation times and can therefore not be applied in an online fashion. After incubation, bioassay readout can be done via a plate reader and the readout of each well can be correlated to a corresponding retention time window. From this a bioassay chromatogram can be reconstructed that can be directly correlated with the parallel obtained MS chromatogram.

not only mixtures can be analyzed, but the MS analysis simultaneously provides structural information regarding the ligands isolated.

A more general approach for measuring label-free product formation from an enzymatic reaction is the RapidFire system [129–131]. This system uses miniaturized and reusable SPE cartridges that allow subsequent and continuous SPE procedures to be performed in 5- to 10-s time intervals. This technology actually performs rapid SPE purifications from assays conducted in wells of 96- or 384-well plates after incubation followed by MS-based analysis. For this, the extracted substrate(s) and product(s) from the incubations are rapidly eluted to a triple quad mass spectrometer after every SPE step in order to quantify the amount of substrate and product in every well. The use of fractionation approaches followed by RapidFire analysis will allow analysis and identification of bioactives for a certain enzymatic reaction.

When a protein is His-tagged, a very high affinity between the His tag and the immobilized metal ion efficiently traps any His-tagged protein. Jonker *et al.* [132] utilized this principle by trapping an in-solution formed His-tagged protein–ligand complex on a small nickel-loaded column. This allows ligand–protein complex formation in solution, using the immobilized metal column merely to separate the bound and unbound fraction followed by release of the ligands from the trapped proteins for MS detection. Another variant of affinity capture methodologies is ligand fishing using magnetic particles. For this so-called magnetic bead dynamic protein-affinity selection, IMAC-coated magnetic beads are used to trap a His-tagged protein–ligand complex from a sample for further processing and final MS [133, 134]. In other examples, Marsza *et al.* [135] employed magnetic beads coated with heat shock protein for ligand fishing, while Hu *et al.* [136] published an enzyme inhibition screening assay utilizing enzymes immobilized on magnetic beads.

A number of groups have developed methodologies using continuous ultrafiltration [137, 138]. In this technique, a fixed amount of protein is injected into an ultrafiltration chamber and the analyte is then pumped through the chamber. Fung *et al.* [139] were one of the first to automate ultrafiltration for high-throughput analysis. Some other recent publications include the work of Comess *et al.* [140], who developed a high-throughput serial ultrafiltration methodology to screen a library of compounds for affinity to a pharmacologically relevant streptococcal enzyme. Finally, Li *et al.* [141] developed an online coupled ultrafiltration–LC-MS methodology that was used to screen natural extracts for α-glucosidase inhibitors.

Some other examples include the quantification of unbound prednisolone, prednisone, cortisol, and cortisone in human plasma by ultrafiltration–LC–MS [142] and the development of an ultrafiltration–LC–MS-based ligand binding assay for *Mycobacterium tuberculosis* shikimate kinase [143], to be used for the development of antimicrobial agents. For screening of potential lead compounds in malaria drug discovery, work was performed for the drug targets Plasmodium falciparum thioredoxin and glutathione reductases [144]. Another study used ultrafiltration-based affinity selection and MS for the kinase Chk1 involved in DNA damage [145]. A slightly different example employed hollow fiber membranes online coupled to MS for continuous affinity selection [146].

The development of pulsed ultrafiltration by van Breemen *et al.* significantly increased the applicability of ultrafiltration [147]. Instead of filtration of an entire sample, a small amount of sample was injected into the ultrafiltration unit. The liquid flow forces the nonbound fraction through the molecular weight cutoff membrane, to waste. Afterward the bound ligand is dissociated from the protein by buffer adjustment. The formerly bound fraction is then pushed through the membrane, and immediately measured and identified by MS. The two main advantages over traditional ultrafiltration are that (i) less analyte is needed and (ii) target proteins can be reused if a nondestructive dissociation buffer is applied. The methodology can be applied to complex mixtures as well as combinatorial libraries, and it functions online and is automated and suitable for HTS. Some successful applications of pulsed ultrafiltration include the work of Shin, Liu, and Cheng, all in the group of van Breemen, who used pulsed ultrafiltration to study metabolic stability [148], inhibitors of protein aggregation [149], and ligands for human retinoid X receptor alpha [150]. Another development involved the discovery of cyclooxygenase inhibitors from medicinal plants used to treat inflammation [151].

A direct successor to the classic GPC is **SpeedScreen technology** [152]. It consists of a double 96-well plate format. The upper part contains a size exclusion gel, and has holes in the bottom of the plate. An in-solution protein–ligand incubation mixture is pipetted on top of the gel, and it is placed on the second plate, a collection plate. The incubate is separated based on centrifugal force, and bound and unbound ligands are separated. The 96-well plate is then analyzed using an LC–MS system. The system screens one 96-well plate within 10 min and has routinely been used by Novartis leading to a number of lead compounds [153].

In 2004, Neogenesis developed an automated HTS assay using online SEC coupled to MS for the assessment of protein–ligand interactions [154]. This method, named the Automated Ligand Identification System (ALIS)), uses a standardized SEC-LC–MS setup, but contains an in-house-developed resin as the size exclusion material. The fast affinity screening results include the discovery of a lipid phosphatase inhibitor involved in diabetes type 2 [155]. Characterization of orthosteric and allosteric ligands for the muscarinic M-2 AChR was also feasible with the system [156]. Furthermore, Whitehurst and Annis [157] describe affinity selection methodologies and their emerging role for screening GPCR (G protein-coupled receptor) ligands.

Following the success of ALIS, a number of groups have developed innovative applications for size exclusion affinity measurements. Flarakos *et al.* [158] have developed a methodology able to assess and rank ligand binding toward human serum albumin (HSA) based on automated SEC coupled online to a two-dimensional LC–MS system. Binding of ligands to Akt-1 and Zap-70 kinases, the muscarinic AChR M2 [127], or the histamine H2 receptor [128] followed by SEC-based purification of the receptor ligands and disruption for release of ligands to MS were performed by Annis *et al.* and Derks *et al.*, respectively.

Phosphorylation of (mostly) proteins activates or deactivates them in global and/or specific signal transduction cascades that are usually initiated by other signaling events. Furthermore, on phosphorylation, changes in cellular location and/or association events of proteins with other proteins (or other biomolecules) can be mechanisms of action for protein kinases. When protein kinase activity in the body is deregulated in an organism, significant effects on cells such as cell death or proliferation can cause many different diseases, especially cancer, but also inflammation. Protein kinases possess similar adenosine tri-phosphate (ATP) binding sites. This implies that for developing new lead compounds, selectivity toward one protein kinase or a subset of relevant protein kinases is extremely difficult.

When using a chemical proteomics approach based on so-named kinobeads, the binding of ligands to hundreds of different protein kinases can be analyzed at once. For this, cells with endogenous protein kinases are lysed (or in theory solutions with all desired protein kinases to be screened can be used). Ligands are then added followed by the addition of the kinobeads. The kinobeads are able to bind many different protein kinases rather nonspecifically. When ligands are bound to a specific protein kinase, however, this protein kinase will no longer bind to the kinobeads, resulting in the detection of a lower end signal for that specific protein kinase. For the end readout, the kinobeads are isolated, the bound protein kinases are digested, labeled with an isobaric tag for relative and absolute quantitation (iTRAQ) reagent [159], and finally the labeled peptides are subjected to MS peptide profiling to quantify the amounts of all protein kinases bound to the kinobeads by comparison of iTRAQ reporter ions relative to the negative control [160, 161]. This methodology simultaneously allows analyzing initial ligand-induced changes in the phosphorylation state of the protein kinases via their respective peptides allowing an analysis of signaling pathways downstream of the target protein kinases as well. Chemical proteomics approaches, chemical cross-linking proteomics, as well as so-called interactomics or interaction proteomics studies, based on protein–protein or protein–ligand interactions, are described by Kool *et al.* [162] in a review paper on protein–protein and ligand–protein affinity interaction studies analyzed by MS. These approaches are very similar to the kinobead trapping approach.

4.6
Bioassay Testing

4.6.1
Environmental Analysis

Various ways exist for the biological analysis of fractions from environmental extracts. *In vitro* techniques are mainly used; however, there are also studies in which *in vivo* assays are employed. Examples of test organisms for *in vivo* assays are zebra fish [46], fresh water snails [43], and algae [45]. Snails have been used to assess responses to chemicals with respect to reproduction. In addition, cell reproduction tests have been done with green algae cells and, furthermore, green algae and zebra fish have been analyzed for growth inhibition and developmental toxicity. These tests are, however, time-consuming and can handle only small numbers of samples and/or fractions. Consequently, in most cases *in vitro* tests are carried out.

In vitro bioassays can easily be classified in fluorimetric, spectrophotometric, luminescent, and radioactivity assays. Recently, the radiolabeled thyroxine (T4)-TTR inhibition assay in which TTR binding of inhibiting compounds is expressed in T4 equivalent concentrations was used for the detection of thyroid hormone-disrupting compounds [37, 114, 163]. The most commonly applied bioassays are, however, spectrophotometric based, as no special laboratory requirements are necessary, except a simple plate reader. Various assays are available such as assays for the detection of endocrine disruption, enzyme activity, mutagenicity, and genotoxicity.

Endocrine disruption mediated by activation/inhibition of estrogen and androgen receptors (ARs) can be measured with the YES [19, 23, 48] and the yeast androgen assay (YAS) [23]. These assays are sensitive and have the advantage that they are relatively simple. Their drawback, however, is that uptake through the cell membrane is in many cases not comparable to mammalian cells. For the latter, mammalian cell-based assays such as the chemically activated luciferase gene expression ER-LUC [43] and CALUX assays [41, 43, 164] are more suitable. These are gene reporter assays typically in 96-well plate format and are analyzed via plate reader instruments. Nevertheless, the principle of both assays is the same and is schematically illustrated in Figure 4.12. When a pollutant binds to the estrogen receptor (ER) or AR, the receptor complex migrates toward the nucleus where it binds to a responsive element in the DNA. After transcription and translation of the reporter gene, an enzyme is formed that either transforms an agent, inducing a change in color, or induces luminescence. Besides the ER-CALUX (estrogen receptor chemically activated luciferase expression) and AR-CALUX the dioxin receptor (DR)-CALUX is also commonly used, which allows detection of dioxin-like compounds [23]. The assay principle is similar to ER- and AR-CALUX; however, here the AhR is the target instead of the ER or AR. In addition, similar gene reporter assays with other end points are available; however, these have thus far not received much attention in EDA studies, potentially because they are rather new and are still being implemented in different laboratories worldwide.

In addition to assays for endocrine disruption, other environmentally relevant assays are available. For example, the 96- or 384-well plate Ames fluctuation test is used for the detection of mutagenic compounds [42, 44]. This assay utilizes *Salmonella typhimurium* with certain mutations that make the bacteria rely on histidine for growth. When the bacteria are placed in low-concentration histidine medium together with a mutagenic compound the bacteria mutates back to a non-histidine-dependent form. This is detected via a pH-sensitive dye that changes color from purple to yellow and can be measured in a plate reader. In addition, the microtiter umuC test can be used for the measurement of genotoxicity [165]. Also, *Salmonella typhimurium* is used, and genotoxicity is detected by dye conversion

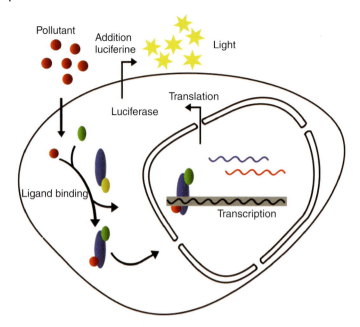

Figure 4.12 Mechanism of the CALUX assays. In addition to AR-CALUX and ER-CALUX for androgenic and estrogenic compounds, respectively, various other assays also are shown such as the CALUX assay for dioxins and thyroid compounds.

induced by the formation of β-galactosidase. Furthermore, the EROD assay is used for the detection of environmental contaminants binding to the AhR and photometric detection is also used [19, 21]. Occasionally, the microtiter MTT assay is used for cytotoxicity analysis [19]. This is a spectrophotometric-based assay in which a purple color is formed when cytotoxicity is absent. However, in the presence of cytotoxic compounds, this color formation is inhibited.

Besides these offline assays, an online EDA approach has been published for the detection of acetyl cholinesterase inhibitors [166]. In this approach, gas is inserted into the column eluate to create air segments to reduce peak broadening. Bioaffinity is assessed by means of an online acetylcholine inhibition assay with colorimetric readout using a UV detector and Ellman's reagent. The platform is schematically illustrated in Figure 4.13.

In addition, an online bioluminescence platform has been published in which the bioluminescent bacteria *Vibrio fischeri* is used for toxicity testing. In the absence of toxicants a constant light intensity is measured; however, when one of the multiple enzymes involved in luminescence is inhibited, a loss of light can be detected indicating toxicity [167]. Unfortunately, only poor sensitivities were obtained most likely because of insufficiently long incubation times. This also points to one of the difficulties of using living organisms in online setups. Incubation times in online systems are generally insufficient for this purpose and many of the applied bioassays in environmental analysis are not suitable for

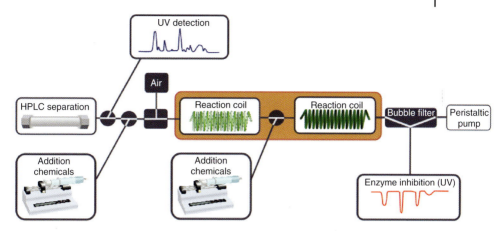

Figure 4.13 Illustration of an online bioluminescence platform. Compounds are separated by chromatography and part of the eluent is guided toward a UV detector via a split. This is followed by the continuous addition of enzyme via a syringe pump and introduction of air to create air segments. Incubation takes place in a reaction coil at a temperature of 37 °C. After the first reaction, the coil substrate and reagents are continuously infused via a second syringe pump and incubated in a second reaction coil also at 37 °C. Finally, for the detection of enzyme inhibition, the inserted gas is removed via a bubble filter and the eluent is guided toward the enzyme inhibition detector.

online analysis. Insertion of gas segments has shown to reduce peak broadening; however, this is only until a small extent and is often still insufficient. Additionally, online assays are less sensitive and, furthermore, efforts are not always taken to transform microtiter assays into an online assay.

For assays in microtiter format, there might, however, be the potential to analyze these in an at-line approach, as has been discussed in the section "fractionation and detection." However, it has to be kept in mind that solvent compatibility issues have to be overcome, either by sufficient dilution or evaporation or by making use of a mixing device/spotter technology as described by Giera *et al.* [168]. Here, the LC eluent is split toward a detector and toward a mixing device. The mixing device mixes essential reagents for the bioanalysis with the LC eluent and fractionates this in a microtiter plate for bioanalysis. After bioassay readout, the results can be directly correlated to the chemical signal. This would aid in the throughput of generally labor-intensive and costly EDA studies and might enable more standardized procedures, which could be useful for application in environmental monitoring laboratories. Alternatively, a direct nanofractionation setup can be used in which the identification of bioactives is performed by MS and the parallel nanofractionation allows for high-resolution fractionation in 96- to 1536-well plates. This is followed by vacuum centrifuge evaporation after which the nanofractionated plates can be stored or directly used in a bioassay of choice. Examples of radioligand binding assays as well as cell-based functional assays using a rather new direct postfractionation cell seeding approach can be named [169, 170]. The high-resolution fractionation allows for reconstructed bioactivity or bioaffinity chromatograms, which can be used for accurate peak shape and retention time correlation of bioactives found in the bioassay used. For this, the bioassay chromatograms are overlaid with the parallel obtained LC–MS data and with help of extracted ion currents (EICs), the retention times, and peak shapes of the EICs that match those of the retention time and peak shape of a bioactive are matched and consequently correlated. Additional MS/MS (tandem mass spectrometry) data enables further elucidation of the bioactive and/or enables database searches. Furthermore, when an accurate mass of a bioactive is known, subsequent straightforward *m/z* guided LC–MS fractionation facilitates collection of enough material for, for example, NMR analysis.

4.6.2
Food Analysis

The bioassays used for BGF-like studies in functional foods are diverse and range from antioxidant and free radical scavenging assays until antimicrobial and enzyme or receptor inhibition assays. With respect to, for example, plant extracts, this can be measured with the DPPH (2,2-diphenyl-1-picrylhydrazyl radical) chemical assay [79, 82, 83]. This compound is colored purple in its radical form. However, when antioxidants are present, a yellow color is formed, which can be detected via spectrophotometric techniques. Applications of antioxidant activity not only in food analysis but also in other research fields were reviewed in 2008 by Niederländer *et al.* [171]. Besides antioxidant assays, the MTT and 3-(4,5-dimethylthiazol-2-yl)-5-(3-carboxymethoxyphenyl)-2-(4-sulfophenyl)-2H-tetrazolium (MTS) assay is used [79, 82, 83]. The latter is similar to the MTT assay but is easier as it requires fewer steps. It is, however, also more susceptible to interferences. Furthermore, inhibition tests with ACE have been carried out [82, 84]. This enzyme is known to be involved in a process that finally leads to elevated blood pressure, and it is therefore of interest to identify compounds that inhibit this enzyme. ACE inhibition can be monitored by HPLC-UV or -MS analysis of the hydrolysis of hippuric acid–histidine–leucine by ACE and the consequent formation of hippuric acid or via fluorescent measurements [172]. Finally, antimicrobial assays have been performed by counting bacterial colonies grown on agar plates or by spectrophotometric absorbance measurements on the individual wells of a microtiter plate [79, 91]. Unfortunately, until now, as in environmental analysis, in this field bioassays are most often performed offline, making the method labor-intensive, time-consuming, and more susceptible to errors.

4.6.3
Drug Discovery

As was already introduced in the fractionation section in the field of drug discovery, advances have been made in the hyphenation of bioassays with HPLC separation and chemical detection [3]. For postcolumn online analysis this is mainly done for bioaffinity studies for enzymes and soluble receptors. The latter are suitable for online analysis as many of these receptors have available assay formats with good dynamic windows and with fast fluorescence responses, which is ideal for online screening approaches. For example, the online affinity analysis of AChBP can be mentioned in this regard [115]. This protein is similar to nicotinic acetylcholine receptors (nAChRs), which are pharmaceutical drug targets for several diseases such as Parkinson's disease, epilepsy, and Alzheimer's disease. Furthermore, method development and application for kinase inhibition has been the subject of several studies [168, 173, 174]. In addition, the online AchBP affinity analysis platform has also been used for natural products such as venom [123]. In contrast, cell-based assays, in which, for example, nuclear receptors are the target, require long incubation times as this time frame is necessary for signaling, DNA binding, transcription, and translation. These could only be performed atline via a microfractionation and nanospotting technique that has potential for not only drug discovery but also other fields. Besides these developments in hyphenation, in drug discovery, also numerous offline techniques are used. Several tests have been carried out with zebra fish embryos [125] but MTT and MTS assays are also used [103]. Furthermore, the antioxidant ferric reducing antioxidant power (FRAP) assay [102] and AChBP [123] and AChR [106] competition assays have been used.

4.7
Identification and Confirmation Process

After evaluation of the active fractions, the next step is to identify the compound(s) responsible for the measured bioactivity. This is the most complex part of an EDA study and various techniques are used to achieve identification of the toxicant. The identification process shares many characteristics with nontarget screening methodologies that are applied for the detection of unknown and emerging compounds. In true nontarget screening, no information is available about which compound to expect and the analysis is generally based on solely chemical techniques. Various reviews have been published about identification and confirmation by means of nontarget screening [12, 175–177] and considerable experience has been obtained in metabolomics and environmental analysis. This subchapter aims to give an overview of the approaches used for identification and confirmation in EDA as well as in nontarget screening.

4.7.1
Instrumentation

4.7.1.1   Gas Chromatography–Mass Spectrometry
A popular technique used for compound identification in EDA studies and untargeted screening is GC–MS [18, 23, 46, 114, 178, 179]. The availability of extensive databases for EI spectra (NIST and Wiley) against which measured spectra can be

matched eases the identification process and does not necessarily demand high-resolution MS. However, when the searched spectrum is absent in the database, which may be the case with a newly emerging contaminant, it is a difficult task to unravel the identity of the compound with a low-resolution instrument. In this case, accurate mass data of the parent ion obtained with a high-resolution MS could provide a solution for compound identification. The accurate mass can be used for the calculation of a molecular formula and this is often the starting point of the identification process when a database search fails [180, 181]. Therefore, it is important that the molecular ion is clearly visible in the spectrum and with EI, due to the extensive fragmentation, the molecular ion is not always present. In this case, soft ionization techniques such as chemical ionization (CI) and atmospheric pressure chemical ionization (APCI) can provide a solution because, with these techniques, a clearly visible peak of the protonated molecule is generated with only minor fragmentation. In addition, the combination of soft ionization with a hybrid TOF instrument provides the possibility of acquiring accurate mass data from the protonated molecule together with its fragments in a single run. This could serve as a complementary tool to EI and provide extra proof for confirmation. Furthermore, for high-resolution separations achieved with GC and multidimensional GC (G×GC) [182] it is important that an MS that is able to rapidly acquire multiple spectra be used. This ensures that enough data points per chromatographic peak are obtained. Despite the high-resolution capabilities of orbitrap [183] and FTICR instruments, these require relatively long cycle times to achieve high-resolution data and are currently not compatible with high-resolution separations. However, TOF and hybrid TOF instruments are excellent for this purpose as high-resolution mass accurate spectra can be acquired rapidly with good sensitivity in full scan mode. In addition, the rapid acquisition rate increases the signal-to-noise ratio because it is possible to apply signal averaging of the multiple spectra collected [184]. This results in cleaner spectra and in turn allows more accurate database searching. Nevertheless, although GC–MS allows robust, high-resolution separations with MS detection, this technique also has its limitations as only volatile, thermally stabile, nonpolar compounds can be measured. Taken into account that polar compounds are gaining interest due to frequently detected bioactivity in polar fractions, it is important that these compounds also be investigated [20, 185, 186].

4.7.1.2 Liquid Chromatography–Mass Spectrometry

A suitable instrument for the analysis of more polar contaminants is LC–MS. The use of LC extends the range of compounds that can be covered with GC and is an important tool in EDA and nontarget screening [22, 114, 180, 181, 187, 188]. NMR analysis and X-ray analysis of crystallized fractions or extracts is mostly not feasible due to trace concentrations and insufficiently pure extracts. The two main ionization techniques applied in LC–MS are electron spray ionization (ESI) and APCI. In contrast to EI in GC–MS, ESI, and APCI are both soft ionization techniques that induce only minor fragmentation and the mass of the intact molecule can be retrieved from the spectrum. It is, however, more difficult to create mass spectral libraries than with EI as fragments created by ion activation will vary between instruments and applied LC–MS conditions. Because only relatively small libraries exist for LC–MS, it is even more important to obtain mass accurate data than with GC–MS as there is a reasonable chance no match will be found in the database search and the remaining option is to start with the calculation of a molecular formula from the accurate mass and isotope pattern. FTICR, orbitrap, as well as TOF instruments enable high-resolution measurements but most likely because of the cost and size of FTICR instruments, these have been used only rarely for nontarget screening in environmental analysis [185]. Orbitrap and and TOF instruments [114, 187, 188], however, have been used frequently for nontarget screening and EDA studies [22, 180, 189]. Besides mass accuracy, fragmentation capabilities are also important for compound identification as this provides structural information that can be useful to narrow down the candidate compounds retrieved after the molecular formula calculation. Because in nontarget screening and EDA there is no selection of masses before the analysis, for a product ion scan, an initial screening has to be performed after which in a second injection the product ions of compounds of interest can be obtained.

Data-dependent acquisition (DDA) allows fragmentation of parent ions without the need for an initial screening. Predefined selection criteria such as ion intensity and exclusion of buffer clusters can be set so that only those compounds that meet these criteria are fragmented. These selection criteria, however, induce bias and may cause compounds to be overlooked. A way to overcome this is to apply so-called MS^E. With MS^E, the energy of the collision cell is alternated between low and high energy. In the low-energy condition, parent ions are mainly observed whereas in the high-energy mode, product ions of all parent compounds are generated. In this way, in a single measurement the parent ions with all their product ions are retrieved [190]. It is important to note that, in contrast to MS–MS, parent and product ions cannot immediately be linked because all parent ions are fragmented simultaneously. To link parent and product ions, their retention times have to be compared. This can be done manually or automatically by software algorithms. When the retention time of a parent ion corresponds with several product ions, these product ions most likely originate from that specific parent ion [190]. Proper separation of compounds is important in this approach as coeluting compounds can complicate the correlation process of parent and product ions. This can be a complication in EDA as generic separation conditions are used and coelution might occur. Nevertheless, MS^E can serve as a rapid screening tool and where complexities arise with regard to parent and product ions this may be resolved with a product ion scan.

4.7.2
Data Analysis

After separation and MS detection, the difficult task of compound identification remains. Data analysis for nontargeted analysis is well investigated and has been the subject of multiple studies [22, 178–181, 188] and review articles [12, 175–177, 191]. The applicability of these data analysis approaches is also recognized for EDA and progressively applied for toxicant identification. For that reason, the coming references will refer to not only EDA but also nontarget screening as several perhaps not yet applied techniques from nontarget analysis could prove useful for EDA as well. Considerable expertise has been obtained in the field of metabolomics, and many techniques are also applied in environmental nontarget screening. In general, once the accurate mass data is obtained, the first step is to apply peak detection to distinguish peaks from noise. Subsequently, a blank subtraction can be performed where the spectrum of a blank or solvent injection is subtracted from the sample spectra. Nowadays, in most of the data analysis software provided with the MS, this can be done with built-in tools. However, other software programs such as MZmine [180, 185] that often have additional features such as the ability to handle data from different instruments are also used [192]. Certain criteria can be set for peak detection such as minimum intensity threshold and signal-to-noise ratio. In addition, spectra alignment can be carried out, which is commonly applied in metabolomics for the correction of minor retention time shifts between samples but has also been used in food and environmental analysis [22, 188]. An example of an alignment program is Metalign [193]. Nevertheless, in EDA, alignment is less useful as fractionation and bioassay testing in principle reduces sample complexity to only a few compounds rather than giving crowded spectra as in metabolomics.

4.7.2.1 Deconvolution

Depending on the ion source, the polarity used during the measurement and mobile-phase composition, different adducts can be formed. Deconvolution is necessary to obtain the compound's pure spectra and to allow determination of the accurate mass of the neutral molecules. This can be done manually or, when the spectrum is complex, software tools can be used to simplify this process. The deconvolution tools nowadays incorporated within MS-accompanied analysis software are mainly for charge deconvolution of multiply charged species. This is only of use for large molecules that can carry multiple charges but is not useful for the small molecules in EDA and environmental nontarget screening. For adduct deconvolution, several MS analysis software programs are available such as CAMERA, ACD IntelliXtract, and HighChem Mass Frontier [176]. However, in-house software and scripts have also been developed [194, 195]. An example is an R package named "nontarget" [181] that has been developed for this purpose [194]. Once the accurate masses of the compounds are obtained, the elemental composition can be calculated with built-in tools in the accompanied MS software or external software programs such as SIRIUS2 [195]. In the majority of cases, a high number of possible molecular formulas is obtained. By application of seven heuristic rules such as isotope pattern matching and rules for element ratios as proposed by Kind and Fiehn [196], the number of candidates can be reduced significantly and some elemental compositions have to be evaluated further.

4.7.2.2 Database Spectrum Searching

After these steps, the next step is to perform database searches for which the measured spectra are compared against mass spectra databases [41, 114, 163, 179, 181]. Especially for EI spectra from GC–MS, this can be a successful approach as there is a viable chance that a match is found in the large databases available for EI spectra (NIST and Wiley). Even if no exact match is found, closely matching compounds can give information about the compound structure. The usefulness of database searching has also been recognized for LC–MS, and although spectra are more dependent on instrument type, instrument settings, and separation conditions, initiatives have been taken to construct databases. MassBank database initiated in Japan [197] for LC–MS but also GC–MS spectra has a rich repository of tools for database searching such as spectrum search, substructure search, a metabolome prediction tool, and various other features. Nowadays a European variant of the Japanese MassBank is available from the NORMAN network, a network of reference laboratories, research centers, and related organizations for monitoring of emerging environmental substances (*www.norman-network.net*). METLIN [198] is another database specific for metabolites. It is, however, developed for metabolomics purposes and most likely stores limited information about metabolites of environmental toxicants. Nevertheless, the existence of such a database for environmental toxicants could prove interesting for the identification of TPs and will perhaps become available in the future.

4.7.2.3 Database Searching on Molecular Formula

The success of spectrum database searching is dependent on the quality and number of spectra stored in the database and is thus far limited for LC–MS. Instead of a spectrum database search another possibility is to generate molecular formulas based on the measured accurate mass [22, 180, 181]. The generated formulas can be searched in compound databases such as PubChem, Chemspider, NIST, and the Merck index. Depending on the database, different search inputs besides that of a molecular formula can be entered. With the Merck index it is possible to search on chemical abstract service (CAS) number, name, molecular formula, and molecular weight. In PubChem, some extra features are included such as substructure/superstructure search and identical/similarity search. After the search, a list of potential compounds with their structures is returned that closely match the calculated molecular formulas; however, in practice this means multiple candidates still exist and further data treatment is required.

4.7.2.4 In Silico Tools

In silico tools have been developed that can aid in further reduction of candidate compounds and their confirmation. AMDIS is a software program for data deconvolution and identification and has been used in EDA studies [40, 46, 114]. Fragmentation prediction software such as Mass Frontier and MetFrag [199] can be used to generate fragmentation spectra of the candidates found with Pubmed or other databases. Subsequently, the measured spectra can be compared with the *in silico* generated spectra for identification [180, 200]. Also, software for the prediction of metabolites formed by microbial transformation is available [201] and it aids in the identification of microbial TP of organic pollutants [202, 203]. Structure generation software is another type of software that is available and can be crucial when the unknown compound is not present in the database. This software is able to calculate all possible structures for a specific molecular formula (MOLGEN-MS). This has been investigated especially for GC–MS, and MOLGEN-MS software allows to combine fragments with the calculated molecular formula to predict possible structures [179].

4.7.2.5 Use of Physicochemical Properties

Besides MS spectra analysis, there are also other parameters that can aid in compound identification. An example is the use of retention indices such as the Kovat or Lee index. In addition, steric energy calculations can help to eliminate *in silico* predicted structures that are energetically unfavorable. Furthermore, the log K_{ow} value can aid in the identification and confirmation process [22, 179, 204]. For that, log K_{ow} values are calculated from the retention time [205] and compared with the log K_{ow} values of the tentatively identified compound. Finally, although software tools for database searching, fragmentation and metabolite prediction, and log p calculations are valuable and necessary in the identification and confirmation process, it is important to realize that they cannot be trusted blindly and manual interpretation and confirmation will always be necessary.

4.7.2.6 Applications in Nontarget Screening and EDA

Multiple papers have focused on compound identification. Schymanski *et al.* [179] used the structure generation software MOLGEN-MS for identification of compounds measured by GC–MS. The calculated molecular formula was searched in NIST, and together with the generated structures from MOLGEN-MS a list of possible compounds was obtained. To reduce the high number of candidates, fragmentation prediction was applied and the *in silico* spectra were compared with the measured spectra. Further candidate reduction was done by comparison of physicochemical properties. Compounds were filtered on boiling point and Lee retention index, log p, steric energy, and, finally, spectral matching. The number of candidate molecules was successfully reduced from over a million down to six structures, making it a promising workflow when no database match is found.

Also, combinations of target, suspect, and nontarget analysis have proven successful. LC–HRMS (liquid chromatography–high-resolution mass spectrometry) analysis was applied for the identification of novel micropollutants in wastewater [180]. In brief, by target analysis a number of peaks were identified and could be excluded from the suspect and nontarget analysis. For the suspect screening, a list of suspect compounds (site specific and known from literature) was constructed. Based on the molecules' functional groups, two new lists were composed with corresponding masses for positive ionization and negative ionization solely for $[M + H]^+$ and $[M - H]^-$. The spectra were searched for the masses in the compound lists with MZmine, and the resulting candidates were further filtered by comparison of measured retention time and linear solvent energy relation (LSER) model calculated retention times. Next, spectra were inspected manually for characteristic isotope patterns and peak shape. Finally, database searching was done on an in-house database and MassBank. For the nontarget approach, peak detection was also by MZmine, and molecular formulas were calculated with MS accompanied software. The number of candidates was reduced by application of seven golden rules as published by Kind and Fiehn [196]. A subsequent database search in PubChem returned possible candidates, and these were further investigated via the fragment prediction software Metfrag. Multiple compounds were successfully identified or tentatively identified; however, the authors do emphasize that more expanded mass spectral databases would be beneficial for nontarget screening. The LSER model proved useful for an initial automated exclusion of candidates but care should be taken with ionic compounds as the model does not take ionic interactions into account. Nevertheless, with a relatively wide retention interval of 5 min, the majority of measured ionic compounds did match the predicted retention time, and multiple candidates could be excluded successfully narrowing down the candidate list.

A combination of target, suspect, and nontarget analysis was also used in another study [181]. For suspect screening the software program Envimass was applied for mass list compilation. With the R package "nontarget," isotope and adduct grouping was applied and a database search with MassBank was performed. With the exact mass and fragmentation data, MOLGEN-MS/MS was used to calculate molecular formulas. Finally, MetFusion was utilized, a software program that allows parallel searches on different databases such as the mass spectral database MassBank, compound databases, as well as the results from *in silico* fragmentation prediction. The extensive information obtained with HRMS (high-resolution mass spectrometry) and the vast number of calculated molecular formulas could be reduced to 18 compounds via this methodology.

Recently Radović *et al.* [178] applied chemometrics in an EDA study on oil by GC×GC TOF analysis. Oil was fractionated by liquid chromatography and tested for AhR receptor agonists and AR agonists and antagonists. Active fractions were analyzed by GC×GC TOF, and the data was processed by MATLAB and partial least squares (PLS) toolbox for N-way PLS (N-PLS) and unfolded PLS (U-PLS) analysis. This is to our knowledge one of the first approaches in EDA where multivariate analysis is applied for toxicant identification. In EDA, however, it is important to reduce sample complexity by fractionation; this can result in many fractions each of which have to be analyzed. In this approach the chemometric reduction of the data in combination with bioactivity data does not demand fine fractionation and allows the identification of compounds in fractions containing multiple compounds.

Liscio [30] used an EDA approach for the detection of antiandrogens in river water. Different passive sampling techniques were used and antiandrogenic activity was measured with the YAS assay. The bioactive fractions were further investigated via GC–MS and LC–TOF-MS for compound identification. Data analysis was done with instrument-accompanied software and other software tools such as MS manager software (ACD labs) for spectra deconvolution. Molecular formula calculation was done based on exact mass and isotope pattern fitting. For compound identification, the NIST, KEGG LIGAND, PubChem, Chemspider METLIN, and a custom-made database for silylated compounds were searched. Over 31 compounds were successfully identified as antiandrogenic, and the activity could be attributed to a number of fungicides, germicides, flame retardants, and pharmaceuticals.

The previous examples illustrate nontarget approaches; however, these methodologies are starting to gain attention in EDA as well. Where previous EDA studies mainly used GC–MS for compound identification via target or nontarget analysis with NIST database searching [46], nowadays LC–MS with HR MS instruments are also being used [22, 114, 185]. Simon *et al.* [114] used a combination of target and nontarget analysis utilizing GC–MS with a low-resolution MS as well as LC–MS with a high-resolution TOF-MS for the identification of thyroid hormone-like compounds in polar bear plasma. Among others, molecular formula calculation, database searching, and isotope clustering led to the identification of several thyroid hormone-disrupting compounds. In another study an LC-orbitrap system was used in a follow-up study that aimed to further identify bioactive compounds [22]. The obtained spectra were aligned, and peaks were identified from the noise by the program SIEVE. Subsequently, only peaks with a certain intensity were further investigated and their peak shape was assessed manually. Molecular formulas were generated and a database search was undertaken. For confirmation, pure standards were purchased (where possible) and analyzed chemically and biologically.

4.8
Conclusion and Perspectives

A wide variety of techniques is used for EDA-like and EDA approaches. Among the different research fields, these techniques are applied in many different ways but all with the same aim: to identify biologically active compounds in complex chemical mixtures and matrices. Unfortunately, thus far the majority of these approaches are labor-intensive and not always successful in identifying the bioactive compounds. Furthermore, a broad knowledge is necessary for the use of biological and chemical techniques. Because of the labor-intensive methodologies and the necessary expertise, thus far EDA is not a frequently used technique in routine analysis and monitoring. It would, however, be beneficial if EDA-like approaches could be implemented in these fields. The use of EDA in, for example, water quality assessment could aid in the identification of CECs. In addition, there are many possible applications in food. For example, in the screening of new compounds that may be used as growth stimulators in livestock and for the identification of new kinds of doping in doping control. In order to implement EDA, it is necessary that robust systems with increased throughput are developed. It is, however, important to keep in mind that depending on the results obtained, adjustments have to be made to the method in order to isolate specific compounds. Furthermore, to capture a wide range of compounds, a combination of GC–MS and LC–MS is very successful to extend the polarity range of compounds that can be identified. In addition, the ionization interface inherently influences the identification results and it is important to use different ionization techniques to cover a broad window of compounds with different physicochemical properties. Furthermore, high-resolution MS was shown to be successful in nontarget analysis and is gaining attention in EDA. The high-resolution data can be used for molecular formula calculation and subsequent database searching. In contrast, spectral databases for LC–MS are thus far still limited in size and there is a demand for larger databases to facilitate the identification process. In the case of emerging compounds, *in silico* structure generation tools can be helpful for identification when no database hit is found. It is important that databases and software tools continue to develop; however, it should be kept in mind that visual confirmation and assessment of spectra will always be necessary. Finally, once a new bioactive compound is identified, faster, more sensitive, and robust traditional techniques have to take over for quantitation and subsequent monitoring. To achieve increased throughput of EDA approaches, challenges have to be overcome in sample preparation, separation, and chemical and biological detection. Recent developments such as AMAC for lipid-rich matrices, SPE for the extraction of a broad range of contaminants from plasma samples, nanospotting technologies, and hyphenation of biological and chemical detection seem promising for future research and might encourage laboratories to apply EDA.

References

1. Weller, M.G. (2012) A unifying review of bioassay-guided fractionation, effect-directed analysis and related techniques. *Sensors (Basel)*, **12** (7), 9181–9209.

2. Schobel, U., Frenay, M., Van Elswijk, D.A., McAndrews, J.M., Long, K.R., Olson, L.M., Bobzin, S.C., and Irth, H. (2001) High resolution screening of plant natural product extracts for estrogen receptor a and f3 binding activity using an online HPLC-MS biochemical detection system. *J. Biomol. Screen.*, **6** (5), 291–303.

3. Kool, J., Giera, M., Irth, H., and Niessen, W.M.A. (2011) Advances in mass spectrometry-based post-column bioaffinity profiling of mixtures. *Anal. Bioanal. Chem.*, **399** (8), 2655–2668.

4. Brack, W. (2003) Effect-directed analysis: a promising tool for the identification of organic toxicants in complex mixtures? *Anal. Bioanal. Chem.*, **377** (3), 397–407.

5. Brack, W., Schmitt-Jansen, M., Machala, M., Brix, R., Barceló, D., Schymanski, E., Streck, G., and Schulze, T. (2008) How to confirm identified toxicants in effect-directed analysis. *Anal. Bioanal. Chem.*, **390** (8), 1959–1973.

6. Brack, W. (ed) (2011) *The Handbook of Environmental Chemistry*, vol. **15**, Springer.

7. Burgess, R.M., Ho, K.T., Brack, W., and Lamoree, M. (2013) Effects-directed analysis (EDA) and toxicity identification evaluation (TIE): complementary but different approaches for diagnosing causes of environmental toxicity. *Environ. Toxicol. Chem.*, **32** (9), 1935–1945.

8. Kokkali, V. and van Delft, W. (2014) Overview of commercially available bioassays for assessing chemical toxicity in aqueous samples. *TrAC, Trends in Analytical Chemistry*, **61**, (2014), 133–155.

9. d'Acampora Zellner, B., Dugo, P., Dugo, G., and Mondello, L. (2008) Gas chromatography-olfactometry in food flavour analysis. *J. Chromatogr. A*, **1186** (1-2), 123–143.

10. Plutowska, B. and Wardencki, W. (2008) Application of gas chromatography–olfactometry (GC–O) in analysis and quality assessment of alcoholic beverages – A review. *Food Chem.*, **107** (1), 449–463.

11. Connolly, L., Ropstad, E., and Verhaegen, S. (2011) In vitro bioassays for the study of endocrine-disrupting food additives and contaminants. *TrAC, Trends Anal. Chem.*, **30** (2), 227–238.

12. Schymanski, E.L., Bataineh, M., Goss, K.-U., and Brack, W. (2009) Integrated analytical and computer tools for structure elucidation in effect-directed analysis. *TrAC, Trends Anal. Chem.*, **28** (5), 550–561.

13. Kazda, R., Hajšlová, J., Poustka, J., and Čajka, T. (2004) Determination of polybrominated diphenyl ethers in human milk samples in the Czech Republic. *Anal. Chim. Acta*, **520** (1-2), 237–243.

14. Olivella, M.A. (2006) Polycyclic aromatic hydrocarbons in rainwater and surface waters of Lake Maggiore, a subalpine lake in Northern Italy. *Chemosphere*, **63** (1), 116–131.

15. Senta, I., Terzić, S., and Ahel, M. (2008) Simultaneous determination of sulfonamides, fluoroquinolones, macrolides and trimethoprim in wastewater and river water by LC-tandem-MS. *Chromatographia*, **68** (9-10), 747–758.

16. López-Blanco, C., Gómez-Alvarez, S., Rey-Garrote, M., Cancho-Grande, B., and Simal-Gándara, J. (2006) Determination of pesticides by solid phase extraction followed by gas chromatography with nitrogen-phosphorous detection in natural water and comparison with solvent drop microextraction. *Anal. Bioanal. Chem.*, **384** (4), 1002–1006.

17. Sandra Babić, M.K.-M., Mutavdžić Pavlović, D., Ašperger, D., Periša, M., Zrnčić, M., and Horvat, A.J.M. (2010) Determination of multiclass pharmaceuticals in wastewater by liquid chromatography–tandem mass spectrometry (LC–MS–MS). *Anal. Bioanal. Chem.*, **398**, 1185–1194.

18. Schulze, T., Weiss, S., Schymanski, E., von der Ohe, P.C., Schmitt-Jansen, M., Altenburger, R., Streck, G., and Brack, W. (2010) Identification of a phytotoxic photo-transformation product of diclofenac using effect-directed analysis. *Environ. Pollut.*, **158** (5), 1461–1466.

19. Smital, T., Terzic, S., Zaja, R., Senta, I., Pivcevic, B., Popovic, M., Mikac, I., Tollefsen, K.E., Thomas, K.V., and Ahel, M. (2011) Assessment of toxicological profiles of the municipal wastewater effluents using chemical analyses and bioassays. *Ecotoxicol. Environ. Saf.*, **74** (4), 844–851.

20. Lübcke-von Varel, U., Machala, M., Ciganek, M., Neca, J., Pencikova, K., Palkova, L., Vondracek, J., Löffler, I., Streck, G., Reifferscheid, G., Flückiger-Isler, S., Weiss, J.M., Lamoree, M., and Brack, W. (2011) Polar compounds dominate in vitro effects of sediment extracts. *Environ. Sci. Technol.*, **45** (6), 2384–2390.

21. Wölz, J., Fleig, M., Schulze, T., Maletz, S., Lübcke-von Varel, U., Reifferscheid, G., Kühlers, D., Braunbeck, T., Brack, W., and Hollert, H. (2010) Impact of contaminants bound to suspended particulate matter in the context of flood events. *J. Soils Sediments*, **10** (6), 1174–1185.

22. Weiss, J.M., Simon, E., Stroomberg, G.J., de Boer, R., de Boer, J., van der Linden, S.C., Leonards, P.E.G., and Lamoree, M.H. (2011) Identification strategy for unknown pollutants using high-resolution mass spectrometry: androgen-disrupting compounds identified through effect-directed analysis. *Anal. Bioanal. Chem.*, **400** (9), 3141–3149.

23. Grung, M., Næs, K., Fogelberg, O., Nilsen, A.J., Brack, W., Lübcke-von Varel, U., and Thomas, K.V. (2011) Effects-directed analysis of

sediments from polluted marine sites in Norway. *J. Toxicol. Environ. Health A*, **74** (7-9), 439–454.

24. Sun, Y., Zhang, W.-Y., Xing, J., and Wang, C.-M. (2011) Solid-phase microfibers based on modified single-walled carbon nanotubes for extraction of chlorophenols and organochlorine pesticides. *Microchim. Acta*, **173** (1-2), 223–229.

25. Giordano, A., Fernández-Franzón, M., Ruiz, M.J., Font, G., and Picó, Y. (2009) Pesticide residue determination in surface waters by stir bar sorptive extraction and liquid chromatography/tandem mass spectrometry. *Anal. Bioanal. Chem.*, **393** (6-7), 1733–1743.

26. Booij, P., Sjollema, S.B., Leonards, P.E.G., de Voogt, P., Stroomberg, G.J., Vethaak, a.D., and Lamoree, M.H. (2013) Extraction tools for identification of chemical contaminants in estuarine and coastal waters to determine toxic pressure on primary producers. *Chemosphere*, **93** (1), 107–114.

27. Booij, P., Vethaak, A.D., Leonards, P.E.G., Sjollema, S.B., Kool, J., de Voogt, P., and Lamoree, M.H. (2014) Identification of photosynthesis inhibitors of pelagic marine algae using 96-well plate microfractionation for enhanced throughput in effect-directed analysis. *Environ. Sci. Technol.*, **48**, 8003.

28. Esteve-Turrillas, F.A., Pastor, A., and de la Guardia, M. (2007) Behaviour of semipermeable membrane devices in neutral pesticide uptake from waters. *Anal. Bioanal. Chem.*, **387** (6), 2153–2162.

29. Koč, V., Ocelka, T., Dragoun, D., Vít, M., Grabic, R., and Šváb, M. (2007) Concentration of organochlorine pollutants in surface waters of the central european biosphere reserve krivoklatsko (8 pp). *Environ. Sci. Pollut. Res.*, **14** (2), 94–101.

30. Liscio, C., Abdul-Sada, A., Al-Salhi, R., Ramsey, M.H., and Hill, E.M. (2014) Methodology for profiling anti-androgen mixtures in river water using multiple passive samplers and bioassay-directed analyses. *Water Res.*, **57**, 258–269.

31. Grung, M., Lichtenthaler, R., Ahel, M., Tollefsen, K.-E., Langford, K., and Thomas, K.V. (2007) Effects-directed analysis of organic toxicants in wastewater effluent from Zagreb, Croatia. *Chemosphere*, **67** (1), 108–120.

32. Urbatzka, R., van Cauwenberge, A., Maggioni, S., Viganò, L., Mandich, A., Benfenati, E., Lutz, I., and Kloas, W. (2007) Androgenic and antiandrogenic activities in water and sediment samples from the river Lambro, Italy, detected by yeast androgen screen and chemical analyses. *Chemosphere*, **67** (6), 1080–1087.

33. Fidalgo-Used, N., Blanco-González, E., and Sanz-Medel, A. (2007) Sample handling strategies for the determination of persistent trace organic contaminants from biota samples. *Anal. Chim. Acta*, **590** (1), 1–16.

34. Suzuki, G., Tue, N.M., van der Linden, S., Brouwer, A., van der Burg, B., van Velzen, M., Lamoree, M., Someya, M., Takahashi, S., Isobe, T., Tajima, Y., Yamada, T.K., Takigami, H., and Tanabe, S. (2011) Identification of major dioxin-like compounds and androgen receptor antagonist in acid-treated tissue extracts of high trophic-level animals. *Environ. Sci. Technol.*, **45** (23), 10203–10211.

35. Streck, H.-G., Schulze, T., and Brack, W. (2008) Accelerated membrane-assisted clean-up as a tool for the clean-up of extracts from biological tissues. *J. Chromatogr. A*, **1196–1197**, 33–40.

36. Simon, E., Lamoree, M.H., Hamers, T., Weiss, J.M., Balaam, J., de Boer, J., and Leonards, P.E.G. (2010) Testing endocrine disruption in biota samples: a method to remove interfering lipids and natural hormones. *Environ. Sci. Technol.*, **44** (21), 8322–8329.

37. Simon, E., Bytingsvik, J., Jonker, W., Leonards, P.E.G., de Boer, J., Jenssen, B.M., Lie, E., Aars, J., Hamers, T., and Lamoree, M.H. (2011) Blood plasma sample preparation method for the assessment of thyroid hormone-disrupting potency in effect-directed analysis. *Environ. Sci. Technol.*, **45** (18), 7936–7944.

38. Hill, E.M., Evans, K.L., Horwood, J., Rostkowski, P., Oladapo, F.O., Gibson, R., Shears, J.a., and Tyler, C.R. (2010) Profiles and some initial identifications of (anti)androgenic compounds in fish exposed to wastewater treatment works effluents. *Environ. Sci. Technol.*, **44** (3), 1137–1143.

39. Rostkowski, P., Horwood, J., Shears, J.a., Lange, A., Oladapo, F.O., Besselink, H.T., Tyler, C.R., and Hill, E.M. (2011) Bioassay-directed identification of novel antiandrogenic compounds in bile of fish exposed to wastewater effluents. *Environ. Sci. Technol.*, **45** (24), 10660–10667.

40. Houtman, C.J., Van Oostveen, A.M., Brouwer, A., Lamoree, M.H., and Legler, J. (2004) Identification of estrogenic compounds in fish bile using bioassay-directed fractionation. *Environ. Sci. Technol.*, **38** (23), 6415–6423.

41. Weiss, J.M., Hamers, T., Thomas, K.V., van der Linden, S., Leonards, P.E.G., and Lamoree, M.H. (2009) Masking effect of anti-androgens on androgenic activity in European river sediment unveiled by effect-directed analysis. *Anal. Bioanal. Chem.*, **394** (5), 1385–1397.

42. Reifferscheid, G., Buchinger, S., Cao, Z., and Claus, E. (2011) Identification of mutagens in freshwater sediments by the ames-fluctuation assay using nitroreductase and acetyltransferase overproducing test strains. *Environ. Mol. Mutagen.*, **52**, 397–408.

43. Schmitt, C., Streck, G., Lamoree, M., Leonards, P., Brack, W., and de Deckere, E. (2011) Effect directed analysis of riverine sediments--the usefulness of Potamopyrgus antipodarum for in vivo effect confirmation of endocrine disruption. *Aquat. Toxicol.*, **101** (1), 237–243.

44. Lübcke-von Varel, U., Bataineh, M., Lohrmann, S., Löffler, I., Schulze, T., Flückiger-Isler, S., Neca, J., Machala, M., and Brack, W. (2012) Identification and quantitative confirmation of dinitropyrenes and 3-nitrobenzanthrone as major mutagens in contaminated sediments. *Environ. Int.*, **44**, 31–39.

45. Bandow, N., Altenburger, R., Streck, G., and Brack, W. (2009) Effect-directed analysis of contaminated sediments with partition-based dosing using green algae cell multiplication inhibition. *Environ. Sci. Technol.*, **43** (19), 7343–7349.

46. Legler, J., van Velzen, M., Cenijn, P.H., Houtman, C.J., Lamoree, M.H., and Wegener, J.W. (2011) Effect-directed analysis of municipal landfill soil reveals novel developmental toxicants in the zebrafish Danio rerio. *Environ. Sci. Technol.*, **45** (19), 8552–8558.

47. Brack, W., Schirmer, K., Lothar, E., and Hollert, H. (2005) Effect-directed analysis of mutagens and ethoxyresofurin-o-deethylase inducers in aquatic sediments. *Environ. Toxicol. Chem.*, **24** (10), 2445–2458.

48. Schmitt, S., Reifferscheid, G., Claus, E., Schlüsener, M., and Buchinger, S. (2012) Effect directed analysis and mixture effects of estrogenic compounds in a sediment of the river Elbe. *Environ. Sci. Pollut. Res. Int.*, **19** (8), 3350–3361.

49. Hawthorne, S.B. and Miller, D.J. (1994) of Organics from environmental solids. *Anal. Chem.*, **66** (22), 4005–4012.

50. Xiong, G., He, X., and Zhang, Z. (2000) Microwave-assisted extraction or saponification combined with microwave-assisted decomposition applied in pretreatment of soil or mussel samples for the determination of polychlorinated biphenyls. *Anal. Chim. Acta*, **413** (1-2), 49–56.

51. Vallecillos, L., Borrull, F., and Pocurull, E. (2012) Determination of musk fragrances in sewage sludge by pressurized liquid extraction coupled to automated ionic liquid-based headspace single-drop microextraction followed by GC-MS/MS. *J. Sep. Sci.*, **35**, 2735–2742.

52. Regueiro, J., Matamoros, V., Thibaut, R., Porte, C., and Bayona, J.M. (2013) Use of effect-directed analysis for the identification of organic toxicants in surface flow constructed wetland sediments. *Chemosphere*, **91** (8), 1165–75.

53. Houtman, C.J., Booij, P., Jover, E., Pascual del Rio, D., Swart, K., van Velzen, M., Vreuls, R., Legler, J.,

Brouwer, A., and Lamoree, M.H. (2006) Estrogenic and dioxin-like compounds in sediment from Zierikzee harbour identified with CALUX assay-directed fractionation combined with one and two dimensional gas chromatography analyses. *Chemosphere*, **65** (11), 2244–2252.

54. Bandow, N., Altenburger, R., Lübcke-Von Varel, U., Paschke, A., Streck, G., and Brack, W. (2009) Partitioning-based dosing: an approach to include bioavailability in the effect-directed analysis of contaminated sediment samples. *Environ. Sci. Technol.*, **43** (10), 3891–3896.

55. Schmitt, C., Vogt, C., Machala, M., and de Deckere, E. (2011) Sediment contact test with Potamopyrgus antipodarum in effect-directed analyses-challenges and opportunities. *Environ. Sci. Pollut. Res. Int.*, **18** (8), 1398–1404.

56. Schwab, K. and Brack, W. (2007) Large volume TENAX extraction of the bioaccessible fraction of sediment-associated organic compounds for a subsequent effect-directed analysis. *J. Soils Sediments*, **7** (3), 178–186.

57. Campo, E., Ferreira, V., Escudero, A., and Cacho, J. (2005) Prediction of the wine sensory properties related to grape variety from dynamic-headspace gas chromatography-olfactometry data. *J. Agric. Food Chem.*, **53** (14), 5682–5690.

58. Ugliano, M. and Moio, L. (2008) Free and hydrolytically released volatile compounds of Vitis vinifera L. cv. Fiano grapes as odour-active constituents of Fiano wine. *Anal. Chim. Acta*, **621** (1), 79–85.

59. Barata, A., Campo, E., Malfeito-Ferreira, M., Loureiro, V., Cacho, J., and Ferreira, V. (2011) Analytical and sensorial characterization of the aroma of wines produced with sour rotten grapes using GC-O and GC-MS: identification of key aroma compounds. *J. Agric. Food Chem.*, **59** (6), 2543–2553.

60. Falcão, L.D., de Revel, G., Rosier, J.P., and Bordignon-Luiz, M.T. (2008) Aroma impact components of Brazilian Cabernet Sauvignon wines using detection frequency analysis (GC–olfactometry). *Food Chem.*, **107** (1), 497–505.

61. Mo, X., Xu, Y., and Fan, W. (2010) Characterization of aroma compounds in Chinese rice wine Qu by solvent-assisted flavor evaporation and headspace solid-phase microextraction. *J. Agric. Food Chem.*, **58** (4), 2462–2469.

62. Niu, Y., Zhang, X., Xiao, Z., Song, S., Eric, K., Jia, C., Yu, H., and Zhu, J. (2011) Characterization of odor-active compounds of various cherry wines by gas chromatography–mass spectrometry, gas chromatography-olfactometry and their correlation with sensory attributes. *J. Chromatogr. B Analyt. Technol. Biomed. Life Sci.*, **879** (23), 2287–2293.

63. Chin, S.-T., Eyres, G.T., and Marriott, P.J. (2012) Cumulative solid phase microextraction sampling for gas chromatography-olfactometry of Shiraz wine. *J. Chromatogr. A*, **1255**, 221–227.

64. Ugliano, M. and Moio, L. (2006) The influence of malolactic fermentation and Oenococcus oeni strain on glycosidic aroma precursors and related volatile compounds of red wine. *J. Sci. Food Agric.*, **86**, 2468–2476.

65. San-Juan, F., Pet'ka, J., Cacho, J., Ferreira, V., and Escudero, A. (2010) Producing headspace extracts for the gas chromatography–olfactometric evaluation of wine aroma. *Food Chem.*, **123** (1), 188–195.

66. Thomsen, M., Martin, C., Mercier, F., Tournayre, P., Berdagué, J.-L., Thomas-Danguin, T., and Guichard, E. (2012) Investigating semi-hard cheese aroma: relationship between sensory profiles and gas chromatography-olfactometry data. *Int. Dairy J.*, **26** (1), 41–49.

67. Cornu, A., Rabiau, N., Kondjoyan, N., Verdier-Metz, I., Pradel, P., Tournayre, P., Berdagué, J.L., and Martin, B. (2009) Odour-active compound profiles in Cantal-type cheese: effect of cow diet, milk pasteurization and cheese ripening. *Int. Dairy J.*, **19** (10), 588–594.

68. Guneser, O. and Yuceer, Y.K. (2011) Characterisation of aroma-active

compounds, chemical and sensory properties of acid-coagulated cheese: circassian cheese. *Int. J. Dairy Technol.*, **64** (4), 517–525.

69. Owusu, M., Petersen, M.A., and Heimdal, H. (2012) Effect of fermentation method, roasting and conching conditions on the aroma volatiles of dark chocolate. *J. Food Process. Preserv.*, **36** (5), 446–456.

70. Afoakwa, E.O., Paterson, A., Fowler, M., and Ryan, A. (2009) Matrix effects on flavour volatiles release in dark chocolates varying in particle size distribution and fat content using GC–mass spectrometry and GC–olfactometry. *Food Chem.*, **113** (1), 208–215.

71. Théron, L., Tournayre, P., Kondjoyan, N., Abouelkaram, S., Santé-Lhoutellier, V., and Berdagué, J.-L. (2010) Analysis of the volatile profile and identification of odour-active compounds in Bayonne ham. *Meat Sci.*, **85** (3), 453–460.

72. Culleré, L., Ferreira, V., Chevret, B., Venturini, M.E., Sánchez-Gimeno, A.C., and Blanco, D. (2010) Characterisation of aroma active compounds in black truffles (Tuber melanosporum) and summer truffles (Tuber aestivum) by gas chromatography–olfactometry. *Food Chem.*, **122** (1), 300–306.

73. Díaz, P., Ibáñez, E., Reglero, G., and Señoráns, F.J. (2009) Optimization of summer truffle aroma analysis by SPME: comparison of extraction with different polarity fibres. *LWT Food Sci. Technol.*, **42** (7), 1253–1259.

74. Akiyama, M., Murakami, K., Hirano, Y., Ikeda, M., Iwatsuki, K., Wada, A., Tokuno, K., Onishi, M., and Iwabuchi, H. (2008) Characterization of headspace aroma compounds of freshly brewed arabica coffees and studies on a characteristic aroma compound of Ethiopian coffee. *J. Food Sci.*, **73** (5), C335–C346.

75. Aceña, L., Vera, L., Guasch, J., Busto, O., and Mestres, M. (2011) Determination of roasted pistachio (Pistacia vera L.) key odorants by headspace solid-phase microextraction and gas chromatography-olfactometry. *J. Agric. Food Chem.*, **59** (6), 2518–2523.

76. Vera, P., Uliaque, B., Canellas, E., Escudero, A., and Nerín, C. (2012) Identification and quantification of odorous compounds from adhesives used in food packaging materials by headspace solid phase extraction and headspace solid phase microextraction coupled to gas chromatography-olfactometry-mass spectrometry. *Anal. Chim. Acta*, **745**, 53–63.

77. Parker, D.B., Wright, D.W., Eaton, D.K., Nielsen, L.T., Kuhrt, F.W., Koziel, J.A., and Spinhirne, J.P. (2005) Multidimensional gas chromatography – Olfactometry for the identification and prioritization of malodors from confined animal feeding operations. *J. Agric. Food Chem.*, **53**, 8663–8672.

78. Tyl, C.E. and Bunzel, M. (2012) Antioxidant activity-guided fractionation of blue wheat (UC66049 Triticum aestivum L.). *J. Agric. Food Chem.*, **60** (3), 731–739.

79. Wang, Y., Bao, L., Yang, X., Li, L., Li, S., Gao, H., Yao, X.-S., Wen, H., and Liu, H.-W. (2012) Bioactive sesquiterpenoids from the solid culture of the edible mushroom Flammulina velutipes growing on cooked rice. *Food Chem.*, **132** (3), 1346–1353.

80. Wetli, H.A., Brenneisen, R., Tschudi, I., Langos, M., Bigler, P., Sprang, T., Schürch, S., and Mühlbauer, R.C. (2005) A gamma-glutamyl peptide isolated from onion (Allium cepa L.) by bioassay-guided fractionation inhibits resorption activity of osteoclasts. *J. Agric. Food Chem.*, **53** (9), 3408–3414.

81. Nitteranon, V., Zhang, G., Darien, B.J., and Parkin, K. (2011) Isolation and synergism of in vitro anti-inflammatory and quinone reductase (QR) inducing agents from the fruits of Morinda citrifolia (noni). *Food Res. Int.*, **44** (7), 2271–2277.

82. Qian, B., Xing, M., Cui, L., Deng, Y., Xu, Y., Huang, M., and Zhang, S. (2011) Antioxidant, antihypertensive, and immunomodulatory activities of peptide fractions from fermented skim milk with Lactobacillus delbrueckii ssp. bulgaricus LB340. *J. Dairy Res.*, **78** (1), 72–79.

83. Kim, A.-R., Shin, T.-S., Lee, M.-S., Park, J.-Y., Park, K.-E., Yoon, N.-Y., Kim, J.-S., Choi, J.-S., Jang, B.-C., Byun, D.-S., Park, N.-K., and Kim, H.-R. (2009) Isolation and identification of phlorotannins from Ecklonia stolonifera with antioxidant and anti-inflammatory properties. *J. Agric. Food Chem.*, **57** (9), 3483–3489.

84. Lagemann, A., Dunkel, A., and Hofmann, T. (2012) Activity-guided discovery of (S)-malic acid 1'-O-β-gentiobioside as an angiotensin I-converting enzyme inhibitor in lettuce (Lactuca sativa). *J. Agric. Food Chem.*, **60** (29), 7211–7217.

85. Nielen, M.W.F., van Bennekom, E.O., Heskamp, H.H., van Rhijn, J.H.A., Bovee, T.F.H., and Hoogenboom, L.R.A.P. (2004) Bioassay-directed identification of estrogen residues in urine by liquid chromatography electrospray quadrupole time-of-flight mass spectrometry. *Anal. Chem.*, **76** (22), 6600–6608.

86. Peters, R.J.B., Rijk, J.C.W., Bovee, T.F.H., Nijrolder, A.W.J.M., Lommen, A., and Nielen, M.W.F. (2010) Identification of anabolic steroids and derivatives using bioassay-guided fractionation, UHPLC/TOFMS analysis and accurate mass database searching. *Anal. Chim. Acta*, **664** (1), 77–88.

87. de Abreu, D.A.P., Rodriguez, K.V., and Cruz, J.M. (2012) Extraction, purification and characterization of an antioxidant extract from barley husks and development of an antioxidant active film for food package. *Innovative Food Sci. Emerg. Technol.*, **13**, 134–141.

88. Ochiai, N. and Sasamoto, K. (2011) Selectable one-dimensional or two-dimensional gas chromatography-olfactometry/mass spectrometry with preparative fraction collection for analysis of ultra-trace amounts of odor compounds. *J. Chromatogr. A*, **1218** (21), 3180–3185.

89. Naudé, Y., van Aardt, M., and Rohwer, E.R. (2009) Multi-channel open tubular traps for headspace sampling, gas chromatographic fraction collection and olfactory assessment of milk volatiles. *J. Chromatogr. A*, **1216** (14), 2798–2804.

90. Ortner, E.K. and Rohwer, E.R. (1996) Trace analysis of semi-volaile organic air pollutants using thick film silicone rubber traps with capillary gas chromatography. *J. High Resolut. Chromatogr.*, **19**, 339–344.

91. Silván, J.M., Mingo, E., Hidalgo, M., de Pascual-Teresa, S., Carrascosa, A.V., and Martinez-Rodriguez, A.J. (2013) Antibacterial activity of a grape seed extract and its fractions against Campylobacter spp. *Food Control*, **29** (1), 25–31.

92. Si, W., Gong, J., Tsao, R., Kalab, M., Yang, R., and Yin, Y. (2006) Bioassay-guided purification and identification of antimicrobial components in Chinese green tea extract. *J. Chromatogr. A*, **1125** (2), 204–210.

93. Harvey, A.L. (2008) Natural products in drug discovery. *Drug Discovery Today*, **13** (19-20), 894–901.

94. Phillipson, D.W., Milgram, K.E., Yanovsky, A.I., Rusnak, L.S., Haggerty, D.A., Farrell, W.P., Greig, M.J., Xiong, X., and Proefke, M.L. (2002) High-throughput bioassay-guided fractionation: a technique for rapidly assigning observed activity to individual components of combinatorial libraries, screened in HTS bioassays. *J. Comb. Chem.*, **4** (6), 591–599.

95. Carpenter, C.D., O'Neill, T., Picot, N., Johnson, J.A., Robichaud, G.A., Webster, D., and Gray, C.A. (2012) Anti-mycobacterial natural products from the Canadian medicinal plant Juniperus communis. *J. Ethnopharmacol.*, **143** (2), 695–700.

96. Ishola, I.O., Agbaje, O.E., Narender, T., Adeyemi, O.O., and Shukla, R. (2012) Bioactivity guided isolation of analgesic and anti-inflammatory constituents of Cnestis ferruginea Vahl ex DC (Connaraceae) root. *J. Ethnopharmacol.*, **142** (2), 383–389.

97. Forgo, P., Zupkó, I., Molnár, J., Vasas, A., Dombi, G., and Hohmann, J. (2012) Bioactivity-guided isolation of antiproliferative compounds from Centaurea jacea L. *Fitoterapia*, **83** (5), 921–925.

98. Han, Q.-B., Zhou, Y., Feng, C., Xu, G., Huang, S.-X., Li, S.-L., Qiao, C.-F., Song, J.-Z., Chang, D.C., Luo, K.Q.,

and Xu, H.-X. (2009) Bioassay guided discovery of apoptosis inducers from gamboge by high-speed counter-current chromatography and high-pressure liquid chromatography/electrospray ionization quadrupole time-of-flight mass spectrometry. *J. Chromatogr. B Analyt. Technol. Biomed. Life Sci.*, **877** (4), 401–407.

99. Xia, Z.-X., Zhang, D.-D., Liang, S., Lao, Y.-Z., Zhang, H., Tan, H.-S., Chen, S.-L., Wang, X.-H., and Xu, H.-X. (2012) Bioassay-guided isolation of prenylated xanthones and polycyclic acylphloroglucinols from the leaves of Garcinia nujiangensis. *J. Nat. Prod.*, **75** (8), 1459–1464.

100. Hou, Y., Cao, X., Dong, L., Wang, L., Cheng, B., Shi, Q., Luo, X., and Bai, G. (2012) Bioactivity-based liquid chromatography-coupled electrospray ionization tandem ion trap/time of flight mass spectrometry for β_2AR agonist identification in alkaloidal extract of Alstonia scholaris. *J. Chromatogr. A*, **1227**, 203–209.

101. Yuliana, N.D., Khatib, A., Verpoorte, R., and Choi, Y.H. (2011) Comprehensive extraction method integrated with NMR metabolomics: a new bioactivity screening method for plants, adenosine A1 receptor binding compounds in orthosiphon stamineus benth. *Anal. Chem.*, **83** (17), 6902–6906.

102. Kelman, D., Posner, E.K., McDermid, K.J., Tabandera, N.K., Wright, P.R., and Wright, A.D. (2012) Antioxidant activity of Hawaiian marine algae. *Mar. Drugs*, **10** (2), 403–416.

103. Zimba, P.V., Moeller, P.D., Beauchesne, K., Lane, H.E., and Triemer, R.E. (2010) Identification of euglenophycin--a toxin found in certain euglenoids. *Toxicon*, **55** (1), 100–104.

104. Teasdale, M.E., Shearer, T.L., Engel, S., Alexander, T.S., Fairchild, C.R., Prudhomme, J., Torres, M., Le Roch, K., Aalbersberg, W., Hay, M.E., and Kubanek, J. (2012) Bromophycoic acids: bioactive natural products from a Fijian red alga Callophycus sp. *J. Org. Chem.*, **77** (18), 8000–8006.

105. Morita, M., Ohno, O., Teruya, T., Yamori, T., Inuzuka, T., and Suenaga, K. (2012) Isolation and structures of biselyngbyasides B, C, and D from the marine cyanobacterium Lyngbya sp., and the biological activities of biselyngbyasides. *Tetrahedron*, **68** (30), 5984–5990.

106. Vulfius, C.A., Gorbacheva, E.V., Starkov, V.G., Osipov, A.V., Kasheverov, I.E., Andreeva, T.V., Astashev, M.E., Tsetlin, V.I., and Utkin, Y.N. (2011) An unusual phospholipase A_2 from puff adder Bitis arietans venom--a novel blocker of nicotinic acetylcholine receptors. *Toxicon*, **57** (5), 787–793.

107. Miyashita, M., Sakai, A., Matsushita, N., Hanai, Y., Nakagawa, Y., and Miyagawa, H. (2010) A novel amphipathic linear peptide with both insect toxicity and antimicrobial activity from the venom of the scorpion isometrus maculatus. *Biosci. Biotechnol. Biochem.*, **74** (2), 364–369.

108. Ständker, L., Harvey, A.L., Fürst, S., Mathes, I., Forssmann, W.G., Escalona de Motta, G., and Béress, L. (2012) Improved method for the isolation, characterization and examination of neuromuscular and toxic properties of selected polypeptide fractions from the crude venom of the Taiwan cobra Naja naja atra. *Toxicon*, **60** (4), 623–631.

109. Diego-García, E., Peigneur, S., Debaveye, S., Gheldof, E., Tytgat, J., and Caliskan, F. (2013) Novel potassium channel blocker venom peptides from Mesobuthus gibbosus (Scorpiones: Buthidae). *Toxicon*, **61**, 72–82.

110. Queiroz, S.C.N., Cantrell, C.L., Duke, S.O., Wedge, D.E., Nandula, V.K., Moraes, R.M., and Cerdeira, A.L. (2012) Bioassay-directed isolation and identification of phytotoxic and fungitoxic acetylenes from Conyza canadensis. *J Agric Food Chem. 2012 Jun 13;60*, (23), 5893–8.

111. Mandalakis, M. and Gustafsson, O. (2003) Optimization of a preparative capillary gas chromatography–mass spectrometry system for the isolation and harvesting of individual polycyclic aromatic hydrocarbons. *J. Chromatogr. A*, **996** (1-2), 163–172.

112. Meinert, C., Moeder, M., and Brack, W. (2007) Fractionation of technical p-nonylphenol with preparative capillary gas chromatography. *Chemosphere*, **70** (2), 215–223.

113. Pieke, E., Heus, F., Kamstra, J.H., Mladic, M., Van Velzen, M., Kamminga, D., Lamoree, M.H., Hamers, T., Leonards, P., Niessen, W.M.A., and Kool, J. (2013) High-resolution fractionation after gas chromatography for effect-directed analysis. *Anal. Chem.*, **85**, 8204–8211.

114. Simon, E., Van Velzen, M., Brandsma, S.H., Lie, E., Løken, K., De Boer, J., Bytingsvik, J., Jenssen, B.M., Aars, J., Hamers, T., and Lamoree, M.H. (2013) Effect-directed analysis to explore the polar bear exposome: identification of thyroid hormone disrupting compounds in plasma. *Environ. Sci. Technol.*, **47**, 8902–8912.

115. Kool, J., Heus, F., de Kloe, G., Lingeman, H., Smit, A.B., Leurs, R., Edink, E., De Esch, I.J.P., Irth, H., and Niessen, W.M.A. (2011) High-resolution bioactivity profiling of mixtures toward the acetylcholine binding protein using a nanofractionation spotter technology. *J. Biomol. Screen.*, **16** (8), 917–924.

116. Müller, M.B., Dausend, C., Weins, C., and Frimmel, F.H. (2004) A new bioautographic screening method for the detection of estrogenic compounds. *Chromatographia*, **60** (3-4), 207–211.

117. Buchinger, S., Spira, D., Der, K.B., Schlu, M., Ternes, T., and Rei, G. (2013) Direct coupling of thin-layer chromatography with a bioassay for the detection of estrogenic compounds: applications for effect-directed analysis. *Anal. Chem.*, **85**, 7248–7256.

118. Lagemann, A. *et al.* (2012) Activity-guided discovery of (S)-malic acid 1′-O-β-gentiobioside as an angiotensin I-converting enzyme inhibitor in lettuce (Lactuca sativa). *J. Agric. Food Chem.*, **60**, 7211–7217.

119. Ma, H. and Horiuchi, K.Y. (2006) Chemical microarray: a new tool for drug screening and discovery. *Drug Discovery Today*, **11** (13-14), 661–668.

120. Mayer, A.M.S., Avilés, E., and Rodríguez, A.D. (2012) Marine sponge Hymeniacidon sp. amphilectane metabolites potently inhibit rat brain microglia thromboxane B2 generation. *Bioorg. Med. Chem.*, **20** (1), 279–282.

121. Wang, L.-W., Xu, B.-G., Wang, J.-Y., Su, Z.-Z., Lin, F.-C., Zhang, C.-L., and Kubicek, C.P. (2012) Bioactive metabolites from Phoma species, an endophytic fungus from the Chinese medicinal plant Arisaema erubescens. *Appl. Microbiol. Biotechnol.*, **93** (3), 1231–1239.

122. Oosterkamp, A.J., Irth, H., Beth, M., Unger, K.K., Tjaden, U.R., and van de Greef, J. (1994) Bioanalysis of digoxin and its metabolites using direct serum injection combined with liquid chromatography and on-line immunochemical detection. *J. Chromatogr. B Biomed. Appl.*, **653** (1), 55–61.

123. Heus, F., Vonk, F., Otvos, R.A., Bruyneel, B., Smit, A.B., Lingeman, H., Richardson, M., Niessen, W.M.A., and Kool, J. (2013) An efficient analytical platform for on-line microfluidic profiling of neuroactive snake venoms towards nicotinic receptor affinity. *Toxicon*, **61**, 112–124.

124. Schenk, T., Breel, G.J., Koevoets, P., van den Berg, S., Hogenboom, A.C., Irth, H., Tjaden, U.R., and van der Greef, J. (2003) Screening of natural products extracts for the presence of phosphodiesterase inhibitors using liquid chromatography coupled online to parallel biochemical detection and chemical characterization. *J. Biomol. Screen.*, **8** (4), 421–429.

125. Crawford, A.D., Liekens, S., Kamuhabwa, A.R., Maes, J., Munck, S., Busson, R., Rozenski, J., Esguerra, C.V., and de Witte, P.A.M. (2011) Zebrafish bioassay-guided natural product discovery: isolation of angiogenesis inhibitors from East African medicinal plants. *PLoS One*, **6** (2), e14694.

126. Jonker, N., Kool, J., Irth, H., and Niessen, W.M.A. (2011) Recent developments in protein-ligand affinity mass spectrometry. *Anal. Bioanal. Chem.*, **399** (8), 2669–2681.

127. Annis, D.A., Nazef, N., Chuang, C., Scott, M.P., and Nash, H.M. (2004) A general technique to rank protein – Ligand binding affinities and determine allosteric versus direct binding site competition in compound mixtures. *J. Am. Chem. Soc.*, **126** (8), 15495–15503.

128. Derks, R.J.E., Letzel, T., Jong, C.F., Marle, A., Lingeman, H., Leurs, R., and Irth, H. (2006) SEC–MS as an approach to isolate and directly identifying small molecular GPCR–Ligands from complex mixtures without labeling. *Chromatographia*, **64** (7-8), 379–385.

129. Quercia, A.K., LaMarr, W.A., Myung, J., Ozbal, C.C., Landro, J.A., and Lumb, K.J. (2007) High-throughput screening by mass spectrometry: comparison with the scintillation proximity assay with a focused-file screen of AKT1/PKB alpha. *J. Biomol. Screen.*, **12** (4), 473–480.

130. Jonas, M., LaMarr, W.A., and Ozbal, C. (2009) Mass spectrometry in high-throughput screening: a case study on acetyl-coenzyme a carboxylase using RapidFire--mass spectrometry (RF-MS). *Comb. Chem. High Throughput Screen.*, **12**, 752–759.

131. Holt, T.G., Choi, B.K., Geoghagen, N.S., Jensen, K.K., Luo, Q., LaMarr, W.A., Makara, G.M., Malkowitz, L., Ozbal, C.C., Xiong, Y., Dufresne, C., and Luo, M.-J. (2009) Label-free high-throughput screening via mass spectrometry: a single cystathionine quantitative method for multiple applications. *Assay Drug Dev. Technol.*, **7** (5), 495–506.

132. Jonker, N., Kool, J., Krabbe, J.G., Retra, K., Lingeman, H., and Irth, H. (2008) Screening of protein-ligand interactions using dynamic protein-affinity chromatography solid-phase extraction-liquid chromatography-mass spectrometry. *J. Chromatogr. A*, **1205** (1-2), 71–77.

133. Pochet, L., Heus, F., Jonker, N., Lingeman, H., Smit, A.B., Niessen, W.M.A., and Kool, J. (2011) Online magnetic bead based dynamic protein affinity selection coupled to LC-MS for the screening of acetylcholine binding protein ligands. *J. Chromatogr. B Analyt. Technol. Biomed. Life Sci.*, **879** (20), 1781–1788.

134. Jonker, N., Kretschmer, A., Kool, J., Fernandez, A., Kloos, D., Krabbe, J.G., Lingeman, H., and Irth, H. (2009) Online magnetic bead dynamic protein-affinity selection coupled to LC-MS for the screening of pharmacologically active compounds. *Anal. Chem.*, **81**, 4263.

135. Marsza, M.P., Moaddel, R., Kole, S., Gandhari, M., Bernier, M., and Wainer, I.W. (2008) Ligand and protein fishing with heat shock protein 90 coated magnetic beads. *Anal. Chem.*, **80** (19), 7571–7575.

136. Hu, F., Deng, C., and Zhang, X. (2008) Development of high performance liquid chromatography with immobilized enzyme onto magnetic nanospheres for screening enzyme inhibitor. *J. Chromatogr. B Analyt. Technol. Biomed. Life Sci.*, **871** (1), 67–71.

137. Zlotos, G., Oehlmann, M., Nickel, P., and Holzgrabe, U. (1998) Determination of protein binding of gyrase inhibitors by means of continuous ultrafiltration. *J. Pharm. Biomed. Anal.*, **18** (4-5), 847–858.

138. Kinawi, A. and Teller, C. (1979) Determination of drug-albumin binding in buffered bovine serum-albumin solutions applying a modifed ultrafiltration process. *Arzneimittelforschung*, **29** (2), 1495–1500.

139. Fung, E.N., Chen, Y.-H., and Lau, Y.Y. (2003) Semi-automatic high-throughput determination of plasma protein binding using a 96-well plate filtrate assembly and fast liquid chromatography–tandem mass spectrometry. *J. Chromatogr. B*, **795** (2), 187–194.

140. Comess, K.M., Schurdak, M.E., Voorbach, M.J., Coen, M., Trumbull, J.D., Yang, H., Gao, L., Tang, H., Cheng, X., Lerner, C.G., McCall, J.O., Burns, D.J., and Beutel, B.A. (2006) An ultra-efficient affinity-based high-throughout screening process: application to bacterial cell wall biosynthesis enzyme MurF. *J. Biomol. Screen.*, **11** (7), 743–754.

141. Li, H., Song, F., Xing, J., Tsao, R., Liu, Z., and Liu, S. (2009) Screening and structural characterization of alpha-glucosidase inhibitors from hawthorn leaf flavonoids extract by ultrafiltration LC-DAD-MS(n) and SORI-CID FTICR MS. *J. Am. Soc. Mass Spectrom.*, **20** (8), 1496–1503.

142. Ionita, I.A. and Akhlaghi, F. (2010) Quantification of unbound prednisolone, prednisone, cortisol and cortisone in human plasma by ultrafiltration and direct injection into liquid chromatography tandem mass spectrometry. *Ann. Clin. Biochem.*, **47** (Pt. 4), 350–357.

143. Mulabagal, V. and Calderón, A.I. (2010) Development of an ultrafiltration-liquid chromatography/mass spectrometry (UF-LC/MS) based ligand-binding assay and an LC/MS based functional assay for Mycobacterium tuberculosis shikimate kinase. *Anal. Chem.*, **82** (9), 3616–3621.

144. Mulabagal, V. and Calderón, A.I. (2010) Development of binding assays to screen ligands for Plasmodium falciparum thioredoxin and glutathione reductases by ultrafiltration and liquid chromatography/mass spectrometry. *J. Chromatogr. B Analyt. Technol. Biomed. Life Sci.*, **878** (13-14), 987–993.

145. Comess, K.M., Trumbull, J.D., Park, C., Chen, Z., Judge, R.A., Voorbach, M.J., Coen, M., Gao, L., Tang, H., Kovar, P., Cheng, X., Schurdak, M.E., Zhang, H., Sowin, T., and Burns, D.J. (2006) Kinase drug discovery by affinity selection/mass spectrometry (ASMS): application to DNA damage checkpoint kinase Chk1. *J. Biomol. Screen.*, **11** (7), 755–764.

146. Jiang, Y. and Lee, C.S. (2001) On-line coupling of hollow fiber membranes with electrospray ionization mass spectrometry for continuous affinity selection, concentration and identification of small-molecule libraries. *J. Mass Spectrom.*, **36** (6), 664–669.

147. van Breemen, R.B., Huang, C.R., Nikolic, D., Woodbury, C.P., Zhao, Y.Z., and Venton, D.L. (1997) Pulsed ultrafiltration mass spectrometry: a new method for screening combinatorial libraries. *Anal. Chem.*, **69** (11), 2159–2164.

148. Geun Shin, Y., Bolton, J.L., and van Breemen, R.B. (2002) Screening drugs for metabolic stability using pulsed ultrafiltration mass spectrometry. *Comb. Chem. High Throughput Screen.*, **5**, 59–64.

149. Cheng, X. and van Breemen, R.B. (2005) Mass spectrometry-based screening for inhibitors of beta-amyloid protein aggregation. *Anal. Chem.*, **77** (21), 7012–7015.

150. Liu, D., Guo, J., Luo, Y., Broderick, D.J., Schimerlik, M.I., Pezzuto, J.M., and van Breemen, R.B. (2007) Screening for ligands of human retinoid X receptor-alpha using ultrafiltration mass spectrometry. *Anal. Chem.*, **79** (24), 9398–9402.

151. Cao, H., Yu, R., Choi, Y., Ma, Z.-Z., Zhang, H., Xiang, W., Lee, D.Y.-W., Berman, B.M., Moudgil, K.D., Fong, H.H.S., and van Breemen, R.B. (2010) Discovery of cyclooxygenase inhibitors from medicinal plants used to treat inflammation. *Pharmacol. Res.*, **61** (6), 519–524.

152. Muckenschnabel, I., Falchetto, R., Mayr, L., and Filipuzzi, I. (2004) SpeedScreen: label-free liquid chromatography–mass spectrometry-based high-throughput screening for the discovery of orphan protein ligands. *Anal. Biochem.*, **324** (2), 241–249.

153. Brown, N., Zehender, H., Azzaoui, K., Schuffenhauer, A., Mayr, L.M., and Jacoby, E. (2006) A chemoinformatics analysis of hit lists obtained from high-throughput affinity-selection screening. *J. Biomol. Screen.*, **11** (2), 123–130.

154. Annis, D.A., Athanasopoulos, J., Curran, P.J., Felsch, J.S., Kalghatgi, K., Lee, W.H., Nash, H.M., Orminati, J.-P.A., Rosner, K.E., Shipps, G.W., Thaddupathy, G.R.A., Tyler, A.N., Vilenchik, L., Wagner, C.R., and Wintner, E.A. (2004) An affinity selection–mass spectrometry method for the identification of small molecule ligands from self-encoded combinatorial libraries. *Int. J. Mass Spectrom.*, **238** (2), 77–83.

155. Annis, D.A., Cheng, C.C., Chuang, C.C., McCarter, J.D., Nash, H.M., Nazef, N., Rowe, T., Kurzeja, R.J., and Shipps, G.W. Jr., (2009) Inhibitors of the lipid phosphatase SHIP2 discovered by high-throughput affinity selection-mass spectrometry screening of combinatorial libraries. *Comb. Chem. High Throughput Screen.*, **12**, 2760–2771.

156. Whitehurst, C.E., Nazef, N., Annis, D.A., Hou, Y., Murphy, D.M., Spacciapoli, P., Yao, Z., Ziebell, M.R., Cheng, C.C., Shipps, G.W., Felsch, J.S., Lau, D., and Nash, H.M. (2006) Discovery and characterization of orthosteric and allosteric muscarinic M2 acetylcholine receptor ligands by affinity selection-mass spectrometry. *J. Biomol. Screen.*, **11** (2), 194–207.

157. Whitehurst, C.E. and Annis, D.A. (2008) Affinity selection-mass spectrometry and its emerging application to the high throughput screening of G protein-coupled receptors. *Comb. Chem. High Throughput Screen.*, **11**, 427–438.

158. Flarakos, J., Morand, K.L., and Vouros, P. (2005) High-throughput solution-based medicinal library screening against human serum albumin. *Anal. Chem.*, **77** (5), 1345–1353.

159. Ross, P.L., Huang, Y.N., Marchese, J.N., Williamson, B., Parker, K., Hattan, S., Khainovski, N., Pillai, S., Dey, S., Daniels, S., Purkayastha, S., Juhasz, P., Martin, S., Bartlet-Jones, M., He, F., Jacobson, A., and Pappin, D.J. (2004) Multiplexed protein quantitation in Saccharomyces cerevisiae using amine-reactive isobaric tagging reagents. *Mol. Cell. Proteomics*, **3** (12), 1154–1169.

160. Bantscheff, M., Eberhard, D., Abraham, Y., Bastuck, S., Boesche, M., Hobson, S., Mathieson, T., Perrin, J., Raida, M., Rau, C., Reader, V., Sweetman, G., Bauer, A., Bouwmeester, T., Hopf, C., Kruse, U., Neubauer, G., Ramsden, N., Rick, J., Kuster, B., and Drewes, G. (2007) Quantitative chemical proteomics reveals mechanisms of action of clinical ABL kinase inhibitors. *Nat. Biotechnol.*, **25** (9), 1035–1044.

161. Bantscheff, M., Hopf, C., Kruse, U., and Drewes, G. (2007) Proteomics-based strategies in kinase drug discovery. *Ernst Schering Found. Symp. Proc.*, **3**, 1–28.

162. Kool, J., Jonker, N., Irth, H., and Niessen, W.M.A. (2011) Studying protein-protein affinity and immobilized ligand-protein affinity interactions using MS-based methods. *Anal. Bioanal. Chem.*, **401** (4), 1109–1125.

163. Suzuki, G., Takigami, H., Watanabe, M., Takahashi, S., Nose, K., Asari, M., and Sakai, S.-I. (2008) Identification of brominated and chlorinated phenols as potential thyroid-disrupting compounds in indoor dusts. *Environ. Sci. Technol.*, **42** (5), 1794–1800.

164. Vrabie, C.M., Sinnige, T.L., Murk, A.J., and Jonker, M.T.O. (2012) Effect-directed assessment of the bioaccumulation potential and chemical nature of Ah receptor agonists in crude and refined oils. *Environ. Sci. Technol.*, **46** (3), 1572–1580.

165. Meinert, C., Schymanski, E., Küster, E., Kühne, R., Schüürmann, G., and Brack, W. (2010) Application of preparative capillary gas chromatography (pcGC), automated structure generation and mutagenicity prediction to improve effect-directed analysis of genotoxicants in a contaminated groundwater. *Environ. Sci. Pollut. Res. Int.*, **17** (4), 885–897.

166. Fabel, S., Niessner, R., and Weller, M.G. (2005) Effect-directed analysis by high-performance liquid chromatography with gas-segmented enzyme inhibition. *J. Chromatogr. A*, **1099** (1-2), 103–110.

167. Stolper, P., Fabel, S., Weller, M.G., Knopp, D., and Niessner, R. (2008) Whole-cell luminescence-based flow-through biodetector for toxicity testing. *Anal. Bioanal. Chem.*, **390** (4), 1181–1187.

168. Giera, M., Heus, F., Janssen, L., Kool, J., Lingeman, H., and Irth, H. (2009) Microfractionation revisited: a 1536 well high resolution screening assay. *Anal. Chem.*, **81** (13), 5460–5466.

169. Kool, J., Rudebeck, A.F., Fleurbaaij, F., Nijmeijer, S., Falck, D., Smits, R.A., Vischer, H.F., Leurs, R., and

Niessen, W.M.A. (2012) High-resolution metabolic profiling towards G protein-coupled receptors: rapid and comprehensive screening of histamine H$_4$ receptor ligands. *J. Chromatogr. A*, **1259**, 213–220.

170. Nijmeijer, S., Vischer, H.F., Rudebeck, A.F., Fleurbaaij, F., Falck, D., Leurs, R., Niessen, W.M.A., and Kool, J. (2012) Development of a profiling strategy for metabolic mixtures by combining chromatography and mass spectrometry with cell-based GPCR signaling. *J. Biomol. Screen.*, **17** (10), 1329–1338.

171. Niederländer, H.A.G., van Beek, T.A., Bartasiute, A., and Koleva, I.I. (2008) Antioxidant activity assays on-line with liquid chromatography. *J. Chromatogr. A*, **1210** (2), 121–134.

172. Sentandreu, M. and Toldra, F. (2006) A rapid, simple and sensitive fluorescence method for the assay of angiotensin-I converting enzyme. *Food Chem.*, **97** (3), 546–554.

173. David Falck, H.I., de Vlieger, J.S.B., Niessen, W.M.A., Kool, J., Honing, M., and Giera, M. (2010) Development of an online p38α mitogen-activated protein kinase binding assay. *Anal. Bioanal. Chem.*, **398**, 1771–1780.

174. Falck, D., de Vlieger, J.S.B., Giera, M., Honing, M., Irth, H., Niessen, W.M.A., and Kool, J. (2012) On-line electrochemistry-bioaffinity screening with parallel HR-LC-MS for the generation and characterization of modified p38α kinase inhibitors. *Anal. Bioanal. Chem.*, **403** (2), 367–375.

175. Hernández, F., Sancho, J.V., Ibáñez, M., Abad, E., Portolés, T., and Mattioli, L. (2012) Current use of high-resolution mass spectrometry in the environmental sciences. *Anal. Bioanal. Chem.*, **403** (5), 1251–1264.

176. Kind, T. and Fiehn, O. (2010) Advances in structure elucidation of small molecules using mass spectrometry. *Bioanal. Rev.*, **2** (1-4), 23–60.

177. Krauss, M., Singer, H., and Hollender, J. (2010) LC-high resolution MS in environmental analysis: from target screening to the identification of unknowns. *Anal. Bioanal. Chem.*, **397** (3), 943–951.

178. Radović, J.R., Thomas, K.V., Parastar, H., Díez, S., Tauler, R., and Bayona, J.M. (2014) Chemometrics-assisted effect-directed analysis of crude and refined oil using comprehensive two-dimensional gas chromatography-time-of-flight mass spectrometry. *Environ. Sci. Technol.*, **48** (5), 3074–3083.

179. Schymanski, E.L., Meringer, M., and Brack, W. (2011) Automated strategies to identify compounds on the basis of GC/EI-MS and calculated properties. *Anal. Chem.*, **83**, 903–912.

180. Hug, C., Ulrich, N., Schulze, T., Brack, W., and Krauss, M. (2014) Identification of novel micropollutants in wastewater by a combination of suspect and non-target screening. *Environ. Pollut.*, **184**, 25–32.

181. Schymanski, E.L., Singer, H.P., Longrée, P., Loos, M., Ruff, M., Stravs, M.A., Ripollés Vidal, C., and Hollender, J. (2014) Strategies to characterize polar organic contamination in wastewater: exploring the capability of high resolution mass spectrometry. *Environ. Sci. Technol.*, **48** (3), 1811–1818.

182. Thomas, K.V., Langford, K., Petersen, K., Smith, A.J., and Tollefsen, K.E. (2009) Effect-directed identification of naphthenic acids as important in vitro xeno-estrogens and anti-androgens in North sea offshore produced water discharges. *Environ. Sci. Technol.*, **43** (21), 8066–8071.

183. Makarov, A. and Scigelova, M. (2010) Coupling liquid chromatography to Orbitrap mass spectrometry. *J. Chromatogr. A*, **1217** (25), 3938–3945.

184. Guilhaus, M. (1995) Principles and instrumentation in time-of-flight mass spectrometry physical and instrumental concepts. *J. Mass Spectrom.*, **30**, 1519–1532.

185. Bataineh, M., Lübcke-von Varel, U., Hayen, H., and Brack, W. (2010) HPLC/APCI-FTICR-MS as a tool for identification of partial polar mutagenic compounds in effect-directed analysis. *J. Am. Soc. Mass Spectrom.*, **21** (6), 1016–1027.

186. Kool, J., van Marle, A., Hulscher, S., Selman, M., van Iperen, D.J., van Altena, K., Gillard, M., Bakker, R.A.,

Irth, H., Leurs, R., and Vermeulen, N.P.E. (2007) A flow-through fluorescence polarization detection system for measuring GPCR-mediated modulation of cAMP production. *J. Biomol. Screen.*, **12** (8), 1074–1083.

187. Gómez, M.J., Gómez-Ramos, M.M., Malato, O., Mezcua, M., and Férnandez-Alba, A.R. (2010) Rapid automated screening, identification and quantification of organic micro-contaminants and their main transformation products in wastewater and river waters using liquid chromatography-quadrupole-time-of-flight mass spectrometry with an accurate-mass. *J. Chromatogr. A*, **1217** (45), 7038–7054.

188. Tengstrand, E., Rosén, J., Hellenäs, K.-E., and Aberg, K.M. (2013) A concept study on non-targeted screening for chemical contaminants in food using liquid chromatography-mass spectrometry in combination with a metabolomics approach. *Anal. Bioanal. Chem.*, **405** (4), 1237–1243.

189. Gallampois, C.M.J., Schymanski, E.L., Bataineh, M., Buchinger, S., Krauss, M., Reifferscheid, G., and Brack, W. (2013) Integrated biological-chemical approach for the isolation and selection of polyaromatic mutagens in surface waters. *Anal. Bioanal. Chem.*, **405** (28), 9101–9112.

190. Díaz, R., Ibáñez, M., Sancho, J.V., and Hernández, F. (2012) Target and non-target screening strategies for organic contaminants, residues and illicit substances in food, environmental and human biological samples by UHPLC-QTOF-MS. *Anal. Methods*, **4** (1), 196.

191. Zedda, M. and Zwiener, C. (2012) Is nontarget screening of emerging contaminants by LC-HRMS successful? A plea for compound libraries and computer tools. *Anal. Bioanal. Chem.*, **403** (9), 2493–2502.

192. Pluskal, T., Castillo, S., Villar-Briones, A., and Oresic, M. (2010) MZmine 2: modular framework for processing, visualizing, and analyzing mass spectrometry-based molecular profile data. *BMC Bioinf.*, **11**, 395.

193. Lommen, A. (2009) MetAlign: interface-driven, versatile metabolomics tool for hyphenated full-scan mass spectrometry data preprocessing. *Anal. Chem.*, **81** (8), 3079–3086.

194. Loos, M. (2014) R Package Nontarget: Detecting, Combining and Filtering Isotope, Adduct and Homologue Series Relations in High-Resolution Mass Spectrometry (HRMS) Data, *http://www.eawag.ch/forschung/uchem/software/R_package_start* (accessed 10 June 2014).

195. Böcker, S., Letzel, M.C., Lipták, Z., and Pervukhin, A. (2009) SIRIUS: decomposing isotope patterns for metabolite identification. *Bioinformatics*, **25** (2), 218–224.

196. Kind, T. and Fiehn, O. (2007) Seven Golden Rules for heuristic filtering of molecular formulas obtained by accurate mass spectrometry. *BMC Bioinf.*, **8**, 105.

197. Horai, H., Arita, M., Kanaya, S., Nihei, Y., Ikeda, T., Suwa, K., Ojima, Y., Tanaka, K., Tanaka, S., Aoshima, K., Oda, Y., Kakazu, Y., Kusano, M., Tohge, T., Matsuda, F., Sawada, Y., Hirai, M.Y., Nakanishi, H., Ikeda, K., Akimoto, N., Maoka, T., Takahashi, H., Ara, T., Sakurai, N., Suzuki, H., Shibata, D., Neumann, S., Iida, T., Tanaka, K., Funatsu, K., Matsuura, F., Soga, T., Taguchi, R., Saito, K., and Nishioka, T. (2010) MassBank: a public repository for sharing mass spectral data for life sciences. *J. Mass Spectrom.*, **45** (7), 703–714.

198. Smith, C.A., O'Maille, G., Want, E.J., Qin, C., Trauger, S.A., Brandon, T.R., Custodio, D.E., Abagyan, R., and Siuzdak, G. (2005) METLIN: a metabolite mass spectral database. *Ther. Drug Monit.*, **27** (6), 747–751.

199. Wolf, S., Schmidt, S., Müller-Hannemann, M., and Neumann, S. (2010) In silico fragmentation for computer assisted identification of metabolite mass spectra. *BMC Bioinf.*, **11**, 148.

200. Schymanski, E.L., Gallampois, C.M.J., Krauss, M., Meringer, M., Neumann, S., Schulze, T., Wolf, S., and Brack, W. (2012) Consensus structure elucidation

combining GC/EI-MS, structure generation, and calculated properties. *Anal. Chem.*, **84** (7), 3287–3295.

201. Gao, J., Ellis, L.B.M., and Wackett, L.P. (2010) The University of Minnesota Biocatalysis/Biodegradation Database: improving public access. *Nucleic Acids Res.*, **38** (Database issue), D488–D491.

202. Helbling, D.E., Hollender, J., Kohler, H.-P.E., Singer, H., and Fenner, K. (2010) High-throughput identification of microbial transformation products of organic micropollutants. *Environ. Sci. Technol.*, **44** (17), 6621–6627.

203. Kern, S., Fenner, K., Singer, H.P., Schwarzenbach, R.P., and Hollender, J. (2009) Identification of transformation products of organic contaminants in natural waters by computer-aided prediction and high-resolution mass spectrometry. *Environ. Sci. Technol.*, **43** (18), 7039–7046.

204. Nurmi, J., Pellinen, J., and Rantalainen, A.-L. (2012) Critical evaluation of screening techniques for emerging environmental contaminants based on accurate mass measurements with time-of-flight mass spectrometry. *J. Mass Spectrom.*, **47** (3), 303–312.

205. Galassi, S. and Benfenati, E. (2000) Fractionation and toxicity evaluation of waste waters. *J. Chromatogr. A*, **889** (1-2), 149–154.

5
MS Binding Assays

Georg Höfner and Klaus T. Wanner

5.1
Introduction

Binding to a defined target is the indispensable prerequisite for endogenous messengers as well as for drugs to exert physiologically or pharmacologically relevant effects. Techniques enabling the estimation of a compound's affinity toward a target of interest are therefore indispensable tools in basic research and even more so in the drug discovery process [1–4]. A huge fund of assay techniques is available for this purpose today, each with its own specific strengths and weaknesses in respect to the subject as well as to the investigated target [1, 3–9]. The number of techniques suitable for integral membrane proteins such as GPCRs (G protein-coupled receptors), ion channels, and transporters, which account for at least two-thirds of the proposed and existing protein drug targets [10], is distinctly limited. Radioligand binding assays developed for this purpose, that is, determination of a compound's affinity toward a target of low natural abundance (subnanomolar concentrations in native target sources), have found widespread use since the mid 1970s [9, 11, 12]. They are characterized by the use of a radioligand (i.e., a ligand labeled with a radioisotope; e.g., ^{3}H or ^{125}I) binding to the desired target with high affinity and selectivity. Following this strategy, the amount of radioligand bound to a target can be easily measured with outstanding sensitivity by means of liquid scintillation or gamma counting. While the use of radioligands has many advantages, it also has considerable inherent drawbacks. These are the result of legal restrictions concerning the handling of radioactive substances, security measures that have to be taken, high costs, the limited availability of radioligands, and, last but not least, the costly and difficult disposure of the radioactive waste produced.

Fluorescence- and luminescence-based ligand binding techniques do not have these disadvantages and offer excellent options regarding miniaturization and throughput [1, 9, 13–17]. Although they also require a label, which due to the bulkiness of fluorophores can cause severe losses of affinity, fluorescent ligands have largely replaced radioligands in binding assays addressing soluble targets [1]. Fluorescence- and luminescence-based binding assays are feasible for integral membrane proteins as well [17–21]. As they often have to be

Analyzing Biomolecular Interactions by Mass Spectrometry, First Edition.
Edited by Jeroen Kool and Wilfried M.A. Niessen.
© 2015 Wiley-VCH Verlag GmbH & Co. KGaA. Published 2015 by Wiley-VCH Verlag GmbH & Co. KGaA.

performed in solution or demand custom-made modification of the target they have not replaced radioligand binding assays as the most popular assays for membrane-bound targets [1].

The fact that radioligands were an indispensable prerequisite in binding assays prompted the development of a variety of label-free approaches. These are mainly based on optical or acoustic biosensing, nuclear magnetic resonance (NMR) spectroscopy, calorimetry, or mass spectrometry (MS) as readout systems [1, 5–8, 22]. Of these, particularly MS is highly attractive as a readout. It is highly versatile, sensitive, and, at the same time, capable of providing structural information on detected binders [5, 23]. With the increasing performance of MS and the access to gentle but at the same time effective ionization techniques such as matrix-assisted laser desorption ionization (MALDI) and electrospray ionization (ESI), the first MS-based methods to measure target–ligand interactions were described at the beginning of the 1990s [24, 25]. In the following decades, MS proved to be an efficient tool in a variety of different approaches monitoring binding of native ligands to defined targets, either directly at the level of the target–ligand complexes or at the level of the ligands interacting with the targets [22, 23, 26–29]. Many examples exist that demonstrate the feasibility of homogenous assay formats coupled with direct measurement of target–ligand-complexes by means of ESI-MS analysis [23, 30–33]. Apart from a few exceptions [34, 35] this approach has so far not been applied to integral membrane proteins. Furthermore, the reliability of affinity data (IC_{50} or K_i) resulting from this direct approach in which noncovalent target–ligand complexes are analyzed, is still a matter of debate, as the method suffers from poor reproducibility and a vast number of possible artifacts [36].

Strategies employing MS at the level of ligands interacting with the target of interest require a more complex setup as they require a separation step (e.g., by filtration) or at least a discrimination (e.g., by frontal chromatography) of bound and nonbound ligands. Methods utilizing frontal affinity chromatography [37, 38], ultrafiltration [39, 40], or size exclusion chromatography [41–43] have gained increasing popularity. They are, however, highly sophisticated and therefore still reserved to specialized academic research groups or the screening facilities of Big Pharma, particularly in assays aiming at the characterization of ligand binding to membrane proteins.

Although the pursuit of MS-based approaches proved to be highly effective in a few screening campaigns, implementation of eligible concepts still requires considerable efforts and skilled specialists. Considering this situation, we came to the conclusion that replacing radiometry as readout in radioligand binding assays by MS quantitation might be particularly attractive as such an approach could benefit from the simple working principle of radioligand binding, a specific strength of this method, while at the same time avoiding the necessity to use radioactivity by making use of the capability of MS for the quantitation of nonlabeled compounds. In this approach, the assay setup of radioligand binding assays could be retained, whereas for the quantitation of the now unlabeled reporter ligand – in this context termed *native marker*, *MS marker*, or simply *marker* – MS could be employed. To give this strategy a concise name, the corresponding assays were termed *MS Binding Assays* [44–46].

In the following sections the concept of MS Binding Assays will be further explained and an overview of already implemented assays presented. It will be demonstrated that MS Binding Assays share the advantages of radioligand binding assays, such as their simple working principle, their robustness, their lack of complex target preparations or modifications, their flexibility, and their universal applicability, while avoiding all disadvantages inherent in the use of radioisotopic labeled ligands.

5.2
MS Binding Assays – Strategy

5.2.1
Analogies and Differences Compared to Radioligand Binding Assays

MS Binding Assays and radioligand binding assays addressing membrane-bound proteins largely share the same workflow. Both comprise the following steps: incubation of the target with a reporter ligand, separation of bound from nonbound reporter ligand, quantitation of bound reporter ligand, and analysis of the resulting data. As the basic requirements in MS Binding Assays are essentially the same as in radioligand binding assays, these steps of assay development can be established with a minimum of effort (see Figure 5.1), independent of which of the three different types of experiments, that is, whether saturation, competition, or kinetic experiments, are performed.

For the incubation step in MS Binding Assays all relevant parameters such as target source, incubation conditions, assay format (tubes or microtiter plates), and determination of nonspecific binding can be adapted or deduced from radioligand binding experiments. The nonlabeled "native marker" to be employed in an MS Binding Assay can often be selected from an extensive set of known ligands. Just as the radioligand, the MS marker should have, first and foremost, a high affinity and selectivity for the target together with a low tendency for nonspecific binding. Second, quantitation of the marker by means of MS should be achievable with as high a sensitivity as possible. High-performance liquid chromatography–mass spectrometry (LC–MS) equipment can be assumed to be in general sufficiently sensitive for this task. Separation of bound from nonbound marker by means of filtration (including washing steps) can be performed in the same way as in radioligand binding assays using filter plates (typically glass fiber plates) in combination with a vacuum manifold. Whereas in radioligand binding assays, the radioligand can be directly quantified by measuring the radioactivity present on the filter, in MS Binding Assays the target-bound MS marker remaining on the filter is not immediately suitable for quantitation by LC–MS.

It is therefore essential that the marker be completely liberated and separated from the target. This is achieved best by eluting the filter with a denaturing solvent

Comparison radioligand versus MS binding assay

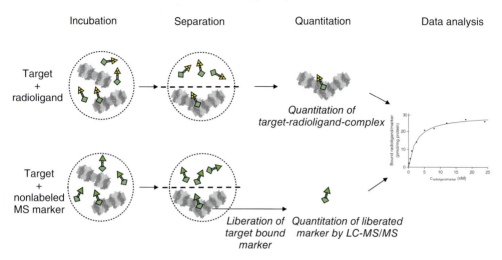

Figure 5.1 Scheme illustrating the analogies and differences of radioligand versus MS Binding Assays. The essential differences between both approaches are accentuated with italic text. Samples are incubated in both assay types typically in 96-well microtiter plates. Separation can be performed in both assay types by means of vacuum filtration over 96-well filter plates. For quantitation of the target-bound radioligand, radioactivity remaining after filtration and washing is measured whereas the target-bound marker remaining on the filter after filtration and washing is liberated in an additional elution step and subsequently quantified by LC–MS/MS. Analysis of binding data is completely identical in both cases.

(e.g., an organic solvent) before the marker is quantified by MS in its free form (Figure 5.1).

The concentration of bound marker that has to be quantified in an MS binding experiment carried out under equilibrium conditions in analogy to a radioligand binding experiment can be calculated from Equation 5.1.

$$K_d = \frac{[T] \cdot [L^*]}{[TL^*]} = \frac{[T] \cdot [M]}{[TM]} \tag{5.1}$$

K_d represents the equilibrium dissociation constant and T the free target; L^* and M represent the free radioligand and the free marker, respectively; and TL^* and TM represent the target–radioligand complex and the target–marker complex, respectively.

For a realistic estimation of the concentration of TM and thus of the amount of bound marker to be quantified in MS Binding Assays, it is considered useful to look first at a typical radioligand binding saturation experiment that features a radioligand with a K_d of 3 nM, a M_r of 300, and two ^{3}H atoms (i.e., a specific activity of 2.1 TBq mmol^{-1}) and is performed in a microtiter plate with a total incubation volume of 250 μl. Following the rules generally accepted for the performance of radioligand binding assays, the radioligand is commonly employed in a concentration range that equals from 0.1 to 10 K_d. At the same time the target is used in a concentration distinctly smaller than K_d to ensure that formation of TL* does not make up more than 5–10% of the nominal radioligand concentration [9, 47–49] and ligand depletion is avoided. In the case of the lowest radioligand concentration employed (i.e., 300 pM) the substance amount of bound radioligand in the final sample (based on a volume of 250 µl) will therefore not exceed 7.5 fmol. Using liquid scintillation counting with an efficiency of about 40% (as it is typical for ^{3}H) a fairly respectable signal for 7.5 fmol bound radioligand (corresponding to 15.8 Bq or 945 dpm) of about 380 cpm can be expected. Since a more detailed discussion of radioligand binding assays is beyond the scope of the present review the interested reader may be referred to comprehensive reviews [9, 46–49]. In the MS Binding Assay the identical substance amount of bound marker has to be quantified in the volume, defined by the eluent required for marker liberation. Assuming a minimum volume of about 250 µl, which in general has been found sufficient for this purpose, the lowest concentration of the marker to be quantified by MS in the final sample can be estimated to be about 30 pM or 9 pg ml^{-1}.

The last step to be accomplished when performing MS Binding Assays, the processing of the obtained MS binding data, is again completely identical to the procedure followed in radioligand binding assays.

5.2.2
Fundamental Assay Considerations

Unlike radioligand binding assays that are restricted to a few commercially available radioligands addressing a target of interest, MS Binding Assays allow a much wider choice of markers. As K_d of the marker is inversely related to the analytical demand on the sensitivity of the MS measurement to be employed, the performance of the available LC–MS instrumentation should be taken into account when choosing the right marker. With respect to the required stability of the separated target–marker complex the situation is completely identical to radioligand binding assays. This means the dissociation rate constant (k_{-1}) of the target–marker complex should be so low that no significant losses of specifically bound marker will occur during the time span required for sufficient washing of the protein retained on the filter [48, 49]. The marker's K_d must therefore not be too high, as k_{-1} typically increases with increasing K_d values. As low target densities in the binding assay can also contribute to the analytical challenge of marker quantitation, appropriate target sources should be employed. Target densities as they are typically obtained when membrane proteins are expressed in mammalian cells are sufficient. Therefore, crude membrane fractions prepared from them (with >1 pmol target per milligram total protein [50]) are in most cases fully suitable.

With the liberation of the bound marker a matrix is generated that influences marker quantitation by MS. During the development of the MS quantitation method, all components that may contribute to the matrix sample, but could be more or less freely chosen, such as the ligand used in excess to determine non-specific binding, the incubation buffer, the employed filter material or reagents to block nonspecific binding to the filter material (e.g., polyethyleneimine) should therefore be taken into account as a source for possible matrix effects. Care should be taken that the solvent used for liberation of the target-bound marker from the filter plate is compatible with the established chromatography and enables injection volumes as large as possible. The total volume of the solvent used for liberation and elution of the marker also has a significant effect on the analytical demand of the quantitation method. It is generally more advantageous to use smaller volumes as this leads to higher marker concentrations in the final sample. This situation may be even further improved, by gently evaporating the eluates (containing the liberated marker) and reconstituting the residues in a milieu optimally suited for LC–MS quantitation.

5.2.3
Fundamental Analytical Considerations

Taking these prerequisites into account, several conclusions can be drawn about MS analytics. An LC–ESI-MS/MS system comprising a triple quadrupole mass spectrometer run in the selected-reaction monitoring (SRM) mode can be assumed to be the best option to achieve the desired sensitivity for marker quantitation in MS Binding Assays [51, 52].

Essentially two aspects should be considered regarding the optimal LC–MS configuration, namely, the sampling efficiency and the capacity of LC columns regarding the injection volume of the sample. Sampling efficiency describes the ratio of the rate of analyte ions entering the vacuum chamber of the mass spectrometer to the rate of analyte molecules entering the ion source dissolved in a liquid solution [53]. The sampling efficiencies of present ESI sources (supported by pneumatic nebulization) coupled to standard LC instruments are distinctly below 1%; nano-ESI sources, however, reach 85% sampling efficiency [54]. Fortunately, this disadvantage of standard LC–ESI-MS instrumentation can be compensated by the much higher sample volume that can be maximally exploited in a single injection into corresponding LC columns as the maximal injection volume is primarily limited by the column's inner diameter (i.e., the maximal sample volume increases with the square of the column inner diameter, [55]).

As nano-LC–ESI-MS/MS is considered to be highly sophisticated, bioanalytical quantitation issues aiming at extremely low quantitation limits are commonly addressed by means of standard LC instrumentation in combination with pneumatically supported ESI-MS/MS systems, if the sample amount is not severely limited [54]. When MS Binding Assays are processed in a 96-well format, the resulting samples typically contain the marker to be quantified in substance amounts down to a low femtomole level in a sample volume in the range of several hundred μl (see above). This means that the sample amount is not limited. A standard LC–ESI-MS/MS instrumentation with a column of about 2-mm inner diameter run with a mobile phase containing a high amount of organic solvent at a flow rate of several hundred microliter per minute can therefore be expected to be a good starting point for highly sensitive marker quantitation.

Owing to the high sensitivity of nano-LC–ESI-MS systems, it can be assumed that MS Binding Assays can be further miniaturized, as marker quantitation in the smaller sample volumes resulting from this can benefit from the extraordinary high sampling efficiency of the corresponding nano-ESI sources.

As ESI-MS can be dramatically affected by matrix effects [51, 56], quantitation according to the internal standard method will facilitate reliable quantitation and improve the robustness of the LC–MS method [52, 57]. Using a marker labeled with a stable isotope as the internal standard – if possible – has proven to be the best choice. Furthermore, chromatography should be fast to enable a reasonable sample throughput and should be robust to guarantee trouble-free routine application. Finally, it should be affected as little as possible by the matrix to avoid the need for any additional efforts for sample preparation.

5.3
Application of MS Binding Assays

5.3.1
MS Binding Assays for the GABA Transporter GAT1

5.3.1.1   GABA Transporters
γ-Aminobutyric acid (GABA) is the most important inhibitory neurotransmitter in the central nervous system. GABAergic neurotransmission is, directly or indirectly, associated with various neurological diseases such as epilepsy, Huntington's chorea, Morbus Parkinson, tardive dyskinesia, schizophrenia, anxiety, depression, and neuropathic pain. GABA transporters (GATs) are – next to GABA receptors ($GABA_A$ and $GABA_B$) – promising drug targets for the development of new drug candidates for basically all the indications listed above [58–60]. Four GAT subtypes, termed GAT1–3 and BGT1, are known. All are members of the Na^+/Cl^--dependent SLC6 (solute carrier family 6) neurotransmitter transporters. Of these transporters, GAT1 and GAT3 are found almost exclusively in the brain, whereas GAT2 and BGT1 are predominantly abundant in peripheral organs such as kidney and liver [61, 62]. Since the function of GAT1 and GAT3 in the brain is to transport GABA released into the synaptic cleft back into neurons or glia cells [59, 63], therapeutic effects of respective GAT inhibitors are based on the extension of the residence time of GABA in the synaptic cleft and thus on increasing its inhibitory effect. GAT1, the GAT subtype found most frequently in the brain, is already a validated target in the search for anticonvulsants. A successful example for a selective inhibitor of GAT1 is tiagabine, which has already been introduced in the therapy of epilepsy [64].

In vitro screening of potential GAT inhibitors is most frequently conducted by uptake assays that characterize the effect of a test compound on [^{3}H]GABA transport into whole cells, synaptosomes, or brain slices [63, 65]. Radioligand binding assays appear to be more advantageous for screening campaigns. In contrast to uptake assays, it is sufficient to use crude membrane fractions of the target material, which can be stored in a freezer and employed on demand. In fact, already in the late 1980 and 1990s, when tiagabine was developed, at Lundbeck a radioligand binding assay based on [³H]tiagabine was established to characterize the binding of this and other test compounds to GAT1 [66–69]. Unfortunately, neither [³H]tiagabine nor any other radioligand for GAT1 is currently commercially available.

With all the relevant data known from radioligand binding, the concept of MS bindings assays can be established in a most efficient manner as will be shown in the following section describing the development of an MS Binding Assay for GAT1 [45].

5.3.1.2 MS Binding Assays for GAT1 – General Setup

One of the central questions for the development of an MS Binding Assay is the selection of a suitable nonlabeled marker. In the present case this task was easy to accomplish, as next to tiagabine several other GAT1 inhibitors displaying high affinity are commercially available. From these, we chose NO 711 (Figure 5.2) as preliminary experiments indicated that this compound can be quantified in very low concentrations (lower limit of quantitation (LLOQ) ≤50 pM) by LC–ESI-MS/MS in the matrix resulting from the binding samples. An essential point was also that the deuterated analog of the marker, [²H₁₀] NO 711 (Figure 5.2), that should be employed as internal standard [45] appeared to be easily accessible.

As it was available to us from previous GAT1 uptake assays a HEK293 cell line stably expressing GAT1 (in this case of the mouse, mGAT1) was selected as a target source. From this appropriate membrane homogenates were prepared. The incubation conditions were adopted from the [³H]tiagabine radioligand binding

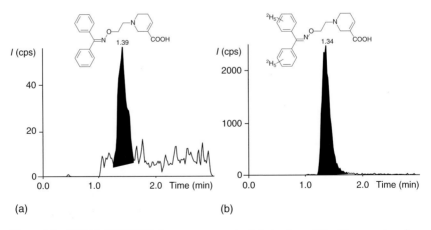

(a) (b)

Figure 5.2 SRM LC–ESI-MS/MS chromatogram of 15 pM NO 711 (351 → 180, *m/z*) and 1 nM [²H₁₀]NO 711 (361 → 190, *m/z*) in matrix; (a) NO 711 and (b) [²H₁₀]NO 711. LC: Purospher STAR RP 18 (55 mm × 2 mm; 3 μm); acetonitrile/methanol/10 mM ammonium formate buffer pH 7 (30 : 20 : 50, v/v/v) as mobile phase at a flow rate of 350 μl min⁻¹; injection volume 50 μl; effluent direct to waste at *t* < 1.0 min. (Adapted from Ref. [45].)

assays developed by Braestrup *et al.* [66] with an incubation buffer consisting of 50 mM Tris citrate and 1 M NaCl. As incubation and analytics are completely decoupled in this setup the high salt content caused no problems in LC–ESI-MS/MS quantitation. Binding assays based on incubation conditions established by Braestrup are highly hyperosmotic due to the high NaCl concentration of 1 M, which diminishes their significance. Therefore, NO 711 binding to mGAT1 was later also investigated at nearly physiological NaCl concentrations (see below). The scheme depicted in Figure 5.3 illustrates the general setup of the GAT1 MS Binding Assays thus established.

This assay is characterized by an incubation step in a conventional 96-well plate (total assay volume 300 μl), followed by filtration (250 μl of each sample, transfer by means of a multichannel pipette) over a 96-well glass fiber plate and subsequent washing of the residue remaining on the filter. After drying of the filter plate, liberation of NO 711 bound to the target is achieved by elution with methanol (added with a multichannel pipette). For elution and final generation of the samples that had now to be quantified by LC–ESI-MS/MS two different procedures were used. In the first procedure, a methanol solution of the internal standard with a defined concentration is employed to liberate the target-bound marker from the filter plate (3 × 100 μl). The eluates obtained are then supplemented with the buffer of the mobile phase used for chromatography containing the internal standard (100 μl). The advantage of this procedure is that elution with a methanolic solution of the internal standard can compensate even for very small volume differences of individual wells in the resulting eluates that may be generated during filtration

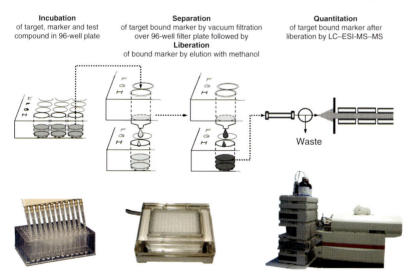

Incubation	**Separation**	**Quantitation**
of target, marker and test compound in 96-well plate	of target bound marker by vacuum filtration over 96-well filter plate followed by	of target bound marker after liberation by LC–ESI-MS–MS
	Liberation	
	of bound marker by elution with methanol	

Figure 5.3 Schematic flowchart for GAT1 MS Binding Assays illustrating incubation in a 96-well plate, separation of bound from nonbound marker, as well as subsequent liberation of bound marker in 96-well filter plates in combination with a vacuum manifold and quantitation of the formerly bound MS marker by LC–ESI-MS/MS. (Adapted from Ref. [45].)

under reduced pressure. In the second procedure, liberation is performed with pure methanol ($3 \times 100\,\mu$l). In this case an internal standard solution in methanol ($200\,\mu$l) is added to the eluate after liberation of the marker. The resulting solutions are then gently dried and the residues subsequently reconstituted in a milieu particularly suited for the following chromatography ($200\,\mu$l 10 mM ammonium formate buffer pH 7.0 and methanol, 95 : 5, v/v). The second procedure has the advantage that the final sample volume and solvent milieu is not dependent on the elution process and that analytical matrix standards are easier to prepare. The greatest advantage is that both strategies avoid elaborate sample preparation while still yielding highly reliable results. The LC–ESI-MS/MS method in combination with the latter liberation/reconstitution procedure has been successfully validated according to the CDER (Center for Drug Evaluation and Research) guidance for bioanalytical methods [70] regarding the parameters' selectivity, calibration, precision, and accuracy [71].

Saturation Assays Saturation assays have been performed complying with the guidelines generally accepted for radioligand binding assays based on the setup described above [9, 47–49]. NO 711 concentrations from 2 to 200 nM (i.e., about 0.1–10 K_d) were investigated, each in two triplicates – one for total binding (incubation in the absence of any other GAT ligand) and the other for the determination of nonspecific binding (incubation in the presence of 10 mM GABA). A representative saturation curve established from the data obtained is shown in Figure 5.4.

The K_d and B_{max} values calculated from the respective saturation curves amounted to 23.4 ± 2.19 nM ($n = 15$) and 34.6 ± 4.02 pmol mg^{-1} protein ($n = 14$, unless stated otherwise all the following results are given as means $\pm$ SEM). The reliability of the established MS Binding Assay was confirmed by means of analogously performed radioligand binding assays employing [^{3}H]NO 711

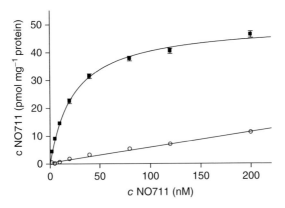

Figure 5.4 Representative saturation experiment characterizing NO 711 binding toward GAT1. Total binding (■) and nonspecific binding (in presence of 10 mM GABA) of NO 711 (○) to a membrane preparation of HEK293 cells stably expressing mGAT1 (means $\pm$ SEM from triplicates). (From Ref. [45].)

(synthesized in a custom labeling project). In $[^3H]$NO 711 saturation experiments performed under incubation conditions identical with those of the MS Binding Assays – with the conventional setup of radioligand binding assays – K_d and B_{max} values of 35.9 ± 1.81 nM ($n = 19$) and 26.0 ± 2.64 pmol mg^{-1} protein ($n = 22$), respectively, were found. Both values were in good accord with the results of the MS Binding Assays [45]. NO 711 binding at mGAT1 was also studied in incubation buffers closer to physiological conditions [72]. Saturation assays conducted in either 50 mM Tris citrate and 120 mM NaCl or Krebs buffer (25 mM Tris, 2.5 mM CaCl$_2$, 1.2 mM MgCl$_2$, 1.2 mM H$_2$KPO$_4$, 4.7 mM KCl, 119 mM NaCl, 11 mM glucose pH 7.2) with the setup identical to that described above yielded distinctly higher K_d values of 143 ± 10 nM ($n = 17$) and 137 ± 30 nM ($n = 4$), respectively. It is interesting to note that MS Binding Assays performed under identical conditions but terminated by centrifugation (in contrast to filtration over glass fiber plates) yielded a similar K_d value of 125 ± 12 nM ($n = 11$). This indicates that the washing procedure after filtration as part of the established setup of the MS Binding Assays does not lead to significant losses of specific marker binding, although the affinity of the MS marker is lowered due to the lower NaCl concentration accompanied by a higher dissociation rate constant (see below).

Kinetic Assays   The general setup of MS Binding Assays is equally applicable to kinetic investigations. As is common in association experiments, the target is incubated with a constant NO 711 concentration for different time spans, after which equilibration is terminated following the described setup [45]. Binding experiments have been performed in this way, for example, employing the original high NaCl incubation buffer (50 mM Tris citrate and 1 M NaCl). These yielded binding curves of which a representative example is shown in Figure 5.5a.

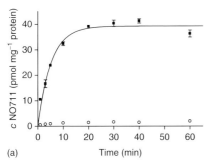
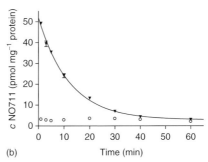
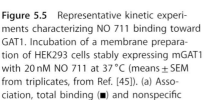

(a)   Time (min)

(b)   Time (min)

Figure 5.5   Representative kinetic experiments characterizing NO 711 binding toward GAT1. Incubation of a membrane preparation of HEK293 cells stably expressing mGAT1 with 20 nM NO 711 at 37 °C (means ± SEM from triplicates, from Ref. [45]). (a) Association, total binding (■) and nonspecific binding (in presence of 10 mM GABA) of NO 711 (○). (b) Dissociation (10 mM GABA was added after 60 min preincubation to initiate dissociation), total binding (▼) and nonspecific binding (in presence of 10 mM GABA) of NO 711 (○).

The observed association rate constant (k_{obs}) calculated from these experiments amounted to 0.0032 ± 0.0002 s^{-1}.

MS Binding Assays for the determination of the dissociation rate constant k_{-1} are also simple to perform [45]. After fully equilibrating a definite amount of NO 711 with mGAT1 (in the original incubation buffer, 50 mM Tris citrate and 1 M NaCl), dissociation is initiated by addition of an excess of GABA. The termination of the experiment by filtration after different periods of time and the subsequent marker quantitation again follows the general setup. A dissociation curve obtained in this way is shown in Figure 5.5b. The dissociation rate constant k_{-1} calculated from these experiments was found to be 0.0015 ± 0.0001 s^{-1}. From k_{-1}, k_{obs}, and the marker concentration employed in the association experiment an association rate constant k_{+1} of 0.15 ± 0.03 10^6 M^{-1} s^{-1} for NO 711 binding to mGAT1 has been derived. The K_d value calculated from the quotient of experimentally determined k_{-1} and k_{+1} (10 nM) was in good agreement with the K_d determined in saturation experiments (23.4 nM, see above) and also confirms the validity of the results derived from kinetic MS binding studies [45, 46, 72]. Corresponding kinetic MS Binding Assays have been performed in the low NaCl incubation buffers as well. As expected, incubation in 50 mM Tris citrate and 120 mM NaCl or Krebs buffer revealed distinctly higher dissociation rate constants (0.0041 ± 0.0004 and 0.0053 ± 0.0001 s^{-1}) as well as lower association rate constants ($0.042 \pm 0.001 \cdot 10^6$ and $0.078 \pm 0.002 \cdot 10^6$ M^{-1} s^{-1}), when compared to the results obtained from the experiments with high NaCl buffer (50 mM Tris citrate and 1 M NaCl) [72].

Competition Assays for mGAT1 The radioligand binding assays most frequently conducted by far in the drug screening process are competition experiments. To demonstrate the feasibility of MS Binding Assays for this purpose, GABA and a series of 21 GAT inhibitors with different lipophilicity and affinity to the target were tested [45]. The experiments were conducted according to the setup described above in the original high NaCl buffer (50 mM Tris citrate and 1 M NaCl) complying again with the guidelines generally accepted for radioligand binding assays [9, 47–49]. Increasing concentrations of the test compounds characteristically decreased NO 711 binding and yielded competition curves such as the example depicted in Figure 5.6.

With the IC$_{50}$ values obtained by nonlinear regression analysis affinity constants of the test compounds (K_i) were calculated using the Cheng–Prusoff equation [74]. A correlation of the results for the whole set of test compounds obtained in competitive MS Binding Assays with results obtained in conventional radioligand binding assays (based on [^{3}H]NO 711) is shown in Figure 5.7. The rise of the graph (1.010 ± 0.016, $R^2 > 0.99$) indicates that the pK_i values resulting from both binding assays are nearly identical, which clearly demonstrates the validity of the data obtained by MS Binding Assays.

Competitive MS Binding Assays were also conducted for a series of 20 GAT inhibitors in low NaCl Tris citrate buffer (Tris citrate and 120 mM NaCl) and Krebs buffer. The K_i values determined were compared to those obtained in assays based

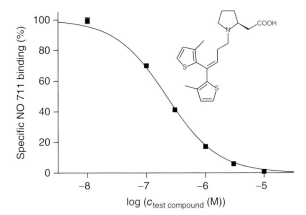

Figure 5.6 Representative competition experiment obtained for a test compound [73]. Data points represent specific binding of NO 711 employed in a nominal concentration of 10 nM (means ± SEM from triplicates). (Adapted from Ref. [46].)

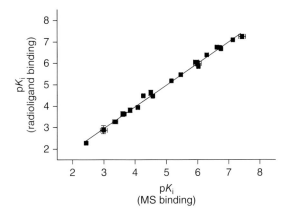

Figure 5.7 Correlation of pK_i values from competitive MS Binding Assays and radioligand binding assays (means ± SEM from triplicates). (From Ref. [46].)

on the original buffer containing 1 M NaCl [72] and indicate that the apparent affinity is noticeably dependent on the incubation milieu.

With respect to the relevance of competitive MS Binding Assays addressing GAT1 it is worth mentioning that this assay proved to be a valuable tool in routine compound testing campaigns in our laboratories. NO 711 binding assays addressing mGAT1 and hGAT1 are performed in a throughput of 10 000 – 20 000 samples per year. Large series of test compounds have been characterized that way although only a part of the results have been published so far [75 – 77].

5.3.1.3 MS Binding Assays for GAT1 – Speeding up Chromatography
LC–ESI-MS/MS quantitation of NO 711 under standard conditions (see Figure 5.1) takes about 3.5 min per sample. The largest part of this time span,

3 min, is the result of chromatography before MS analysis (0.5 min for injection). The established LC–MS method therefore requires considerable more time than liquid scintillation counting for a [³H] labeled radioligand, which typically requires about 1 min per sample. There are numerous options to shorten the time necessary for chromatography, the most time-consuming step of LC–ESI-MS/MS quantitation. Popular approaches are based on UPLC (ultra performance liquid chromatography), monolithic stationary phases, multiplexed electrospray interfaces (MUX) sources, or column switching techniques [52, 78–80]. We focused on a rather simple and straightforward option, that is, the reduction of column length in combination with increasing flow rates to keep additional efforts as small as possible [71]. Employing RP-18 columns with the dimensions of 20 mm × 2 mm and 10 mm × 2 mm (instead of our original 55 mm × 2 mm column) at flow rates of up to 1000 µl min⁻¹, retention times of 8–9 s for NO 711 and chromatographic cycle times of 18 s were enabled (see Figure 5.8).

Although under the fast chromatography condition the matrix from the binding samples leads to a distinctly higher suppression of the NO 711 signal, quantitation of the marker was still achieved with a sensitivity high enough to record even the lowest concentrated samples reliably. It is worth mentioning that the methods established employing columns of 10 and 20 mm length could also be successfully validated with respect to linearity, intra- and interbatch accuracy, and precision in a range from 50 pM to 5 nM according to the FDA (Food and Drug Administration) guideline for bioanalytical methods [70]. Employing a high-performance triple quadrupole mass spectrometer (ABSciex API 5000) allowed quantitation even down to an LLOQ of 20 pM [71]. Application of these short column methods to the quantification of NO 711 in mGAT1 saturation experiments revealed K_d and B_{max} values almost identical to those obtained in MS Binding Assays performed, with the LC being based on 50 mm standard columns [71].

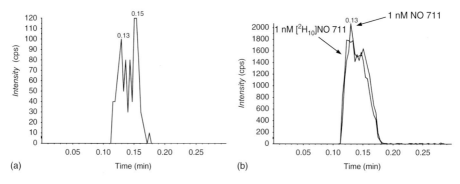

(a) (b)

Figure 5.8 SRM LC–ESI-MS/MS chromatograms of NO 711 (351 → 180, m/z) and [²H₁₀]NO 711 (361 → 190, m/z) in matrix standards; (a) 50 pM NO 711 in matrix, (b) 1 nM NO 711 and 1 nM [²H₁₀]NO 711 in matrix. LC: Luna C18 (10 mm × 2 mm; 3 µm); acetonitrile/10 mM ammonium formate buffer pH 7.0 (95/5, v/v) as mobile phase at a flow rate of 800 µl min⁻¹; injection volume 5 µl, effluent direct to waste at t < 0.1 min. (From Ref. [71].)

5.3.1.4 MS Binding Assays for GAT1 with MALDI-MS/MS for Quantitation

As outlined in the chapter, marker quantitation in MS Binding Assays by LC–ESI-MS/MS can be achieved with an expenditure of time markedly below 1 min per sample, the duration approximately necessary for liquid scintillation counting of [³H] labeled radioligands. An even faster option for marker quantitation by MS appeared to be feasible by completely omitting the chromatographic separation step. Automated infusion devices for nano-ESI-MS systems, which perform data acquisition within 10–30 s per sample [81], are certainly to be considered a promising alternative. The best approach to shorten the time required for quantitation was seen in the possibility to benefit from the unsurpassed speed of MALDI sources. MS systems employing MALDI sources are commonly equipped with TOF (time of flight) analyzers. TOF analyzers are, however, not the first choice when it comes to highly sensitive quantitation of analytes [82]. Furthermore, MALDI is rarely used for the quantification of small molecules as typical matrices often interfere with analyte signals. After a crucial breakthrough regarding sensitivity and throughput when combining a MALDI source with a triple quadrupole analyzer [83], these MALDI-MS/MS systems fortunately became commercially available under the brand name FlashQuant in 2007. Using the FlashQuant MS system quantitation of low molecular weight analytes in the SRM mode is possible within a few seconds per sample even in biological matrices. Compared to previously used MALDI-TOF combinations, this new type of instrument offers a distinct enhancement with respect to sensitivity, precision, and dynamic range [84–90]. This development prompted us to investigate the potential of the FlashQuant system for marker quantitation in MS Binding Assays, which was exemplarily studied with GAT1 MS Binding Assays [91]. Employing α-cyano-4-hydroxy cinnamic acid (CHCA) as MALDI matrix it was found that the most prominent mass transition of NO 711 (m/z 351 → 180) could be recorded almost without any matrix interferences. This was also the case when using [²H₁₀]NO 711 (monitored at m/z 361 → 154) the internal standard used for the MALDI-MS/MS quantitation of NO 711. NO 711 could therefore be reliably quantified in a range from 208 pM to 16.7 nM with the MALDI-MS/MS system (Figure 5.9).

The established MALDI-MS/MS method appeared to be well suited for the quantitation of NO 711 in saturation assays for mGAT1. For two series of identically performed saturation experiments ($n = 3$) – one quantified by means of FlashQuant and the other quantified by means of the original LC–ESI-MS/MS method – K_d and B_{max} values (19.5 ± 1.2 nM and 36.9 ± 6.8 pmol mg⁻¹ protein for MALDI-MS/MS and 21.4 ± 1.8 nM and 45.9 ± 9.0 pmol mg⁻¹ protein for LC–ESI-MS/MS) that were in excellent accord with each other were obtained. Competition experiments produced equally positive results. Two series of competition experiments for 13 test compounds were carried out: one quantified by means of FlashQuant and the other quantified by means of the original LC–ESI-MS/MS method [91]. Figure 5.10 illustrates the excellent correlation of the pK_i values that resulted from the MALDI-MS–MS-based study compared to these based on the LC–ESI-MS/MS quantitation method. The best evidence of this is the slope

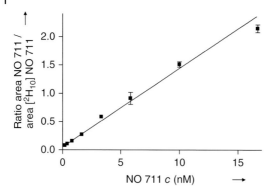

Figure 5.9 Calibration function ($y = 0.1387$ $x + 0.04549$; $R^2 = 0.984$) for NO 711 in the range from 0.208 nM to 16.7 nM obtained by MALDI-MS/MS quantitation of NO 711 standards in matrix. MALDI SRM (NO 711: $351 \rightarrow 180$, m/z; $[^2H_{10}]$NO 711: $361 \rightarrow 154$, m/z) data were collected on a FlashQuant work station with a laser raster speed across the MALDI plate of 1 mm s^{-1} and a dwell time of 20 ms (mean area ratios $\pm$ SD). (From Ref. [91].)

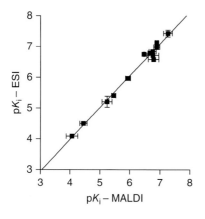

Figure 5.10 Correlation of pK_i values for a series of 12 GAT inhibitors and GABA as determined in competition experiments based on MALDI-MS/MS and LC–ESI-MS/MS quantitation (means $\pm$ SD, $n = 4$ for MALDI-MS/MS, $n = 3$ for LC–ESI-MS/MS). (From Ref. [91].)

of the correlation line depicted in Figure 5.10 that amounts to a value of 1.036 ($R^2 = 0.986$).

Marker quantitation based on FlashQuant for a MALDI target plate with 96 samples requires 22 min, which comes to 14 s as the mean time for the measurement of a single spot. In fact, data acquisition for a single spot is even shorter, taking only 1.7 s. The difference is due to the fact that the laser beam of the FlashQuant system moves with a constant speed over the whole MALDI target plate. The time the laser beam travels from spot to spot is significantly longer than the time for data acquisition. Nonetheless, this study illustrates the potential for MS Binding Assays to be carried out in a high-throughput manner when appropriate MS instruments are employed. It is finally worth mentioning as well that all steps of MS Binding Assays, beginning with incubation until the quantitation of the marker by being executed in microtiter plate formats, may easily be automated.

5.3.1.5   Library Screening by Means of MS Binding Assays

Hit identification represents an essential part in the modern drug discovery process and is often based on screening campaigns of large compound collections, referred to as libraries that either are generated in-house or originate from commercial sources. Competitive MS Binding Assays in the original setup (described in the Sections 5.3.1.2–5.3.1.4) have proved to be an effective substitute for radioligand binding assays for the affinity determination of single test compounds toward a target of interest. The concept of MS Binding Assays can be easily extended to the screening of libraries as well. Instead of single compounds, collections of several test compounds are incubated at a defined concentration together with the target and the marker. In this case, however, instead of recording a full competition curve only a measurement at a single test concentration is performed. Active libraries are detected by their inhibition of marker binding in relation to control samples without any inhibitor. The contained hits can then be identified by means of deconvolution experiments.

Dynamic combinatorial chemistry (DCC) represents a promising strategy for a highly efficient generation of libraries [92–94]. Suitable dynamic combinatorial libraries (DCLs) are built up from appropriate building blocks under equilibrium conditions allowing the formation of all possible library members in a thermodynamically controlled distribution. For the screening process, the library is generated once in the absence of the target (untemplated library) and once in its presence (templated library). In the presence of the target, acting as a template, the equilibrium of the formed compounds is shifted toward the best binders. They are identified by monitoring this shift induced by the target by means of a suitable readout. Until now a successful application of this concept could only be shown for a few targets [92, 95], mostly enzymes or carbohydrates, presumably due to their availability in pure form, their high solubility in an aqueous milieu, and, following from this, their applicability in quite high (i.e., micromolar) concentrations.

Membrane proteins, such as GAT1, unfortunately do not possess these favorable features and are poorly suited for such an application, as possible template effects resulting from the target present in the low nanomolar concentration range could hardly be detected by means of the analytical tools presently available. Nevertheless, it seemed worthwhile to consider the use of libraries generated by means of reactions known from DCC for drug screening even in those cases where the target is available only in low nanomolar concentrations as these reactions belong to the most fundamental and efficient transformations in organic chemistry. In the search of new GAT1 inhibitors an exploratory study was undertaken in which hydrazone libraries were generated and screened against GAT1 employing again MS Binding Assays as readout (see Figure 5.11).

Hydrazone libraries were generated by reacting various sets of aromatic aldehydes with a single hydrazine derivative delineated from piperidine-3-carboxylic acid (nipecotic acid), a common fragment of known *GAT1 inhibitors*. This was accomplished in the presence of GAT1 to ease the screening process as will be outlined below (see Figure 5.12). The hydrazine derivative was used in large excess to shift the equilibrium toward the product side. This ensures that a defined

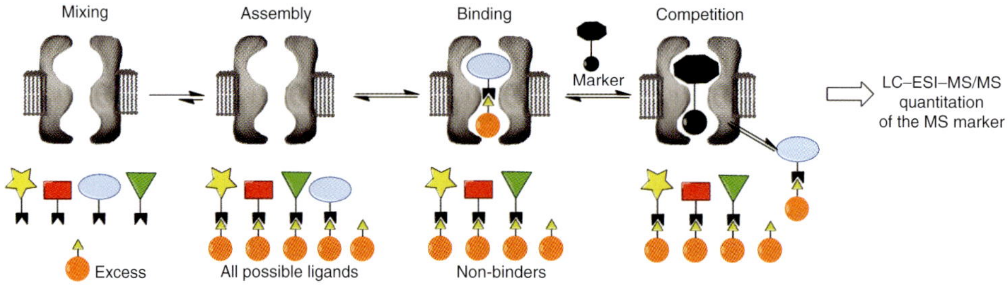

Mixing   Assembly   Binding   Competition

Marker

LC–ESI–MS/MS
quantitation
of the MS marker

Excess   All possible ligands   Non-binders

Figure 5.11 Scheme for generation of pseudo-static libraries by DCC and analysis of biological activity by competitive MS Binding Assays. At first pseudo-static libraries are generated by combining a diverse set of building blocks in equal amounts with a large excess of one compound of complementary chemical reactivity. After equilibration of the mixture a competitive GAT1 MS binding assay is performed by addition of NO 711. Finally, target-bound NO 711 is quantified to unravel the presence of hits. (From Ref. [96].)

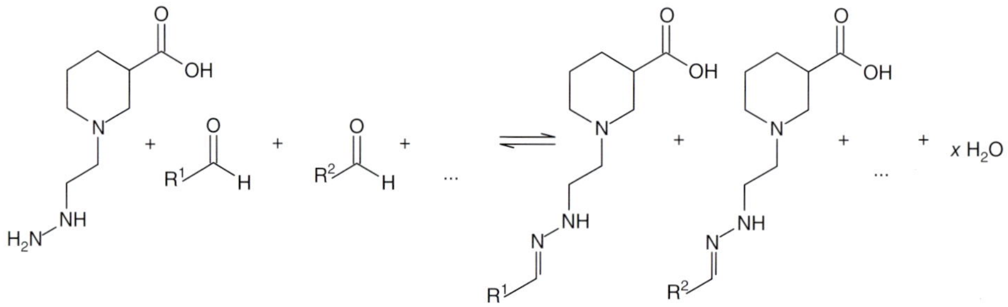

Figure 5.12 Condensation of nipecotic acid-derived hydrazine building block with diverse aldehydes to generate pseudo-static compound libraries. (From Ref. [96].)

concentration of the reaction products equal to the concentration of the starting aldehydes will be achieved. Accordingly, by employing the latter in identical concentrations, the reaction products must also form in equal amounts [96].

The libraries generated in this way can be referred to as *pseudostatic libraries* as their composition is largely fixed although they are still dynamic. The GAT1 MS binding assay was directly implemented into the process of library generation in terms of a one-pot approach. A particular benefit of this approach is that several working steps, which would otherwise have been required in a procedure comprising conventional synthesis and activity determination, can be avoided. When library formation and affinity determination are combined in a single-stage process no efforts regarding separation, isolation, purification, dissolution, and dosing of library components have to be made. However, such a strategy requires a buffer equally suited for library formation and library screening by means of MS Binding Assays. Since the original 50 mM Tris, 1 M NaCl buffer pH 7.1 used for GAT1 MS Binding Assays could possibly interfere with hydrazone formation and due to its limited buffer capacity it was substituted by 25 mM sodium phosphate, 1 M

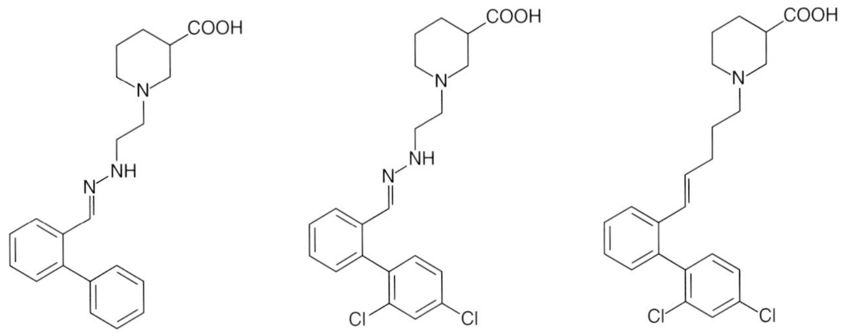

Figure 5.13 GAT1 inhibitor hits obtained by means of library screening. The substitution pattern of hit (1) obtained in the first library screening was optimized by screening-focused libraries leading to (2). (3) Carba analog of (2).

NaCl buffer pH 7.1. This buffer appeared to be a good compromise with respect to a reasonable speed for hydrazone formation and the intention to keep the binding characteristics of NO 711 toward mGAT1 nearly unchanged in comparison to the original conditions [96]. Under the conditions developed in this way, library formation in the presence of mGAT1 was allowed to proceed for 4 h, a period of time that had been found to be sufficient for an almost complete conversion of the educts into the products. The activity of the libraries are then indicated by MS competition experiments started by addition of NO 711 and performed in the same way as described above ("Competition Assays for mGAT1" section).

Nine libraries, each based on the same hydrazine derivative, and nine sets of four different aromatic aldehydes were generated this way [96]. After deconvolution of the libraries identified as most active, two potent binders with pK_i values in the nanomolar range could be identified [96]. Employing the same screening strategy, further attempts were made in a subsequent study to optimize one of the biaryl carrying hits ((1), Figure 5.13) as a lead structure [97]. When focused hydrazone libraries based on the same hydrazine derivative and diversely substituted biphenylcarbaldehydes were generated and analyzed, a highly potent binder ((2), Figure 5.13) with low nanomolar affinity was found. Since hydrazones are unsuitable for drug development, the stable carba analog ((3), Figure 5.13) was synthesized as a substitute for this hydrazone hit and could be shown to be a highly potent GAT1 binder as well.

5.3.2
MS Binding Assays for the Serotonin Transporter

5.3.2.1 Serotonin Transporter

The serotonin transporter (SERT) belongs – like the GATs – to the Na^+/Cl^--dependent SLC6 neurotransmitter transporter family. It is responsible for the reuptake of serotonin (5-hydroxytryptamine; 5-HT) into presynaptic neurons and glia cells, thus regulating 5-HT levels in the synaptic cleft in the brain. Representing the primary target in the treatment of emotional disorders such as depression, obsessive–compulsive disorder, and anxiety, it is of great pharmacological interest [98]. Although inhibitors increasing synaptic levels of 5-HT, such as nonselective tricyclic antidepressants (TCAs) and selective 5-HT reuptake inhibitors (SSRIs), are widespread in clinical use, the search for SERT inhibitors for the development of new drugs with improved therapeutic profiles is still of enormous pharmacological relevance. Screening for SERT inhibitors is typically based on radioligand binding assays, for example, employing [³H]citalopram, [³H]imipramine, or [³H]paroxetine [99–101]. However, MS Binding Assays developed for SERT as a label-free substitute for radioligand binding assays can be considered as a powerful and at the same time a more environment-friendly alternative.

5.3.2.2 MS Binding Assays for SERT – General Setup

Fluoxetine (Figure 5.14) was found well suited as a SERT MS marker due to its high affinity to the transporter and the fact that it can be quantified with satisfying sensitivity by MS [102]. A further bonus is that enantiomers of fluoxetine as well the deuterated analog of the racemate, which is ideally suited to serve as internal standard for MS quantitation, are commercially available. The procedure and the general performance of the MS Binding Assays follows the setup already described for GAT1 almost exactly (see Figure 5.3). A crude membrane preparation, this time obtained from HEK293 cells stably expressing the human serotonin transporter (hSERT), was used as a target source. An LC–ESI-MS/MS method with a sensitivity down to 50 pM (LLOQ) at an attractive speed and, again without the need for sample preparation, was employed for quantitation of fluoxetine (Figure 5.14).

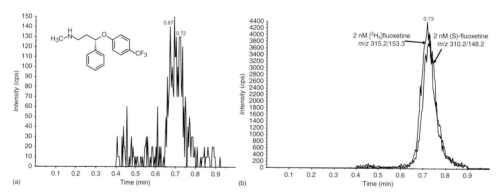

(a) (b)

Figure 5.14 SRM LC-ESI-MS/MS chromatograms of (S)-fluoxetine (310 → 148, *m/z*) and [²H₅]fluoxetine (315 → 153, *m/z*) in matrix standards; (a) 50 pM (S)-fluoxetine and (b) 2 nM (S)-fluoxetine and 2 nM [²H₅]fluoxetine. LC: Luna C18 (20 mm × 2 mm; 3 μm); acetonitrile/5 mM ammoniumbicarbonate buffer pH 9.5 (80/20, v/v) as mobile phase at a flow rate of 800 μl min⁻¹; injection volume 5 μl; effluent direct to waste at *t* < 0.4 min and *t* > 0.9 min. (Adapted from Ref. [102].)

Before its application to SERT MS Binding Assays, this method had been validated according to the CDER guidance for bioanalytical methods of the FDA [70].

Saturation Assays Although fluoxetine is one of most frequently prescribed
antidepressants worldwide, hardly any literature concerning radioligand binding
assays using a radiolabeled fluoxetine derivative such as [³H]fluoxetine as radioligand exists [103–107]. This may be due to the known radiochemical instability of
[³H]fluoxetine, which presumably is also the case with other analogs. It is likely
the reason why such ligands are not commercially available. It is in situations such
as this that the value of MS Binding Assays becomes apparent. They were found
to be also well applicable to the characterization of both fluoxetine enantiomers
in saturation experiments [102].

For the saturation experiments, quantitation of fluoxetine had to accomplished
even below the LLOQ of 50 pM. A lowering of the quantitation limit was achieved
by simply performing binding experiments at a larger scale (1000 µl assay volume,
900 µl employed for filtration) while leaving unchanged the volume for denaturing
and eluting the bound marker from the protein–marker complexes leading to the
final analytical sample (280 µl). By increasing the concentration by a factor of 3.2
in this way, it was possible to quantify fluoxetine in the binding assay (at its LLOQ
of 50 pM in the final sample) down to 15.6 pM.

Following this modification saturation binding experiments were performed
with (S)-fluoxetine (eutomer, 0.3–75 nM) or (R)-fluoxetine (distomer, 0.5–75 nM)
in the presence of a hSERT membrane preparation in 50 mM Tris–HCl, 120 mM
NaCl, 5 mM KCl pH 7.4 (a buffer known from SERT radioligand binding assays).
The results obtained in a representative experiment with (S)-fluoxetine are shown
in Figure 5.15.

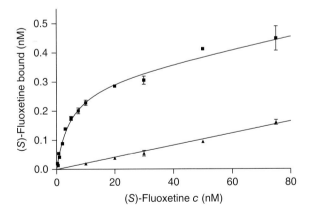

Figure 5.15 Representative saturation
experiment characterizing (S)-fluoxetine
binding toward SERT. Total binding (■) and
nonspecific binding (in presence of 10 µM
imipramine of (S)-fluoxetine, ▲) to a membrane preparation of HEK293 cells stably
expressing hSERT (means ± SD from triplicates). (From Ref. [102].)

When nonspecific binding data are not directly accessible by MS quantitation, as they are below the LLOQ, they can also be delineated from data for nonspecific binding that apply to higher concentrations as nominal marker concentrations and nonspecific binding are known to be linked in a linear manner. This approach was also used in the present case to determine nonspecific fluoxetine binding at low nominal concentrations. The data points for nonspecific binding defined as binding in the presence of 10 μM imipramine at nominal fluoxetine concentrations of <10 nM (the limit from which the concentration of nonspecifically bound fluoxetine was below 15.6 pM, the LLOQ in the binding sample) were therefore calculated by extrapolation from a linear regression curve based on experimental data obtained for nonspecific binding of fluoxetine at nominal concentrations of ≥10 nM. The K_d values obtained for both enantiomers from the saturation isotherms amounting to 4.4 ± 0.4 nM ($n = 13$, B_{max} 26.9 ± 5.7 pmol mg^{-1} protein) for (*S*)-fluoxetine and 5.2 ± 0.9 nM ($n = 5$, B_{max} 18.8 ± 3.8 pmol mg^{-1} protein) for (*R*)-fluoxetine [108] were in good agreement with the K_i values of 6.1 ± 0.3 and 7.7 ± 1.0 nM for (*S*)- and (*R*)-fluoxetine, respectively, reported by Koch *et al.* [101], again demonstrating the reliability of MS binding data.

Kinetic Assays   It is surprising that only a few studies investigating binding kinetics of SERT inhibitors are known so far. Fluoxetine was studied by Martin *et al.* [109] in an indirect approach, determining the dissociation rate constant in a competitive manner. The MS binding assay setup for SERT now allowed us to study (*S*)-fluoxetine kinetics at the target in a direct manner. In contrast to the procedure used for saturation experiments, it had to be taken into account that the sample volume must not be too large to allow termination of equilibration of target marker binding by transfer of aliquots from the incubation microtiter plate to the filter plate (by means of a 300 μl multichannel pipette) even after very short periods of time. The marker concentration also plays an important role as higher concentrations lead to faster association kinetics whereas lower concentrations enhance the analytical demand for bound marker quantitation. An assay volume of 300 μl and a concentration of 5 nM (*S*)-fluoxetine was found to be a good compromise for association experiments in this regard. Analysis of the resulting association curves yielded a k_{obs} value of 0.0078 ± 0.0009 s^{-1} ($n = 3$) and dissociation experiments performed with the same setup a k_{-1} of 0.0032 ± 0.0002 s^{-1} ($n = 3$). The k_{+1} value calculated from k_{obs} and k_{-1} amounted to 0.92 ± 0.17 10^6 M^{-1} s^{-1} for (*S*)-fluoxetine. Both rate constants, k_{+1} and $k_{-1,}$ are in very good accord with data published by Martin *et al.* [109]. The K_d value of 3.5 nM for (*S*)-fluoxetine calculated from these rate constants matched the K_d determined in saturation experiments almost perfectly as well.

Competition Assays   SERT inhibitors with a broad spectrum of potency were studied in competition experiments with (*S*)-fluoxetine performed at a concentration of 5 nM in a total assay volume of 300 μl. Competition curves, IC$_{50}$ values, and K_i values were obtained in this way for a series of 14 known SERT inhibitors [102]. For 10 of the investigated compounds a direct comparison of the K_i values with

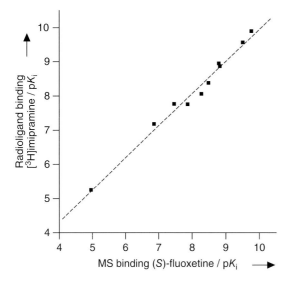

Figure 5.16 Correlation of pK_i values from competitive MS Binding Assays and radioligand binding assays (means, $n = 3$–6, from Ref. [108]). The straight line is characterized by $y = 0.9393x - 0.5527$, $R^2 = 0.9841$.

published data obtained from [³H]imipramine binding [100, 102] was possible. Figure 5.16 shows a correlation between pK_i values derived from the MS binding assay for SERT and those from radioligand binding assays with [³H]imipramine.

Although the affinities determined in both binding assays cannot be expected to be identical, as different markers ((S)-fluoxetine versus [³H]imipramine) and slightly different assay conditions are used to label SERT, the differences for the obtained K_i values are almost negligible as illustrated in Figure 5.16.

5.3.3
MS Binding Assays Based on the Quantitation of the Nonbound Marker

As described in the previous sections, radioligand binding assays are generally based on the quantitation of the radioligand bound to the target. It is, however, also possible to quantify the amount of nonbound instead of bound marker to monitor binding. The feasibility of this concept has been shown for fluorescent markers addressing nicotinic acetylcholine receptors as well as benzodiazepine receptors [110, 111]. Examples for quantitation of nonbound ligands by MS assay formats distinctly differing from conventional binding assays are also already known, such as frontal affinity chromatography or "continuous-flow" ligand binding assays [37, 112].

MS Binding Assays making use of the quantitation of a nonbound marker addressing a desired target are also possible. However, in this case, the typical conditions in radioligand binding assays with target concentrations far below K_d are not suitable, as the differences between the concentrations of the free marker

M caused by changes in the concentrations of the bound marker TM would be almost imperceptible or, at best, extremely hard to detect. To overcome this problem the concentration of TM has to be increased considerably compared to the nominal marker concentration (M_{tot}) applied. This can be achieved by distinctly raising the nominal target concentration (T_{tot}) compared to the concentration used in radioligand binding assays. Favorable conditions are especially ensured when the concentrations of M_{tot} and T_{tot} are in the range of K_d (for a more detailed discussion of this topic see reference [46]). It has to be taken into account, however, that for membrane-bound proteins as targets it may be difficult to reach high enough concentrations. The free marker concentration that has to be quantified in this approach ranges from the concentration of M_{tot} to the difference of the concentrations of M_{tot} and TM. As long as the concentration of TM does not come close to the concentration of M_{tot} (as this is the case when M_{tot} and T_{tot} are in the range of K_d, [46]), the analytical demand on the LLOQ is distinctly lower than in the approach based on quantitation of the bound marker (see Section 5.2.1). Another important parameter is the buffer employed in the binding assay. As the nonbound marker is present in the matrix of the binding assay (e.g., in a supernatant after centrifugation or in a filtrate), the compatibility of the incubation buffer with the MS detection has to be ensured or, alternatively, a sample preparation step may be required to remove the buffer salts.

The applications described below will demonstrate that competitive MS Binding Assays even for membrane targets are feasible following this strategy [44, 46, 113].

5.3.3.1 Competitive MS Binding Assays for Dopamine D_1 and D_2 Receptors

Dopamine receptors are integral membrane proteins belonging to the GPCR family, which is currently the most frequently addressed class of targets in drug development [2]. Binding assays for dopamine receptors play a crucial role in the development of new drugs for several indications such as Morbus Parkinson or schizophrenia. Five different subtypes of dopamine receptors (D_{1-5}) are known. They can be classified into two groups according to their signal transduction and pharmacology: D_1 and D_5, on one side and D_{2-4} on the other [114].

Competitive MS Binding Assays for the subtypes D_1 und D_2 occurring most frequently in the brain have been established following the setup shown in Figure 5.17, employing SCH 23390 (Figure 5.18) as marker for D_1 and spiperone (Figure 5.18) for D_2, respectively [44, 113].

As nonbound marker is quantified in this approach, compounds inhibiting binding to the target are detected by greater intensities of the marker in the presence of increasing concentrations of the test compound. The effect of (+)-butaclamol (a nonselective dopamine antagonist) on binding of SCH 23390 to D_1 and of spiperone to D_2 is shown in Figure 5.18.

The setup described above was successfully used in competitive binding experiments for a series of known dopamine antagonists leading to binding curves that allowed the calculation of the affinities of the test compounds. The determined rank orders of potency for the compounds investigated in the D_1 as well as in the

Incubation
of target, marker, and
test compound in
single tubes

Separation
of non bound marker
by centrifugation

Quantitation
of non bound maker
by LC–MS–MS

Supernatant

Waste

Figure 5.17 Schematic flowchart of the competitive MS binding assay quantifying the nonbound marker employed for dopamine D_1 and D_2 receptors. After incubation of the target in presence of the marker (D_1: SCH 23390, D_2: spiperone) and a test compound the binding samples are centrifuged to separate bound from nonbound marker. The nonbound marker in the resulting supernatant is quantified by LC–ESI-MS/MS without further sample preparation (D_1) or after SPE (D_2). (Adapted from Ref. [46].)

D_2 assay were found to be in good accord with those obtained in conventional radioligand binding assays based on [^{3}H]SCH 23390 and [^{3}H]spiperone, respectively [44, 113].

5.3.4
Other Examples Following the Concept of MS Binding Assays

In 2007, de Jong *et al.* published a study that aimed to apply the principle of MS Binding Assays simultaneously to benzodiazepine and β-adrenergic receptors as well as SERT in a rat cortical membrane fraction. They employed flunitrazepam (benzodiazepine receptor), pindolol (β-adrenergic receptor), and 2-({2-[(dimethylamino)methyl]phenyl}sulphanyl)-5-methylaniline) (MADAM) (SERT), respectively, as markers [115]. Unfortunately, only preliminary binding experiments have been published. The study indicates, however, that measuring specific binding of the employed markers toward the addressed targets by means of LC–ESI-MS/MS is feasible. Although the setup described in this report calls for further improvement as far as measurement of binding constants is concerned, the idea to address several targets simultaneously by means of MS Binding Assays seems promising [46].

A group at Lilly Research Laboratories followed the concept of MS Binding Assays and established LC–MS-based assays to assess receptor binding *in vivo* in animals employing nonlabeled markers instead of radioligands [116, 117]. In contrast to conventional binding assays, the markers used to address the corresponding targets (i.e., raclopride for dopamine D_2 receptors, MDL-100907 for serotonin 5-HT$_{2A}$ receptors, GR205171 for neurokinin 1 receptors, naltrexone for μ-opioid receptors, GR103345 for κ-kappa opioid receptors, and naltriben for δ-opioid receptors) were administered intravenously into mice, rats, and gerbils. After sacrification of the animals, corresponding brain regions were rapidly

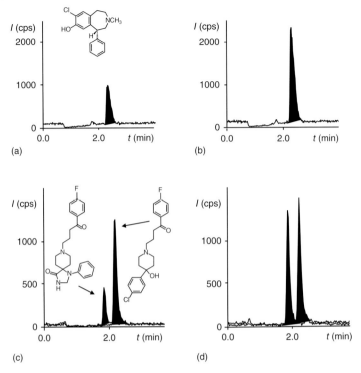

Figure 5.18 SRM LC–ESI-MS/MS chromatograms for nonbound SCH 23390 (288 → 91, m/z) and spiperone (396 → 123, m/z) in competitive MS binding experiments addressing dopamine D$_1$ and D$_2$ receptors, respectively, in pig striatum. SCH 23390 in supernatants of binding samples (a) without or (b) with 10 μM (+)-butaclamol. LC: Superspher 60 RP-select B (125 mm × 2 mm; 4 μm); 0.1% formic acid in water/acetonitrile (50/50, v,v) as mobile phase; flow rate 300 μl min^{-1}; injection volume 50 μl; external standard calibration. Spiperone in supernatants of binding samples (c) without or (d) with 10 μM (+)-butaclamol. LC: Luna C8 (50 mm × 2 mm; 3 μm); 0.1% formic acid in water/acetonitrile (70/30, v,v) as mobile phase; flow rate 150 μl min^{-1}; injection volume 25 μl; internal standard (haloperidol, 376 → 123, m/z) calibration. (Adapted from Refs. [46, 113].)

removed and the bound markers liberated with acetonitrile containing formic acid and finally quantified by LC–MS/MS.

All the examples presented so far demonstrate that the concept of MS Binding Assays provides a simple option to transform radioligand binding assays addressing membrane-bound proteins into a label-free assay format.

Radioligand binding assays are, however, not restricted to this target category they are also in widespread use to study soluble proteins such as nuclear receptors [118]. Recently, Aqai et al. [119] described a bioaffinity mass spectrometry (BioMS) concept employing rapid and radiolabel-free screening based on competitive MS Binding Assays for a soluble serum protein. They established MS Binding Assays for transthyretin (TTR) that functions as carrier of the thyroid hormone ɪ-thyroxine (T4). Incubation of the target (recombinant His-tagged transthyretin, rTTR) with $^{13}C_6$-l-thyroxine (marker) and test compounds was performed in ultrafiltration devices. After removing the unbound marker by centrifugation, subsequent liberation of the target-bound marker by means of a water methanol mixture containing formic acid and centrifugation of the ultrafiltration units, samples containing the liberated marker as well as test compounds inhibiting $^{13}C_6$-l-thyroxine binding were obtained. The amount of marker in this filtrate was quantified by means of a fast LC–MS method employing an UPLC coupled to an ESI–triple quadrupole mass spectrometer. For model test compounds such as the natural thyroid hormone T4, dose–response curves could be generated, and IC_{50} values calculated from them. Additionally, identification of unknown ligands with rTTR bioaffinity was possible in the filtrate containing the liberated $^{13}C_6$-l-thyroxine by means of a nano-UPLC–ESI-Q-TOF-MS-system. In a technical improvement rTTR was immobilized on paramagnetic microbeads enabling a 96-well assay format and an elegant separation of the nonbound marker from the marker bound to the target. In this way, it was possible to identify and characterize endocrine disruptors (i.e., compounds that can interfere with the endocrine system) in urine addressing rTTR.

In a follow-up paper, Aqai *et al.* [120] extended this approach to the screening and identification of known and unknown designer steroids in dietary supplements with affinity toward the recombinant human sex hormone binding globulin (rhSHBG). Here D_3-17β-testosterone was used as marker and the target immobilized on paramagnetic microbeads. The whole assay could be performed in a 96-well plate and reached a reasonable throughput enabling measurements of 96 samples within 4 h.

Assays addressing soluble enzymes by MS quantitation of substrates of the enzymes or the products formed by them or both (for a review, see [121]) show some degree of similarity to MS Binding Assays but also distinct differences when compared to the latter. As enzyme assays analyzed by MS do not follow the concept of MS Binding Assays they have not been included in this review.

5.4
Summary and Perspectives

MS Binding Assays represent a universal label-free screening technique that is ideally suited for affinity determination of small molecule ligands toward membrane-bound targets. As shown by the applications presented here they provide all the information that had to be deduced previously from the different types of radioligand binding assays. MS Binding Assays employing nonlabeled markers instead of radioligands can be conducted in an analogous experimental setup as conventional radioligand binding assays and can attain a throughput at least as high as these. In general, almost any target can be studied in MS Binding Assays as long as suitable markers are available. The most demanding task in MS Binding Assays, the reliable mass spectrometric quantitation of the marker, can be achieved by means of standard LC–MS/MS equipment and is, furthermore, continuously made easier by the progressively improving sensitivity of modern mass spectrometers. It can be assumed that the efforts necessary to establish such an MS-based quantitation method for a suitable marker are more than compensated by the amount of time, work, and money saved, which radiolabeling would require.

To sum up, the concept of MS Binding Assays can be easily and universally applied without the inherent disadvantages of radioligands. As MS Binding Assays have already reached the status of a mature technique that guarantees results just as reliable as radioligand binding assays, their potential can be expected to be increasingly exploited in future in basic research as well as in the drug discovery process.

Acknowledgments

We thank K. Heimberger and Dr. M. Simon for editing and proofreading and Dr. J. Pabel for providing graphical material.

References

1. Fang, Y. (2012) Ligand receptor interaction platforms and their applications for drug discovery. *Expert Opin. Drug Discovery*, **7** (10), 969–988.
2. Bleicher, K.H., Böhm, H.-I., Müller, K., and Alanine, A.I. (2003) A guide to drug discovery: hit and lead generation: beyond high throughput screening. *Nat. Rev. Drug Discovery*, **2** (5), 369–378.
3. Höfliger, M. M., Beck-Sickinger, A. G. (2003) in *Protein-Ligand Interactions. From Molecular Recognition to Drug Design*, (ed. H.-J. Böhm, G. Schneider), Wiley-VCH Verlag GmbH, Weinheim, pp. 107–135.
4. Sweetnam, P.M., Price, C.H., and Ferkany, J.W. (1995) in *Mass Ligand Screening as a Tool for Drug Discovery and Development in Burger's Medicinal Chemistry and Drug Discovery*, Vol. I: Principles and Practice (ed M.E. Wolff), John Wiley & Sons, Inc., New York, pp. 697–731.
5. Shiau, A.K., Massari, M.E., and Ozbal, C.C. (2008) Back to basics: label-free technologies for small molecule screening. *Comb. Chem. High Throughput Screening*, **11** (3), 231–237.
6. Zhu, Z. and Cuozzo, J. (2009) High-throughput affinity-based technologies for small-molecule drug discovery. *J. Biomol. Screening*, **14** (10), 1157–1164.
7. Lundqvist, T. (2005) The devil is still in the details – driving early drug discovery forward with biophysical experimental methods. *Curr. Opin. Drug Discovery Dev.*, **8** (4), 513–519.
8. Cooper, M.A. (2004) Advances in membrane receptor screening and analysis. *J. Mol. Recognit.*, **17**, 286–315.
9. Lammertsma, A.A., Leysen, J.E., Heylen, L., and Langlois, X. Receptors: Binding Assays, in *Encyclopedia of Psychopharmacology* (ed. I.P. Stolerman), http://www.springerreference.com/index/chapterdbid/169194 (accessed 4 February 2012).
10. Shukla, H.D., Vaitiekunas, P., and Cotter, R.J. (2012) Advances in membrane proteomics and cancer biomarker discovery: current status and future, perspective. *Proteomics*, **12** (19-20), 3085–3104.
11. Lefkowitz, R.J. (2004) Historical review: a brief history and personal retrospective of seven-transmembrane

receptors. *Trends Pharmacol. Sci.*, **25** (8), 413–422.

12. Hulme, E.C. (1992) Preface, in *Receptor-Ligand Interactions A Practical Approach* (ed E.C. Hulme), IRL Press, Oxford.

13. Jameson, D.M. and Ross, J.A. (2010) Fluorescence polarization/anisotropy in diagnostics and imaging. *Chem. Rev.*, **110** (5), 2685–2708.

14. Lam, M.Y.H. and Stagljar, I. (2012) Strategies for membrane interaction proteomics: no mass spectrometry required. *Proteomics*, **12** (10), 1519–1526.

15. Pope, A.J., Haupts, V.M., and Moore, K.J. (1999) Homogenous fluorescence readouts for miniaturized high-throughput screening: theory and practice. *Drug Discovery Today*, **4** (8), 350–362.

16. Wienken, C.J., Baaske, P., Rothbauer, U., Braun, D., and Duhr, S. (2010) Protein-binding assays in biological liquids using microscale thermophoresis. *Nat. Commun.*, **1**, 100. doi: 10.1038/ncomms1093.

17. Handl, H.L. and Gillies, R.J. (2005) Lanthanide-based luminescent assays for ligand-receptor interactions. *Life Sci.*, **77** (4), 361–671.

18. Carter, A., Kevin Thompson, K., and Wark, C. (2012) Quantifying Fluorescent Ligand Binding to GPCRs in Live Cells Using the PHERAstar FS – A New Format for HTS, BMG Labtech, Application Note 227, Rev. 06/2012.

19. Zemanova, L., Schenk, A., Valler, M.J., Nienhaus, G.U., and Heilker, R. (2005) in *Protein-Ligand Interactions*, Methods in Molecular Biology, vol. **305** (ed G.U. Nienhaus), Humana Press, Totowa, pp. 365–383.

20. Zwier, J.M., Roux, T., Cottet, M., Durroux, T., Douzon, S., Bdioui, S., Gregor, N., Bourrier, E., Oueslati, N., Nicolas, L., Tinel, N., Boisseau, C., Yverneau, P., Charrier-Savournin, F., Fink, M., and Trinquet, E. (2010) A fluorescent ligand-binding alternative using tag-lite® technology. *J. Biomol. Screening*, **15** (10), 1248–1259.

21. Janssen, M.J., Ensing, K., and de Zeeuw, R.A. (2001) A fluorescent receptor assay for benzodiazepines using coumarin labeled desethylflumazenil as ligand. *Anal. Chem.*, **73** (13), 3168–3173.

22. Holdgate, G.A., Anderson, M., Edfeldt, F., and Geschwindner, S. (2010) Affinity-based, biophysical methods to detect and analyze ligand binding to recombinant proteins: matching high information content with high throughput. *J. Struct. Biol.*, **172** (1), 142–157.

23. Siegel, M.M. (2005) in *Integrated Strategies for Drug Discovery Using Mass Spectrometry* (ed M.S. Lee), John Wiley & Sons, Inc., New York, pp. 27–70.

24. Katta, V. and Chait, B.T. (1991) Observation of the heme-globin complex in native myoglobin by electrospray-ionization mass spectrometry. *J. Am. Chem. Soc.*, **113** (22), 8535–8537.

25. Ganem, B., Li, Y.-T., and Henion, J.D. (1991) Detection of noncovalent receptor-ligand complexes by mass spectrometry. *J. Am. Chem. Soc.*, **113** (16), 6294–6296.

26. Annis, D.A., Nickbarg, E., Yang, X., Ziebelland, M.R., and Whitehurst, C.E. (2007) Affinity selection-mass spectrometry screening techniques for small molecule drug discovery. *Curr. Opin. Chem. Biol.*, **11** (5), 518–526.

27. Jonker, N., Kool, J., Irth, H., and Niessen, W.M.A. (2011) Recent developments in protein–ligand affinity mass spectrometry. *Anal. Bioanal.Chem.*, **399** (8), 2669–2681.

28. Geoghegan, A.F. and Kelly, M.A. (2005) Biochemical applications of mass spectrometry in pharmaceutical drug discovery. *Mass Spectrom. Rev.*, **24** (3), 347–366.

29. Schermann, S., Simmons, D.A., and Konermann, L. (2005) Mass spectrometry based approaches to protein-ligand interactions. *Expert Rev. Neurother.*, **2** (4), 475–485.

30. Hofstadler, S.A. and Sannes-Lowery, K.A. (2007) in *Mass Spectrometry in Medicinal Chemistry* (eds K.T. Wanner

and G. Höfner), Wiley-VCH Verlag GmbH, Weinheim, pp. 321–338.

31. Akashi, S. (2006) Investigation of molecular interaction within biological macromolecular complexes by mass spectrometry. *Med. Res. Rev.*, **26** (3), 339–368.

32. Yin, S. and Loo, J.A. (2009) in *Mass Spectrometry of Proteins and Peptides*, Methods in Molecular Biology, vol. **492** (eds M.S. Lipton and L. Paša-Tolic), Humana Press, Totowa, pp. 273–282.

33. Hofstadler, S.A. and Sannes-Lowery, K.A. (2006) Applications of ESI-MS in drug discovery: interrogation of noncovalent complexes by ESI-MS. *Nat. Rev. Drug Discovery*, **5** (7), 585–595.

34. Barrera, N.P., Isaacson, S.C., Zhou, M., Bavro, V.N., Welch, A., Schaedler, T.A., Seeger, M.A., Núñez Miguel, R., Korkhov, V.M., van Veen, H.W., Venter, H., Walmsley, A.R., Tate, C.G., and Robinson, C.V. (2009) Mass spectrometry of membrane transporters reveals subunit stoichiometry and interactions. *Nat. Methods*, **6** (8), 585–589.

35. Vivat Hannah, V., Atmanene, C., Zeyer, D., Van Dorsselaer, A., and Sanglier-Cianférani, S. (2010) Native MS: an 'ESI' way to support structure- and fragment-based drug discovery. *Future Med. Chem.*, **2** (1), 35–50.

36. Kitova, E.N., El-Hawiet, A., Schnier, P.D., and Klassen, J.S. (2012) Reliable determinations of protein–ligand interactions by direct ESI-MS measurements, are we there yet? *J. Am. Soc. Mass. Spectrom.*, **23** (3), 431–441.

37. Chan, N., Lewis, D., Kelly, M., Ng, E.S.M., and Schriemer, D.C. (2007) in *Mass Spectrometry in Medicinal Chemistry* (eds K.T. Wanner and G. Höfner), Wiley-VCH Verlag GmbH, Weinheim, pp. 217–246.

38. Calleri, E., Temporini, C., and Massolini, G. (2011) Frontal affinity chromatography in characterizing immobilized receptors. *J. Pharm. Biomed. Anal.*, **54** (5), 911–925.

39. Clouthier, T.E. and Comess, K.M. (2007) in *Mass Spectrometry in Medicinal Chemistry* (eds K.T. Wanner and G. Höfner), Wiley-VCH Verlag GmbH, Weinheim, pp. 157–183.

40. Van Breemen, R.B. (2003) in *Burger's Medicinal Chemistry and Drug Discovery*, Vol. I: Drug Discovery (ed D.J. Abraham), John Wiley & Sons, Inc., New York, pp. 583–610.

41. Siegel, M.M. (2007) in *Mass Spectrometry in Medicinal Chemistry* (eds K.T. Wanner and G. Höfner), Wiley-VCH Verlag GmbH, Weinheim, pp. 65–120.

42. Annis, A., Chuang, C.-C., and Nazef, N. (2007) in *Mass Spectrometry in Medicinal Chemistry* (eds K.T. Wanner and G. Höfner), Wiley-VCH Verlag GmbH, Weinheim, pp. 121–156.

43. Zehender, H., Le Goff, F., Lehmann, N., Filipuzzi, I., and Mayr, L.M. (2004) SpeedScreen: the "missing link" between genomics and lead discovery. *J. Biomol. Screening*, **9** (6), 498–505.

44. Höfner, G. and Wanner, K.T. (2003) Competitive binding assays simply done with a native marker and Mass spectrometric quantification. *Angew. Chem. Int. Ed.*, **42** (42), 5235–5237; *Angew. Chem.*, **115** (42), 5393–5395.

45. Zepperitz, C., Höfner, G., and Wanner, K.T. (2006) MS-binding assays: kinetic, saturation, and competitive experiments based on quantitation of bound marker as exemplified by the GABA transporter mGAT1. *ChemMedChem*, **1** (2), 208–217.

46. Höfner, G., Zepperitz, C., and Wanner, K.T. (2007) in *Mass Spectrometry in Medicinal Chemistry* (eds K.T. Wanner and G. Höfner), Wiley-VCH Verlag GmbH, Weinheim, pp. 247–283.

47. Bylund, D.B., Deupree, I.D., and Toews, M.L. (2004) in *Receptor Signal Transduction Protocols*, Methods in Molecular Biology, vol. **259** (eds G.B. Willars and R.A.I. Challiss), Humana Press, Totowa, pp. 1–28.

48. Hulme, E.C. and Birdsall, N.J.M. (1992) in *Receptor-Ligand Interactions. A Practical Approach* (ed E.C. Hulme), IRL Press, Oxford, pp. 63–176.

49. Bennett, J.P. Jr., and Yamamura, H.I. (1985) in *Neurotransmitter Receptor Binding* (eds H.I. Yamamura, S.J. Enna, and M.J. Kuhar), Raven Press, New York, pp. 61–89.

50. Fraser, C.M. (1990) in *Receptor Biochemistry. A Practical Approach* (ed

E.C. Hulme), IRL Press, Oxford, pp. 236–275.

51. Hopfgartner, G. (2007) in *Mass Spectrometry in Medicinal Chemistry* (eds K.T. Wanner and G. Höfner), Wiley-VCH Verlag GmbH, Weinheim, pp. 3–62.

52. Novakova, L. (2013) Challenges in the development of bioanalytical liquid chromatography–massspectrometry method with emphasis on fast analysis review. *J. Chromatogr. A*, **1292**, 25–37.

53. Schneider, B.B., Javaheri, H., and Covey, T.R. (2006) Ion sampling effects under conditions of total solvent consumption. *Rapid Commun. Mass Spectrom.*, **20** (10), 1538–1544.

54. Covey, T.R., Schneider, B.B., Javaheri, H., LeBlanc, J.C.Y., Ivosev, G., Corr, J.J., and Kovarik, P. (2010) in *Electrospray and MALDI Mass Spectrometry*, 2nd edn (ed R.B. Cole), John Wiley & Sons, Inc., Hoboken, NJ, pp. 443–490.

55. Snyder, L.R., Kirkland, J.J., and Donland, J.W. (2010) *Introduction to Modern Liquid Chromatography*, 3rd edn, John Wiley & Sons, Inc., Hoboken, NJ, pp. 70–86.

56. Trufelli, H., Palma, P., Famiglini, G., and Cappiello, A. (2011) An overview of matrix effects in liquid chromatography–mass spectrometry. *Mass Spectrom. Rev.*, **30** (3), 491–509.

57. Van Eeckhaut, A., Lanckmans, K., Sarre, S., Smolders, I., and Michotte, Y. (2009) Validation of bioanalytical LC–MS/MS assays: evaluation of matrix effects. *J. Chromatogr. B*, **877** (23), 2198–2207.

58. Beleboni, R.O., Carolino, R.O., Pizzo, A.B., Castellan- Baldan, L., Coutinho-Netto, J., dos Santos, W.F., and Coimbra, N.C. (2004) Pharmacological and biochemical aspects of GABAergic neurotransmission: pathological and neuropsychobiological relationships. *Cell. Mol. Neurobiol.*, **24** (6), 707–728.

59. Bröer, S. and Gether, U. (2012) The solute carrier 6 family of transporters. *Br. J. Pharmacol.*, **167** (2), 256–278.

60. Iversen, L. (2004) GABA pharmacology—what prospects for the future? *Biochem. Pharmacol.*, **68** (8), 1537–1540.

61. Zhou, Y., Holmseth, S., Hua, R., Lehre, A.C., Olofsson, A.M., Poblete-Naredo, I., Kempson, S.A., and Danbolt, N.C. (2012) The betaine-GABA transporter (BGT1, slc6a12) is predominantly expressed in the liver and at lower levels in the kidneys and at the brain surface. *Am. J. Physiol. Renal Physiol.*, **302**, F316–F328.

62. Zhou, Y., Holmseth, S., Guo, C., Hassel, B., Höfner, G., Huitfeldt, H.S., Wanner, K.T., and Danbolt, N.C. (2012) Deletion of the γ-aminobutyric acid transporter 2 (GAT2 and SLC6A13) gene in mice leads to changes in liver and brain taurine contents. *J. Biol. Chem.*, **287** (42), 35733–35746.

63. Borden, L.A. (1996) GABA transporter heterogeneity: pharmacology and cellular localization. *Neurochem. Int.*, **29** (4), 335–356.

64. Stefan, H. and Feuerstein, T.J. (2007) Novel anticonvulsant drugs. *Pharmacol. Ther.*, **113** (1), 165–183.

65. Høg, S., Greenwood, J.R., Madsen, K.B., Larsson, O.M., Frølund, B., Schousboe, A., Krogsgaard-Larsen, P., and Clausen, R.P. (2006) Structure-activity relationships of selective GABA uptake inhibitors. *Curr. Top. Med. Chem.*, **6** (17), 1861–1882.

66. Braestrup, C., Nielsen, E.B., Sonnewald, U., Knutsen, L.S., Andersen, K.E., Jansen, I.A., Frederiksen, K., Andersen, P.H., Mortensen, A., and Suzdak, P.D. (1990) (*R*)-*N*-[4,4-bis(3-methyl-2-thienyl)but-3-en-1- yl]nipecotic acid binds with high affinity to the brain γ-aminobutyric acid uptake carrier. *J. Neurochem.*, **54** (2), 639–647.

67. Suzdak, P.D., Frederiksen, K., Andersen, K.E., Sorensen, P.O., Knutsen, L.I.S., and Nielsen, E.B. (1992) NNC-711, a novel potent and selective γ-aminobutyric acid uptake inhibitor: pharmacological characterization. *Eur. J. Pharmacol.*, **224** (2-3), 189–198.

68. Suzdak, P.D., Foged, C., and Andersen, K.E. (1994) Quantitative autoradiographic characterization of the binding of [³H]tiagabine (NNC 05-328) to the

GABA uptake carrier. *Brain Res.*, **647** (2), 231–241.

69. Eriksson, I.S., Allard, P., and Marcusson, J. (1999) [³H]Tiagabine binding to GABA uptake sites in human brain. *Brain Res.*, **851** (1-2), 183–188.

70. United States Food and Drug Administration (2001) *Guidance for Industry*, Bioanalytical Methods Validation, Washington, DC.

71. Höfner, G. and Wanner, K.T. (2010) Using short columns to speed up LC–MS quantification in MS Binding Assays. *J. Chromatogr. B*, **878** (17-18), 1356–1364.

72. Zepperitz, C., Höfner, G., and Wanner, K.T. (2008) Expanding the scope of MS Binding Assays to low-affinity markers as exemplified for mGAT1. *Anal. Bioanal. Chem.*, **391** (1), 309–316.

73. Wanner, K., Fülep, G., and Höfner, G. (2000) GABA-uptake-Inhibitoren mit Pyrrolidinstruktur, WO Patent 0014064; *Chem. Abstr.*, **132** (2000), 194656.

74. Cheng, Y.-C. and Prusoff, W.H. (1973) Relationship between the inhibition constant (K_i) and the concentration of inhibitor which causes 50 per cent inhibition (I_{50}) of an enzymatic reaction. *Biochem. Pharmacol.*, **22** (23), 3099–3108.

75. Kulig, K., Wićckowska, A., Wićckowski, K., Gajda, J., Pochwat, B., Höfner, G., Wanner, K.T., and Malawska, B. (2011) Synthesis and biological evaluation of new derivatives of 2-substituted 4-hydroxybutanamides as GABA uptake inhibitors. *Eur. J. Med. Chem.*, **46** (1), 183–190.

76. Kowalczyk, P., Höfner, G., Wanner, K.T., and Kulig, K. (2012) Synthesis and pharmacological evaluation of new 4,4-diphenylbut-3-enyl derivatives Of 4-hydroxybutanamides As GABA uptake inhibitor. *Acta Poloniae Pharm.*, **69**, 157–160.

77. Salat, K., Wieckowska, A., Wieckowski, K., Höfner, G., Kaminski, J., Wanner, K.T., Malawska, B., Filipek, B., and Kulig, K. (2012) Synthesis and pharmacological properties of new GABA uptake inhibitors. *Pharmacol. Rep.*, **64** (4), 817–833.

78. Hopfgartner, G. and Bourgogne, E. (2003) Quantitative high-throughput analysis of drugs in biological matrices by mass spectrometry. *Mass Spectrom. Rev.*, **22** (3), 195–214.

79. Jerkovich, A.D. and Vivilecchia, R.V. (2007) in *HPLC for Pharmaceutical Scientists* (eds Y. Kazakevich and R. Lobrutto), John Wiley & Sons Inc., Hoboken, NJ, pp. 765–810.

80. Hsieh, Y., Fukuda, E., Wingate, J., and Korfmacher, W.A. (2006) Fast mass spectrometry-based methodologies for pharmaceutical analyses. *Comb. Chem. High Throughput Screening*, **9** (1), 3–8.

81. Wickremsinhe, E.R., Singh, G., Ackermann, B.L., Gillespie, T.A., Todd, A.C., and Chaudhary, A.K. (2006) A review of nanoelectrospray ionization applications for drug metabolism and pharmacokinetics. *Curr. Drug Metab.*, 7 (8), 913–928.

82. Cohen, L.H. and Gusev, A.I. (2002) Small molecule analysis by MALDI mass spectrometry. *Anal. Bioanal. Chem.*, **373** (7), 571–586.

83. Corr, J.J., Kovarik, P., Schneider, B.B., Hendrikse, J., Loboda, A., and Covey, T.R. (2006) Design considerations for high speed, quantitative mass spectrometry, with MALDI ionization. *J. Am. Soc. Mass. Spectrom.*, **17** (8), 1129–1141.

84. Sleno, L. and Volmer, D.A. (2005) Toxin screening in phytoplankton: detection and quantitation using maldi triple quadrupole mass spectrometry. *Anal. Chem.*, **77** (5), 1509–1517.

85. Gobey, J., Cole, M., Jaiszewski, J., Covey, T., Chau, T., Kovarik, P., and Con, J. (2005) Characterization and performance of MALDI on a triple quadrupole mass spectrometer for analysis and quantification of small molecules. *Anal. Chem.*, **77** (17), 5643–5654.

86. Kovarik, P., Grivet, C., Bourgogne, E., and Hopfgartner, G. (2007) Method development aspects for the quantitation of pharmaceutical compounds in human plasma with a matrix-assisted laser desorption/ionization source in the multiple reaction monitoring mode.

Rapid Commun. Mass Spectrom., **21** (6), 911–919.

87. Vollmer, D.A., Sleno, L., Bateman, K., Sturino, C., Oballa, R., Mauriala, T., and Corr, J. (2007) Comparison of MALDI to ESI on a triple quadrupole platform for pharmacokinetic analyses. *Anal. Chem.*, **79** (23), 9000–9006.

88. Hatsis, P., Brombacher, S., Corr, J., Kovarik, P., and Volmer, D.A. (2003) Quantitative analysis of small pharmaceutical drugs using a high repetition rate laser matrix-assisted laser/desorption ionization source. *Rapid Commun. Mass Spectrom.*, **17** (20), 2303–2309.

89. Hopfgartner, G. and Varesio, E. (2005) New approaches for quantitative analysis in biological fluids using mass spectrometric detection. *Trends Anal. Chem.*, **24** (7), 583–589.

90. van Kampen, J.J.A., Burgers, P.C., de Groot, R., and Luider, T.M. (2006) Qualitative and quantitative analysis of pharmaceutical compounds by MALDI-TOF mass spectrometry. *Anal. Chem.*, **78** (15), 5403–5411.

91. Höfner, G., Merkel, D., and Wanner, K.T. (2009) MS Binding Assays - with MALDI towards high throughput. *ChemMedChem*, **4** (9), 1523–1528.

92. Hochgürtel, M. and Lehn, J.-M. (2006) in *Fragment-Based Approaches in Drug Discovery* (eds W. Jahnke and D.A. Erlanson), Wiley-VCH Verlag GmbH, Weinheim, pp. 341–364.

93. Ramström, O. and Lehn, J.-M. (2002) Drug discovery by dynamic combinatorial libraries. *Nat. Rev. Drug Discovery*, **1**, 26–36.

94. Furlan, R.L.E., Otto, S., and Sanders, J.M.K. (2002) Supramolecular templating in thermodynamically controlled synthesis. *Proc. Natl. Acad. Sci. U.S.A.*, **99** (8), 4801–4804.

95. Sakai, S., Shigemasa, Y., and Sasaki, T. (1997) A self-adjusting carbohydrate ligand for GalNAc specific lectins. *Tetrahedron Lett.*, **38** (47), 8145–8148.

96. Sindelar, M. and Wanner, K.T. (2012) Library screening by means of MS Binding Assays - exemplarily demonstrated for a pseudo-static library

addressing GAT1. *ChemMedChem*, **7** (9), 1678–1690.

97. Sindelar, M., Lutz, T.A., Petrera, M., and Wanner, K.T. (2013) Focused pseudostatic hydrazone libraries screened by mass spectrometry binding assay – optimizing affinities towards γ-aminobutyric acid transporter 1. *J. Med. Chem.*, **56** (3), 1323–1340.

98. Murphy, D.L., Lerner, A., Rudnick, G., and Lesch, K.P. (2004) Serotonin transporter: gene, genetic disorders, and pharmacogenetics. *Mol. Interventions*, **4** (2), 109–123.

99. Owens, M.J., Morgan, W.N., Plott, S.J., and Nemeroff, C.B. (1997) Neurotransmitter receptor and transporter binding profile of antidepressants and their metabolites. *J. Pharmacol. Exp. Ther.*, **283** (3), 1305–1322.

100. Tatsumi, M., Groshan, K., Blakely, R.D., and Richelson, E. (1997) Pharmacological profile of antidepressants and related compounds at human monoamine transporters. *Eur. J. Pharmacol.*, **340** (2-3), 249–258.

101. Koch, S., Perry, K.W., Nelson, D.L., Conway, R.G., Threlkeld, P.G., and Bymaster, F.P. (2002) R-fluoxetine increases extracellular DA, NE,As Well As 5-HT in Rat prefrontal cortex and hypothalamus: an in vivo microdialysis and receptor binding study. *Neuropsychopharmacology*, **27** (6), 949–959.

102. Hess, M., Höfner, G., and Wanner, K.T. (2011) Development and validation of a rapid LC-ESI-MS/MS method for quantification of fluoxetine and its application to binding assays. *Anal. Bioanal. Chem.*, **400** (10), 3505–3515.

103. Wong, D.T., Bymaster, F.P., and Engleman, E.E. (1995) Prozac (Fluoxetine, Lilly 110140) the first selective serotonin uptake inhibitor and antidepressant drug: twenty years since its first publication. *Life Sci.*, **57** (5), 411–441.

104. Chen, F., Larsen, M.B., Sánchez, C., and Wiborg, O. (2005) The S-enantiomer of *R,S*-citalopram, increases inhibitor binding to the human serotonin transporter by an allosteric mechanism. Comparison with other serotonin

transporter inhibitors. *Eur. Neuropsychopharmacol.*, **15** (2), 193–198.

105. Wong, D.T., Bymaster, F.P., Reid, L.R., Fuller, R.W., and Perry, K.W. (1985) Inhibition of serotonin uptake by optical isomers of fluoxetine. *Drug Dev. Res.*, **6** (4), 398–403.

106. Elfving, B., Madsen, J., and Knudsen, G.M. (2007) Neuroimaging of serotonin uptake sites requires high-affinity ligands. *Synapse*, **61** (11), 882–888.

107. Wong, D.T. and Bymaster, F.P. (1983) Serotonine (5HT) neuronal uptake binding sites in rat brain membranes labelled with [³H] fluoxetine. *Fed. Proc.*, **42**, 1164.

108. Hess, M., Höfner, G., and Wanner, K.T. (2011) (S)- and (R)-fluoxetine as native markers in MS Binding Assays addressing the serotonin transporter. *ChemMedChem*, **6** (10), 1900–1908.

109. Martin, R.S., Henningsen, R.A., Suen, A., Apparsundaram, S., Leung, B., Jia, Z., Kondru, R.K., and Milla, M.E. (2008) Kinetic and thermodynamic assessment of binding of serotonin transporter inhibitors. *J. Pharmacol. Exp. Ther.*, **327** (3), 991–1000.

110. Takeuchi, T. and Rechnitz, G.A. (1991) Nonisotopic receptor assay for benzodiazepine receptors for utilizing fluorophore labelled ligand. *Anal. Biochem.*, **194** (2), 250–255.

111. Chen, L., Takeuchi, T., and Rechnitz, G.A. (1992) Development of a nonisotopic acetylcholine receptor assay for the investigation of cholinergic ligands. *Anal. Chim. Acta*, **267** (1), 55–62.

112. Irth, H. (2007) in *Mass Spectrometry in Medicinal Chemistry* (eds K.T. Wanner and G. Höfner), Wiley-VCH Verlag GmbH, Weinheim, pp. 185–215.

113. Niessen, K.V., Höfner, G., and Wanner, K.T. (2005) Competitve MS Binding Assays for dopamine O₂ receptors employing spiperone as a native marker. *ChemBioChem*, **6** (10), 1769–1775.

114. Missale, C., Nash, R., Robinson, S.W., Jaber, M., and Caron, M.O. (1998) Dopamine receptors: from structure to function. *Physiol. Rev.*, **78** (1), 189–225.

115. de Jong, L.A.A., Jeronimus-Stratingh, C.M., and Cremers, T.I.F.H. (2007) Development of a multiplex non-radioactive receptor assay: the benzodiazepine receptor, the serotonin transporter and the β-adrenergic receptor. *Rapid Commun. Mass Spectrom.*, **21** (4), 567–572.

116. Chernet, E., Martin, L.J., Li, D., Need, A.B., Barth, V.N., Rash, K.S., and Phebus, L.A. (2005) Use of LC/MS to assess brain tracer distribution in preclinical, in vivo receptor occupancy studies: dopamine D2, serotonin 2A and NK-1 receptors as examples. *Life Sci.*, **78** (4), 340–346.

117. Need, A.B., McKinzie, J.H., Mitch, C.H., Statnick, M.A., and Phebus, L.A. (2007) In vivo rat brain opioid receptor binding of LY255582 assessed with a novel method using LC/MS/MS and the administration of three tracers simultaneously. *Life Sci.*, **81** (17-18), 1389–1396.

118. Jones, S.A., Parks, D.J., and Kliewer, S.A. (2003) Cell-free ligand binding assays for nuclear receptors. *Methods Enzymol.*, **364**, 53–71.

119. Aqai, P., Fryganas, C., Mizuguchi, M., Haasnoot, W., and Nielen, M.W.F. (2012) Triple bioaffinity mass spectrometry concept for thyroid transporter ligands. *Anal. Chem.*, **84** (15), 6488–6493.

120. Aqai, P., Cevik, E., Gerssen, A., Haasnoot, W., and Nielen, M.W.F. (2013) High-throughput bioaffinity mass spectrometry for screening and identification of designer anabolic steroids in dietary supplements. *Anal. Chem.*, **85** (6), 3255–3262.

121. Greis, K.D. (2007) Mass spectrometry for enzyme assays and inhibitor screening: an emerging application in pharmaceutical research. *Mass Spectrom. Rev.*, **26** (4), 324–339.

6
Metabolic Profiling Approaches for the Identification of Bioactive Metabolites in Plants

Emily Pipan and Angela I. Calderón

6.1
Introduction to Plant Metabolic Profiling

The study of plant metabolites and their bioactive properties has long captivated mankind. Some of the earliest known investigations of bioactive plants and herbs date back to 3000 BC with Egyptian scrolls detailing a variety of plants and their medicinal properties [1]. The modern study of bioactive metabolites began ~200 years ago with the isolation of morphine by F.W. Sertürner [2]. Since then, great advances have been made in the field of phytochemistry. Bioactive metabolites are defined as secondary plant metabolites that elicit pharmacological or toxicological effects in humans or animals and are produced within the plant outside of primary biosynthetic and metabolic routes [3]. That is not a narrow definition, as it includes a plethora of compound classes.

With that in mind, it is unsurprising that plant metabolomics plays an integral role in both scientific and commercial settings, including crop improvement, such as in the improvement in concentrations of carotenoids, proteins, and oils [4], drug discovery [5], food safety [6], and elucidation of the interconnection of the plant metabolome [7]. All of these pursuits are unquestionably connected to plant metabolite bioactivity, and their importance cannot be overstated. However, there is no method that allows the profiling of the complete array of a single plant's natural products. In addition to this, only an estimated 10% of higher plants have been investigated for these metabolites, and many of them not in great detail [1]. This underlines the importance and promise of phytochemistry to grow rapidly and with much success. The challenges faced by analytical methods include the vast differences in concentrations of metabolites and their extremely different chemical properties [8], such as size, polarity, and stability.

Mass spectrometry (MS) is the widely accepted choice for metabolic profiling of bioactive compounds due to its high sensitivity, resolution, and dynamic range of detection [9]. Nuclear magnetic resonance (NMR) spectroscopy, the competing metabolite determination technique, is limited by detection sensitivity [10, 11], although it is not uncommon to use both MS and NMR in plant metabolite profiling studies [12]. There are two common approaches for plant

Analyzing Biomolecular Interactions by Mass Spectrometry, First Edition.
Edited by Jeroen Kool and Wilfried M.A. Niessen.

metabolite profiling: targeted and nontargeted. Targeted approaches are used when the researcher would like to focus on a particular metabolite/metabolite class whereas nontargeted approaches seek to gather data on the entire range of metabolites. Targeted approaches are more popular in profiling studies whereas nontargeted approaches are the first step in generating data for biomarker discovery [10]. Both approaches are used regularly, depending on independent research goals. Metabolic profiling relies heavily on separation techniques before mass spectrometric measurement, a process different from metabolic fingerprinting that often has no separation and seeks a chemical "fingerprint" instead of quantitative studies [10].

6.2
Sample Collection and Processing

Bioactive plant metabolites are inherently part of a mixture that has a wide range of physical and chemical properties. The preparation of samples can be divided into three steps: sample collection, extraction, and further preparation.

An important requirement of scientific inquiry is the reproducibility of results. Plant samples present a challenge in this regard as there is a great variety of metabolites even among plants of the same species grown in different areas. Because of this, the procedure must also include a specific method of plant breeding [13]. This procedure often includes controlled temperature, humidity, small soil area, and meticulous measurement of water and fertilizer [14–16]. In addition, the time of the day when plants are harvested can change the metabolic profile; so this must also be taken into consideration [17]. As an additional step after harvesting, plants can be rinsed with deionized water in addition to other solvents that contain complexing agents such as EDTA, dilute HCl solutions, or organic solvents to remove unwanted soil or other material [13]. Freezing of the plant material should take place immediately after harvesting it to avoid enzymatic changes. This is most commonly done in liquid nitrogen [18]. The amount of water present in samples can greatly affect spectra, so a method of drying must be used. The most popular method is freeze-drying, or lyophilization [16, 18]. Other methods include the use of ventilated ovens at a low drying temperature to reduce the loss of volatile or subliming substances [13]. Substances are often homogenized to a fine powder before drying to improve reduction of water [14, 18]. Extraction of plant metabolites falls into two categories: traditional techniques and newer, less conventional procedures. The wide array of extraction options is due to the fact that no single solvent system will extract all metabolites. Traditional procedures include Soxhlet extraction, decoction, maceration, and steam distillation. There are a variety of resources available outlining these techniques [13, 18–21] that have become less ideal because of cleanup and concentration often required before analysis, difficulty in extracting low concentrations or thermally labile compounds, the use of large amounts of solvent, and time consumption [22]. More recently, modern extraction techniques have replaced the traditional solid–liquid extraction procedures. They are widely considered ideal because of the shorter time, ease of automation, and precision compared to traditional procedures. However, there is currently no single technique favored over the others and even traditional techniques are still used. The decision must rely on what compounds one wishes to extract and at what cost and time.

Microwave-assisted extraction (MAE) uses the absorption of microwave energy by the molecules of polar chemical compounds to generate heat. The combination of microwaves and traditional solvent extraction enhances penetration of solvent and, therefore, promotes the faster dissolution of bioactive compounds such alkaloids, terpenoids, and phenolic compounds [19]. The procedure is carried out at high temperatures (150–190 °C). The hot solvent allows rapid isolation of thermally stable analytes [13] but excludes isolation of temperature-sensitive analytes. MAE is a reliable method with good stability and reproducibility, moderate costs, and few safety issues [23]. An integrated version of this procedure, dynamic MAE, is worthy of note because of its coupling to chromatographic systems for online sample preparation and analysis [24, 25]. Traditional MAE separates the steps of extraction and analysis, while dynamic MAE combines the two and allows the extracted compounds to be analyzed in an online HPLC (high performance liquid chromatography) separation. Further information on this technique and its derivatives are covered in detail in two recent reviews [23, 26]. Supercritical fluid extraction (SFE) is another variation on solvent extraction that uses supercritical fluids as the extracting solvent. Carbon dioxide is the most commonly used solvent (SC-CO_2). SFE is a versatile technique that has applications in food processing, small- and large-scale extraction, and natural products research [27]. It is particularly popular in the isolation of essential oils investigated for their antioxidant and anti-inflammatory activity [28] and other bioactive compounds such as carotenoids and polyphenols [29]. However, solvents can be used to extract polar analytes as well [29]. The main advantage of this technique is that active compounds avoid long exposure to heat and atmospheric oxygen [30]. SFE is notable for its compatibility with green chemistry [31].

Ultrasound-assisted extraction (UAE) is a technique that capitalizes on the energy of ultrasonic waves to induce the phenomenon of cavitation. This process creates a dynamic equilibrium of bubble cavities produced by the energy applied, which in turn interacts with the liquid/solid boundary surface to erode solids [13]. The main advantages of this technique are lower solvent consumption, speed, potential to uncover tightly bound residues, and room temperature operation [32]. It is a well-established technique that has successfully been used to extract a variety of bioactive metabolites–polyphenols, anthocyanins, and aromatic compounds, to name a few [33]. Pressurized liquid extraction (PLE) and pressurized hot water extraction (PHWE) both belong to the class of "green" extraction techniques. These techniques use solvents (and in the latter case, water, specifically) at a high temperature and pressure, rendering them more efficient than techniques carried out at room temperature and atmospheric pressure. PLE has been used to extract phenolic compounds, lignans, carotenoids, oils, lipids, and essential oils, highlighting the versatility of this technique in the extraction of bioactive compound classes [34]. PLE and PHWE improve extraction yield and solvent consumption, protect oxygen and light-sensitive compounds, and decrease time. The drawbacks of these techniques are cost and difficulty with thermolabile compounds [34, 35].

Another green extraction technique is enzyme-assisted extraction (EAE). This technique uses cell wall degrading enzymes such as glucanases and pectinases to make the intracellular bioactive compounds accessible for extraction [36]. This technique has been shown to have high extraction yields for a variety of bioactive compounds such as polysaccharides, oils, flavors, and other natural products [37, 38]. EAE minimizes the use of solvents, decreases extraction time, and provides good yield and quality of product. Its drawbacks are the cost of enzymes, difficulty of scale-up because of the irreproducibility of enzymes in different conditions [37], and the need to find enzymes for specific substrates, requiring further studies on this technique [19]. Table 6.1 summarizes the key points of the above techniques.

Before chemical analysis, a pretreatment of crude extracts is often necessary to remove common interfering metabolites such as lipids, pigment, chlorophyll, and tannins [19]. Liquid–liquid extraction, solid-phase extraction (SPE), and gel filtration are often the methods of choice. These techniques are not new and the authors point to two often-cited reviews for further perusal [39, 40]. Recent advances in SPE include the use of dual-column SPE for improved extraction with minimal loss

Table 6.1 Summary of extraction techniques.

MAE	SFE	UAE	PLE/PHWE	EAE
Uses microwave energy to generate heat, with increased penetration of solvent and faster dissolution of bioactive compounds	Uses supercritical fluids as the extracting solvent	Uses ultrasonic waves to induce cavitation	Uses traditional solvents at a high temperature and pressure to increase efficiency	Uses cell wall degrading enzymes to extract intracellular bioactives
Best for thermally stable analytes	Can be used for polar and nonpolar analytes	Can be used for polar and nonpolar analytes	Can be used for polar and nonpolar analytes	Can be used for polar and nonpolar compounds
—	Avoids long analyte exposure to heat and oxygen	Avoids analyte exposure to heat	Green chemistry technique	Fast, low solvent consumption
—	Green chemistry technique	Fast, low solvent consumption	—	Green chemistry technique

of analyte as well as improved time of extraction when coupled to an online system [41]. Another new and emerging technique in extraction is the use of ionic liquids (ILs). These organic and inorganic salts in the liquid state have gained popularity in recent years due to their use as "green" solvents and success in extraction techniques. These "designer" solvents offer the user many tunable elements in terms of polarity, hydrophobicity, viscosity, and other chemical and physical properties [42]. They have been used in liquid–liquid extraction, UAE, MAE, and SPE, especially in extraction of bioactive phytochemicals such as tanshinones in *Salvia miltiorrhiza* [24], caffeine, phenolic acids, and a variety of other known bioactive natural products [43, 44]. This newer technique is expected to rise in popularity as more research is carried out. In the field of metabolomics attention has recently been turned to not only which metabolites are present, but also where they are in the cell. Many of the newer soft ionization techniques (e.g., MALDI (matrix-assisted laser desorption ionization), DESI (desorption electrospray ionization), LAESI (laser ablation electrospray ionization); see Section 6.5) take samples from a solid analyte with a picoliter emitting device. The sample is typically pressed against a Teflon or polytetrafluoroethylene (PTFE) surface immediately after harvest and followed by a brief rinse with deionized water to remove dirt and dust [45–48]. These emerging techniques for single, whole cell analysis will be discussed later in the chapter. Figure 6.1 summarizes the steps for sample preparation and subsequent analysis.

6.3
Hyphenated Techniques

6.3.1
Liquid Chromatography–Mass Spectrometry

Liquid chromatography–mass spectrometry (LC–MS) remains the most commonly used technique in plant metabolomics because of its high sensitivity and selectivity [49]. This versatile technique pairs an atmospheric pressure ionization method to a mass spectrometer. Electrospray ionization (ESI) is best suited to high-polarity compounds while atmospheric pressure chemical ionization (APCI) is better for low- and medium-polarity compounds. The workflow and objective outcomes follow a path similar to profiling by GC–MS (gas chromatography–mass spectrometry) (Figure 6.2 [50]). LC separation is highly dependent on the mobile phase and column used. Normal-phase (NP), reverse-phase (RP), and hydrophilic liquid interaction columns (HILICs) are commonly used in metabolomics. RP columns (such as C8 or C18) are highly popular for global profiling studies. Most modern RP columns have sub-2 μm particles. These small, robust, reproducible stationary phases combined with higher pressure pumping systems form the techniques of ultra-performance liquid chromatography (UPLC) and ultra-high performance liquid chromatography (UHPLC). The small-diameter stationary phase provides higher separation power

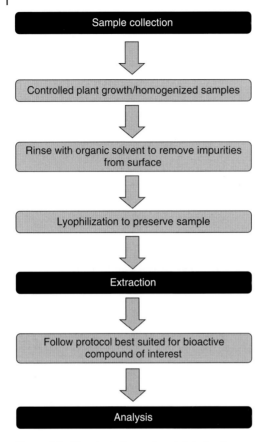

Figure 6.1 Flowchart of experimental design for preparation and analysis of bioactive natural products.

and peak capacity over traditional HPLC column materials but produces higher back pressure. Smaller-diameter columns and high-pressure pumps allowing for faster linear velocities yield faster elution, with less solvent waste and faster analysis [51, 52]. Recent work has shown that an increase in temperature of an RP column, which decreases solvent viscosity and therefore backpressure, allows for faster flow rates, which can decrease analysis time by two to threefold without affecting peak capacities or resulting in degradation of natural products over a large polarity range [53]. The main drawback of RP columns is that they do not have good separation or retention of polar metabolites [54]. This drawback has led to the rise of supplementary HILIC columns. These columns solve the problem of polar, hydrophilic compounds being unretained or eluting too quickly from traditional RP columns. This technique is well suited for predominately aqueous metabolomes such as biofluids [55]. This technique has recently been used to quantify polar sugars, amino acids, organic acids, and amines in grapes [10] and free amino acids in *Ginkgo biloba* seeds [56]. An RP column was also used with

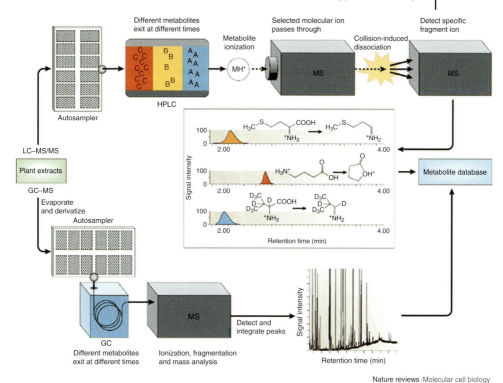

Figure 6.2 Small-molecule analysis by mass spectrometry with chromatographic separation [50]. (Reprinted by permission from Macmillan Publishers Ltd.: Nature Reviews Molecular Cell Biology, Copyright 2007.)

tandem MS/MS to monitor antioxidant activity of chlorogenic acid, forsythiaside, rutin, phillyrin, and phillygenin in *Forsythia suspensa* leaves [57]. Of course, LC is not limited to only these column types; HPLC in combination with thin-layer chromatography was used to investigate the antibacterial properties in *Ficus coronata* with the isolation of suberenol, dihydrocoumarin, chalepin, rutamarin, bergapten, skimminaine, 7-(1,1-dimethylallyloxy)coumarin, 7-hydroxycoumarin, and 7-(1,1-dimethylallyloxy)coumarin, along with 28 unidentified compounds [58]. LC–MS is well suited for targeted and nontargeted analysis, with global profiling as the most popular application. LC–MS has been used to successfully identify phenols, flavonoids, alkaloids, terpenoids, steroids, coumarins, and lignans [59, 60]. LC has additional benefits over GC–MS in that analytes can be recovered by fraction collection or concentration [54], that there is measurement of a broader range of metabolites [4], and that many metabolites can be detected without prior derivatization steps [61]. One of the current challenges of LC techniques is accurate identification of metabolite peaks. Although databases and other resources are outlined later in the chapter, recent studies used improved LC retention time prediction on an HILIC column, which allowed identification of organic acids, amino acids, amines, purines and pyrimidines, phosphates, alcohols, and more in *Trypanosoma brucei* when combined with accurate mass [62]. This is an example of an accurate mass–time tag, in which masses are compared to control substrates that have a predetermined mass and elution time. This allows for faster identification of predetermined substances without the need for additional MS/MS analysis [63]. Accurate mass–time tags [60] or simply accurate mass tags (AMTs) are also commonly used for better identification of peptides [64, 65] with possible applications to the metabolome. Another relatively recent improvement in LC techniques is the use of two-dimensional liquid chromatography (LC × LC). The major benefit of this method is the multiplication of separation capacities, resulting in high peak capacity and resolving power for complex samples [66]. Some recent uses in plant metabolomics include NP and RP columns (together NP × RP) to quantify polyphenols in red wines [67] and to analyze polyphenols and stevioside extracts in *Stevia rebaudiana* [68]. Hydrophilic interaction chromatography and an RP column (HILIC × RP) were used to observe pesticides in foods [69] and dihydrochalcones, flavanones, and flavones in the plant *Aspalathus linearis* [70]. A heart-cutting method (where only the interesting portion moves to the second dimension versus "comprehensive" LC × LC used in the previous examples) used size exclusion chromatography and an RP column (SEC × RP) to analyze flavonol glycosides and their health-protective properties in *Maytenus ilicifolia*, a traditional Brazilian medicinal plant [71].

6.3.2
Gas Chromatography–Mass Spectrometry

Gas chromatography is another popular technique for profiling that is relatively low cost and provides high separation efficiencies [49], effective protocols, and faster scan times [4]. It is well-suited for volatile compounds, although nonvolatile compounds with proper derivatization can also be detected. Only 20% of organic compounds can be detected by GC–MS without prior treatment [72]. In order to make compounds volatile, a two-step derivatization procedure is most commonly applied to polar functional groups [73]. Methoxyamination followed by silylation is the most acceptable form for GC–MS [73]. This additional preparation adds a layer of complexity and room for experimental error not seen in other techniques [61]. However, ion suppression seen from coeluting compounds in LC–MS is "virtually absent" and assignment of the identity of peaks with databases is many times more direct because of "extensive and reproducible" fragmentation patterns when used in full-scan mode [74]. Electron ionization is the most commonly paired ionization technique because it is robust, highly reproducible, and less susceptible to matrix effects [75]. GC is commonly used in metabolic profiling of bioactives, for example, in the investigation of medicinal properties of pungent alkamides in *Spilanthes acmella* [76]. Just as with LC, GC can also be used in a multidimensional manner. GC × GC is an orthogonal technique [77]. Most metabolomics studies use a nonpolar stationary phase in the first dimension followed by a midpolarity stationary phase in the second dimension although other combinations can be used [78]. GC×GC has been used to profile rice and was shown to have improved resolution over single-dimension GC [79]. Another example of GC×GC was the analysis of Salvinorin A, a hallucinogen, from *Salvia divinorum* [80]. Another newer technique using retention time locking GC–MS-selected ions monitoring was shown to have higher sensitivity, linearity, and data quality compared to traditional GC–MS when used in a pseudo-targeted metabolomics study [81].

6.3.3
Capillary Electrophoresis–Mass Spectrometry

Capillary electrophoresis–mass spectrometry (CE–MS) is another well-known separation technique often used for polar and charged compounds [77]. CE–MS has very high resolution, good throughput, and can simultaneously identify all charged molecular weight compounds in a sample [77, 78]. APCI is the most commonly paired technique [75], although DESI, ESI, and direct analysis in real time (DART) ionization are also used [79]. CE–MS has begun to gain even more popularity in plant metabolomics, with studies of phenylpropanoids, flavonoids, phenolic acids, and iridoid glycosides in lemon verbena (*Lippia citriodora*) extract [80], alkaloids, and other metabolites of traditional herbal medicine Toki-Shakuyaku-San [81] and amino acid content in fruit development of *Elaeis guineensis* [82]. CE–MS has also been used to target alkaloids in a variety of psychoactive plant extracts that make up the recreational drug kratom and alkaloids in *Nelumbo nucifera* [83, 84]. It has also been used to investigate the antioxidant, antimicrobial, and other health-beneficial effects likely tied to metabolites such as phenolic acids, flavonoids, organic acids, vitamins, and phytohormones in avocado [85]. CE–MS is more difficult to interface with MS because of low flow rates (nanoliter per minute) and the need to maintain a closed electrical circuit [86].

6.4
Mass Spectrometry

MS is the dominant technique in the study of bioactive plant metabolites. Mass spectrometers produce mass spectra representative of the elemental composition of an analyte presented as the mass-to-charge ratio (m/z). The field of MS continues to grow and develop with the advent of new machines and techniques for metabolite identification. Mass spectrometers can be run as single MS, tandem MS, or MS^n studies. These techniques vary depending on the goal of the research and the machines available. Tandem MS (MS/MS) is a popular technique that in itself is insufficient for metabolite identification but, combined with manual interpretation and growing MS/MS databases, can give a general idea of structure [87]. The mass analyzer has the ability to separate selected ions and further fragment them as a way to elucidate the structure. This technique can be used with a pair of magnetic sector or quadrupole instruments, a single ion cyclotron resonance or quadrupole ion trap [88]. MS^n furthers this technique, in which multiple stages of mass spectrometric analyses are carried out with events occurring between each stage. MS/MS and MS^n are preferred for metabolite identification because of the variety of tandem MS databases and greater identification success through fragmentation patterns.

6.4.1
Time of Flight

The time-of-flight mass (TOF) spectrometer measures the time it takes for ions to move from a fixed point to the detector when accelerated with a high-voltage pulse. The flight time over that distance is dependent on the particle's mass. Linear TOF instruments have a defined starting time for particles, while reflectron instruments utilize an ion optic device to reverse particles' flight to minimize the distribution time and energy for improved mass accuracy and spectra resolution. The TOF instrument is often used in metabolomics applications, such as in the profiling of *Panax notoginseng* [89], and is often coupled with a chromatography system.

The quadrupole time-of-flight (Q-TOF) mass spectrometer differs from single-stage mass spectrometers in that it is capable of tandem MS experiments and additional scanning types such as product ion and selected reaction monitoring [90]. The hybrid electrospray ionization-quadrupole time-of-flight (ESI-Q-TOF) was used, for example, to globally profile glycosides, phenylpropanoids, isothiocyanates, nitriles, indoles, and isoflavones in *Arabidopsis thaliana* [91]. This technique has high resolution and sensitivity and therefore is well-suited to non-targeted metabolomic studies. Another example of this work is the profiling of the tomato metabolome by LC–ESI-Q-TOF [92]. A comparison of triple quadrupole (QqQ), TOF, and Q-TOF instruments in identification of pesticide residues in a variety of fruits favored the TOF instrument because of its *sensitivity* in full spectrum mode and accurate mass measurements compared to the other instruments [93]. A comparison of quadrupole, TOF, and Fourier transform mass analyzers showed that the TOF and Fourier transform obtained high-resolution mass spectra through a broader mass range with greater sensitivity than the quadrupole for LC–MS applications [94].

6.4.2
Quadrupole Mass Filter

The quadrupole mass spectrometer is a mass filter that has the capability to filter masses with combined direct current and radiofrequency potentials on the quadrupole rods. Single-quadrupole machines can do selected ion monitoring (SIM) to reduce chemical noise and improve detection limits of targeted analytes, but this pales in comparison with the multiple reaction monitoring (MRM) capabilities of the QqQ mass filter. Triple quadrupoles are highly popular for MS/MS analyses because of their ease of coupling to LC [95] for MS/MS studies [56]. Their advantages over the single-mass filter include an improved signal-to-noise ratio, more reliable identification of analytes, and better accuracy and reproducibility [96]. The first and last quadrupole work as mass filters while the middle quadrupole, or multipole, functions as a collision cell to induce fragmentation of the molecular ion. QqQ mass analyzers are also often used because of the elimination of time delays often found in chromatographic techniques [97]. QqQ have been used to identify a variety of bioactive phytochemicals. Examples of these include flavonoids in the pericarp of *Citrus reticulata* [98], glycosides and furfural derivatives in *Rehmanniae Radix*, the root of *Rehmannia glutinosa* [99], and several phenolic compounds, cyclic peptides, and amides in *Cortex Lycii*, the root bark of *Lycium barbarum* [100].

6.4.3
Ion Traps (Orbitrap and Linear Quadrupole (LTQ))

Conventional ion trap mass spectrometers differ from quadrupoles in that they produce a three-dimensional quadrupole field, making them highly efficient [101]. These 3D ion traps are physically small and relatively inexpensive but are highly sensitive and capable of MS^n analysis [102]. The orbitrap instrument traps ions radially along a central spindle electrode surrounded by an outer electrode, which allows the machine to identify m/z values from the harmonic ion oscillations [103]. These instruments can be used alone or hybridized for greater resolution and MS^n capabilities [104]. A quadrupole linear ion trap (LTQ) was recently used in conjunction with an FTICR (Fourier transform ion cyclotron resonance) setup to identify glucosinolates in broccoli (*Brassica oleracea* L. var. *italica*), cauliflower (*Brassica oleracea* L. var. *botrytis*), and rocket salad (*Eruca sativa*) with high mass accuracy and sensitivity [105]. Recently a quadrupole orbitrap instrument was designed with improvement in resolution for proteomics studies [106]. A study comparing the orbitrap to the Q-TOF instrument found that both machines were highly efficient with comparable sensitivity and mass accuracy. The orbitrap provided slightly lower detection limits for some compounds while the Q-TOF had better spectral accuracy and lower analytical variability but both were found comparable for plant metabolomics analysis [107]. Unlike the Q-TOF, ion trap machines can measure MS^n data tandem in time instead of space. The limiting factor for this technique is residual ion intensity [108]. Mycotoxins in wheat flour, barley flour, and crisp bread were successfully identified with LC−high-energy collision dissociation (HCD)−high-resolution mass spectrometry (HRMS) with an orbitrap. The protocol was shown to have similar detection limits, recovery, and repeatability when compared to a QqQ method [109].

6.4.4
Fourier Transform Mass Spectrometry

Fourier transform ion cyclotron resonance mass spectrometry (FT-ICR-MS) is one of the most powerful instruments available for mass accuracy, detection sensitivity, and resolution [110, 111]. This instrument measures m/z ratios through the cyclotron frequency of ions in a fixed magnetic field. The above characteristics make it an ideal instrument for the complex mixtures often dealt with in metabolomics applications [112]. Its mass accuracy (potentially in excess of 10^6) can often allow identification of small molecules by accurate mass alone [113]. FT-ICR is also compatible with ESI, APCI, atmospheric pressure photoionization (APPI), MALDI, electron ionization (EI), and chemical ionization (CI), increasing its versatility [113]. FT-ICR has recently been used to find biologically significant metabolites containing heteroatoms in onion bulbs through isotopic labeling [114] and for profiling jasmonate-treated plant tissues [115]. FT-ICR is overall highly ideal for high-throughput analysis of metabolites in plants because of its high mass accuracy over a relatively short data acquisition time [116].

6.4.5
Ion Mobility Mass Spectrometry

Ion mobility mass spectrometry (IMMS) has only recently become more utilized in analytical experiments because of two-dimensional separation with the use of an electrodynamic field instead of the previous electrostatic field [117]. IMMS separates ions by size/charge ratio in a buffer gas in a charged electric field, which makes it a powerful tool for separation of complex structures and metabolite identification [118]. IMMS has recently been used to globally profile metabolites such as inorganic ions, volatile alcohols, ketones, and amino and nonamino organic acids in *Escherichia coli* [119] and human blood [120]. These techniques will hopefully translate rapidly to plant metabolome profiling. Recently a new ion mobility quadrupole time-of-flight mass spectrometer (IM-Q-TOF) specifically designed for biological sample analyses was introduced [121, 122]. High-field asymmetric waveform ion mobility mass spectrometry (FAIMS) showed somewhat improved sensitivity and better signal-to-noise ratios over traditional ESI-MS for amino acid determination of 20 pure samples of proteinogenic amino acids [123]. FAIMS also shows promise in recognizing the chirality of small molecules [124]. This technique will continue to develop and may add new experimental direction in the profiling of bioactive compounds.

6.5
Mass Spectrometric Imaging

Mass spectrometric imaging is an increasingly popular technique in the field of plant metabolomics. The focus on plants as living, dynamic tissues has

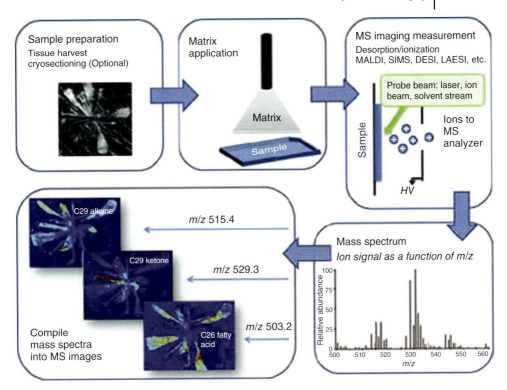

Figure 6.3 Experimental scheme for mass spectrometry imaging experiments [48]. (Reprinted by permission from John Wiley and Sons: Plant Journal, Copyright 2012.)

necessitated a movement toward investigation of different concentrations of metabolites within solid plant material. Even a small disturbance of a plant can cause a cascade of metabolic activity that inherently changes the plant's profile [125]. The workflow of a mass spectrometric imaging experiment is shown in Figure 6.3 [48]. All ionization sources work the same way: the source moves across the sample in an X,Y coordinate manner while the mass spectrometer records the mass spectrum for each laser-sized "pixel" [126]. Ions observed in the spectra can be assigned color and spatial visualization on an optical image of the sample [127]. There are several ionization techniques currently employed in mass spectrometric imaging that vary in their method and applications.

6.5.1
MALDI-MS

MALDI is one of the most popular imaging techniques. It relies on a UV or IR laser to ionize molecules [128]. It produces singly charged ions with a spatial resolution between 10 and 100 µm [48]. Sample preparation for MALDI involves covering the tissue uniformly with a selected matrix by robotic spotting, nebulization, or sublimation [129]. The most common matrices used for small molecules are α-cyano-4-hydroxycinnamic acid (CHCA) or 2,5-dihydroxybenzoic acid (2,5-DHB), but other matrices (such as ionic liquid matrices) exist [130]. These ionic liquid matrices were shown in one study to provide better spectral quality, crystallization on tissues, analysis duration, and resistance to laser irradiation [131]. Sample preparation is of utmost importance in reproducibility and optimization. TOF-MS has high throughput and is often used in combination with MALDI [132, 133]. However, its mass resolving power can result in losses of important chemical and spatial information [134]. Other options include the use of FT-ICR-MS and orbitrap instruments for better sensitivity and resolution [132, 135]. Another improvement needed is an increase in spatial resolution, which is limited by the size of matrix crystals (often close in size to the analyte of interest) and laser focus [129, 136]. MALDI has been used in the investigation of bioactive compounds to identify spatial placement of amino acids and sugars in wheat (*Triticum aestivum* var. axona) [137], the distribution of flavanols and dihydrochalcones in golden delicious apples [46], and the distribution of toxic glycoalkaloids in potato tuber tissue (*Solanum tuberosum*) [6]. It has also been used in the rough visualization of phospholipids in barley (*Hordeum vulgare*) and tobacco roots (*Nicotiana tabacum*) [138] and the identification of isoflavones and other unidentified metabolites in the leaves of medicinal plant *Leptoderris fasciculata* [139].

6.5.2
SIMS-MS

Secondary ion mass spectrometry (SIMS) uses a primary ion beam to desorb and ionize ambient molecules (secondary ions) [48]. The greatest advantage of SIMS over other ionization sources is its spatial resolution and that it has no need for sample pretreatment. However, because of its high fragmentation and low ionization efficiency, it is compatible with small-molecule research only [140]. SIMS produces singly charged ions with a spatial resolution of $0.2-3\,\mu m$ [48]. SIMS can be run in static or dynamic mode. TOF-SIMS is able to localize molecules as large as m/z $1000-1500$ with a routine resolution from 400 to $1-2\,\mu m$ [141], making it a very attractive tool for mass spectrometric imaging. Recent reports estimate that this quickly evolving technique can have a spatial resolution as low as 500 nm when not limited by static analysis conditions [142]. SIMS has been used to investigate silicon transport in root and shoot tissues of annual blue grass and orchard grass [143], imaging of flavonoids in *Pisum sativum* and *Arabidopsis thaliana* seeds [141], and localization of iron in durum wheat grain (*Triticum durum*) [144].

6.5.3
DESI-MS

In DESI-MS, a spray nozzle provides a high-pressure nebulizing gas to create charged microdroplets that are then directed toward the analyte. The collision induces secondary droplets combined with desorbed analyte that are suctioned into an extended atmospheric pressure ion transfer capillary that moves analytes into the vacuum region [145]. DESI can produce singly or multiply charged ions with a resolution of 200 μm [48]. DESI and LAESI (see below) are most appropriate when the surface chemistry of the analyte is of interest [140]. The TOF-MS analyzer remains the mass analyzer of choice for both DESI and SIMS techniques because of its speed, sensitivity, and broad mass range detection. The Q-TOF mass spectrometer is also used for its ability to obtain MS/MS mass spectral data [134]. DESI has been used to visualize hydroxynitrile glucoside levels, a key component in growth of potato tubers [146]. DESI has also been used to image hyperforin, hypericin, pseudohypericin, protopseudohypericin, and protohypericin, small molecules that possibly promote the antibacterial, antiviral, or antidepressant activity in *Hypericum perforatum* to great reproducibility and penetration of plant tissue [45].

6.5.4
LAESI-MS

LAESI is a soft ionization method that uses focused mid-IR laser radiation for ambient biological samples of sufficient water content [147, 148]. This atmospheric pressure technique (along with DESI) allows for less sample pretreatment and can study living plants [149]. The IR laser beam can produce singly or multiply charged analytes with a spatial resolution of 300–500 μm [48]. LAESI has been used to image cyanidins in onion cells and single cells on sour orange leaf [150], color variegation in zebra plant (*Aphelandra squarrosa*) leaf [147], onion (*Allium cepa*) and daffodil (*Narcissus pseudonarcissus*) epidermal cells [148], and leaf tissues in zebra plant and peace lily (*Spathiphyllum cochlearispathum*) [149].

6.5.5
LDI-MS and Others for Imaging

Laser desorption imaging (LDI)-MS is a relatively new technique that attempts to capitalize on the protocol of MALDI-MS but without the use of interfering matrices. This technique is still limited to low weight molecules [151]. LDI-MS has been used to find the distribution of naphthodianthrones and biflavanoids within the plants of the genus *Hypericum* as well as model plant *Arabidopsis* without the assistance of a matrix. LDI-MS was tentatively shown to have better spatial resolution than traditional MALDI imaging protocols [152] for flavonoids, xanthones, biflavanoids, naphthodianthrones, and prenylated phloroglucinols in *Hypericum*. LDI-MS was also used for the imaging of biofuels in *Miscanthus giganteus*, but was shown to be limited in providing complete chemical information and good spatial resolution with a single sample preparation method [153]. A popular subset of this technique, colloidal graphite-assisted LDI (GALDI), successfully imaged saccharides, glycoalkaloids, flavonoids, organic acids, and glucosinolates from model plant *Arabidopsis thaliana* [154].

There are other innovative methods to define metabolites spatially, although with more time-consuming data collection. One study simply used laser microdissection and UHPLC–MS to identify 38 chromatographic peaks of small bioactive molecules such as anthraquinones, phenylbutanone, glucopyranosides, stilbene and its glycosides, tannins, and procyanidins and their gallates from a range of single-cell tissue preparations in *Rheum palmatum* [155] without the use of ambient ionization. Another recent technique developed for *in vivo* studies is leaf spray MS. This technique ionizes chemicals directly from plant tissue with minimal sample preparation and disturbance to the plant [156].

6.6
Data Analysis

6.6.1
Data Processing

After data are obtained, they must be exported. MS gathers an extremely large amount of raw data, especially if combined with a hyphenated technique such as LC–MS. Automated software is needed to take the raw data and interpret them for peak alignment, metabolite identity, and quantification. This is done in order to compare mass and intensity as well as retention time of metabolites in the case of chromatography or CE [157]. The typical steps of data processing are filtering (removal of measurement noise), feature detection (finding representation of ions from a raw signal), alignment (clustering of measurements across samples), and normalization (removal of unwanted systematic variation. [158]. Clearly, good data processing is of utmost importance. Many mass spectrometers come with their own company's version of a data processor. However, there is also a variety of freely available metabolomic data processors such as MetAlign [158, 159]. Most raw data need to be converted to the software file types netCDF or mzXML before processing by the chosen software [160, 161].

6.6.2
Data Analysis Methods

As the data obtained become more complex, scientists find that they have to adapt to a more multidisciplinary approach to interpret results. Multivariate data analysis is popular as a way to visualize not only information on a single metabolite, but also how metabolites compare in a sample. A good beginner's review on the subject is covered by Jansen *et al.* [162]. Principal component analysis (PCA) is commonly used because it can provide an overview of all samples by projecting and clustering samples and highlighting differences [75]. PCA is an unsupervised, orthogonal technique that converts a set of possibly correlated variables into a set of linearly uncorrelated values (principal components). PCA can be used alone or in conjunction with other statistical methods. Recent applications include quantification of sugars, amino acids, organic acids, and amines in eight red and green grape varieties [10], comparison of amino acids in *Ginkgo biloba* seed [56], investigation of regional differences of primary and secondary metabolites such as nicotine in Chinese tobacco leaf (*Nicotiana tabacum*) [163], study of cold tolerance and associated flavonoids and unknown small molecules produced in *Arabidopsis thaliana* [14], and profiling of wheat lines *Triticum turgidum and T. aestivum* [164]. Analysis of variance (ANOVA) is a variation of PCA that has been used as a supplemental technique in nontargeted metabolomics profiling [165]. Another prevalent data analysis technique is orthogonal partial least squares discriminant analysis (OPLS-DA). The main advantage of this technique is to correctly focus data analysis to the question at hand. PCA will occasionally overfocus and divert the model away from the information of interest [166]. OPLS-DA is often coupled with PCA as a following technique [167]. OPLS-DA has been used to investigate metabolic variances across a range of factors in *Allium cepa* and *Narcissus pseudonarcissus* cells [150]. It has also been used for genetically modified metabolome equivalence assessment in tomatoes [168], for determining changes in phenolic metabolism through the shikimate, phenylpropanoid, and flavonoid pathways of *Nicotiana tabacum* cells [169], and to determine novel functions of *Arabidopsis thaliana* proteins [170]. Linear discriminant analysis (LDA) is also used in metabolomics studies. This is a supervised method that attempts to distinguish different groups based on the treatment they received, making it highly applicable to metabolomics studies [162]. This technique is often paired with PCA [171] and has recently been used to discriminate attributes in wine [172] and *in vivo* plant tissue typing [173].

6.6.3
Databases

A significant pitfall of the multitude of techniques available for metabolomic analysis is the wide variability in data reporting. The data themselves are often not enough; metadata, or data about the data, is also needed to accurately compare experiments. ArMet is a freely available data model designed specifically for plant metabolomics. Its guidelines for formal reporting encompass the entire experimental timeline and provide solid standards for data sharing across laboratories [174]. The Metabolomics Standards Initiative (MSI) [175] follows the work of ArMet and continues to homogenize reporting standards [176]. This work is far from complete and continues to be a bottleneck in phytochemical research. METLIN (*http://metlin.scripps.edu*) is the largest curated database of high-resolution tandem MS, with over 10 000 metabolites covered [177]; MassBank is another public repository for small molecule life sciences data [178]. PRIMe (Platform for RIKEN Metabolomics; *http://prime.psc.riken.jp/*) is another oft-used database. This includes MS2T, a tandem MS database that has over 1 million untargeted plant metabolomics entries [179], ReSpect, another plant-specific MS/MS resource, as well as a variety of other useful tools. Other databases often used include NIST (National Institute of Science and

Technology Database; *http://www.nist.gov/srd/nist1a.htm*), the Golm Metabolite Database (*http://gmd.mpimp-golm.mpg.de/*), and the Madison Metabolomics Consortium Database (MMCD; *http://mmcd.nmrfam.wisc.edu/*). FiehnLib also has a good assortment of small-molecule data for GC/MS [180]. META-PHOR is a European-based platform for plant metabolite analysis with an emphasis on species that are of interest for food quality and safety (*http://www.meta-phor.eu*). Another valuable resource available to the plant metabolomics researcher includes databases outlining metabolic pathways. These include the Kyoto Encyclopedia of Genes and Genomes Pathway Database, abbreviated KEGG (*http://www.genome.jp/kegg/pathway.html*), PlantCyc (*http://www.plantcyc.org/*), and Arabidopsis Reactome (*http://arabidopsisreactome.org/*). MetMask is recent software that successfully consolidated KEGG, PlantCyc, and ChEBI databases to create a more complete in-house reference library from external resources [181]. This consolidation would potentially allow greater ease of cross-referencing metadata among a large variety of resources.

6.7
Future Perspectives

Rapid improvements in all aspects of plant metabolomics in combination with the vast amount of unexplored plant material ensures further growth in the profiling of bioactive compounds in plants. The next step in metabolomics research is to provide a single database and comprehensive reporting standards. The completion of this herculean task will represent the threshold of a new era for researchers attempting to elucidate the possible bioactivity of secondary compounds. However, even in the absence of standardized databases and data reporting, the enormous complexity of the metabolome combined with research interests as diverse as the plant kingdom guarantees that the investigation into plant bioactive metabolites will be a new frontier for years to come.

References

1. Paulsen, B.S. (2010) in *Bioactive Compounds in Plants-Benefits and Risks for Man and Animals, 2010* (ed. A. Bernhoft), Norwegian Academy of Science and Letters, Oslo, pp. 18–29.
2. Sertürner, F.W. (2007) From waste products to ecochemicals: fifty years research of plant secondary metabolism. *Phytochemistry*, **68** (22–24), 2831–2846.
3. Bernhoft, A. (2010) A brief review on bioactive compounds in plants, in *Bioactive Compounds in Plants- Benefits and Risks for Man and Animals,* 2010 (ed. A. Bernhoft), Norwegian Academy of Science and Letters, Oslo.
4. Fernie, A.R. and Schauer, N. (2009) Metabolomics-assisted breeding: a viable option for crop improvement? *Trends Genet.*, **25** (1), 39–48.
5. Ara, I., Bukhari, N.A., Solaiman, D., and Bakir, M.A. (2012) Antimicrobial effect of local medicinal plant extracts in the Kingdom of Saudi Arabia and search for their metabolites by gas chromatography-mass spectrometric (GC-MS) analysis. *J. Med. Plant Res.*, **6** (45), 5688–5694.

6. Ha, M. *et al.* (2012) Direct analysis for the distribution of toxic glycoalkaloids in potato tuber tissue using matrix-assisted laser desorption/ionization mass spectrometric imaging. *Food Chem.*, **133** (4), 1155–1162.

7. Lorenzo Tejedor, M. *et al.* (2012) In situ molecular analysis of plant tissues by live single-cell mass spectrometry. *Anal. Chem.*, **84** (12), 5221–5228.

8. Kueger, S. *et al.* (2012) High-resolution plant metabolomics: from mass spectral features to metabolites and from whole-cell analysis to subcellular metabolite distributions. *Plant J.*, **70** (1), 39–50.

9. Gholipour, Y. *et al.* (2013) Living cell manipulation, manageable sampling, and shotgun picoliter electrospray mass spectrometry for profiling metabolites. *Anal. Biochem.*, **433** (1), 70–78.

10. Gika, H.G. *et al.* (2012) Quantitative profiling of polar primary metabolites using hydrophilic interaction ultrahigh performance liquid chromatography-tandem mass spectrometry. *J. Chromatogr. A*, **1259**, 121–127.

11. Lei, Z., Huhman, D.V., and Sumner, L.W. (2011) Mass spectrometry strategies in metabolomics. *J. Biol. Chem.*, **286** (29), 25435–25442.

12. Nakabayashi, R. *et al.* (2009) Metabolomics-oriented isolation and structure elucidation of 37 compounds including two anthocyanins from *Arabidopsis thaliana*. *Phytochemistry*, **70** (8), 1017–1029.

13. Romanik, G. *et al.* (2007) Techniques of preparing plant material for chromatographic separation and analysis. *J. Biochem. Biophys. Methods*, **70** (2), 253–261.

14. Vaclavik, L. *et al.* (2013) Mass spectrometry-based metabolomic fingerprinting for screening cold tolerance in *Arabidopsis thaliana* accessions. *Anal. Bioanal. Chem.*, **405** (8), 2671–2683.

15. Troncoso, C. *et al.* (2012) Induction of defensive responses in *eucalyptus globulus (Labill)* plants, against *Ctenarytaina eucalypti* (Maskell) (Hemiptera: Psyllidae). *Am. J. Plant Sci.*, **03** (05), 589–595.

16. Tsikou, D. *et al.* (2013) Cessation of photosynthesis in Lotus japonicus leaves leads to reprogramming of nodule metabolism. *J. Exp. Bot.*, **64** (5), 1317–1332.

17. Urbanczyk-Wochniak, E. *et al.* (2005) Profiling of diurnal patterns of metabolite and transcript abundance in potato (*Solanum tuberosum*) leaves. *Planta*, **221** (6), 891–903.

18. Kim, H.K. and Verpoorte, R. (2010) Sample preparation for plant metabolomics. *Phytochem. Anal.*, **21** (1), 4–13.

19. Brusotti, G. *et al.* (2014) Isolation and characterization of bioactive compounds from plant resources: the role of analysis in the ethnopharmacological approach. *J. Pharm. Biomed. Anal.*, **87**, 218–228.

20. Luque de Castro, M.D. and García-Ayuso, L.E. (1998) Soxhlet extraction of solid materials an outdated technique with a promising innovative future. *Anal. Chim. Acta*, **369**, 1–10.

21. Tiwari, P. *et al.* (2011) Phytochemical screening and extraction: a review. *Int. Pharm. Sci.*, **1** (1), 98–106.

22. Wang, L. and Weller, C.L. (2006) Recent advances in extraction of nutraceuticals from plants. *Trends Food Sci. Technol.*, **17** (6), 300–312.

23. Chan, C.H. *et al.* (2011) Microwave-assisted extractions of active ingredients from plants. *J. Chromatogr. A*, **1218** (37), 6213–6225.

24. Gao, S. *et al.* (2012) On-line ionic liquid-based dynamic microwave-assisted extraction-high performance liquid chromatography for the determination of lipophilic constituents in root of *Salvia miltiorrhiza* Bunge. *J. Sep. Sci.*, **35** (20), 2813–2821.

25. Tong, X., Xiao, X., and Li, G. (2011) On-line coupling of dynamic microwave-assisted extraction with high-speed counter-current chromatography for continuous isolation of nevadensin from *Lyeicnotus pauciflorus* Maxim. *J. Chromatogr. B Anal. Technol. Biomed. Life Sci.*, **879** (24), 2397–2402.

26. Zhang, H.-F., Yang, X.-H., and Wang, Y. (2011) Microwave assisted extraction of secondary metabolites from plants: current status and future directions. *Trends Food Sci. Technol.*, **22** (12), 672–688.

27. Henry, M.C. and Yonker, C.R. (2006) Supercritical fluid chromatography, pressurized liquid extraction, and supercritical fluid extraction. *Anal. Chem.*, **78**, 3909–3915.

28. Fornari, T. *et al.* (2012) Isolation of essential oil from different plants and herbs by supercritical fluid extraction. *J. Chromatogr. A*, **1250**, 34–48.

29. Wijngaard, H. *et al.* (2012) Techniques to extract bioactive compounds from food by-products of plant origin. *Food Res. Int.*, **46** (2), 505–513.

30. Huie, C.W. (2002) A review of modern sample-preparation techniques for the extraction and analysis of medicinal plants. *Anal. Bioanal. Chem.*, **373** (1–2), 23–30.

31. Herrero, M. *et al.* (2010) Supercritical fluid extraction: recent advances and applications. *J. Chromatogr. A*, **1217** (16), 2495–2511.

32. Picó, Y. (2013) Ultrasound-assisted extraction for food and environmental samples. *TrAC, Trends Anal. Chem.*, **43**, 84–99.

33. Vilkhu, K. *et al.* (2008) Applications and opportunities for ultrasound assisted extraction in the food industry — A review. *Innovative Food Sci. Emerg. Technol.*, **9** (2), 161–169.

34. Mustafa, A. and Turner, C. (2011) Pressurized liquid extraction as a green approach in food and herbal plants extraction: a review. *Anal. Chim. Acta*, **703** (1), 8–18.

35. Teo, C.C. *et al.* (2010) Pressurized hot water extraction (PHWE). *J. Chromatogr. A*, **1217** (16), 2484–2494.

36. Li, B.B., Smith, B., and Hossain, M.M. (2006) Extraction of phenolics from citrus peels. *Sep. Purif. Technol.*, **48** (2), 189–196.

37. Puri, M., Sharma, D., and Barrow, C.J. (2012) Enzyme-assisted extraction of bioactives from plants. *Trends Biotechnol.*, **30** (1), 37–44.

38. Sowbhagya, H.B. and Chitra, V.N. (2010) Enzyme-assisted extraction of flavorings and colorants from plant materials. *Crit. Rev. Food Sci. Nutr.*, **50** (2), 146–161.

39. Vas, G. and Vekey, K. (2004) Solid-phase microextraction: a powerful sample preparation tool prior to mass spectrometric analysis. *J. Mass Spectrom.*, **39** (3), 233–254.

40. Smith, R.M. (2003) Before the injection—modern methods of sample preparation for separation techniques. *J. Chromatogr. A*, **1000** (1-2), 3–27.

41. Enders, J.R. *et al.* (2012) A dual-column solid phase extraction strategy for online collection and preparation of continuously flowing effluent streams for mass spectrometry. *Anal. Chem.*, **84** (20), 8467–8474.

42. Han, D. and Row, K.H. (2010) Recent applications of ionic liquids in separation technology. *Molecules*, **15** (4), 2405–2426.

43. Tang, B. *et al.* (2012) Application of ionic liquid for extraction and separation of bioactive compounds from plants. *J. Chromatogr. B Anal. Technol. Biomed. Life Sci.*, **904**, 1–21.

44. Fontanals, N., Borrull, F., and Marcé, R.M. (2012) Ionic liquids in solid-phase extraction. *TrAC, Trends Anal. Chem.*, **41**, 15–26.

45. Thunig, J., Hansen, S.H., and Janfelt, C. (2011) Analysis of secondary plant metabolites by indirect desorption electrospray ionization imaging mass spectrometry. *Anal. Chem.*, **83** (9), 3256–3259.

46. Franceschi, P. *et al.* (2012) Combining intensity correlation analysis and MALDI imaging to study the distribution of flavonols and dihydrochalcones in Golden Delicious apples. *J. Exp. Bot.*, **63** (3), 1123–1133.

47. Vaikkinen, A. *et al.* (2012) Infrared laser ablation atmospheric pressure photoionization mass spectrometry. *Anal. Chem.*, **84** (3), 1630–1636.

48. Lee, Y.J. *et al.* (2012) Use of mass spectrometry for imaging metabolites in plants. *Plant J.*, **70** (1), 81–95.

49. Pereira, D.M. *et al.* (2010) Metabolomic analysis of natural products. *Rev. Pharm. Biomed. Anal.*, **01** (19), 19.

50. Last, R.L., Jones, A.D., and Shachar-Hill, Y. (2007) Towards the plant metabolome and beyond. *Nat. Rev. Mol. Cell Biol.*, **8**, 167–174.

51. Theodoridis, G., Gika, H.G., and Wilson, I.D. (2011) Mass spectrometry-based holistic analytical approaches for metabolite profiling in systems biology studies. *Mass Spectrom. Rev.*, **30** (5), 884–906.

52. Allwood, J.W. and Goodacre, R. (2010) An introduction to liquid chromatography-mass spectrometry instrumentation applied in plant metabolomic analyses. *Phytochem. Anal.*, **21** (1), 33–47.

53. Grata, E. *et al.* (2009) Metabolite profiling of plant extracts by ultra-high-pressure liquid chromatography at elevated temperature coupled to time-of-flight mass spectrometry. *J. Chromatogr. A*, **1216** (30), 5660–5668.

54. Bedair, M. and Sumner, L.W. (2008) Current and emerging mass-spectrometry technologies for metabolomics. *TrAC, Trends Anal. Chem.*, **27** (3), 238–250.

55. Cubbon, S. *et al.* (2010) Metabolomic applications of HILIC-LC-MS. *Mass Spectrom. Rev.*, **29** (5), 671–684.

56. Zhou, G. *et al.* (2013) Hydrophilic interaction ultra-performance liquid chromatography coupled with triple-quadrupole tandem mass spectrometry for highly rapid and sensitive analysis of underivatized amino acids in functional foods. *Amino Acids*, **44** (5), 1293–1305.

57. Jiao, J. *et al.* (2012) Comparison of main bioactive compounds in tea infusions with different seasonal Forsythia suspensa leaves by liquid chromatography–tandem mass spectrometry and evaluation of antioxidant activity. *Food Res. Int.*, **53**, 857–863.

58. Smyth, T.J. *et al.* (2012) Investigation of antibacterial phytochemicals in the bark and leaves of *Ficus coronata* by high-performance liquid chromatography-electrospray ionization-ion trap mass spectrometry (HPLC-ESI-MS(n)) and ESI-MS(n). *Electrophoresis*, **33** (4), 713–718.

59. Wu, H. *et al.* (2013) Recent developments in qualitative and quantitative analysis of phytochemical constituents and their metabolites using liquid chromatography-mass spectrometry. *J. Pharm. Biomed. Anal.*, **72**, 267–291.

60. Cuthbertson, D.J. *et al.* (2013) Accurate mass-time tag library for LC/MS-based metabolite profiling of medicinal plants. *Phytochemistry*, **91**, 187–197.

61. Viant, M.R. and Sommer, U. (2012) Mass spectrometry based environmental metabolomics: a primer and review. *Metabolomics*, **9** (S1), 144–158.

62. Creek, D.J. *et al.* (2011) Toward global metabolomics analysis with hydrophilic interaction liquid chromatography-mass spectrometry: improved metabolite identification by retention time prediction. *Anal. Chem.*, **83** (22), 8703–8710.

63. Pasa-Tolic, L. *et al.* (2004) Proteomic analyses using an accurate mass and time tag strategy. *Biotechniques*, **37** (4), 621–639.

64. Jaitly, N. *et al.* (2006) Robust algorithm for alignment of liquid chromatography-mass spectrometry analyses in an accurate mass and time tag data analysis pipeline. *Anal. Chem.*, **78**, 7397–7409.

65. Whitelegge, J.P. (2004) Mass spectrometry for high throughput quantitative proteomics in plant research: lessons from thylakoid membranes. *Plant Physiol. Biochem.*, **42** (12), 919–927.

66. Dugo, P. *et al.* (2008) Comprehensive multidimensional liquid chromatography: theory and applications. *J. Chromatogr. A*, **1184** (1-2), 353–368.

67. Dugo, P. *et al.* (2009) Comprehensive two-dimensional liquid chromatography to quantify polyphenols in red wines. *J. Chromatogr. A*, **1216** (44), 7483–7487.

68. Cacciola, F. *et al.* (2011) Employing ultra high pressure liquid chromatography as the second dimension in a comprehensive two-dimensional system for analysis of *Stevia rebaudiana* extracts. *J. Chromatogr. A*, **1218** (15), 2012–2018.

69. Kittlaus, S. *et al.* (2013) Development and validation of an efficient automated method for the analysis of 300 pesticides in foods using two-dimensional liquid chromatography-tandem mass spectrometry. *J. Chromatogr. A*, **1283**, 98–109.

70. Beelders, T. *et al.* (2012) Comprehensive two-dimensional liquid chromatographic analysis of rooibos *(Aspalathus linearis)* phenolics. *J. Sep. Sci.*, **35** (14), 1808–1820.

71. de Souza, L.M. *et al.* (2009) Heart-cutting two-dimensional (size exclusion x reversed phase) liquid chromatography-mass spectrometry analysis of flavonol glycosides from leaves of *Maytenus ilicifolia*. *J. Chromatogr. A*, **1216** (1), 99–105.

72. Meyer, V. (2013) *Practical High Performance Liquid Chromatography*, John Wiley & Sons, Inc., New York, p. 426.

73. Osorio, S., Do, P.T., and Fernie, A.R. (2012) Profiling primary metabolites of tomato fruit with gas chromatography mass spectrometry. *Methods Mol. Biol.*, **860**, 101–109.

74. Koek, M.M. *et al.* (2011) Quantitative metabolomics based on gas chromatography mass spectrometry: status and perspectives. *Metabolomics*, **7** (3), 307–328.

75. Okazaki, Y. and Saito, K. (2012) Recent advances of metabolomics in plant biotechnology. *Plant Biotechnol. Rep.*, **6** (1), 1–15.

76. Leng, T.C. *et al.* (2011) Detection of bioactive compounds from *Spilanthes acmella* (L.) plants and its various in vitro culture products. *J. Med. Plants Res.*, **5** (3), 371–378.

77. Urakami, K. *et al.* (2010) Quantitative profiling of illicium anisatum by capillary electrophoresis time of flight mass spectrometry. *Biomed. Res.*, **31**, 161–163.

78. Soga, T. and Heiger, D.N. (2000) Amino acid analysis by capillary electrophoresis electrospray ionization mass spectrometry. *Anal. Chem.*, **72**, 1236–1241.

79. Chang, C. *et al.* (2013) Online coupling of capillary electrophoresis with direct analysis in real time mass spectrometry. *Anal. Chem.*, **85** (1), 170–176.

80. Quirantes-Pine, R. *et al.* (2010) Characterization of phenolic and other polar compounds in a lemon verbena extract by capillary electrophoresis-electrospray ionization-mass spectrometry. *J. Sep. Sci.*, **33**, 2818–2827.

81. Iino, K. *et al.* (2011) Profiling of the charged metabolites of traditional herbal medicines using capillary electrophoresis time-of-flight mass spectrometry. *Metabolomics*, **8** (1), 99–108.

82. Teh, H.F. *et al.* (2013) Differential metabolite profiles during fruit development in high-yielding oil palm mesocarp. *PLoS One*, **8** (4), e61344.

83. Posch, T.N. *et al.* (2012) Nonaqueous capillary electrophoresis-mass spectrometry: a versatile, straightforward tool for the analysis of alkaloids from psychoactive plant extracts. *Electrophoresis*, **33** (11), 1557–1566.

84. Do, T.C.M.V. *et al.* (2013) Analysis of alkaloids in Lotus *(Nelumbo nucifera Gaertn.)* leaves by non-aqueous capillary electrophoresis using ultraviolet and mass spectrometric detection. *J. Chromatogr. A*, **1302**, 174–180.

85. Contreras-Gutiérrez, P. *et al.* (2013) Determination of changes in the metabolic profile of avocado fruits *(Persea americana)* by two capillary electrophoresis-mass spectrometry approaches (targeted and non-targeted). *Electrophoresis*, **34** (19), 2928–2942.

86. Seger, C., Sturm, S., and Stuppner, H. (2013) Mass spectrometry and NMR spectroscopy: modern high-end detectors for high resolution separation techniques – state of the art in natural product HPLC-MS, HPLC-NMR, and CE-MS hyphenations. *Nat. Prod. Rep.*, **30** (7), 970–987.

87. Matsuda, F. *et al.* (2009) MS/MS spectral tag-based annotation of non-targeted profile of plant secondary metabolites. *Plant J.*, **57** (3), 555–577.

88. de Hoffmann, E. (1996) Tandem mass spectrometry: a primer. *J. Mass Spectrom.*, **31**, 129–137.

89. Chan, E.C. *et al.* (2007) Ultra-performance liquid

chromatography/time-of-flight mass spectrometry based metabolomics of raw and steamed *Panax notoginseng*. *Rapid Commun. Mass Spectrom.*, **21** (4), 519–528.

90. Choi, B.K. *et al.* (2003) Comparison of Quadrupole, Time-of-Flight, and Fourier Transform Mass Analyzers for LC–MS Applications. *Current Trends in Mass Spectrom.*, **18** (5S), S24–S31.

91. von Roepenack-Lahaye, E. *et al.* (2004) Profiling of Arabidopsis secondary metabolites by capillary liquid chromatography coupled to electrospray ionization quadrupole time-of-flight mass spectrometry. *Plant Physiol.*, **134** (2), 548–559.

92. Moco, S. *et al.* (2006) A liquid chromatography-mass spectrometry-based metabolome database for tomato. *Plant Physiol.*, **141** (4), 1205–1218.

93. Grimalt, S. *et al.* (2010) Quantification, confirmation and screening capability of UHPLC coupled to triple quadrupole and hybrid quadrupole time-of-flight mass spectrometry in pesticide residue analysis. *J. Mass Spectrom.*, **45** (4), 421–436.

94. Choi, B.K. *et al.* (2003) Comparison of quadrupole, time-of-flight, and fourier-transform mass analyzers for LC-MS applications. *Curr. Trends Mass Spectrom.*, **18**, 524–531.

95. Brunnee, C. (1987) The ideal mass analyzer: fact or fiction? *Int. J. Mass Spectrom.*, **76**, 125–237.

96. Schreiber, A. (2010) *Advantages of Using Triple Quadrupole over Single Quadrupole Mass Spectrometry to Quantify and Identify the Prescence of Pesticides in Water and Soil Samples*, ABSCIEX Food & Environmental.

97. Yost, R.A. and Enke, C.G. (1979) Triple quadrupole mass spectrometry for direct mixture analysis and structure elucidation. *Anal. Chem.*, **51** (12), 1251–1264.

98. Liu, E.H. *et al.* (2013) Simultaneous determination of six bioactive flavonoids in Citri Reticulatae Pericarpium by rapid resolution liquid chromatography coupled with triple quadrupole electrospray tandem mass spectrometry. *Food Chem.*, **141** (4), 3977–3983.

99. Xu, J. *et al.* (2012) Simultaneous determination of iridoid glycosides, phenethylalcohol glycosides and furfural derivatives in *Rehmanniae Radix* by high performance liquid chromatography coupled with triple-quadrupole mass spectrometry. *Food Chem.*, **135** (4), 2277–2286.

100. Zhang, J.X. *et al.* (2013) Simultaneous determination of 24 constituents in *Cortex Lycii* using high-performance liquid chromatography-triple quadrupole mass spectrometry. *J. Pharm. Biomed. Anal.*, **77**, 63–70.

101. Hager, J.W. (2002) A new linear ion trap mass spectrometer. *Rapid Commun. Mass Spectrom.*, **16**, 512–526.

102. Hopfgartner, G. *et al.* (2004) Triple quadrupole linear ion trap mass spectrometer for the analysis of small molecules and macromolecules. *J. Mass Spectrom.*, **39** (8), 845–855.

103. Hu, Q. *et al.* (2005) The orbitrap: a new mass spectrometer. *J. Mass Spectrom.*, **40** (4), 430–443.

104. Makarov, A. *et al.* (2006) Performance evaluation of a hybrid linear ion trap/orbitrap mass spectrometer. *Anal. Chem.*, **78**, 2113–2120.

105. Lelario, F. *et al.* (2012) Establishing the occurrence of major and minor glucosinolates in Brassicaceae by LC-ESI-hybrid linear ion-trap and Fourier-transform ion cyclotron resonance mass spectrometry. *Phytochemistry*, **73** (1), 74–83.

106. Michalski, A. *et al.* (2011) Mass spectrometry-based proteomics using q exactive, a high-performance benchtop quadrupole orbitrap mass spectrometer. *Mol. Cell Proteomics*, **10**, M111.011015.

107. Glauser, G. *et al.* (2013) Ultra-high pressure liquid chromatography-mass spectrometry for plant metabolomics: a systematic comparison of high-resolution quadrupole-time-of-flight and single stage Orbitrap mass spectrometers. *J. Chromatogr. A*, **1292**, 151–159.

108. March, R.E. (1992) Ion trap mass spectrometry. *Int. J. Mass Spectrom. Ion Process.*, **118/119**, 71–135.

109. Lattanzio, V.M. *et al.* (2011) Quantitative analysis of mycotoxins in cereal foods by collision cell fragmentation-high-resolution mass spectrometry: performance and comparison with triple-stage quadrupole detection. *Food Addit. Contam. Part A Chem. Anal. Control Exposure Risk Assess.*, **28** (10), 1424–1437.

110. Forcisi, S. *et al.* (2013) Liquid chromatography-mass spectrometry in metabolomics research: mass analyzers in ultra high pressure liquid chromatography coupling. *J. Chromatogr. A*, **1292**, 51–65.

111. Han, J. *et al.* (2008) Towards high-throughput metabolomics using ultrahigh-field Fourier transform ion cyclotron resonance mass spectrometry. *Metabolomics*, **4** (2), 128–140.

112. Brown, S.C., Kruppa, G., and Dasseux, J.L. (2005) Metabolomics applications of FT-ICR mass spectrometry. *Mass Spectrom. Rev.*, **24** (2), 223–231.

113. Allwood, J.W. *et al.* (2012) Fourier transform ion cyclotron resonance mass spectrometry for plant metabolite profiling and metabolite identification. *Methods Mol. Biol.*, **860**, 157–176.

114. Nakabayashi, R. and Saito, K. (2013) Metabolomics for unknown plant metabolites. *Anal. Bioanal. Chem.*, **405**, 5005–5011.

115. Pollier, J. and Goossens, A. (2013) Metabolite profiling of plant tissues by liquid chromatography Fourier transform ion cyclotron resonance mass spectrometry. *Methods Mol. Biol.*, **1011**, 277–286.

116. Aharoni, A. *et al.* (2002) Nontargeted metabolome analysis by use of fourier transform ion cyclotron mass spectrometry. *OMICS*, **6** (3), 217–234.

117. Ridenour, W. *et al.* (2010) Structural characterization of phospholipids and peptides directly from tissue sections by MALDI traveling-wave ion mobility-mass spectrometry. *Anal. Chem.*, **82**, 1881–1889.

118. Kanu, A.B. *et al.* (2008) Ion mobility-mass spectrometry. *J. Mass Spectrom.*, **43** (1), 1–22.

119. Dwivedi, P. *et al.* (2007) Metabolic profiling by ion mobility mass spectrometry (IMMS). *Metabolomics*, **4** (1), 63–80.

120. Dwivedi, P., Schultz, A.J., and Hill, H.H. (2010) Metabolic profiling of human blood by high resolution ion mobility mass spectrometry (IM-MS). *Int. J. Mass Spectrom.*, **298** (1-3), 78–90.

121. Kurulugama, R. *et al.* (2013) *Development of a New Ion Mobility-Quadrupole Time-of-Flight Mass Spectrometer for High-Resolution and High Throughput Biological Sample Analyses*, Agilent Technologies.

122. Klein, C. *et al.* (2013) *Structural Analysis of Transmembrane Spanning Peptides by Drift Tube Based Ion Mobility-Mass Spectrometry*, Agilent Technologies, pp. 1–4.

123. McCooeye, M. and Mester, Z. (2006) Comparison of flow injection analysis electrospray mass spectrometry and tandem mass spectrometry and electrospray high-field asymmetric waveform ion mobility mass spectrometry and tandem mass spectrometry for the determination of underivatized amino acids. *Rapid Commun. Mass Spectrom.*, **20** (11), 1801–1808.

124. Wu, L. and Vogt, F.G. (2012) A review of recent advances in mass spectrometric methods for gas-phase chiral analysis of pharmaceutical and biological compounds. *J. Pharm. Biomed. Anal.*, **69**, 133–147.

125. Glauser, G. *et al.* (2011) Induction and detoxification of maize 1,4-benzoxazin-3-ones by insect herbivores. *Plant J.*, **68** (5), 901–911.

126. Weaver, E.M. and Hummon, A.B. (2013) Imaging mass spectrometry: from tissue sections to cell cultures. *Adv. Drug Deliv. Rev.*, **65**, 1039–1055.

127. Yang, Y.L. *et al.* (2009) Translating metabolic exchange with imaging mass spectrometry. *Nat. Chem. Biol.*, **5** (12), 885–887.

128. Li, Y., Shrestha, B., and Vertes, A. (2007) Atmospheric pressure molecular imaging by infrared MALDI mass spectrometry. *Anal. Chem.*, **79**, 523–532.

129. Balluff, B. *et al.* (2011) MALDI imaging mass spectrometry for direct tissue analysis: technological advancements and recent applications. *Histochem. Cell Biol.*, **136** (3), 227–244.

130. Kaspar, S. *et al.* (2011) MALDI-imaging mass spectrometry – an emerging technique in plant biology. *Proteomics*, **11** (9), 1840–1850.

131. Lemaire, R. *et al.* (2006) Solid ionic matrixes for direct tissue analysis and MALDI imaging. *Anal. Chem.*, **78**, 809–819.

132. Yukihira, D. *et al.* (2010) MALDI-MS-based high-throughput metabolite analysis for intracellular metabolic dynamics. *Anal. Chem.*, **82**, 4278–4282.

133. Miura, D. *et al.* (2010) Highly sensitive matrix-assisted laser desorption ionization-mass spectrometry for high-throughput metabolic profiling. *Anal. Chem.*, **82**, 498–504.

134. Amstalden van Hove, E.R., Smith, D.F., and Heeren, R.M. (2010) A concise review of mass spectrometry imaging. *J. Chromatogr. A*, **1217** (25), 3946–3954.

135. Wang, H.Y. *et al.* (2011) Analysis of low molecular weight compounds by MALDI-FTICR-MS. *J. Chromatogr. B Anal. Technol. Biomed. Life Sci.*, **879** (17–18), 1166–1179.

136. Esquenazi, E. *et al.* (2009) Imaging mass spectrometry of natural products. *Nat. Prod. Rep.*, **26** (12), 1521–1534.

137. Burrell, M., Earnshaw, C., and Clench, M. (2007) Imaging matrix assisted laser desorption ionization mass spectrometry: a technique to map plant metabolites within tissues at high spatial resolution. *J. Exp. Bot.*, **58** (4), 757–763.

138. Peukert, M. *et al.* (2012) Spatially resolved analysis of small molecules by matrix-assisted laser desorption/ionization mass spectrometric imaging (MALDI-MSI). *New Phytol.*, **193** (3), 806–815.

139. Kouamé, F.P.B.K. *et al.* (2013) Phytochemical investigation of the leaves of *Leptoderris fasciculata*. *Phytochem. Lett.*, **6** (2), 253–256.

140. Matros, A. and Mock, H.P. (2013) Mass spectrometry based imaging techniques for spatially resolved analysis of molecules. *Front Plant Sci.*, **4**, 89.

141. Seyer, A. *et al.* (2010) Localization of flavonoids in seeds by cluster time-of-flight secondary ion mass spectrometry imaging. *Anal. Chem.*, **82**, 2326–2333.

142. Fletcher, J.S. and Vickerman, J.C. (2013) Secondary ion mass spectrometry: characterizing complex samples in two and three dimensions. *Anal. Chem.*, **85** (2), 610–639.

143. Sparks, J.P. *et al.* (2010) Subcellular localization of silicon and germanium in grass root and leaf tissues by SIMS: evidence for differential and active transport. *Biogeochemistry*, **104** (1–3), 237–249.

144. Moore, K.L. *et al.* (2012) Localisation of iron in wheat grain using high resolution secondary ion mass spectrometry. *J. Cereal Sci.*, **55** (2), 183–187.

145. Monge, M.E. *et al.* (2013) Mass spectrometry: recent advances in direct open air surface sampling/ionization. *Chem. Rev.*, **113** (4), 2269–2308.

146. Li, B. *et al.* (2013) Visualizing metabolite distribution and enzymatic conversion in plant tissues by desorption electrospray ionization mass spectrometry imaging. *Plant J.*, **74** (6), 1059–1071.

147. Nemes, P. *et al.* (2008) Ambient molecular imaging and depth profiling of live tissue by infrared laser ablation electrospray ionization mass spectrometry. *Anal. Chem.*, **80**, 4575–4582.

148. Shrestha, B. and Vertes, A. (2009) In situ metabolic profiling of single cells by laser ablation electrospray ionization mass spectrometry. *Anal. Chem.*, **81** (20), 8265–8271.

149. Nemes, P., Barton, A.A., and Vertes, A. (2009) Three-dimensional imaging of metabolites in tissues under ambient conditions by laser ablation electrospray ionization mass spectrometry. *Anal. Chem.*, **81** (16), 6668–6675.

150. Shrestha, B., Patt, J.M., and Vertes, A. (2011) In situ cell-by-cell imaging and analysis of small cell populations by mass spectrometry. *Anal. Chem.*, **83** (8), 2947–2955.

151. Peterson, D.S. (2007) Matrix-free methods for laser desorption/ionization

mass spectrometry. *Mass Spectrom. Rev.*, **26** (1), 19–34.

152. Holscher, D. *et al.* (2009) Matrix-free UV-laser desorption/ionization (LDI) mass spectrometric imaging at the single-cell level: distribution of secondary metabolites of *Arabidopsis thaliana* and *Hypericum species*. *Plant J.*, **60** (5), 907–918.

153. Li, Z., Bohn, P.W., and Sweedler, J.V. (2010) Comparison of sample pre-treatments for laser desorption ionization and secondary ion mass spectrometry imaging of *Miscanthus x giganteus*. *Bioresour. Technol.*, **101** (14), 5578–5585.

154. Cha, S. *et al.* (2008) Direct profiling and imaging of plant metabolites in intact tissues by using colloidal graphite-assisted laser desorption ionization mass spectrometry. *Plant J.*, **55** (2), 348–360.

155. Liang, Z. *et al.* (2013) Profiling of secondary metabolites in tissues from *Rheum palmatum* L. using laser microdissection and liquid chromatography mass spectrometry. *Anal. Bioanal. Chem.*, **405** (12), 4199–4212.

156. Liu, J. *et al.* (2011) Leaf spray: direct chemical analysis of plant material and living plants by mass spectrometry. *Anal. Chem.*, **83** (20), 7608–7613.

157. Dettmer, K., Aronov, P.A., and Hammock, B. (2007) Mass spectrometry-based metabolomics. *Mass Spectrom. Rev.*, **26**, 51–78.

158. Katajamaa, M. and Oresic, M. (2007) Data processing for mass spectrometry-based metabolomics. *J. Chromatogr. A*, **1158** (1-2), 318–328.

159. Lommen, A. and Kools, H.J. (2012) MetAlign 3.0: performance enhancement by efficient use of advances in computer hardware. *Metabolomics*, **8** (4), 719–726.

160. Pedrioli, P.G. *et al.* (2004) A common open representation of mass spectrometry data and its application to proteomics research. *Nat. Biotechnol.*, **22** (11), 1459–1466.

161. Sugimoto, M. *et al.* (2012) Bioinformatics tools for mass spectroscopy-based metabolomic data processing and analysis. *Curr. Bioinform.*, **7**, 96–108.

162. Jansen, J.J. *et al.* (2010) The photographer and the greenhouse: how to analyse plant metabolomics data. *Phytochem. Anal.*, **21** (1), 48–60.

163. Zhang, L. *et al.* (2013) Metabolic profiling of chinese tobacco leaf of different geographical origins by GC-MS. *J. Agric. Food Chem.*, **61**, 2597–2605.

164. Matthews, S.B. *et al.* (2012) Metabolite profiling of a diverse collection of wheat lines using ultraperformance liquid chromatography coupled with time-of-flight mass spectrometry. *PLoS One*, **7** (8), e44179.

165. Li, Y. *et al.* (2012) A novel approach to transforming a non-targeted metabolic profiling method to a pseudo-targeted method using the retention time locking gas chromatography/mass spectrometry-selected ions monitoring. *J. Chromatogr. A*, **1255**, 228–236.

166. Wiklund, S. *et al.* (2008) Visualization of GC TOF-MS-based metabolomics data for identification of biochemically interesting compounds using OPLS class models. *Anal. Chem.*, **80**, 115–122.

167. Zhang, A. *et al.* (2013) Metabolomics study on the hepatoprotective effect of scoparone using ultra-performance liquid chromatography/electrospray ionization quadruple time-of-flight mass spectrometry. *Analyst*, **138** (1), 353–361.

168. Kusano, M. *et al.* (2011) Covering chemical diversity of genetically-modified tomatoes using metabolomics for objective substantial equivalence assessment. *PLoS One*, **6** (2), e16989.

169. Madala, N.E. *et al.* (2013) Metabolomic analysis of isonitrosoacetophenone-induced perturbations in phenolic metabolism of *Nicotiana tabacum* cells. *Phytochemistry*, **94**, 82–90.

170. Diaz, C. *et al.* (2011) Determining novel functions of *Arabidopsis* 14-3-3 proteins in central metabolic processes. *BMC Syst. Biol.*, **5**, 192.

171. Allwood, J.W. *et al.* (2006) Metabolomic approaches reveal that phosphatidic and phosphatidyl glycerol phospholipids are major discriminatory non-polar metabolites in responses by

Brachypodium distachyon to challenge by *Magnaporthe grisea*. *Plant J.*, **46** (3), 351–368.

172. Cuadros-Inostroza, A. *et al.* (2010) Discrimination of wine attributes by metabolome analysis. *Anal. Chem.*, **82**, 3573–3580.

173. Judge, E.J. *et al.* (2011) Nonresonant femtosecond laser vaporization with electrospray postionization for ex vivo plant tissue typing using compressive linear classification. *Anal. Chem.*, **83** (6), 2145–2151.

174. Jenkins, H. *et al.* (2004) A proposed framework for the description of plant metabolomics experiments and their results. *Nat. Biotechnol.*, **22** (12), 1601–1606.

175. Fiehn, O. *et al.* (2007) The metabolomics standards initiative (MSI). *Metabolomics*, **3** (3), 175–178.

176. Fiehn, O. *et al.* (2008) Quality control for plant metabolomics: reporting MSI-compliant studies. *Plant J.*, **53** (4), 691–704.

177. Zhu, Z.J. *et al.* (2013) Liquid chromatography quadrupole time-of-flight mass spectrometry characterization of metabolites guided by the METLIN database. *Nat. Protoc.*, **8** (3), 451–460.

178. Horai, H. *et al.* (2010) MassBank: a public repository for sharing mass spectral data for life sciences. *J. Mass Spectrom.*, **45** (7), 703–714.

179. Sakurai, T. *et al.* (2013) PRIMe Update: innovative content for plant metabolomics and integration of gene expression and metabolite accumulation. *Plant Cell Physiol.*, **54** (2), e5.

180. Kind, T. *et al.* (2009) FiehnLib: mass spectral and retention index libraries for metabolomics based on quadrupole and time-of-flight gas chromatography/mass spectrometry. *Anal. Chem.*, **81**, 10038–10048.

181. Redestig, H. *et al.* (2010) Consolidating metabolite identifiers to enable contextual and multi-platform metabolomics data analysis. *BMC Bioinf.*, **11**, 214.

7
Antivenomics: A Proteomics Tool for Studying the Immunoreactivity of Antivenoms

Juan J. Calvete, José María Gutiérrez, Libia Sanz, Davinia Pla, and Bruno Lomonte

7.1
Introduction

Envenoming following snakebite is largely a neglected threat to public health in tropical and subtropical regions of Africa, Asia, Latin America, and Oceania, affecting some of the world's poorest rural communities. An estimated 5.5 million people are bitten by snakes each year, resulting in about 400 000 amputations, and between 20 000 and 125 000 deaths; however, the true scale of this "disease of poverty" may be much greater than these hospital-based statistics [1–5]. Despite these figures, which affect mainly those involved in subsistence farming activities in vast regions of the world, the burden of human suffering caused by snakebites was added to the World Health Organization's (WHO's) list of neglected tropical diseases only in April 2009 (*http://www.who.int/neglected_diseases/diseases/snakebites/en/*).

The timely parenteral administration of an appropriate antivenom remains, more than a century after the development of the first *serum antivenimeux* by Calmette [6, 7], and Phisalix and Bertrand [8–11], the only currently effective treatment for snakebite envenomings [12, 13]. Poor access to health services in these settings and, in some instances, a scarcity of antivenom, often leads to poor outcomes and considerable morbidity and mortality.

7.2
Challenge of Fighting Human Envenoming by Snakebites

More than 45 commercial or government antivenom producers are located around the world (*http://apps.who.int/bloodproducts/snakeantivenoms/database*) [14]. However, the lack of financial incentives in a technology that has remained relatively unchanged for the better part of the second half of the twentieth century, along with dwindling markets and stagnant leadership from global public health organizations, has made antivenom production a field of modest improvements and very little innovation. Snake antivenoms became scarce or

Analyzing Biomolecular Interactions by Mass Spectrometry, First Edition.
Edited by Jeroen Kool and Wilfried M.A. Niessen.
© 2015 Wiley-VCH Verlag GmbH & Co. KGaA. Published 2015 by Wiley-VCH Verlag GmbH & Co. KGaA.

nonexistent as poor commercial incentives forced some manufacturers to leave the market, and others to downscale production or increase the price, leading to an incongruous decline in affordability for these crucial antidotes to the millions of rural poor most at risk from snakebites in low- and middle-income countries [15]. To raise awareness of public health authorities on the relevance of the snakebite problem, and ensure supplies of effective antivenoms in deficitary parts of the world, several initiatives have emerged in the last decade ([16]; the Global Snakebite Initiative (GSI, *http://www.snakebiteinitiative.org*), [3, 14]). The GSI is an international collaborative project aimed at, among other goals, developing new regional polyvalent antivenoms for Asia and Africa [15]. A key technical issue of the GSI-promoted generation of novel antivenoms is the design of optimized immunization venom mixtures that ensure that the resulting antidotes will be effective against the highest number of venoms from snakes of medical concern across the geographical range where they will be used. This purpose is not trivial given (i) the well-documented occurrence of intraspecific individual venom variability, which is particularly relevant for highly adaptable and widely distributed species, in which allopatric venom variability represents a source of diversity of the pathological effects of envenoming [17, 18] and (ii) the emerging view that venom variation is evolutionarily a highly labile trait even among very closely related taxa [19]. For instance, rattlesnakes (*Crotalus* and *Sistrurus*) show variation in the presence of type I (high levels of metalloprotease and low toxicity) versus type II (low metalloprotease, high toxicity) venoms that shows no strong association with phylogeny [20]. These observations, and the finding of different evolutionary solutions within arboreal *Bothriechis* taxa for the same trophic purpose [21], strengthen the view that phylogeny cannot be invoked as a criterion for species selection for antivenom production. Clearly, the design of venom mixtures should be based on a rigorous analysis of epidemiological, clinical, proteomic, immunological, and toxicological information. Such methodologies include the biochemical and proteomic analysis of venoms, the toxicological profile of venom activities, and the analysis of immunological cross-reactivity of antivenoms toward homologous and heterologous venoms [22, 23]. Knowledge on the paraspecificity of antivenoms is of applied importance not only to optimize the production strategy of a novel antivenom, but also for defining the full clinical range of existing antivenoms.

7.3
Toolbox for Studying the Immunological Profile of Antivenoms

The analysis of the ability of an antivenom to neutralize the most relevant toxic activities of the snake venoms toward which it was designed is a preclinical requisite before it can be approved for medical use. Simple experimental protocols have been developed to assess the antivenom's neutralization efficacy of the most relevant pathophysiological effects of snakebite envenomings [23, 24]. Many of the

tests used for the preclinical evaluation of antivenoms involve the use of experimental animals, mostly mice. Since venoms induce pain and other deleterious
effects in these animals, there is a growing concern about this issue and interest
in finding suitable surrogate *in vitro* tests that can be used in the evaluation of
antivenoms [23]. As a complement to the in vivo neutralization assays and traditional immunological methods for assessing the preclinical neutralizing spectrum
of antivenoms, in 2008, the proteomics-centered protocol dubbed "antivenomics"
was developed [25]. Its major aim was to gain qualitative and quantitative information on both the set of toxins bearing antivenom-recognized epitopes and those
toxins exhibiting poor or null immunoreactivity. The combination of antivenomics
and neutralization assays provides a powerful toolbox for analyzing at the molecular level the preclinical efficacy of antivenoms. In addition, and assuming that the
degree of immunorecognition of a toxin by the antivenom's antibodies represents
a measure of the capability of this antivenom to neutralize the toxic activity of that
toxin, antivenomics analysis may assist in assessing the range of clinical application of currently commercial or experimental antivenoms, and in the development
of improved antivenoms on an immunologically sound basis [26, 27]. Conceptual
and operational principles of antivenomics are discussed below.

7.4
First-Generation Antivenomics

First-generation antivenomics (Figure 7.1) consisted of the immunodepletion of
antivenom-binding toxins on incubation of whole venom (WV) with antivenom
followed by precipitation of antigen–antibody complexes (Ag–Ab) out of
solution by the addition of a secondary antibody (e.g., rabbit anti-horse IgG),
or depletion of Ag–Ab from the reaction mixture using IgG-binding protein
A coupled to Sepharose beads [25]. Antigen–antibody complexes immunodepleted from the reaction mixture contain the toxins against which antibodies
in the antivenom are directed. By contrast, venom components that remain
in the supernatant are those that failed to raise antibodies in the antivenom,
or which triggered the production of low-affinity antibodies. The fraction of
non-immunodepleted molecules (%NR) can be estimated as the relative ratio of
the chromatographic areas (CA) of the toxin peaks in the supernatant and in WV:
%NR = (NR/WV) × 100. Based on the extent of immunoprecipitation, toxins were
classified as non-immunoprecipitated (proteins present in the supernatant, which
failed to raise antibodies in the antivenom, or which triggered the production
of very low-affinity antibodies), toxins exhibiting variable degree of immunoprecipitation (proteins that generated low- to high-affinity antibodies), and
toxins quantitatively immunodepleted from the venom (highly antigenic proteins
bearing antigenic determinants for very high-affinity antivenom antibodies).
These components can be easily identified by comparison of a reverse-phase
high-performance liquid chromatography (RP-HPLC) separation of the non-
precipitated fraction (supernatant of immunoprecipitation) with that of the WV

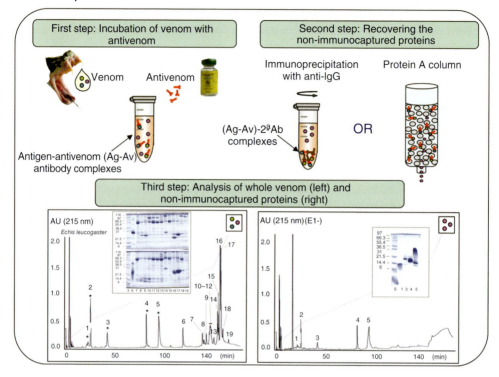

Figure 7.1 First-generation antivenomics. Scheme of the first-generation "antivenomics" analytical strategy. Left display, depletion of venom proteins immunoreactive with ICP polyvalent (Crotalinae) antivenom antibodies. Right display, panel A, reverse-phase-HPLC separations of whole *Echis leucogaster* (El) venom proteins, and panel B, the non-antivenom-binding proteins recovered after incubation of the venom–antivenom mixture (step 1) with rabbit antihorse IgG or protein A–Sepharose beads (step 2) [28]. Column eluates were monitored at 215 nm. Step 3, the non-immunocaptured proteins were identified using the snake venomics approach described in the text and schematized in Figure 7.2. Notice that the excess of rabbit antihorse IgGs used to deplete venom–antivenom complexes in step 2 elutes at the end of the chromatogram, interfering with the detection of certain venom components.

whose toxin composition had been previously characterized by a snake venomics approach (Figure 7.1) [22, 29].

7.5
Snake Venomics

The last decade has witnessed the development of techniques and strategies for assessing the toxin composition of snake venoms ("venomics"), directly (through proteomics-centered approaches) or indirectly (via high-throughput venom gland transcriptomics and bioinformatic analysis) in a relatively rapid and cost-effective manner. Bottom-up proteomics is the workhorse for venomic analysis [18].

Our snake venomics approach [22] (Figure 7.2) includes an initial step of fractionation of the crude venom by RP-HPLC followed by the characterization of each protein fraction by a combination of N-terminal sequencing and SDS-PAGE (sodium dodecyl sulfate polyacrylamide gel electrophoresis) (in nonreduced and reduced conditions to assess the occurrence of multi-subunit toxins). We favor RP-HPLC over two-dimensional electrophoresis because chromatographic separation allows the quantitative recovery of all the components present in the venom in an ideal solution (aqueous acetonitrile containing TFA (trifluoroacetic acid) or formic acid) for subsequent analysis by mass spectrometric determination of toxin-specific features, such as the native molecular mass, the occurrence of oligomeric toxin arrangements, and the number of sulfhydryl groups and

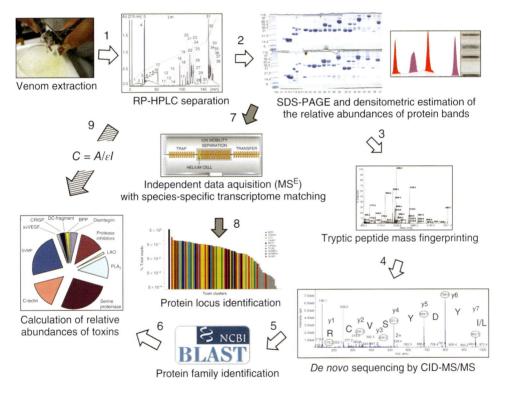

Figure 7.2 Snake venomics. Venom proteins are separated by reverse-phase-HPLC (1) and analyzed by SDS-PAGE (2) followed by in-gel tryptic digestion. Venom proteins are then identified by tryptic peptide mass fingerprinting (3), data-dependent de novo sequencing (4), and BLAST search (5), or data-independent MS^E (7) and product ion spectra matching to a species-specific venom gland transcriptomic database (8). Combining qualitative (protein family or locus identifications) (6) and quantitative (relative abundances derived from absorption data by the Lambert–Beer law, $c = A/\varepsilon L$) (9) data allows an accurate overview of the toxin composition of the venom.

disulfide bonds [29]. Furthermore, the initial part of the acetonitrile gradient of the RP chromatography resolves peptides and small proteins (0.4–7 kDa) that could not be recovered from a 2DE (two-dimensional electrophoresis) separation. Monitoring the eluate at the absorption wavelength of the peptide bond (190–230 nm) represents a reliable method for quantifying the relative abundance of the different pure venom components in the RP chromatogram. Thus, according to the Lambert–Beer law ($c[M] = A/\varepsilon l$), the calculated figures correspond to the weight% (g/100 g) of peptide bonds. To estimate the relative contribution of each toxin family expressed as protein molecules/100 molecules of total venom proteins, the weight percentages of peptide bonds of each family should be normalized for the number of peptide bonds (amino acids) in the full-length sequence of a representative member of the protein family. In the case of chromatographic peaks containing a mixture of proteins (as judged by SDS-PAGE and molecular mass determination by ESI-MS (electrospray-ionization mass spectrometry)), the relative contributions of the different proteins can be estimated by densitometry after SDS-PAGE separation.

For the characterization of the venom proteome, the protein(s) eluting in each RP-HPLC fraction are separated by SDS-PAGE and the Coomassie Brilliant Blue–stained protein bands are submitted to in-gel digestion using sequencing-grade trypsin. Due to the lack of reference genomes, and slow pace in venom gland transcriptomes sequencing, tryptic peptide mass fingerprinting matches very few, if any, proteins in the publically available databases (SwissProt or NCBI). Thus, most snake venomics projects require de novo sequencing of the in-gel-generated tryptic peptide (which in our laboratory involves data-dependent collision-induced dissociation (CID)–tandem mass spectrometry (MS/MS) in an ABI Sciex QTrap2000 or a Waters Synapt G2 system) followed by a database search using BLAST (basic local alignment search tool). This approach allows the elucidation of complete proteomes at gene family resolution. However, if a species-specific venom gland transcriptome comprising full-length toxin-coding sequences (or the species' genome sequence) is available, locus-specific resolution can be achieved using a data-independent LC–MS^E (liquid chromatography–mass spectrometry in an alternating energy mode) shotgun protocol [30]. The quantitative character of the venomics analysis can be preserved by performing the shotgun approach on the tryptic digests of the protein bands rather than on the whole, unfractionated venom.

7.6
Second-Generation Antivenomics

The first-generation antivenomics protocol (Figure 7.1) suffered from several shortcomings. Antivenoms are whole immunoglobulin molecules purified by fractional precipitation of plasma collected from hyperimmunized large animals, mainly horses and to a minor extent also donkeys, sheep, llamas, and camels, or proteolytic antigen-binding site fragments containing the variable regions of

H and L antibody chains generated by limited proteolysis of immunoglobulins with pepsin (F(ab')$_2$ and Fab' after partial reduction) or papain (Fab) [31] (Figure 7.3). Due to the absence in the market of molecules able to specifically and efficiently immunoprecipitate F(ab')$_2$ and Fab antivenoms produced by many manufacturers, first-generation antivenomics cannot be employed to investigate the immunological profile of these antibody fragments. Another limitation of the first-generation antivenomics protocol relates to the fact that the solubility and viscosity of the reaction mixture do not allow using molar ratios of antivenom:venom proteins higher than ~6−8, thus precluding the application of immunoprecipitation-based antivenomics to analyze low and very low affinity Ag−Ab complexes. Increasing the concentration of antivenom in the reaction mixture greatly increases the nonspecificity of the antivenomic assay [32]. In addition, the immunoprecipitated fraction contained predominantly IgG molecules from the primary and secondary antibodies, and is thus essentially not suitable for downstream venomics analysis.

Immunoaffinity chromatography followed by the proteomic analysis of the nonimmunocaptured (flow-through fraction) and the immunocaptured fractions constitute the basis of the "second-generation antivenomics" workflow [33] (Figure 7.4). Second-generation antivenomics was designed to overcome the

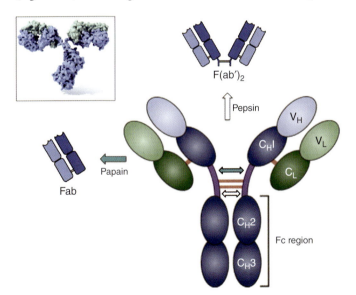

Figure 7.3 Structure of an IgG molecule and its proteolytically generated fragments typically comprising antivenoms. Structural domains of the variable and constant regions of the light chain (VL, CL) and heavy chain (VH, CH1, CH2, CH3) of an immunoglobulin molecule, as well as interchain disulfide bonds (S-S), are indicated. Bivalent F(ab')$_2$ (110 000 Da) fragments generated by proteolysis with pepsin contain two antigen-binding regions joined at the hinge through disulfide linkages. Fab (50 000 Da) is a monovalent fragment generated by digestion of IgG with papain, consisting of the VH, CH1 and VL, CL regions, linked by an intramolecular disulfide bond.

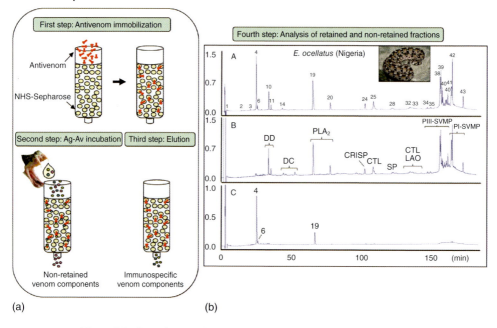

Figure 7.4 Second-generation antivenomics. Scheme of the "immunoaffinity capturing antivenomics" protocol developed by Pla *et al.* [33]. (a) WV is applied to an immunoaffinity column (step 2) packed with antivenom antibodies immobilized onto Sepharose beads (step 1). After eluting the nonretained venom components, the column is thoroughly washed and the immunocaptured proteins eluted (step 3). In (b), panels A–C show, respectively, reverse-phase chromatographic separations of the components of whole *Echis ocellatus* venom [28]; the fraction retained and subsequently recovered from an affinity column of anti-EchiTAb-Plus-ICP antivenom; and the nonimmunocaptured venom fraction. Column eluates were monitored at 215 nm. Proteins within the immunocaptured and the flow-through fractions were identified by the venomics approach described in the text and schematized in Figure 7.2, and quantified by comparing the areas of homologous peaks in the two fractions. Proteins within the different chromatographic fractions of panels A–C were identified by the venomics approach described in the text and schematized in Figure 7.2.

above-mentioned shortcomings of the in-solution immunodepletion protocol. Most important, IgG, F(ab′)$_2$, Fab′, or Fab molecules can be covalently coupled to the activated matrix, making second-generation antivenomics the method of choice for determining the immunoreactivity of any type of antivenom [34–36]. Whereas in the first-generation antivenomics evidence of the immunoreactivity of an IgG antivenom was inferred indirectly through the proteomic characterization of the toxin fraction that remains in solution after immunoprecipitation, in the immunoaffinity-based antivenomics method, the fraction of non-immunodepleted molecules (%NR) is experimentally derived as $100 - ([R/(R + NR)] \times 100)$, where R corresponds to the area of the same protein in the chromatogram of the fraction eluted from the affinity column. The inclusion of R in the calculation compensates for possible losses during sample handling and chromatographic analysis. Furthermore, analysis of the two (R and NR) chromatographic fractions is important because certain venom components (particularly the high molecular mass metalloproteinases) bind with very high avidity to antivenom antibodies and are not recovered from the immunoaffinity column. An analysis of only the retained fraction led to the erroneous conclusion that these components were not immunoreactive. Experiments in which venom is incubated with mock matrix and with matrix-coupled preimmune antibodies run in parallel to immunoaffinity antivenomics analysis and serve as matrix and immunospecificity controls, respectively (Figure 7.5). The smoother baseline in

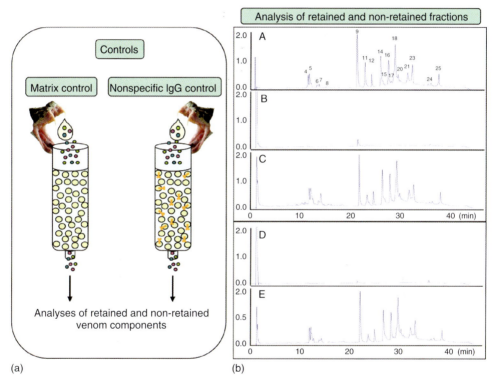

(a) (b)

Figure 7.5 Immunoaffinity controls. (a) Cartoon of the specificity controls of the immunoaffinity-based antivenomics protocol schematized in Figure 7.3. Sepharose 4 Fast Flow matrix, without (matrix control) or with (immunospecificity control) immobilized preimmune IgGs, were incubated with venom and developed in parallel to the immunoaffinity column. (b) Panel A, reverse-phase HPLC separations of the components of the whole venom of *Cerrophidion sasai* [37]. Panels B and C display, respectively, RP-HPLC separations showing the venom proteins recovered in the bound and the flow-through fractions of the Sepharose 4 Fast Flow matrix with immobilized control IgGs. Panels D and E show, respectively, RP-HPLC separations of the venom components recovered in the bound and the flow-through fractions of a mock Sepharose 4 Fast Flow matrix column. Column eluates were monitored at 215 nm. Proteins within the different chromatographic fractions were identified by the venomics approach described in the text and schematized in Figure 7.2.

chromatograms of the affinity column fractions allowed better resolution and a more accurate quantification of the antivenomic results than with the immuno-precipitation methodology. The superior chromatographic performance results are particularly evident in the case of low abundance peaks that become masked in the more noisy chromatograms of supernatants of immunoprecipitation assays, and in the region of high percentage of acetonitrile of the elution gradient of the RP-HPLC run, where some venom proteins (mainly metalloproteinases) and IgG molecules (i.e., primary and secondary antibodies used for immunoprecipitation) coelute (compare panels 1B and 4B).

7.7
Concluding Remarks

The last decade has witnessed the development of techniques and strategies for assessing the toxin composition of snake venoms ("venomics") through proteomics-centered approaches [22, 29, 38, 39]. Antivenomics is translational venomics: a proteomics-based protocol to quantify the extent of cross-reactivity of antivenoms against homologous and heterologous venoms. The capability of this simple knowledge-based analytical method to formulate hypotheses as to how improved venom mixtures might be designed or redesigned for the manu-facture of improved therapeutic antivenoms has been documented in a number of investigations in recent years (reviewed by Calvete [18] and Gutiérrez *et al.* [40]). Its quantitative character and molecular resolution suggest the possibility for antivenomics of supplanting the use of immunoassays and Western blots, the most popular techniques for assessing the immunoreactivity of antibodies. A further advantage of antivenomics lies in the reusability of the affinity columns, which contributes to the economy and the reproducibility of the method.

Acknowledgments

Funding for the research described in this chapter was provided by grants BFU2010-17373 from the Ministerio de Ciencia é Innovación (currently, Minis-terio de Economía y Competitividad), Madrid; PROMETEO/2010/005 from the Generalitat Valenciana; CRUSA-CSIC (2009CR0021); CYTED (project BIOTOX P211RT0412); Vicerrectoría de Investigación, UCR (project 741-B2-652), and FEES-CONARE (Costa Rica).

References

1. Kasturiratne, A., Wickremasinghe, A.R., de Silva, N., Gunawardena, N.K., Pathmeswaran, A., Premaratna, R., Savioli, L., Lalloo, D.G., and de Silva, H.J. (2008) The global burden of snakebite: a literature analysis and modeling based on regional estimates of envenoming and deaths. *PLoS Negl. Trop. Dis.*, **5**, e218.

2. Harrison, R.A., Hargreaves, A., Wagstaff, S.C., Faragher, B., and Lalloo, D.G. (2009) Snake envenoming: a disease

of poverty. *PLoS Negl. Trop. Dis.*, **3**, e569.

3. Williams, D.J., Gutierrez, J.M., Harrison, R., Warrell, D.A., White, J., Winkel, K.D., and Gopalakrishnakone, P. (2010) The global snake bite initiative: an antidote for snake bite. *Lancet*, **375**, 89–91.

4. Gutiérrez, J.M., Williams, D., Fan, H.W., and Warrell, D.A. (2010) Snakebite envenoming from a global perspective: towards an integrated approach. *Toxicon*, **56**, 1223–1235.

5. Gutiérrez, J.M. (2012) in *Public Health-Methodology, Environmental and Systems Issues* (ed J. Maddock), InTech, Rojeka, pp. 131–162.

6. Calmette, A. (1894) L'immunisation artificielle des animaux contre le venin des serpents, et la thérapeutic expérimentale des morsures venimeuses. *C.R. Soc. Biol.*, **46**, 120–124.

7. Calmette, A. (1894) Contribution à l'étude du venin des serpents. Immunisation des animaux et traitement de l'envenimation. *Ann. Inst. Pasteur*, **8**, 275–291.

8. Phisalix, C. and Bertrand, G. (1894) Sur la propriété antitoxique du sang des animaux vaccinée contre le venin de vipére. *C.R. Soc. Biol.*, **46**, 111–113.

9. Phisalix, C. and Bertrand, G. (1894) Propriétés antitoxique du sang des animaux vaccineé contre le venin de vipére. Contribution à l'étude du mécanisme de la vaccination contre ce venin. *Arch. Physiol.*, **6**, 611–619.

10. Hawgood, B.J. (1999) Doctor Albert Calmette 1863-1933: founder of antivenomous serotherapy and of antituberculous BCG vaccination. *Toxicon*, **37**, 1241–1258.

11. Lalloo, D.G. and Theakston, R.D.G. (2003) Snake antivenoms. *J. Toxicol. Clin. Toxicol.*, **41**, 277–290.

12. Gutiérrez, J.M., León, G., and Burnouf, T. (2011) Antivenoms for the treatment of snakebite envenomings: the road ahead. *Biologicals*, **39**, 129–142.

13. Harrison, R.A., Cook, D.A., Renjifo, C., Casewell, N.R., Currier, R.B., and Wagstaff, S.C. (2011) Research strategies to improve snakebite treatment: challenges and progress. *J. Proteomics*, **74**, 1768–1780.

14. Gutiérrez, J.M. (2012) Improving antivenom availability and accessibility: science, technology, and beyond. *Toxicon*, **60**, 676–687.

15. Williams, D.J., Gutiérrez, J.M., Calvete, J.J., Wüster, W., Ratanabanangkoon, K., Paiva, O., Brown, N.I., Casewell, N.R., Harrison, R.A., Rowley, P.D., O'Shea, M., Jensen, S.D., Winkel, K.D., and Warrell, D.A. (2011) Ending the drought: new strategies for improving the flow of affordable, effective antivenoms in Asia and Africa. *J. Proteomics*, **74**, 1735–1767.

16. World Health Organization (2007) Rabies and Envenomings. A Neglected Public Health Issue, World Health Organization, Geneva, *http://www.who.int/bloodproducts/animal_sera/Rabies.pdf* (accessed 27 September 2014).

17. Massey, D.J., Calvete, J.J., Sánchez, E.E., Sanz, L., Richards, K., Curtis, R., and Boesen, K. (2012) Venom variability and envenoming severity outcomes of the *Crotalus scutulatus scutulatus* (Mojave rattlesnake) from Southern Arizona. *J. Proteomics*, **75**, 2576–2587.

18. Calvete, J.J. (2013) Snake venomics: from the inventory of toxins to biology. *Toxicon*, **75**, 44–62.

19. Gibbs, H.L., Sanz, L., Sovic, M.G., and Calvete, J.J. (2013) Phylogeny-based comparative analysis of venom proteome variation in a clade of rattlesnakes (*Sistrurus* sp.). *PLoS One*, **8**, e67220.

20. Mackessy, S.P. (2008) Venom composition in rattlesnakes: trends and biological significance, in *The Biology of Rattlesnakes* (eds H.K. Hayes, K.R. Beaman, M.D. Cardwell, and S.P. Bush), Loma Linda University Press, Loma Linda, CA.

21. Fernández, J., Lomonte, B., Sanz, L., Angulo, Y., Gutiérrez, J.M., and Calvete, J.J. (2010) Snake venomics of *Bothriechis nigroviridis* reveals extreme variability among palm pitviper venoms: different evolutionary solutions for the same trophic purpose. *J. Proteome Res.*, **9**, 4234–4241.

22. Calvete, J.J. (2011) Proteomic tools against the neglected pathology of snake bite envenoming. *Expert Rev. Proteomics*, **8**, 739–758.

23. Gutiérrez, J.M., Solano, G., Pla, D., Herrera, M., Segura, A., Villalta, M., Vargas, M., Sanz, L., Lomonte, B., Calvete, J.J., and León, G. (2013) Assessing the preclinical efficacy of antivenoms: from the lethality neutralization assay to antivenomics. *Toxicon*, **69**, 168–179.

24. World Health Organization (2010) Guidelines for the Production, Control and Regulation of Snake Antivenom Immunoglobulins, WHO, Geneva, *www.who.int/bloodproducts/snakeantivenoms* (accessed 27 September 2014).

25. Lomonte, B., Escolano, J., Fernández, J., Sanz, L., Angulo, Y., Gutiérrez, J.M., and Calvete, J.J. (2008) Snake venomics and antivenomics of the arboreal neotropical pitvipers *Bothriechis lateralis* and *Bothriechis schlegelii*. *J. Proteome Res.*, **7**, 2445–2457.

26. Calvete, J.J., Sanz, L., Angulo, Y., Lomonte, B., and Gutiérrez, J.M. (2009) Venoms, venomics, antivenomics. *FEBS Lett.*, **583**, 1736–1743.

27. Warrell, D.A., Gutiérrez, J.M., Calvete, J.J., and Williams, D. (2013) New approaches and technologies of venomics to meet the challenge of human envenoming by snake-bites in India. *Indian J. Med. Res.*, **138**, 38–59.

28. Calvete, J.J., Cid, P., Sanz, L., Segura, A., Villalta, M., Herrera, M., León, G., Harrison, R., Durfa, N., Nasidi, A., Theakston, R.D., Warrell, D.A., and Gutiérrez, J.M. (2010) Antivenomic assessment of the immunological reactivity of EchiTAb-Plus-ICP, an antivenom for the treatment of snakebite envenoming in sub-Saharan Africa. *Am. J. Trop. Med. Hyg.*, **82**, 1194–1201.

29. Calvete, J.J., Juárez, P., and Sanz, L. (2007) Snake venomics. Strategies and applications. *J. Mass Spectrom.*, **42**, 1405–1414.

30. Margres, M.J., McGivern, J.J., Wray, K.P., Seavy, M., Calvin, K., and Rokyta, R.R. (2014) Linking the transcriptome and proteome to characterize the venom of the eastern diamondback rattlesnake (*Crotalus adamanteus*). *J. Proteomics*, **96**, 145–158. doi: 10.1016/j.jprot.2013.11.001.

31. Andrew, S.M. and Titus, J.A. (2000) Fragmentation of immunoglobulin G. *Curr. Protoc. Cell Biol.*, (Supplement 5), 16.4.1–16.4.10, John Wiley & Sons, Inc.

32. Petras, D., Sanz, L., Segura, A., Herrera, M., Villalta, M., Solano, D., Vargas, M., León, G., Warrell, D.A., Theakston, R.D.G., Harrison, R.A., Durfa, N., Nasidi, A., Gutiérrez, J.M., and Calvete, J.J. (2011) Snake venomics of African spitting cobras: toxin composition and assessment of congeneric cross-reactivity of the Pan-African EchiTAb-Plus-ICP® antivenom by antivenomics and neutralization approaches. *J. Proteome Res.*, **10**, 1266–1280.

33. Pla, D., Gutiérrez, J.M., and Calvete, J.J. (2012) Second generation snake antivenomics: comparing immunoaffinity and immunodepletion protocols. *Toxicon*, **60**, 688–699.

34. Fahmi, L., Makran, B., Pla, D., Sanz, L., Lkhider, M., Harrison, R.A., Ghalim, N., and Calvete, J.J. (2012) Venomics and antivenomics profiles of North African *Cerastes cerastes* and *C. vipera* populations reveals a potentially important therapeutic weakness. *J. Proteomics*, **75**, 2442–2453.

35. Makran, B., Fahmi, L., Pla, D., Sanz, L., Lkhider, M., Ghalim, N., and Calvete, J.J. (2012) Snake venomics of *Macrovipera mauritanica* from Morocco, and assessment of the para-specific immunoreactivity of an experimental monospecific and a commercial antivenoms. *J. Proteomics*, **75**, 2431–2441.

36. Pla, D., Sanz, L., Molina-Sánchez, P., Zorita, V., Madrigal, M., Flores-Díaz, M., Alape-Girón, A., Núñez, V., Andrés, V., Gutiérrez, J.M., and Calvete, J.J. (2013) Snake venomics of *Lachesis muta rhombeata* and genus-wide antivenomics assessment of the paraspecific immunoreactivity of two antivenoms evidence the high compositional and immunological conservation across Lachesis. *J. Proteomics*, **89**, 112–123.

37. Gutiérrez, J.M., Tsai, W.C., Pla, D., Solano, G., Lomonte, B., Sanz, L., Angulo, Y., and Calvete, J.J. (2013) Preclinical assessment of a polyspecific antivenom against the venoms of

Cerrophidion sasai, Porthidium nasutum and *Porthidium ophryomegas*: insights from combined antivenomics and neutralization assays. *Toxicon*, **64**, 60–69.

38. Georgieva, D., Arni, R.K., and Betzel, C. (2008) Proteome analysis of snake venom toxins: pharmacological insights. *Expert Rev. Proteomics*, **5**, 787–797.

39. Fox, J.W. and Serrano, S.M.T. (2008) Exploring snake venom proteomes: multifaceted analyses for complex toxin mixtures. *Proteomics*, **8**, 909–920.

40. Gutiérrez, J.M., Lomonte, B., Sanz, L., Calvete, J.J., and Pla, D. (2014) *J. Proteomics*, **105**, 340–350.

Part III
Direct Pre- and On-Column Coupled Techniques

Analyzing Biomolecular Interactions by Mass Spectrometry, First Edition.
Edited by Jeroen Kool and Wilfried M.A. Niessen.
© 2015 Wiley-VCH Verlag GmbH & Co. KGaA. Published 2015 by Wiley-VCH Verlag GmbH & Co. KGaA.

8
Frontal and Zonal Affinity Chromatography Coupled to Mass Spectrometry

Nagendra S. Singh, Zhenjing Jiang, and Ruin Moaddel

8.1
Introduction

Cell surface transmembrane proteins, including G-protein coupled receptors (GPCRs), ligand gated ion channels (LGICs), and transporters, are key therapeutic targets in drug discovery and development [1–3]. In fact, over 30% of modern drugs target GPCRs [4]. The ATP-binding cassette (ABC) superfamily consists of 48 members, of which P-glycoprotein (Pgp), multidrug resistance protein 1 (MRP1), MRP2, and breast cancer resistant protein (BCRP) are the best characterized members. These transporters are targeted due to their role in drug resistance [5]. Currently, the predominant method for screening transmembrane receptors is through classic binding assays, functional assays, calcium fluorescent assays, monolayer efflux assays for transporters, and surface plasmon resonance [6]. While all these assays work, these methods are laborious and time-consuming and in some cases have weak inter-laboratory reproducibility. In addition, there are very limited assays to study allosteric sites, for example, no standard screen exists for the study of noncompetitive inhibitors (NCIs) of the nicotinic receptors. Another method to study protein properties has been recently developed using nanotechnology, where the isolation of a single biomolecule is carried out to study specific properties such as ligand-induced mechanical folding of the protein [7]. Specifically, Yadavalli *et al.* [8] isolated a single protein to study its structure, interactions, and nanomechanical properties. While each of these methods has its advantages, all these assays are limited in their capabilities of characterizing orphan receptors.

An alternative to study ligand–protein interactions is bioaffinity chromatography. With bioaffinity chromatography, the targeted protein is immobilized onto a chromatographic support covalently or through adsorption, and the immobilized protein can be used as a tool to study ligand–protein interactions. While covalent immobilizations are preferred, for transmembrane proteins, as the integrity of the boundary lipids is important, adsorption is desirable. The resulting protein-based

Analyzing Biomolecular Interactions by Mass Spectrometry, First Edition.
Edited by Jeroen Kool and Wilfried M.A. Niessen.
© 2015 Wiley-VCH Verlag GmbH & Co. KGaA. Published 2015 by Wiley-VCH Verlag GmbH & Co. KGaA.

liquid chromatography (LC) stationary phases (SPs) can be used to determine and characterize ligand–protein interactions [9, 10].

The theory and applicability of using immobilized proteins (protein-based SPs) to explore the interactions between ligands/substrates and an immobilized protein were initially described by Chaiken [11] and Wade *et al.* [12]. The earliest application involving a drug–protein system was by Lagercrantz *et al.* [13] in 1979, where they used low-performance Sepharose columns to measure the interactions of fatty acids, steroids, and various drugs with immobilized bovine serum albumin (BSA). Protein-based SPs have been used for the following purposes: to study protein–ligand interactions [14]; as chiral-based stationary phases (CSPs) [15], for example, α-glycoprotein (AGP) [16], BSA [17], and ovomucoid [18] based CSPs; as chiral bioreactors [19]; as immobilized enzyme reactors such as trypsin/chymotrypsin [20]; and for the determination of drug–plasma protein binding by using a human serum albumin (HSA) column [21]. The experimental approach, including the development of these protein-based SPs, is not discussed here, as it has already been extensively reviewed [22].

The first protein-based SP developed with a transmembrane protein was carried out by Per Lundahl and coworkers from Uppsala University [23–25], where they immobilized the glucose transporter, GLUT1, through the entrapment of red blood cell membranes in proteoliposomes [24]. They demonstrated that the immobilized GLUT1 retained functional activity [24, 25].

Following the pivotal work on the GLUT1 transporter, Irving Wainer and coworkers expanded the work to include the immobilization of cellular membrane fragments from the α3β4 nicotinic receptors and several other receptors [22, 26–28] onto the surface of the immobilized artificial membrane (IAM) SP (12 μ, 300 Å pore) developed by Pidgeon and Venkataram [29]. To date, a large number of transmembrane proteins have been immobilized by several groups [30] including GPCR proteins: $β_2$-adrenergic [31], cannabinoid receptors [32], and P2Y1 receptor [33]; LGICs: multiple subtypes of the nicotinic receptors, γ-amino-butyric acid receptors, and N-methyl-d-aspartate receptors [34–36]; ABC transporters: Pgp, BCRP, and MRP1 [37, 38, 28]; and solute carrier transmembrane proteins: OAT, OCT1, OCT1R488M, and OCT1G465R [39, 40]. An added advantage of these protein-based SPs compared to functional-based screenings is that while the latter require the structure or function of an orphan receptor, whereas in affinity-based screening this is not required. These columns were characterized using frontal and zonal chromatographic techniques.

8.2
Frontal Affinity Chromatography

Frontal chromatography is a quantitative chromatographic technique that can be used to characterize interactions between the SP and the solute that was initially

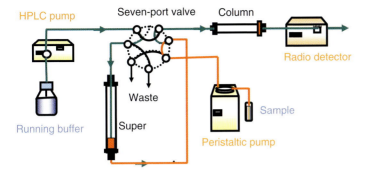

Figure 8.1 Radioflow detector system with superloop used for frontal chromatography. (Reprinted by permission from Macmillan Publishers Ltd.: (Nature Protocol) Ref. [22], Copyright (2009).)

described by Kasai and Ishii [41]. This is carried out using a frontal chromatographic system (Figure 8.1), which includes an LC pump, a superloop, and a detector. The detector depends on the marker compound used and could include a radioflow detector, UV detector, fluorescence detector, as well as a mass spectrometer, with mass spectrometry (MS) being the method of choice. Frontal chromatography is carried out under dynamic equilibrium conditions and determines the binding of the ligand in the mobile phase and the immobilized biopolymer by determining the change in retention volume resulting from an increase in ligand concentration.

As shown in Figure 8.2, in frontal affinity chromatography (FAC), a marker ligand is placed in the mobile phase and passed through the column in the presence or absence of serial concentrations of a displacing ligand under dynamic equilibrium conditions [42]. The frontal regions are composed of the relatively flat initial portions of the chromatographic traces, which represent the nonspecific binding of the marker ligand to the SP, followed by a vertical breakthrough, which reflects the specific interactions with the immobilized protein, ending in a plateau, which corresponds to the saturation of the binding sites on the immobilized protein. The inflection point of the breakthrough curves is directly related to the concentration

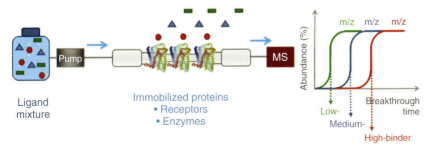

Figure 8.2 Illustration showing the basic principle of frontal chromatography. (Reprinted with permission of John Wiley & Sons, Inc. Ref. [42] Copyright (2009).)

of the applied ligand and the association equilibrium constants between the ligand and immobilized protein. In addition to specific interactions, the presence of nonspecific interactions must be considered as well. While nonspecific binding can be well tolerated in FAC as nonspecific interactions are typically concentration-independent, on occasion it can become problematic, and in some cases, mask the specific interactions. In this case, organic modifiers can be added to reduce nonspecific interactions, or a change can be effected in the immobilization protocol.

In frontal chromatography, increasing concentrations of the applied ligand results in shorter breakthrough times, and allows for the determination of the binding affinity (K_d) of the applied ligand and the number of active binding sites (B_{max}) for the immobilized protein. This is determined using Equation 8.1:

$$(V - V_{min}) = P[D](K_d + [D])^{-1} \qquad (8.1)$$

where [D] is the concentration of the displacer ligand, V is the retention volume of the displacer ligand, and V_{min} is the retention volume of the displacer ligand when the specific interaction is completely suppressed and P is the product of the B_{max} (the number of active binding sites) and K_d/K_{dM}. From the plot of [D] $(V - V_{min})$ versus [D], dissociation constant values (K_d) for the displacer ligand can be obtained. The data are analyzed using a nonlinear regression with a rectangular hyperbolic curve using Prism 4 software (Graph Pad Software Inc., San Diego, CA, USA). For the results to be quantitative, Equation 8.1 assumes that the concentration of the marker ligand is significantly smaller than its K_d for the immobilized protein.

With the introduction of frontal and zonal analysis, protein-based SPs became an attractive option for determining the binding affinity (K_d) of a target protein. The chromatographically determined affinities using a protein-based SP were comparable to affinities obtained using standard membrane-binding techniques, such as filtration assays, demonstrating that the FAC could be used to study binding interactions [22]. For example, a linear correlation $(r = 0.99)$ was observed for the human organic cation transporter (hOCT1) protein between the K_d values obtained by frontal chromatography when compared to that reported in the literature [39].

Conventional frontal affinity and zonal affinity chromatographic systems are typically used with a single-wavelength UV/vis spectrophotometer [43] or radio-flow as the detector for the trace component [35]. In the former case, all the solution constituents can contribute to the signal intensity, making it difficult to distinguish between the desired component and the background. In the latter case, the desired marker compound is radiolabeled and therefore selectively monitored. The latter method has many advantages; however, as it requires the use of radioactive labels, the markers are at times expensive or unavailable, for example, in the case of an orphan receptor. Coupling FAC to MS provides much more versatility. It allows the monitoring of the trace component through its m/z value. As long as the trace component has an ionizable group it can be usually detected by MS. A potential limiting use of FAC with MS lies in the requirement of FAC for the trace component to be run at concentrations significantly lower than its binding affinity (K_d) for the immobilized protein. Therefore, if the marker ligand has nanomolar affinity for the protein, it may be difficult to detect the trace component unless it is readily ionizable. In the following sections we discuss different applications of FAC including the staircase method, which allows the rapid determination of a binding affinity for the immobilized protein; the simultaneous monitoring of multiple breakthrough curves, allowing the identification and ranking of the active components for a single target, based on their m/z and K_d values, in a single chromatogram; and the analysis of a single component binding characteristics to multiple protein columns. Subsequently, zonal chromatography and nonlinear chromatography (NLC) are discussed.

8.3
Staircase Method

In FAC, in order to determine the binding affinity of a ligand for the immobilized protein, at least five separate concentrations of the displacer should be carried out, with a wash time in between each injection. While this is a proven method, it can become time-consuming based on the wash time. In order to circumvent the wash time, Schreimer *et al.* configured a novel high-precision nano-FAC system [44] and established a binding assay called the *staircase method*, which voids the wash time. The method was utilized for high-throughput screening of various ligands against immobilized proteins [44–48]; however, it has been limited to cytosolic receptors.

Only recently has the staircase approach been successfully applied to membrane receptors immobilized on an open tube capillary [49]. In this approach the target protein is immobilized onto the SP directly or through a streptavidin–biotin system. The resulting micro-affinity column (250 µM i.d. × 1 cm) (Figure 8.3) can be connected to the mass spectrometer for frontal analysis. Using the staircase method, the binding affinity of a ligand is determined by infusing the ligand and

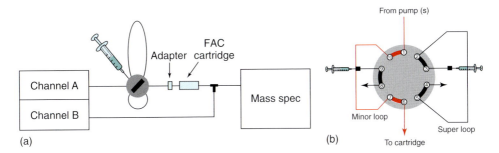

(a) (b)

Figure 8.3 (a) Schematic representation of high-precision nano-FAC system. (b) Ten port valve configuration. (Reprinted by permission from Macmillan Publishers Ltd.: (Nature Protocol) Ref. [46], Copyright (2007).)

the void marker simultaneously through the column, under frontal analysis conditions, and tracked using its respective m/z value on a mass spectrometer. Starting at the lowest concentration, the ligand is sequentially infused with increasing concentrations with a constant concentration of void marker. The breakthrough volumes of the successive ligand concentrations relative to their respective void volumes are then used for the determination of B_{max} and K_d by linear regression analysis. The use of the staircase method has been extensively reviewed [46], including the assembled system required for frontal analysis. For calculating the K_d of the test ligand, a solution of the void marker and a low concentration of the test ligand (e.g., 0.1 K_d) in an assay buffer is prepared and loaded on the minor loop. When the baseline is stable, the contents of the minor loop are injected and monitored by MS. A series of concentrations are subsequently injected until saturation is observed. An optimally functioning system will demonstrate breakthrough curves for the two compounds that are overlapped and symmetrical (Figure 8.4). The K_d is calculated using Equation 8.2:

$$[A]_0 + y = B_{max} \left[\frac{1}{V - V_0} \right] - K_d \tag{8.2}$$

where $V - V_0$ is the corrected breakthrough volume for the ligand and B_{max} is the active capacity of the cartridge. A_0 refers to the ligand infusion concentration and the summed concentrations ($[A]_0 + y$) refer to the initial concentration of the ligand for the first step of the staircase; however, for the subsequent steps it will be the sum of that current step and all its predecessors. A plot of $[A]_0 + y$ versus

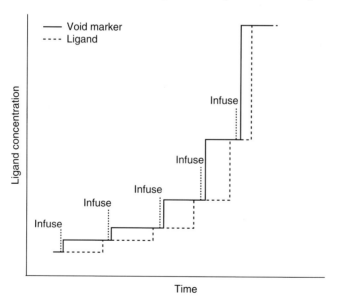

Figure 8.4 Diagram of the modified staircase approach, used to measure binding data without washing steps. (Reprinted from Ref. [47], Copyright (2003), with permission from Elsevier.)

reciprocal $(V - V_0)$ supports the determination of B_{max} and K_d by linear regression analysis.

The approach has demonstrated that there is a direct correlation between the amount of analyte required in a frontal chromatographic assays and the capacity of the affinity column. Ng *et al.* demonstrated that binding affinities for compounds could be determined with even picomoles of immobilized proteins. This was clearly shown when only 1 pmol of cholera toxin B was immobilized and the K_d of the pentasaccharide GM1 was determined ($K_d = 0.44\,\mu M$) [46].

On the basis of the proposed approach, Schreimer *et al.* successfully calculated K_d for ligands of the estrogen receptor-β (ERB), sorbitol dehydrogenase, human α-thrombin, cholera toxin B subunit, β-galactosidase, and *Griffonia simplicifolia* isolectin B4 [47]. In addition, the method can be utilized for high-throughput screening of various ligands against immobilized proteins [42, 44–48] using an indicator ligand to discriminate between mixtures that contain high levels of weak ligands and those that contain single strong ligands. This approach was utilized for screening ligands for ERB and human α-thrombin [47].

8.4
Simultaneous Frontal Analysis of a Complex Mixture

In the previous sections, the use of frontal chromatography was discussed for the determination of binding affinities, with some discussion of their application for screening. Massolini and coworkers [50] developed a novel screening method, where they could simultaneously screen 18 compounds against the GPR17 using FAC with MS.

GPR17, a dual uracil nucleotide/cysteinyl leukotriene receptor, is believed to be a novel target for neuroprotective agents to counteract damage resulting from stroke. As a result, the GPR17 SP was synthesized and characterized by screening three known receptor ligands with different potencies: the antagonists cangrelor (IC$_{50}$ = 0.7 nM) and MRS 2179 (IC$_{50}$ = 508 nM) and the agonist UDP (EC$_{50}$ = 1.14 μM). The three ligands were infused in tandem with equimolar mixture and monitored using FAC-MS by monitoring the individual m/z values. It was shown that the compounds separated based on affinity, that is, cangrelor was the most strongly retained, while UDP had the lowest retention. The percentage of breakthrough time with respect to cangrelor for UDP was ~10% [50]. Therefore, FAC-MS is an ideal and rapid method capable of simultaneously ranking ligands based on their affinity for the immobilized target by monitoring the individual m/z values.

In an extension of this work, a library of 18 nucleotide derivatives were selected for screening against the GPR17, and ranked based on their retention on the GPR17-IAM column (Figure 8.5). These compounds were divided into two groups: a group consisting of analogs of nucleotides (library A) and a group consisting of acyclic nucleotides (library B) (Table 8.1). Two separate solutions containing the ligands from the two libraries and the three known

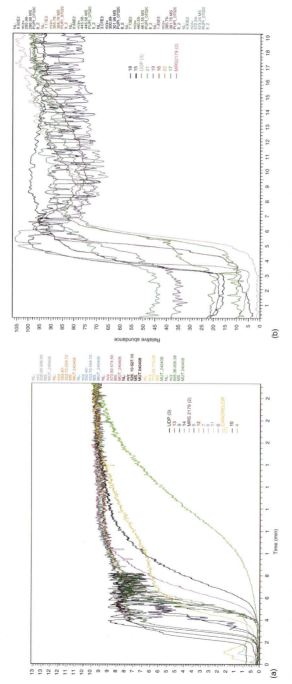

Figure 8.5 Frontal affinity chromatography–mass spectrometry of (a) library A and (b) library B. Extracted breakthrough curves for each analyte of library A and library B are shown. Mixtures of ligands in the presence of the three reference compounds each at 1 μM were infused through the GPR17-IAM-I column using the mass spectrometer in negative mode. (Reprinted with permission from Ref. [50]. Copyright (2010) American Chemical Society.)

Table 8.1 Chemical structures of the compounds selected for FAC-MS studies.

Compound	Structure	Pharmacology	Compound	Structure	Pharmacology
	Reference compounds			Reference compounds	
Cangrelor (1)		P2Y$_{12}$ antagonist IC$_{50}$ = 0.7 nM	UDP (3)		P2Y agonist EC$_{50}$ = 1.14 μM
MRS 2179 (2)		P2Y$_{1}$ competitive antagonist IC$_{50}$ = 508 nM			

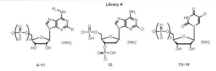

Library A

4^{18}	R = C≡C-Ph R$_1$ = H X = H n = 3, m = 4	P2Y$_{1,12}$ antagonist	8	R = Cl R$_1$ = CH$_3$ X = CH n = 3, m = 4	
5^{19,20}	R = H R$_1$ = CH$_3$ X = H n = 3, m = 4		9^{22}	R = Cl R$_1$ = CH$_3$ X = N n = 2, m = 3	
6^{21}	R = H R$_1$ = cC$_5$H$_9$ X = H n = 3, m = 4		10^{21}	R = C≡C-Ph R$_1$ = H X = N n = 2, m = 3	P2Y$_{1,12}$ antagonist
7	R = Cl R$_1$ = CH$_3$ X = H n = 3, m = 4		11^{18}	R = C≡C-nC$_4$H$_9$ R$_1$ = H X = N n = 2, m = 3	P2Y$_{1,12}$ agonist
			12^{23,24}		MRS 2179 analog
			13^{25,26}	R = I	P2Y$_4$ agonist
			14	R = C≡C-nC$_4$H$_9$	P2Y$_4$ agonist

Library B

15^{27}	R = H n = 1, m = 1		19^{27}	R = NH$_2$ n = 2, m = 3	
16^{27}	R = H n = 2, m = 3		20^{27}	R = NH$_2$ n = 3, m = 4	
17^{27}	R = H n = 3, m = 4		21^{27}	R = I n = 3, m = 4	
18^{27}	R = NH$_2$ n = 1, m = 1				

Source: Reprinted with permission from Ref. [50]. Copyright (2010) American Chemical Society.

GPR17 ligands (UDP, MRS 2179, and cangrelor) all at 1 μM were prepared in mobile phase buffer and infused through the GPR17-IAM column connected to an ESI-MS. Frontal elution of each of the compounds was monitored based on individual *m/z* values in negative mode. Overlayed frontal chromatograms (Figure 8.5) indicate the retention of each compound relative to each other and the known GPR17 ligands (UDP, MRS 2179, and cangrelor). The percentage of breakthrough time with respect to cangrelor was calculated for all the analytes. In order to determine the accuracy of the ranking obtained using FAC-MS, the ranking was compared to the results obtained using the $[^{35}S]GTP\gamma S$ binding assay. The binding data were consistent with the ranking order obtained using the FAC-MS approach. The total run time and ranking of the 18 compounds was 55 min, while the equivalent information obtained by $[^{35}S]GTP\gamma S$ binding assay took over a week. Further, Massolini *et al.* demonstrated that the FAC-MS screening data in combination with molecular modeling stimulation could be utilized to predict structure–affinity relationships that would help the synthesis/designing of selective GPR17 ligands (agonists/antagonists). Overall, the FAC-MS ranking data suggested that lipophilic substituents on the C-2 and on the N6 amino group of purine ring can lead to a tight interaction with the hydrophobic residues in the binding pocket. The application of simultaneous frontal analysis has been carried out on a larger scale. Chan *et al.* used an FAC column containing β-galactopyranoside enzyme to screen a library of 356 modified β-galactopyranosides. Ten different sets of mixtures containing 24–40 β-galactopyranosides were infused through the FAC-MS system. Overall 34 entries containing isomers were identified with K_d values $\leq$10 μM, based on one single injection of each mixture. Further, it was demonstrated that the approach was able to distinguish differences in activity of four diastereomers. In a similar kind of approach, Ng *et al.* demonstrated that a high-throughput affinity screening of upto 200 ligands could be carried out by coupling FAC-MS with a docking approach. This was demonstrated with a mixture of 100 compounds. They were simultaneously infused through the hERβ column and the breakthrough time for each compound was determined relative to a known marker ligand. This led to the identification of 16 compounds with submicromolar affinity for the hERβ.

While this method has been shown to be very effective, no method is foolproof, and this method is no exception. Complex mixtures contain large numbers of compounds and therefore can result in false negatives or missed binders, in addition to the presence of allosteric modifiers or NCIs eliciting a conformational change in the immobilized protein and thus changing the binding characteristics of the immobilized protein. While these are limitations and should always be considered, the identification of several compounds from a complex mixture makes it worthwhile.

8.5
Multiprotein Stationary Phase

While the previous sections have described the screening of multiple ligands for a single receptor, we have shown that a single compound can be used for the characterization of multiple receptors. Wainer and coworkers [35] demonstrated that the immobilization of solubilized rat forebrain membrane fragments resulted in the co-immobilization of nicotinic receptors (nAChR (nicotinic acetylcholine receptor)), γ-amino-butyric acid (GABA) receptors, and *N*-methyl d-aspartate (NMDA). In these studies, in order to study one receptor, for example, the nAChR, a saturating concentration of a known marker for another receptor was added, such as flunitrazepam for $GABA_A$. This allowed the characterization of a specific receptor in the case that the marker ligand was not selective. In the aforementioned study, these experiments did not result in any changes in the elution volumes with or without the inhibitor for the other receptor, indicating that the marker ligands were specific for the immobilized receptors and that the immobilized receptors were independent of each other. Further, the K_d values obtained for the ligands for each of these receptors correlated with previously reported values.

The characterization of multiple receptors was also carried out using cellular membranes from the 1321N1 and A172 astrocytoma cell lines [36]. In this study, in addition to the presence of multiple LGICs, including the $GABA_A$ and NMDA receptors, the presence of multiple subtypes of the nAChR was demonstrated. Specifically, homomeric ($α_7$ nAChR) (α-Bungarotoxin (α-Bgt) sensitive nAChRs) and heteromeric nicotinic receptors ($α_xβ_y$ nAChRs) (κ-Bgt sensitive nAChRs) were co-immobilized. The addition of these subtype-specific inhibitors to the mobile phase allows the user to study each subtype independently using the same marker ligand. Of interest is the fact that, in the absence of any subtype-specific inhibitors, the frontal elution curve of epibatidine resulted in a biphasic curve (Figure 8.6), indicating the presence of multiple subtypes of the nicotinic receptor. The addition of the subtype-specific inhibitor at saturating concentrations allows for the characterization of a single subtype. This demonstrates that the co-immobilization of multiple subtypes of a receptor can be characterized using frontal chromatography. However, it should also be considered whether a marker ligand has affinity for multiple proteins, and this will exhibit itself as a multiphasic curve on multiprotein SP.

8.6
Zonal Chromatography

Zonal affinity chromatography is a method where a small quantity of ligand is injected in the presence or absence of the competing ligand in the mobile phase. The retention time of the ligand is determined at various concentrations of the competing ligand. This technique has been most frequently used to study the binding of drugs and other solutes with protein immobilized on a column [51]; however, it has also been used to examine the effect of various solvent conditions, such as pH, composition, and temperature, which affect the drug–protein interaction. In addition, zonal chromatography can also be used to determine the kinetics and the thermodynamics of the binding process using temperature-dependent experiments and functional parameters such as IC_{50} (inhibitory concentration) and EC_{50} (electrochemical conversion). These studies are typically carried out using nonlinear chromatographic studies, and the chromatographically

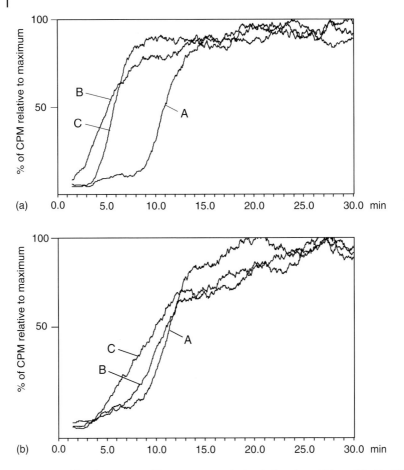

Figure 8.6 The chromatographic traces obtained for 60 pM [³H]-EB on the CMAC(1321N1) column (a) and CMAC(A172) column (b) where A is the trace obtained using a mobile phase composed of ammonium acetate (10 mM, pH 7.4); B is the trace obtained after the addition of 5 nM α-BTx to the mobile phase; and C is the trace obtained after the addition of 1 nM κ-BTx to the mobile phase. (Reprinted with permission from Ref. [36]. Copyright (2008) American Chemical Society.)

determined values are comparable to those obtained using standard bio-chemical and pharmacological techniques [30, 52], as discussed in this section.

By the late 1980s and early 1990s, reports began to appear in which high-performance affinity chromatography (HPAC) and zonal elution were used in quantitative studies of drug–protein interactions [53–59]. Zonal chromatographic studies can be carried out on any HPLC (high performance liquid chromatography) system and the mode of detection is dependent on the characteristics of the ligand; however, MS is the preferred choice.

In zonal affinity chromatography, the elution time is the summation of specific and nonspecific interaction, while in FAC the equilibrium is under dynamic equation. Hage and colleagues demonstrated how the ligand's overall retention factor is related to the number of binding sites [51] and the equilibrium constants of the ligand for the immobilized protein [30]:

$$\kappa = \frac{(K_{A1}n_1 + \cdots + K_{An}n_n)m_L}{V_M} \tag{8.3}$$

In Equation 8.3, the retention factor is calculated by using $k = (t_R - t_M)/t_M$ or $k = (V_R - V_M)/V_M$, where t_R is the retention time of the injected compound, V_R is the corresponding retention volume, t_M is the column void time, and V_M is the void volume. Other terms in Equation 8.3 include the total moles of analyte binding sites in the column (m_L), the association equilibrium constants for the analyte at the individual sites in this population (K_{A1} through K_{AN}), and the fraction of each type of site in the column (n_1 through n_n). This demonstrates that the retention factors of the solute reflect their relative binding strength to the immobilized protein. By manipulating the assay conditions, including pH, ionic strength, polarity of the mobile phase, and temperature, zonal chromatography can determine which factors play a role in the formation of the protein–drug complex. For example, if the retention is dependent on the pH of the mobile phase, this would be indicative that there is considerable contribution of Coulomb interactions in the binding of a drug or that the protein underwent a conformational change as a result of the change in the net charges; the presence of organic modifiers plays a role in disturbing hydrophobic interactions [60]. Another advantage of zonal chromatography is that you can study the changes in protein–drug interactions as a result of pathological lesions [61].

A typical zonal elution experiment is shown in Figure 8.7, where the change in retention of *cis*-clomiphene to HSA is being determined in the presence of increasing concentrations of (*R,S*) warfarin [63]. As can be seen in Figure 8.7, there is an increase in the retention time of clomiphene with increasing concentrations of warfarin, indicative of positive allosteric effects [63].

The use of zonal chromatography for determining the binding affinity of a drug to the immobilized protein can be carried out as the retention factor is determined under equilibrium conditions, meaning the bound and free forms of the ligand are equal. Further, the retention factor is equal to the moles of an injected solute that is bound to the ligand (*b*) divided by the moles of the solute that remain free in the surrounding mobile phase (*f*), that is, $k = b/f$. Then the equation can be rearranged to calculate the bound fraction, such that $b = k/k + 1$ [64]. Noctor *et al.* [64] utilized this approach to compare the binding percentage of HSA with various benzodiazepines, coumarins, and triazole derivatives. Thus, these compounds could be ranked using their relative retention for their affinity toward HSA (Figure 8.8).

In addition to determining the relative binding affinity, the competitive displacement of a known ligand can also be calculated using zonal chromatography.

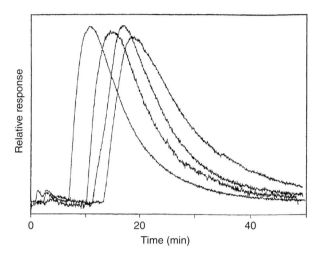

Figure 8.7 Chromatograms obtained in a zonal elution study for the injection of *cis*-clomiphene in the presence of racemic warfarin as a mobile phase additive. The mobile phase concentrations of warfarin (from left to right) were 0, 0.5, 1.0, and 7.5 mM. All of these studies were performed at 37.8 °C in a pH 7.4, 0.067 M phosphate buffer. (Reprinted with permission from Ref. [62]. Copyright (1999) American Chemical Society.)

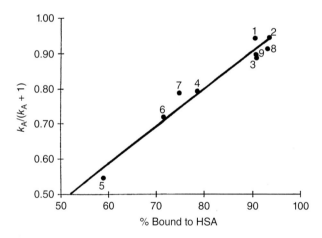

Figure 8.8 Relationship between the binding predicted by retention measurements on an immobilized HSA column, $k/(k11)$, and the binding measured by ultrafiltration for a series of coumarins with soluble HSA. The compounds studied in this experiment were as follows: (1) *R*-warfarin, (2) *S*-warfarin, (3) 7-hydroxycoumarin-4-acetic acid, (4) coumarin, (5) coumarin, (6) 4-hydroxycoumarin, (7) umbelliferone, (8) 4-trifluoromethylumbelliferone, and (9) 7-amino-4-methylcoumarin. (Reprinted with permission of John Wiley & Sons, Inc. Ref. [64] Copyright (1993).)

This is carried out by injecting the drug as the analyte with the presence of varying concentrations of the displacer in the mobile phase. This is also the type of experiment illustrated in Figure 8.7, wherein *cis*-clomiphene was shown to compete with racemic warfarin on HSA, demonstrated by the shift in clomiphene retention as the mobile phase concentration of warfarin was varied. Zonal displacement chromatography can therefore be used to determine whether or not the drug is competing or interacting with another ligand. However, in order to better quantify the competitive displacement, a series of zonal chromatographic studies need to be carried out examining the various types of interaction models. Various equations and models that have been developed are shown in Table 8.2 [65].

In addition to relative binding affinities and displacement studies, the zonal approach can be used to observe drug–protein interactions using solvent and temperature parameters. As expected, the formation and stabilization of the drug–protein complex is dependent on a number of factors including the mobile phase composition, pH, and ionic strength, as mentioned earlier. Serum proteins, for example, elicit their strongest binding at around pH 7.4. A small change in the pH, however, will result in a significant change in the binding of drugs. While a change in pH can result in changes in binding as a result of changes in the net charge and coulombic interactions, an increase in ionic strength results in a decrease of coulombic interactions through a shielding effect, but at the same time may cause an increase in nonpolar solute adsorption. The addition of an organic modifier, for example, 1- or 2-propanol, changes the solvent's polarity and thereby can alter solute–protein binding by disrupting nonpolar interactions or causing a change in the protein's structure. This usually decreases the protein's binding to solutes, along with decreasing the width and tailing of the injected solute peaks [66]. One of the examples demonstrating the effect of solvent composition is reported by Yang and Hage [67], wherein increasing concentrations of organic modifier are used in the interaction of *R*- and *S*-warfarin with HSA. Although *R*- and *S*-warfarin have a common binding site on HSA, it was observed that an increase in the organic modifier concentration decreased the binding of *R*-warfarin while it increased the binding of *S*-warfarin to HSA, indicating that *R*- and *S*-warfarin might be binding to different sites, with respect to the specific HSA binding site.

Zonal chromatography can also be used for the determination of the location and structure of the binding site. This can be done through competitive displacement studies or by quantitative structure–retention relationships (QSRRs). In the former case, this would be carried out with a drug whose binding characteristics are well known. Therefore, displacement of this drug will allow the identification of the region to which the tested ligand binds. For example, Sengupta and Hage [62] generated a map that shows the relationship between the binding regions on a ligand (Figure 8.9), based on competitive displacement studies for the multiple binding regions of HSA. In the latter approach, QSRR, the interactions of a series of structurally related ligands with the immobilized protein are affected under constant temperature and mobile phase conditions. The factors that play

Table 8.2 Relationships used to describe the retention of injected solutes during zonal elution.

Type of system [Reference]	Model	Predicted response
Self-competition of an injected analyte, A, with itself as a competing agent at a single type of binding site, L [46, 59]	$A + L \underset{}{\overset{K_A}{\rightleftharpoons}} A - L$	$K_A = \dfrac{K_A m_L}{V_M(1+K_A[A])}$ or $\dfrac{1}{k_A} = \dfrac{V_M[A]}{m_L} + \dfrac{V_M}{K_A m_L}$
Self-competition of an injected analyte, A, with itself as a competing agent at two types of binding sites, L_1 and L_2 [46, 55, 59]	$A + L_1 \underset{}{\overset{K_{A1}}{\rightleftharpoons}} A - L_1$ $A + L_2 \underset{}{\overset{K_{A2}}{\rightleftharpoons}} A - L_2$	$K_A = \dfrac{K_{A1} m_{L1}}{V_M(1+K_{A1}[A])} + \dfrac{K_{A2} m_{L2}}{V_M(1+K_{A2}[A])} = k_{Site1} + k_{Site2}$ or $\dfrac{1}{(k_A - k_{Site2})} = \dfrac{V_M[A]}{m_{L1}} + \dfrac{V_M}{K_{A1} m_{L1}}$
Direct competition of an injected analyte, A and competing agent, I, at a single common binding site, L, where A has no other binding sites and is present in a small amount versus L [65]	$A + L \underset{}{\overset{K_A}{\rightleftharpoons}} A - L$ $I + L \underset{}{\overset{K_I}{\rightleftharpoons}} A - L$	$k_A = \dfrac{K_A m_L}{V_M(1+K_I[I])}$ or $\dfrac{1}{k_A} = \dfrac{V_M K_I[I]}{K_A m_L} + \dfrac{V_M}{K_A m_L}$
Direct competition of an injected analyte, A, and competing agent, I, at a single common binding site, L; A has one or more additional sites, $L_{N1}, \dots, L_{Nt}$, that produce no competition, and A is present in a small amount versus L [28]	$A + L \underset{}{\overset{K_A}{\rightleftharpoons}} A - L$ $I + L \underset{}{\overset{K_I}{\rightleftharpoons}} I - L$ $A + L_{N1} \underset{}{\overset{K_{AN1}}{\rightleftharpoons}} A - L_{N1}$ $A + L_{Nt} \underset{}{\overset{K_{ANt}}{\rightleftharpoons}} A - L_{Nt}$	$k_A = \dfrac{K_A m_L}{V_M(1+K_I[I])} + \dfrac{(K_{AN1} m_{LN1} + \dots + K_{ANt} m_{LNt})}{V_M} = k_{AL} + X$ or $\dfrac{1}{(k_A - X)} = \dfrac{V_M K_I[I]}{K_A m_L} + \dfrac{V_M}{K_A m_L}$

Source: Reprinted from Ref. [65], Copyright (2002), with permission from Elsevier.

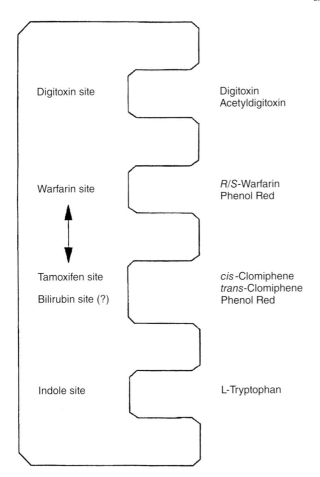

| Digitoxin site | Digitoxin |
| | Acetyldigitoxin |

| Warfarin site | R/S-Warfarin |
| | Phenol Red |

Tamoxifen site	cis-Clomiphene
	trans-Clomiphene
Bilirubin site (?)	Phenol Red

| Indole site | L-Tryptophan |

Figure 8.9 Relative arrangement of the major and minor binding sites of HSA based on competition studies between various probe compounds. The compounds that were examined in this work are shown on the right and their binding regions are shown on the left. The arrow between the warfarin and tamoxifen sites indicates the presence of allosteric effects between these two regions. (Reprinted with permission from Ref. [62]. Copyright (1999) American Chemical Society.)

the most important role in the protein–drug binding interactions are determined, thus allowing the approximate description of the sites that are involved in these interactions [68].

In addition to studying the factors that play a role in the formation of the protein–drug complex, conformational changes can also be studied on columns using zonal affinity chromatography. While there are several examples in the literature [27, 64, 57, 69, 70], only Pgp transporter is discussed here. The Pgp transporter undergoes an ATP-dependent conformational change, which results in the efflux of substrates into the extracellular fluid. Pgp is believed to cycle through four ATP-catalyzed phases. In the initial phase (absence of ATP),

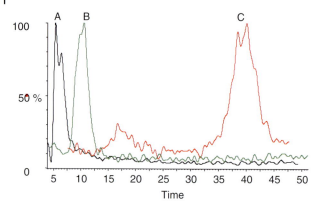

Figure 8.10 PGP-SP (IAM) was packed into peek tubing (762 µm id × 40 cm), with a flow rate of 50 µl/min. Zonal injections of 1 nM vinblastin were given in the presence of 1 µM vinblastin (A), on the column trapped in the first ATP cycle in the presence of vanadate in the mobile phase and alone (C). (Reprinted from Ref. [26], Copyright (2006), with permission from Elsevier.)

Pgp has a high affinity for the substrates; upon binding of ATP (phase II), a conformational change takes place that results in a substantial reduction in the affinity of substrates and an increase in the affinity of inhibitors for the Pgp transporter. The conformational mobility of the immobilized Pgp transporter is clearly illustrated in Figure 8.10 [26]. Vinblastin, a known Pgp substrate that binds to the Pgp transporter in the absence of ATP, was retained for 40 min. Exposure of the column to 3 mM ATP with 1 mM $MgCl_2$ resulted in a conformational change to phase II, which was maintained by the addition of sodium vanadate, causing a significant decrease in vinblastin retention to 8 min. Once the vanadate and ATP were washed out of the column, vinblastin retention increased to 40 min. The same change in retention was not observed on the Pgp(−) column, indicating that the change in retention volume resulted only from the conformational change of the immobilized Pgp.

8.7
Nonlinear Chromatography

The shape of a chromatographic peak is the function of the specific and non-specific interactions between the solute and the SP. When the SP contains an immobilized protein, the mass transfer process defined by the dissociation and association of a ligand–protein complex is usually slow, producing broad, non-Gaussian chromatographic peaks with significant tailing. The degree of deviation from a Gaussian distribution is a function of applied ligand concentration, and the concentration-dependent asymmetry can be used with NLC techniques to characterize the separation processes occurring on the column, including the kinetics involved in the formation and dissolution of the solute–SP complex, the association (k_a) and dissociation (k_d) rate constants, and the equilibrium constant (K_a, calculated as k_a/k_d) [71, 72]. The observed peak asymmetries were analyzed using impulse input solution (Equation 8.4) as described previously [72, 12, 73]. PeakFit software was used to perform the calculations.

$$y = \frac{a_0}{a_3}\left[1 - \exp\left(-\frac{a_3}{a_2}\right)\right]\left[\frac{\sqrt{a_1/x}I_1\left[2\sqrt{a_1x}/a_2\right]\exp[(-x-a_1)/a_2]}{1 - T[(a_1/a_2),(x/a_2)][1-\exp[-a_3/a_2]]}\right]$$

$$T(u,v) = \exp(-v)\int_0^u \exp(-t)I_0(2\sqrt{vt})dt \tag{8.4}$$

The T function acts as a "switching" function that produces the skew in the peak profile when the column is overloaded; y is intensity of signal, x is reduced retention time, $I_0(x)$ and $I_1(x)$ are modified Bessel functions, a_0 is the area parameter, and a_1 is the center parameter (which determines the true thermodynamic capacity factor k'), a_2 is the width parameter, and a_3 is the distortion parameter. Kinetic parameters can be calculated as follows: $k_d = 1/a_2/t_0$, $K_a = a_3/C_0$, and $k_a = K_a/k_d$, where t_0 is the dead time of the column and C_0 is the concentration of ligand injected multiplied by the width of the injection pulse [12].

Several subtypes of the nicotinic receptors have been immobilized, including the α3β2, α3β4, α4β2, and α4β4 nicotinic receptors [74]. While the agonist binding sites were characterized using several techniques including FAC, the same is not true for the NCI binding sites on the inner lumen of the nicotinic receptor [74]. The noncompetitive binding sites of the nicotinic receptor have gained increasing importance in drug discovery, with several drugs developed specifically to target this site. For example, mecamylamine (MCM), a known NCI was developed as a drug for nicotine cessation. NCIs have a moderate affinity for the nicotinic receptor in the resting state, while the binding of an agonist to the nicotinic receptor induces a conformational change that results in an increase in the affinity. NCIs inhibit the function of the nicotinic receptor by a reversible channel blockade or shorten channel opening time in a voltage-sensitive manner [75]. Therefore, NCIs prevent the receptor from undergoing agonist-induced gating in a noncompetitive manner.

To date there is no rapid screening technique available for the NCIs of the nicotinic receptor. Currently, NCIs are primarily identified and characterized using efflux functional assays, as well as electrophysiological experiments, both of which are time-consuming and laborious and require extensive training in the latter method. While FAC has been used to determine the binding affinities of agonist/antagonist for the agonist binding site of the nicotinic receptor, we demonstrated that the NCI binding site of the nicotinic receptor can be studied using NLC using LC–MS. In addition to determining the association and dissociation rate constants, using NLC, we were able to determine the thermodynamic retention factors, the association constant K_a, and demonstrate that NLC could be used to assess the length of inhibition of NCIs for the nicotinic receptor. Specifically, using dextromethorphan and levomethorphan, the length of inhibition of the NCI was determined by measuring 7 min recovery using the Rb^{+86} efflux assay and it was demonstrated that NLC could be used to predict the relative length of inhibition of the NCI. While no NLC parameter was found for the quantitative correlation with the IC_{50} values of the NCIs, it was determined that NLC could qualitatively rank the NCIs into weak inhibitors or strong inhibitors. In this study, the chromatographic retention factors and the molecular descriptors describing these retentions were used for the qualitative assessment of the IC_{50} values (Table 8.3) and this parameter is described in Equation 8.5 [76]:

$$Log\,k' = 5.255(\pm0.942) + 0.491(\pm0.092)E_{HOMO} + 0.012(\pm0.005)S_{YZ}$$
$$r = 0.984, \quad s = 0.168, \quad F = 27.929, \quad n = 17 \qquad\qquad (8.5)$$

Table 8.3 IC_{50} values, logarithm of thermodynamic retention factors (log k) and molecular descriptors (E_{HOMO} and S_{YZ}), and assigned clusters for the compounds used in this study.

	Compound	log k	E_{HOMO}	S_{YZ}	Cluster	IC_{50} (µM)
1	Methadone	1.65	−9.20	66.6	1	1.9
2	Diltiazem	1.64	−8.66	62.3	1	2.3
3	Norverapamil	1.99	−9.12	69.1	1	2.6
4	Verapamil	1.98	−9.04	69.4	1	3
5	Dextromethorphan	1.79	−8.72	51.1	2	10.1
6	Levomethorphan	1.55	−8.74	52.7	2	10.9
7	Dextrorphan	1.43	−8.77	50.9	2	29.6
8	(+)-3-Hydroxy-morphinan	1.75	−8.78	49.8	2	10.3
9	(+)-3-Methoxy-morphinan	1.42	−8.83	49.3	2	54.2
10	Clozapine	2.19	−7.69	44.5	2	28
11	Laudanosine	1.36	−8.54	50.1	2	139.7
12	Phencyclidine	1.38	−9.05	51.6	2	7
13	Adamantadine	0.95	−9.71	35.4	3	3.4
14	Bupropion	1.11	−9.51	34.2	3	1.4
15	Ketamine	0.92	−9.49	42.5	3	1.4
16	Mecamylamine	1.04	−9.22	40.2	3	1
17	Memantine	1.22	−9.71	42.3	3	6.6
18	Methamphetamine	0.92	−9.39	27.8	3	401.2
19	MK-801	1.28	−9.11	43	3	26.6
20	N-Demethyl-diltiazem	1.61	−8.65	66.5	1	4.2
21	Deacetyl-diltiazem	1.6	−8.58	62.4	1	30.4
22	N-Demethyl-deacetyl-diltiazem	1.61	−8.58	57.4	2	77.6
23	O-Demethyl-deacetyl-diltiazem	1.45	−8.41	61.8	1	73.2
24	N,O-Didemethyl-deacetyl-diltiazem	1.48	−8.65	58.1	2	63.1
25	Galapamil	1.88	−9.05	66.3	1	6.4
26	D-620 (verapamil metabolite)	1.25	−9.35	48.5	2	48.9
27	Nicardipine	2.33	−8.84	65.2	1	2.5
28	Amlodipine	2	−8.72	62.9	1	5.8
29	Nifedipine	1.27	−8.63	58.4	2	24.7

where the molecular descriptor E_{HOMO} is the energy of the highest occupied molecular orbital and reflects the ability of the compound to transfer a proton to the carboxylate moiety; S_{YZ} is the surface area of the molecular projection onto the YZ plane and is associated with the entrance of the compound into a defined steric environment [76]; k' is the thermodynamic retention factor that reflects the equilibrium between the initial binding of the compound to the nicotinic receptor, k_{on}, and the stability of the final NCI–nicotinic receptor complex, k_{off}.

While no linear relationship was observed between the IC_{50} values and the NLC parameters in this study, a multidimensional cluster analysis of the 3D scatterplot of the chromatographic and molecular parameters demonstrated that a relationship between the IC_{50} values and the NLC parameters was found by subdividing them into three clusters (Figure 8.11; Table 8.4): Cluster 1 (compounds 1–4), which contained compounds with IC_{50} values ranging from 1.9 to 3.0 μM and had the following ranges for the molecular parameters: log k' values ranging from 1.3 to 2.2, E_{HOMO} from −9.2 to −8.6, and SYZ from 60 to 70; Cluster 2 (compounds 5–12) which contained compounds with IC_{50} values ranging from 7.0 to 139.7 μM and had the following ranges for the molecular parameters: log k' ranging from 1.3 to 2.2, E_{HOMO} from −9.0 to −7.7, and SYZ from 45 to 60; Cluster 3 (compounds 13–19), which contained compounds with IC_{50} values ranging from 1.0 to 6.6 μM and had the following ranges for the molecular parameters: log k', ranging from 0.9 to 1.3, E_{HOMO} from −9.8 to −9.1, and SYZ from 25 to 45.

As the NCIs binding to the inner lumen is a multistep process, it is not surprising that cluster analysis using all multiple factors is required to qualitatively identify compounds with similar IC_{50} values. This experimental approach placed 25 of the

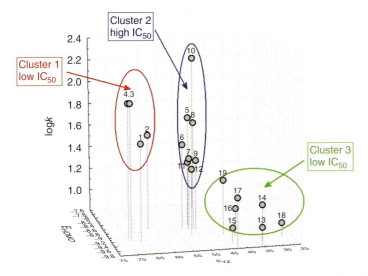

Figure 8.11 The results from the cluster analysis of the compounds used in this study using the log k, E_{HOMO}, and S_{YZ} parameters. (Reprinted from Ref. [76], Copyright (2005), with permission from Elsevier.)

Table 8.4 Binding affinities (nanomolar) calculated by frontal affinity chromatography using the immobilized nicotinic receptor columns, compared to binding affinities calculated by filtration assays.

	α3β2	α3β4	α4β2	α4β4
Epibatidine	0.001 (0.035)	0.086 (0.57)	0.005 (0.061)	0.042 (0.16)
Nicotine	2.18 (47)	80 (440)	16.4 (10)	0.387 (40)
Cytisine	2.43 (37)	7.6 (220)	6.09 (1.5)	0.237 (2.1)

Source: Reprinted with permission from Ref. [74]. Copyright (2005) American Chemical Society.

29 compounds tested in the correct IC_{50} clusters and was carried out over a period of 2 days.

The conformational mobility of the immobilized α3β4 nicotinic receptor was also studied. The nicotinic receptor is known to exist in multiple states, including a resting state, active state, intermediate state, and desensitized state [77–79], and is believed to primarily exist in the resting state. The introduction of an agonist results in conformational change, which in turn results in ion flux followed by desensitization. In order to determine whether the immobilized nAChR could undergo a conformational change, the binding characteristics of four known NCIs, namely, MCM, dextromethorphan (DM), levomethorphan (LM), and phencyclidine (PCP), were studied prior to and after exposure to epibatidine, a known agonist for all subtypes of nAChR. The resting state is known to have a moderate affinity for NCIs, while the desensitized state has a higher affinity for NCIs. The introduction of epibatidine resulted in an increase in retention time for MCM, PCP, and LM for multiple subtypes of nAChR including the α3β2, α3β4, α4β2, and α4β4 receptors, indicating that the immobilized nicotinic receptors underwent agonist-induced conformational change. However, while LM resulted in an increased retention time for all four subtypes tested, DM retention only increased for the α3β4 nAChR and α4β4 nAChR columns. As DM and LM are enantiomers, the change in retention volume between the enantiomers can only result from the immobilized nAChR. Further, the increased enantioseparation resulting from the introduction of an agonist clearly indicates that the immobilized nAChR underwent a conformational change. Prior to exposure to epibatidine, a bimodal peak profile was observed for both DM and LM, with a dominant fronting peak (Figure 8.12a). After exposure to epibatidine, resulting in a conformational change, the bimodal peak collapsed into a single, asymmetric peak (Figure 8.12b) for both DM and LM, indicating that multiple receptor populations exists for DM and LM. The initial frontal peak can therefore be considered to be the resting state of nAChR, while the second peak could be a result of the desensitized state. This indicates that when the nAChRs were immobilized they primarily existed in the resting state with a small population existing in the desensitized state. Nonlinear chromatographic analysis of the data from the pre-EB and post-EB forms of the α3β4 nAChR column indicated that the post-EB dissociation rate constants, k_d, significantly decreased from 6.95

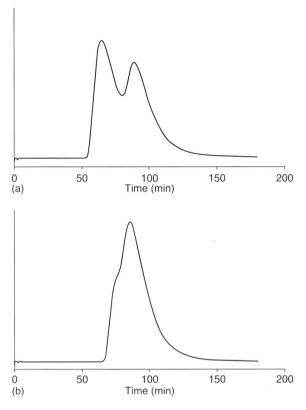

(a)

(b)

Figure 8.12 Chromatographic trace produced by dextromethorphan on an immobilized α3β4 nAChR column: the peak profile produced (a) before exposure of the column to the agonist epibatidine and (b) after exposure to the agonist epibatidine. (Reprinted with permission from Ref. [74]. Copyright (2005) American Chemical Society.)

($\pm$0.01) to 5.68 ($\pm$0.01) s^{-1} (PCP), from 1.92 ($\pm$0.01) to 1.69 ($\pm$0.01) s^{-1} (LM), and from 1.84 ($\pm$0.01) to 1.67 ($\pm$0.01) s^{-1} (DM), indicating an increased stability of the NCI–nAChR complexes on the post-EB column.

Acknowledgments

This research was supported in part by the Intramural Research Program of the NIH, National Institute on Aging.

References

1. Hutchings, C.J., Koglin, M., and Marshall, F.H. (2010) Therapeutic antibodies directed at G protein-coupled receptors. *MAbs*, **6**, 594–606.

2. Bianchi, M.T. and Botzolakis, E.J. (2010) Targeting ligand-gated ion channels in neurology and psychiatry: is pharmacological promiscuity an obstacle or

an opportunity? *BMC Pharmacol.*, **10**, 1–8.

3. Pierce, K.L., Premont, R.T., and Lefkowitz, R.J. (2002) Seven-transmembrane receptors. *Nat. Rev. Mol. Cell Biol.*, **9**, 639–650.

4. Tang, X.L., Wang, Y., Li, D.L., Luo, J., and Liu, M.Y. (2012) Orphan G protein-coupled receptors (GPCRs): biological functions and potential drug targets. *Acta Pharmacol. Sin.*, **33**, 363–371.

5. Dean, M., Rzhetsky, A., and Allikmets, R. (2001) The human ATP-binding cassette (ABC) transporter superfamily. *Genome Res.*, **7**, 1156–1166.

6. Cooper, M.A. (2004) Advances in membrane receptor screening and analysis. *J. Mol. Recognit.*, **17**, 286–315.

7. Zhuang, X. and Rief, M. (2003) Single-molecule folding. *Curr. Opin. Struct. Biol.*, **13**, 88–97.

8. Yadavalli, V.K., Forbes, J.G., and Wang, K. (2006) Functionalized self-assembled monolayers on ultraflat gold as platforms for single molecule force spectroscopy and imaging. *Langmuir*, **22**, 6969–6976.

9. Moaddel, R., Hamid, R., Patel, S., Wainer, I.W., and Bullock, P. (2006) Comparison of an open tubular column containing immobilized P-glycoprotein with Caco-2 cell monolayers for the purpose of investigating interactions between drug candidates and Pgp. *Anal. Chim. Acta*, **578**, 25–30.

10. Moaddel, R., Bullock, P.I., and Wainer, I.W. (2004) Development and characterization of an open tubular column containing immobilized P-glycoprotein for rapid on-line screening for P-glycoprotein substrates. *J. Chromatogr. B*, **799**, 255–263.

11. Chaiken, I.M. (1986) Analytical affinity chromatography in studies of molecular recognition in biology: a review. *J. Chromatogr.*, **376**, 11–32.

12. Wade, J.L., Bergold, A.F., and Carr, P.W. (1987) Theoretical description of nonlinear chromatography, with applications to physicochemical measurements in affinity chromatography

and implications for preparative-scale separations. *Anal. Chem.*, **59**, 1286–1295.

13. Lagercrantz, C., Larsson, T., and Karlsson, H. (1979) Binding of some fatty acids and drugs to immobilized bovine serum albumin studied by column affinity chromatography. *Anal. Biochem.*, **99**, 352–364.

14. Hage, D.S. (1999) Affinity chromatography: a review of clinical applications. *Clin. Chem.*, **45**, 593–615.

15. Haginaka, J. (2008) Recent progresses in protein-based chiral stationary phases for enantioseparations in liquid chromatography. *J. Chromatogr. B*, **875**, 12–19.

16. Haupt, D. (1996) Determination of citalopram enantiomers in human plasma by liquid chromatographic separation on a Chiral-AGP column. *J. Chromatogr. B*, **685**, 299–305.

17. Silan, L., Jadaud, P., Whitfield, L.R., and Wainer, I.W. (1990) Determination of low levels of the stereoisomers of leucovorin and 5-methyltetrahydrofolate in plasma using a coupled chiral-achiral high-performance liquid chromatographic system with post-chiral column peak compression. *J. Chromatogr.*, **532**, 227–236.

18. Kelly, J.W., Stewart, J.T., and Blanton, C.D. (1994) HPLC separation of pentazocine enantiomers in serum using an ovomucoid chiral stationary phase. *Biomed. Chromatogr.*, **8**, 255–257.

19. Bhatia, P., Moaddel, R., and Wainer, I.W. (2010) Application of cellular membrane affinity chromatography: in determining stereoselective interactions with ATP-binding cassette transporters. *Chim. Oggi*, **28**, 32–36.

20. Massolini, G. and Calleri, E. (2005) Immobilized trypsin systems coupled on-line to separation methods: recent developments and analytical applications. *J. Sep. Sci.*, **28**, 7–21.

21. Beaudry, F., Coutu, M., and Brown, N.K. (1999) Determination of drug-plasma protein binding using human serum albumin chromatographic column and multiple linear regression model. *Biomed. Chromatogr.*, **13**, 401–406.

22. Moaddel, R. and Wainer, I.W. (2009) The preparation and development of cellular membrane affinity chromatography columns. *Nat. Protoc.*, **4**, 197–205.

23. Gottschalk, I., Lagerquist, C., Zuo, S.S., Lundqvist, A., and Lundahl, P. (2002) Immobilized-biomembrane affinity chromatography for binding studies of membrane proteins. *J. Chromatogr. B Analyt. Technol. Biomed. Life Sci.*, **768**, 31–40.

24. Brekkan, E., Lundqvist, A., and Lundahl, P. (1996) Immobilized membrane vesicle or proteoliposome affinity chromatography. Frontal analysis of interactions of cytochalasin B and D-glucose with the human red cell glucose transporter. *Biochemistry*, **35**, 12141–12145.

25. Gottschalk, L., Li, Y.M., and Lundahl, P. (2000) Chromatography on cells: analyses of solute interactions with the glucose transporter Glut1 in human red cells adsorbed on lectin-gel beads. *J. Chromatogr. B*, **739**, 55–62.

26. Moaddel, R. and Wainer, I.W. (2006) Development of immobilized membrane-based affinity columns for use in the online characterization of membrane bound proteins and for targeted affinity isolations. *Anal. Chim. Acta*, **564**, 97–105.

27. Moaddel, R. and Wainer, I.W. (2007) Conformational mobility of immobilized proteins. *J. Pharm. Biomed. Anal.*, **43**, 399–406.

28. Zhang, Y., Leonessa, F., Clarke, R., and Wainer, I.W. (2000) Development of an immobilized P-glycoprotein stationary phase for on-line liquid chromatographic determination of drug-binding affinities. *J. Chromatogr. B Biomed. Sci. Appl.*, **739**, 33–37.

29. Pidgeon, C. and Venkataram, U.V. (1989) Immobilized artificial membrane chromatography: supports composed of membrane lipids. *Anal. Biochem.*, **176**, 36–47.

30. Moaddel, R., Lu, L., Baynham, M., and Wainer, I.W. (2002) Immobilized receptor- and transporter-based liquid chromatographic phases for on-line pharmacological and biochemical studies: a mini-review. *J. Chromatogr. B Analyt. Technol. Biomed. Life Sci.*, **768**, 41–53.

31. Beigi, F., Chakir, K., Xiao, R.P., and Wainer, I.W. (2004) G-protein-coupled receptor chromatographic stationary phases. 2. Ligand-induced conformational mobility in an immobilized beta2-adrenergic receptor. *Anal. Chem.*, **76**, 7187–7193.

32. Moaddel, R., Rosenberg, A., Spelman, K., Frazier, J., Frazier, C., Nocerino, S., Brizzi, A., Mugnaini, C., and Wainer, I.W. (2011) Development and characterization of immobilized cannabinoid receptor (CB1/CB2) open tubular column for on-line screening. *Anal. Biochem.*, **412**, 85–91.

33. Temporini, C., Pochetti, G., Fracchiolla, G., Piemontese, L., Montanari, R., Moaddel, R., Laghezza, A., Altieri, F., Cervoni, L., Ubiali, D., Prada, E., Loiodice, F., Massolini, G., and Calleri, E. (2013) Open tubular columns containing the immobilized ligand binding domain of peroxisome proliferator-activated receptors α and γ for dual agonists characterization by frontal affinity chromatography with mass spectrometry detection. *J. Chromatogr. A*, **1284**, 36–43.

34. Moaddel, R., Jozwiak, K., Yamaguchi, R., and Wainer, I.W. (2005) Direct chromatographic determination of dissociation rate constants of ligand-receptor complexes: assessment of the interaction of noncompetitive inhibitors with an immobilized nicotinic acetylcholine receptor-based liquid chromatography stationary phase. *Anal. Chem.*, **77**, 5421–5426.

35. Moaddel, R., Cloix, J.-F., Ertem, G., and Wainer, I.W. (2002) Multiple receptor liquid chromatographic stationary phases: the co-immobilization of nicotinic receptors, γ-amino-butyric acid receptors, and N-methyl D-aspartate receptors. *Pharm. Res.*, **19**, 104–107.

36. Kitabatake, T., Moaddel, R., Cole, R., Gandhari, M., Frazier, C., Hartenstein, J., Rosenberg, A., Bernier, M., and Wainer, I.W. (2008) Characterization of a multiple ligand-gated ion channel

cellular membrane affinity chromatography column and identification of endogenously expressed receptors in astrocytoma cell lines. *Anal. Chem.*, **80**, 8673–8680.

37. Bhatia, P., Bernier, M., Sanghvi, M., Moaddel, R., Schwarting, R., Ramamoorthy, A., and Wainer, I.W. (2012) Breast cancer resistance protein (BCRP/ABCG2) localises to the nucleus in glioblastoma multiforme cells. *Xenobiotica*, **42**, 748–755.

38. Habicht, K.L., Frazier, C., Singh, N., Shimmo, R., Wainer, I.W., and Moaddel, R. (2013) The synthesis and characterization of a nuclear membrane affinity chromatography column for the study of human breast cancer resistant protein (BCRP) using nuclear membranes obtained from the LN-229 cells. *J. Pharm. Biomed. Anal.*, **72**, 159–162.

39. Moaddel, R., Patel, S., Jozwiak, K., Yamaguchi, R., Ho, P.C., and Wainer, I.W. (2005) Enantioselective binding to the human organic cation transporter-1 (hOCT1) determined using an immobilized hOCT1 liquid chromatographic stationary phase. *Chirality*, **17**, 501–506.

40. Moaddel, R., Ravichandran, S., Bighi, F., Yamaguchi, R., and Wainer, I.W. (2007) Pharmacophore modelling of stereoselective binding to the human organic cation transporter (hOCT1). *Br. J. Pharmacol.*, **151**, 1305–1314.

41. Kasai, K.-I. and Ishii, S.-I. (1975) Quantitative analysis of affinity chromatography of trypsin – A new technique for investigation of protein-ligand interaction. *J. Biochem.*, **77**, 261–264.

42. Calleri, E., Temporini, C., Caccialanza, G., and Massolini, G. (2009) Target-based drug discovery: the emerging success of frontal affinity chromatography coupled to mass spectrometry. *ChemMedChem*, **4**, 905–916.

43. Zeng, A., Yuan, B., Wang, C., Yang, G., and He, L. (2009) Frontal analysis of cell-membrane chromatography for the determination of drug-α_{1D} adrenergic receptor affinity. *J. Chromatogr. B*, **877**, 1833–1837.

44. Ng, E.S., Yang, F., Kameyama, A., Palcic, M.M., Hindsgaul, O., and Schriemer, D.C. (2005) High-throughput screening for enzyme inhibitors using frontal affinity chromatography with liquid chromatography and mass spectrometry. *Anal. Chem.*, **77**, 6125–6133.

45. Schriemer, D.C. (2004) Biosensor alternative: frontal affinity chromatography. *Anal. Chem.*, **76**, 440A–448A.

46. Ng, E.S., Chan, N.W., Lewis, D.F., Hindsgaul, O., and Schriemer, D.C. (2007) Frontal affinity chromatography-mass spectrometry. *Nat. Protoc.*, **2**, 1907–1917.

47. Chan, N.W., Lewis, D.F., Rosner, P.J., Kelly, M.A., and Schriemer, D.C. (2003) Frontal affinity chromatography-mass spectrometry assay technology for multiple stages of drug discovery: applications of a chromatographic biosensor. *Anal. Biochem.*, **319**, 1–12.

48. Schriemer, D.C., Bundle, D.R., Li, L., and Hindsgaul, O. (1998) Micro-scale frontal affinity chromatography with mass spectrometric detection: a new method for the screening of compound libraries. *Angew. Chem. Int. Ed. Engl.*, **37**, 3383–3387.

49. Serra, I., Ubiali, D., Cecchini, D.A., Calleri, E., Albertini, A.M., Terreni, M., and Temporini, C. (2013) Assessment of immobilized PGA orientation via the LC-MS analysis of tryptic digests of the wild type and its 3 K-PGA mutant assists in the rational design of a high-performance biocatalyst. *Anal. Bioanal. Chem.*, **405**, 745–753.

50. Calleri, E., Ceruti, S., Cristalli, G., Martini, C., Temporini, C., Parravicini, C., Volpini, R., Daniele, S., Caccialanza, G., Lecca, D., Lambertucci, C., Trincavelli, M.L., Marucci, G., Wainer, I.W., Ranghino, G., Fantucci, P., Abbracchio, M.P., and Massolini, G. (2010) Frontal affinity chromatography-mass spectrometry useful for characterization of new ligands for GPR17 receptor. *J. Med. Chem.*, **53**, 3489–3501.

51. Allenmark, S., Andersson, S., and Bojarski, J. (1988) Direct liquid chromatographic separation of enantiomers

on immobilized protein stationary phases. VI. Optical resolution of a series of racemic barbiturates: studies of substituent and mobile phase effects. *J. Chromatogr.*, **436**, 479–483.

52. Wong, L.S., Khan, F., and Micklefield, J. (2009) Selective covalent protein immobilization: strategies and applications. *Chem. Rev.*, **109**, 4025–4053.

53. Loun, B. and Hage, D.S. (1992) Characterization of thyroxine-albumin binding using high-performance affinity chromatography. I. Interactions at the warfarin and indole sites of albumin. *J. Chromatogr.*, **579**, 225–235.

54. Fitos, I., Visy, J., Simonyi, M., and Hermansson, J. (1992) Chiral high-performance liquid chromatographic separations of vinca alkaloid analogues on α_1-acid glycoprotein and human serum albumin columns. *J. Chromatogr.*, **609**, 163–171.

55. Noctor, T.A.G., Wainer, I.W., and Hage, D.S. (1992) Allosteric and competitive displacement of drugs from human serum albumin by octanoic acid, as revealed by high-performance liquid affinity chromatography, on a human serum albumin-based stationary phase. *J. Chromatogr.*, **577**, 305–315.

56. Noctor, T.A.G., Pham, C.D., Kaliszan, R., and Wainer, I.W. (1992) Stereochemical aspects of benzodiazepine binding to human serum albumin. I. Enantioselective high performance liquid affinity chromatographic examination of chiral and achiral binding interactions between 1,4-benzodiazepines and human serum albumin. *Mol. Pharmacol.*, **42**, 506–511.

57. Domenici, E., Bertucci, C., Salvadori, P., and Wainer, I.W. (1991) Use of a human serum albumin-based high-performance liquid chromatography chiral stationary phase for the investigation of protein binding: detection of the allosteric interaction between warfarin and benzodiazepine binding sites. *J. Pharm. Sci.*, **80**, 164–166.

58. Dalgaard, L., Hansen, J.J., and Pedersen, J.L. (1989) Resolution and binding site determination of d,l-thyronine by high-performance liquid chromatography using immobilized albumin as chiral stationary phase. Determination of the optical purity of thyroxine in tablets. *J. Pharm. Biomed. Anal.*, **7**, 361–368.

59. Kaliszan, R., Noctor, T.A.G., and Wainer, I.W. (1992) Stereochemical aspects of benzodiazepine binding to human serum albumin. II. Quantitative relationships between structure and enantioselective retention in high performance liquid affinity chromatography. *Mol. Pharmacol.*, **42**, 512–517.

60. Hage, D.S. and Chen, J. (2006) Quantitative affinity chromatography: practical aspect, in *Handbook of Affinity Chromatography* (ed D.S. Hage), CRC Press, Boca Raton, FL, pp. 595–628, ISBN: 978-0-8247-4057-3.

61. Basiaga, S.B. and Hage, D.S. (2010) Chromatographic studies of changes in binding of sulfonylurea drugs to human serum albumin due to glycation and fatty acids. *J. Chromatogr. B Analyt. Technol. Biomed. Life Sci.*, **878**, 3193–3197.

62. Sengupta, A. and Hage, D.S. (1999) Characterization of minor site probes for human serum albumin by high-performance affinity chromatography. *Anal. Chem.*, **71**, 3821–3827.

63. Hage, D.S. and Austin, J. (2000) High-performance affinity chromatography and immobilized serum albumin as probes for drug- and hormone-protein binding. *J. Chromatogr. B*, **739**, 39–54.

64. Noctor, T.A.G., Diaz-Perez, M.J., and Wainer, I.W. (1993) Use of a human serum albumin-based stationary phase for high-performance liquid chromatography as a tool for the rapid determination of drug-plasma protein binding. *J. Pharm. Sci.*, **82**, 675–676.

65. Hage, D.S. (2002) High-performance affinity chromatography: a powerful tool for studying serum protein binding. *J. Chromatogr. B*, **768**, 3–30.

66. Hage, D.S. (2001) Chromatographic and electrophoretic studies of protein binding to chiral solutes. *J. Chromatogr. A*, **906**, 459–481.

67. Yang, J. and Hage, D.S. (1996) Role of binding capacity versus binding strength in the separation of chiral compounds on protein-based

high-performance liquid chromatography columns Interactions of d- and l-tryptophan with human serum albumin. *J. Chromatogr. A*, **725**, 273–285.

68. Wainer, I.W. (1994) Enantioselective high-performance liquid affinity chromatography as a probe of ligand-biopolymer interactions: an overview of a different use for high-performance liquid chromatographic chiral stationary phases. *J. Chromatogr. A*, **666**, 221–234.

69. Chen, J. and Hage, D.S. (2004) Quantitative analysis of allosteric drug-protein binding by biointeraction chromatography. *Nat. Biotechnol.*, **22**, 1445–1448.

70. Chen, J. and Hage, D.S. (2006) Quantitative studies of allosteric effects by biointeraction chromatography: analysis of protein binding for low-solubility drugs. *Anal. Chem.*, **78**, 2672–2683.

71. Jozwiak, K., Haginaka, J., Moaddel, R., and Wainer, I.W. (2002) Displacement and nonlinear chromatographic techniques in the investigation of interaction of noncompetitive inhibitors with an immobilized alpha3beta4 nicotinic acetylcholine receptor liquid chromatographic stationary phase. *Anal. Chem.*, **74**, 4618–4624.

72. Moaddel, R., Jozwiak, K., and Wainer, I.W. (2007) Allosteric modifiers of neuronal nicotinic acetylcholine receptors: new methods, new opportunities. *Med. Res. Rev.*, **27**, 723–753.

73. Jaulmes, A. and Vidal-Madjar, C. (1989) in *Advances in Chromatography*, vol. **28**

(eds J. Giddings, E. Grushka, and P.R. Brown), Marcel Dekker, New York, pp. 1–64.

74. Moaddel, R., Jozwiak, K., Whittington, K., and Wainer, I.W. (2005) Conformational mobility of immobilized alpha3beta2, alpha3beta4, alpha4beta2, and alpha4beta4 nicotinic acetylcholine receptors. *Anal. Chem.*, **77**, 895–901.

75. Xiao, Y., Smith, R.D., Caruso, F.S., and Kellar, K.J. (2001) Blockade of rat α3β4 nicotinic receptor function by methadone, its metabolites, and structural analogs. *J. Pharmacol. Exp. Ther.*, **299**, 366–371.

76. Jozwiak, K., Moaddel, R., Yamaguchi, R., Ravichandran, S., Collins, J.R., and Wainer, I.W. (2005) Qualitative assessment of IC50 values of inhibitors of the neuronal nicotinic acetylcholine receptor using a single chromatographic experiment and multivariate cluster analysis. *J. Chromatogr. B Analyt. Technol. Biomed. Life Sci.*, **819**, 169–174.

77. Auerbach, A. (2003) Life at the top: the transition state of AChR gating. *Sci. STKE*, **188**, re11–re24.

78. Quick, M.W. and Lester, R.A. (2002) Desensitization of neuronal nicotinic receptors. *J. Neurobiol.*, **53**, 457–478.

79. Gentry, C.L., Wilkins, L.H. Jr.,, and Lukas, R.J. (2003) Effects of prolonged nicotinic ligand exposure on function of heterologously expressed, human alpha4beta2- and alpha4beta4-nicotinic acetylcholine receptors. *J. Pharmacol. Exp. Ther.*, **304**, 206–216.

9
Online Affinity Assessment and Immunoaffinity Sample Pretreatment in Capillary Electrophoresis–Mass Spectrometry

Rob Haselberg and Govert W. Somsen

9.1
Introduction

Capillary electrophoresis (CE) is a microscale liquid separation technique that is highly suitable for the analysis of complex samples. Typically, CE is characterized by high separation efficiencies, relatively short analysis times, and low sample and solvent consumption. Since the separation mechanism of CE is based on differential mobility of charged species in an electrical field, it encompasses unique selectivity with respect to most chromatographic techniques. Furthermore, selectivity in CE can be adjusted in a straightforward manner by simply changing the composition of the separation medium – the so-called background electrolyte (BGE) - providing a versatile means to efficiently separate ionogenic compounds. Consequently, CE has found widespread application in many fields, such as (bio)pharmaceutical [1] and food analysis [2], forensic science [3], proteomics [4], and metabolomics [5].

CE analyses are normally carried out under aqueous conditions in the absence of a stationary phase, and near-physiological separation conditions can be chosen. In principle these properties make CE a very suitable technique for the investigation of biomolecular affinity interactions. First attempts of such an approach were reported in 1992 by Whitesides *et al.* [6], who coined the technique affinity capillary electrophoresis (ACE). Around the same time, research groups started to investigate the use of affinity materials for selective in-line extraction of compounds before CE analysis [7, 8]. This immunoaffinity capillary electrophoresis (IA-CE) approach aimed for the highly specific isolation of target analytes from complex biological matrices. Meanwhile, both ACE and IA-CE have developed into mature fields of research, with close to 900 papers published on these topics since 1990 and 50 new publications appearing annually over the last decade (Figure 9.1). First successful attempts to couple CE to mass spectrometry (MS) were reported more than 25 years ago. Since then, CE–MS has gradually developed into a routine analytical technique particularly useful for peptides, proteins, and highly polar metabolites. This chapter deals with the incorporation of MS into ACE and IA-CE schemes achieving mass selective detection and/or identification

Analyzing Biomolecular Interactions by Mass Spectrometry, First Edition.
Edited by Jeroen Kool and Wilfried M.A. Niessen.
© 2015 Wiley-VCH Verlag GmbH & Co. KGaA. Published 2015 by Wiley-VCH Verlag GmbH & Co. KGaA.

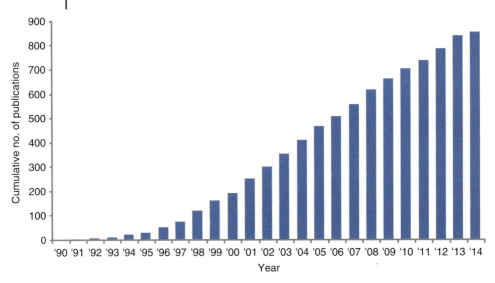

Figure 9.1 Cumulative number of publications between January 1990 and June 2014 having the key terms *affinity capillary electrophoresis* or *immunoaffinity capillary electrophoresis* in the title, abstract, or keywords according to *www.scopus.com*.

of analyzed compounds. In order to understand and appreciate the topic, first the basic aspects of both ACE and IA-CE will be treated in the next sections. Subsequently, the principles of the hyphenation of CE and MS detection are discussed, highlighting factors that are important for affinity determinations. The setup and performance of reported ACE – MS and IA-CE – MS are treated systematically and their applications are outlined giving typical examples.

9.2
Capillary Electrophoresis

First endeavors in electrophoresis began as early as the nineteenth century. In 1937, Nobel Prize winner Arne Tiselius' moving boundary electrophoresis was a landmark, achieving separation of proteins using a U-shaped tank with an electrode at either end [9]. On voltage application, proteins migrated in a direction and at a rate determined by their charge and size. However, voltage-induced separation in free solution, as performed by Tiselius, suffered from limited efficiency due to thermal diffusion and convection caused by the intrinsic heat related to the produced current. Use of supporting media, such as paper and different types of gels, was needed to attain improved electrophoretic separation of high molecular weight compounds. High-efficiency free-zone electrophoresis was explored by Hjertén [10], and later Virtanen [11], Mikkers and Everaerts [12], who used narrow-bore tubes ("capillaries") exhibiting increased surface-to-volume ratios

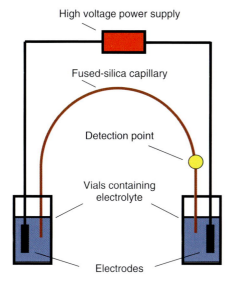

High voltage power supply

Fused-silica capillary

Detection point

Vials containing electrolyte

Electrodes

Figure 9.2 Schematic setup of the basic components of CE instrumentation.

that facilitate heat dissipation. The practical introduction and breakthrough of CE was initiated in the early 1980s by Jorgenson and Lukacs [13]. They used even narrower (<100 µm inner diameter (ID)) capillaries of fused silica in combination with voltages as high as 30 kV to achieve high-resolution separations without experiencing detrimental Joule heating effects. Jorgenson also clarified the underlying theory, described the relationships between operational parameters and separation quality, and demonstrated the actual potential of CE as an analytical tool.

An attractive feature of CE is the overall simplicity of the required instrumentation. A schematic diagram of a generic CE system is shown in Figure 9.2. The ends of a narrow-bore, fused-silica capillary filled with BGE are placed in reservoirs containing the same BGE. The reservoirs also comprise the electrodes that are connected to a high-voltage power supply. A small volume of sample is loaded into the capillary at the inlet end by replacing one of the reservoirs with a sample vial and applying either an electric field or an external pressure. After replacing the BGE reservoir, an electric field is applied performing electrophoretic analysis. A small optical window near the outlet end of the capillary allows in-capillary absorbance or fluorescence detection of passing analytes.

Separation by electrophoresis is based on differences in analyte velocity on application of an electric field [14]. The velocity of an ion in an electric field is given by Equation 9.1, where v is the ion velocity, μ_e is the ion electrophoretic mobility, and E is the applied electrical field.

$$v = \mu_e E \tag{9.1}$$

The electric field (V m^{-1}) is the quotient of the applied voltage and capillary length. For a given BGE and temperature, the electrophoretic mobility is a characteristic analyte property. The velocity of an ionic analyte in an electric field is the resultant of the electric force (F_E) the ion experiences and the counteracting frictional drag by the medium (F_F). Equations 9.2 and 9.3 describe these forces, where q is the ion charge, η is the solution viscosity, and r is the ion radius.

$$F_E = qE \tag{9.2}$$

$$F_F = -6\pi\eta rv \tag{9.3}$$

During electrophoresis $F_E = F_F$, yielding the following equation for the electrophoretic mobility of an ion:

$$\mu_e = \frac{q}{6\pi\eta r} \tag{9.4}$$

Equation 9.4 indicates that analyte separation in CE (i.e., the difference in electrophoretic mobility) relates to differences in charge-to-size (q/r) ratio of the respective analyte ions. The actual charge of an analyte depends on its pK_a and the pH of the selected BGE.

CE is normally performed in fused-silica capillaries of which silanol groups on the inner surface can be deprotonated. This results in a negative charge that attracts cations from the BGE forming a diffuse double layer and creating a so-called zeta potential (Figure 9.3a). When high voltage is applied across the capillary, the cations of the double layer move toward the cathode drawing the rest of the BGE with them, thereby generating an electroosmotic flow (EOF) (Figure 9.3b). The magnitude of the EOF depends on the degree of deprotonation of the silanol groups, and, thus, on the pH of the applied BGE.

The observed or apparent electrophoretic mobility (μ_{app}) of analytes is the sum of the electrophoretic mobility (μ_e) and electroosmotic mobility (μ_{eof}). As the electroosmotic mobility usually is larger than the electrophoretic mobility, all sample compounds will migrate toward the detector at the site of the cathode. Positively charged compounds will reach the detector first, followed by neutral compounds (migrating with the EOF) and negatively charged compounds (Figure 9.3c). The time of an ion to migrate to the detector is a characteristic for that particular sample ion under the applied separation conditions. The migration time allows the calculation of the apparent mobility according to

$$\mu_{app} = \frac{L_{tot}L_{det}}{Vt} \tag{9.5}$$

where L_{tot} is the total capillary length, L_{det} is the capillary length until the detection window, V is the applied voltage, and t is the migration time of an analyte. By subtracting the mobility of the EOF from that of the analytes, their μ_e can be obtained.

Important analytical features of CE are the absence of a stationary phase, the uniform flow profile of the EOF (i.e., no radial velocity differences), and the zero pressure drop across the capillary. As a consequence, in CE, peak broadening predominantly relies on longitudinal diffusion of the analytes. Because of relatively high electric resistance of the narrow capillary, high voltages (10–30 kV) can be

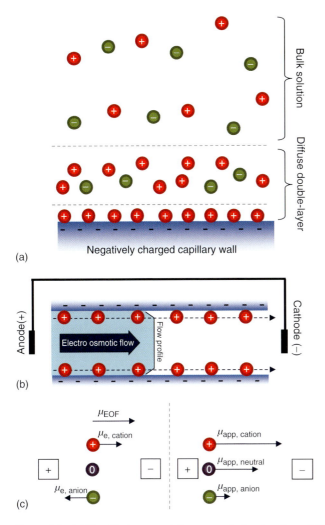

Figure 9.3 (a) Simplified representation of the diffuse double layer formed close to the silica capillary wall. (b) Direction and flow profile of the electroosmotic flow in a fused-silica capillary. (c) Direction of the electrophoretic mobilities of the EOF, cations, neutrals, and anions under normal CE separation conditions (left) and the resulting apparent mobility – and thus migration order – of the respective compounds (right).

applied, which overall can lead to favorable peak efficiencies (plate numbers of 100 000 – 1 000 000).

For CE-based affinity determinations coupled to MS, capillary zone electrophoresis (CZE) is mainly used. In this mode, the sample is applied as a narrow zone surrounded by BGE. As the electric field is applied, each component in the sample zone starts to migrate according to its apparent mobility. Ideally, the sample components will fully separate to form individual zones before detection.

The BGE usually consists of a simple buffer providing conductivity and pH stability needed for reproducible CZE.

9.3
Affinity Capillary Electrophoresis

ACE is a useful technique for measuring association constants of receptor–ligand binding in aqueous solutions, and can be used as a tool for studying the physical interactions that determine binding. ACE is an attractive technique due to its low sample and ligand consumption, relatively short analysis times, and high efficiency and suitability for probing high and weak affinity interactions [15]. Several reviews on the theoretical and practical aspects of ACE methods, including their advantages and limitations, have appeared over time [15–20].

This section provides a general description of ACE modes that have been used in combination with MS detection. The various ACE modes originate from the difference in equilibrium kinetics of the probed affinity binding [21, 22]: dynamic equilibrium ACE, pre-equilibrium ACE, and nonequilibrium ACE. In dynamic equilibrium ACE, the association–dissociation times of the receptor–ligand complex (RL) are very short with respect to the CE separation time. In this mode, receptor, ligand, and complex are in dynamic equilibrium during the complete CE analysis and ligand mobility shifts are measured as function of receptor concentration. When complexation kinetics are slow with respect to the CE separation time, ligand, and receptor can be pre-equilibrated before analysis by CE. In pre-equilibrium CE, the receptor, ligand, and complex will present separate bands in the electropherogram. The so-called kinetic or nonequilibrium ACE methods take a position between dynamic equilibrium and pre-equilibrium approaches. In nonequilibrium CE, the complex, association/dissociation time is in the same order as the CE separation time and perturbations of the ligand and receptor peak shapes are correlated to binding characteristics.

9.3.1
Dynamic Equilibrium ACE (Fast Complexation Kinetics)

Mobility-shift ACE is the most frequently used mode of dynamic equilibrium ACE. Binding constants are determined from the change in ligand electrophoretic mobility observed when the ligand is successively analyzed with BGEs containing increasing concentrations of receptor [16, 21–23]. Prerequisite for this mode is that the equilibrium between receptor and the ligand is fast (many on/off events per second). Most studies also assume that the stoichiometry of the binding between the receptor and the ligand is $1:1$ in order to establish a simple model. Complex formation by the receptor (R) and the ligand (L) is then described by:

$$R + L \leftrightarrows RL \tag{9.6}$$

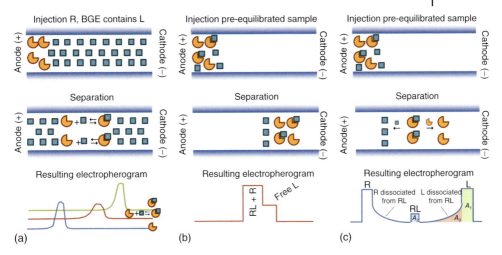

Figure 9.4 Schematic representation of (a) mobility-shift ACE, (b) frontal analysis, and (c) NECEEM.

with the association constant (K_a):

$$K_a = \frac{[RL]}{[R][L]} \tag{9.7}$$

where [R], [L], and [RL] are the respective concentrations of the free receptor, free ligand, and receptor–ligand complex in equilibrium. In mobility-shift ACE, a capillary is filled with BGE containing R in varying concentrations and a small amount of L is injected. In the absence of R, the migration time of L depends solely on its electrophoretic mobility (Figure 9.4a). When R is added to the BGE, and providing that R and L are in rapid equilibrium, a single peak for L is still observed by CE for any given concentration of R (Figure 9.4a). However, the observed migration time of L will be affected by the R–L complex formation. The resulting electrophoretic mobility of L will be a weighted average of the electrophoretic mobilities of L and the complex RL according to

$$\mu_{observed} = L_f \mu_L + (1 - L_f) \mu_{RL} \tag{9.8}$$

where L_f is the free fraction of L, μ_{RL} is the electrophoretic mobility of the complex RL, and μ_L is the electrophoretic mobility of L.

The observed shifts in the effectively mobility of L as function of concentration of R can be used to calculate the K_a of the RL complex. In order to properly calculate K_a values from mobility shifts, boundary conditions should be met. The ligand and receptor should have significantly different electrophoretic mobilities so that binding induces a change of the electrophoretic mobility of L. Moreover, there should be no interactions of L other than with R during the analysis, as this would bias the measured electrophoretic mobility. For example, proteins may not adsorb to the fused-silica capillary wall, as this may cause migration time shifts and peak distortion. Many approaches have been described to establish K_a values

Table 9.1 Plotting methods of different mathematical models used in mobility-shift ACE to calculate association constants (K_a) and the mobility of the complex (μ_c).

Mathematical model	Plotting method	K_a	$\mu_c - \mu_f$
Nonlinear regression	$\frac{(\mu_f - \mu_i)}{(\mu_i - \mu_c)}$ vs [R][a]	Slope	μ_c must be determined experimentally
x-Reciprocal	$\frac{(\mu_i - \mu_f)}{[R]}$ vs $(\mu_i - \mu_f)$	−Slope	1/slope
y-Reciprocal	$\frac{[R]}{(\mu_i - \mu_f)}$ vs [R]	Slope/intercept	−(Intercept/slope)
Double reciprocal	$\frac{1}{(\mu_i - \mu_f)}$ vs $\frac{1}{[R]}$	Intercept/slope	1/intercept

a) μ_f, μ_c, and μ_i are the electrophoretic mobilities of the free ligand, the complexed ligand, and the ligand measured at a specific receptor concentration, [R], respectively.

(Table 9.1) [15, 16, 20, 24]. Most commonly, nonlinear regression is used in which K_a is expressed as

$$K_a[R] = \frac{(\mu_f - \mu_i)}{(\mu_i - \mu_c)} \tag{9.9}$$

Here, [R] is the equilibrium concentration of free receptor, μ_f and μ_c are the electrophoretic mobilities of the free and complexed ligand, and μ_i is the ligand mobility measured at a specific receptor concentration, [R]. With the assumption that at any given time [R] > [L], a plot of $(\mu_f - \mu_i)/(\mu_i - \mu_c)$ versus [R] will yield a linear relationship of which the slope is equal to K_a (Table 9.1). Other mathematical models that have been introduced to calculate K_a are x-reciprocal, y-reciprocal, and double-reciprocal regression. Similar terms as for nonlinear regression are being used in these models, but the way the data is plotted and, consequently K_a is derived, is different (Table 9.1). Although nonlinear regression is most commonly used, it has some drawbacks. The curve should be forced through the origin since the intercept does not have physical meaning. In addition, the electrophoretic mobility of the complex (μ_c) can be difficult or even impossible to determine. The other models do not require knowledge of μ_c (Table 9.1), but also have their limitations. The double-reciprocal plot tends to emphasize on the data points taken at the lowest receptor concentrations, whereas the x-reciprocal plot has been criticized for using the dependent variable on both the x- and y-axes, thus complicating statistical analysis of the data. In general, linear plotting methods give outcomes that are comparable to those obtained from nonlinear curve fitting, provided that the data is properly weighted for linear regression. It is recommended to do a careful evaluation of the mathematical models before choosing one, also taking into account reference values of the studied interaction, preferably obtained with other techniques. It should be noted that currently there is no dedicated software for ACE data evaluation. In most reported ACE studies, commercially available data analysis software (e.g., Microsoft Excel or GraphPad Prism) has been used for fitting the data and perform calculations.

9.3.2
Pre-Equilibrium ACE (Slow Complexation Kinetics)

The most straightforward way to measure interactions governed by slow kinetics is by following a direct separation approach. The receptor and ligand are preincubated and subsequently subjected to CE separation. Because of the slow complexation kinetics discrete peaks for L, R, and RL will be obtained and their respective concentrations can be derived using an external calibration. Another mode of pre-equilibrium ACE uses frontal analysis (FA). In FA-ACE (frontal analysis-affinity capillary electrophoresis), first reported by Kraak *et al.* in 1992 [25], ligands are incubated with a receptor before analysis [17, 18]; however, now a relatively large volume of the mixture is injected. Depending on the dimension of the capillary, the sample volume typically is 20–200 nl (i.e., 5–20% of the total capillary volume). In FA it is assumed that two of the three components (usually R and RL) in the sample have approximately the same mobility and that the mobility of the third (often L) differs significantly. On application of the electrical field, the free L starts to separate from the mixture and two plateaus can be detected (Figure 9.4b). The first plateau corresponds to RL and R, whereas the second plateau corresponds to the free fraction L. Hence, the concentration of free L can be determined from its plateau height through external calibration. An approach similar to FA is frontal analysis continuous capillary electrophoresis (FACCE). It differs from FA in the sample introduction step. The capillary is filled with BGE before sample introduction. Subsequently, the inlet of the capillary is immersed in the sample vial and a voltage is applied to start the injection and separation process. Hence, sample introduction and separation are integrated and separated analytes appear as progressive plateaus in the electropherogram. Similar to FA, FACCE requires free L to migrate faster than RL and the plateau height corresponding to free L can be correlated to its concentration. Compared to FA, FACCE requires larger sample volumes (about 500 nL). Consequently, FACCE usually provides a broader plateau than FA, which is easier for quantitation.

In order to calculate K_a from data obtained with direct separation ACE, FA-ACE, or FACCE, Scatchard analysis can be used. Here, the bound fraction of L is plotted versus the ratio of the bound and free fraction of L [17, 18, 21]. This should provide a linear curve with a slope of $-K_a$. In order to construct such a plot, the concentration of free L should be measured at different concentrations of L, while keeping the concentration of R constant in all samples. Special care must be taken to assure that the total solute concentration is known accurately, since the bound ligand concentration is usually calculated as the difference between the total and free concentrations of the ligand. It is therefore important to ensure that no ligand is lost due to capillary-wall adsorption or any other nonspecific phenomena (e.g., precipitation). Moreover, the concentration in the sample should correspond to the concentration during CE (i.e., no stacking should occur), or at least any loss should be corrected for based on a standard curve obtained under identical conditions. An attractive feature of both FA-ACE and FACCE is that they are insensitive to changes or fluctuations in migration times, EOF, and applied voltage. In addition, lower detection limits are obtained with these two methods.

9.3.3
Kinetic ACE (Intermediate Complexation Kinetics)

In the ACE methods described above, the complex dissociation time is either much shorter or longer than the residence time of the sample in the capillary. In case the timescales for complex dissociation and analysis are similar, this is referred to as kinetic ACE. Early work by Whitesides [26] and Heegaard [27] showed that the change in electrophoretic mobility and shape (but not area) of the sample peak observed could be used for the determination of rate constants of complex formation and dissociation (k_{on} and k_{off}, respectively). From these values K_a follows according to

$$K_a = \frac{k_{on}}{k_{off}} \tag{9.10}$$

More recently, Krylov *et al.* [28] have developed a series of kinetic ACE methods by applying specific experimental settings and data analysis approaches. On the basis of various initial and boundary conditions – the way interacting species enter and exit the capillary – a number of kinetic ACE methods were designed. Here, only the approach used in combination with MS will be treated.

In nonequilibrium capillary electrophoresis of equilibrium mixtures (NECEEM), a short plug of the equilibrium mixture of R and L is injected at the inlet of the capillary filled with BGE (Figure 9.4c) [29, 30]. On application of the voltage, R and L will migrate according to their own electrophoretic mobility, disturbing the complexation equilibrium. When the separation of R and L is efficient, the previously formed RL will start to dissociate and reassociation of R and L can be neglected. The obtained electropherogram contains three peaks (Figure 9.4c), representing R, L, and RL, and two exponential "smears" of R and L, which result from the dissociation of RL. From the electropherogram, the dissociation constant (K_d) and k_{off} can be calculated according to

$$K_d = \frac{1}{K_a} = \frac{[R]_0(1 + A_1/(A_2 + A_3)) - [L]_0}{1 + (A_2 + A_3)/A_1} \tag{9.11}$$

$$k_{off} = \ln\left(\frac{A_2 + A_3}{A_1}\right)/t_{RL} \tag{9.12}$$

where A_1 is the peak area of L, A_2 is the peak area of intact RL, A_3 is the area of the exponential smear caused by L dissociating from RL during the analysis, t_{RL} is the migration time of the intact RL, and $[R]_0$ and $[L]_0$ are the total concentrations of R and L in the equilibrium mixture [21, 29].

For NECEEM, as well as other kinetic CE methods, it is important to accurately distinguish the boundaries of the peaks. Two approaches have been proposed: a control experiment and mathematical modeling [28, 31]. The boundary between A_1 and A_3 can be found by comparing the peaks of free L in the presence and absence of R. However, experimental errors in peak areas larger than 10% can easily occur. In the mathematical approach, the different parameters of Equations 9.11 and 9.12 are adjusted in regression analysis to minimize the deviation between experimental and simulated traces using the least-squares method. In contrast to dynamic equilibrium and pre-equilibrium ACE methods, NECEEM in principle needs one single run only to determine binding constants [21]. However, care should be taken as the results derived from a single injection lack an estimate of error. Moreover, establishing models at one set of initial concentrations only may not be sufficiently reliable. Therefore, multiple measurements may be justified in NECEEM.

9.4
Immunoaffinity Capillary Electrophoresis

IA-CE is a hybrid technique that combines selective immunocapturing of target analytes with CE separation [32–36]. In online IA-CE, a microextractor or concentrator is introduced near the inlet of the separation capillary. The microextraction column contains a sorbent-immobilized antibody that specifically retains the target antigen (Figure 9.5a). A relatively large volume of sample is injected, and target analytes are selectively captured by the antibody. Washing and clean-up procedures are subsequently integrated online to remove excess sample and non-specifically bound interfering components. The bound analytes are then eluted by an appropriate solvent and separated by CE.

In order to introduce the immunoaffinity (IA) sorbent into the capillary, several approaches have been followed (Figure 9.5b) [32–34, 37]. The original concept employs a piece of capillary containing sorbent particles with an immobilized affinity phase. The particles are retained within a restricted space by two frits or porous plugs fabricated from polymeric or glass materials. Alternatively, the sorbent beads can be chemically linked to each other and to the inner wall of the capillary so that no frits are required. Another approach to avoid frits within the capillary is to use the physical restriction of the separation capillary itself. Particles can be packed in a small piece of capillary with a relatively large ID that is placed between a short and a long piece of capillary with considerably smaller ID. If the ID of the longer separation capillary is smaller than that of the size of the sorbent particles, these will be retained. Rather than using particles, solid supports can also be used. Often these are monolithic materials that are polymerized in situ in a specific section of the capillary. After polymerization, antibodies or other affinity ligands can be immobilized on the material. Affinity ligands can also be immobilized on paramagnetic beads, using one or two magnets to hold the beads in place.

IA-CE is a means to selectively capture ligands of interest rather than a way to determine the affinity of a ligand for the immobilized antibody. Crucial for efficient IA-CE is the availability of an affinity capture molecule, such as an antibody,

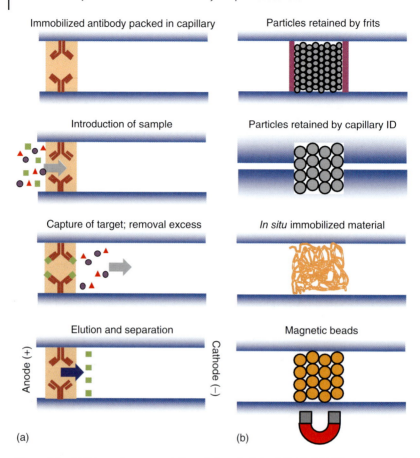

Immobilized antibody packed in capillary

Particles retained by frits

Introduction of sample

Particles retained by capillary ID

Capture of target; removal excess

In situ immobilized material

Elution and separation

Magnetic beads

Anode (+)

Cathode (−)

(a)

(b)

Figure 9.5 (a) Schematic representation of the principle of IA-CE. (b) Different approaches used to introduce IA material in the CE capillary.

with high affinity for the target molecule(s). As a technology for quantitative bio-analysis, IA-CE shows resemblance with immunoassays, such as, for example, the enzyme-linked immunosorbent assays (ELISAs). The main advantages of IA-CE are its high speed of analysis, automation, and low sample consumption [33, 34]. The IA sorbent can also be used for analyte preconcentration and increased sample volumes (typically from several tenths of nanoliters to several microliters) can be used to achieve improved sensitivity. Additionally, the CE separation step after IA trapping decreases the probability of false-positive results. The antibodies used in immunoassays typically cannot distinguish between posttranslationally modified proteins or degraded proteins (aggregated, fragmented, clipped or truncated, glycosylated, glycated, oxidized, deamidated, phosphorylated, etc.) when the epitope is not affected. That is, an overall signal is obtained without information on the identity of the binding component. In IA-CE, the CE separation increases the specificity and selectivity of the method. Furthermore, since IA-CE can be coupled to MS, direct identification of the captured antigen through MS analysis is possible.

9.5
Capillary Electrophoresis–Mass Spectrometry

9.5.1
General Requirements for Effective CE–MS Coupling

When coupling CE to MS, the outlet of the separation capillary has to be connected to the ion source of the mass spectrometer, and obviously, no outlet buffer vial can be used. Still, CE requires a closed electrical circuit and, therefore, a CE–MS interface should provide a means to apply voltage to the capillary outlet. The development of online interfaces for the combination of CE with MS detection started more than 25 years ago, when Smith *et al.* [38, 39] introduced the first CE–MS solutions. Since then, various interfaces for several types of ionization sources have been developed [40]. For the coupling of ACE and IA-CE to MS, electrospray ionization (ESI) has mainly been used.

Coupling CE with MS via an ESI source can roughly be performed in two different ways [40, 41]. In the first and most widely used approach, an additional liquid is used to aid application of CE voltage to the BGE. In this so-called sheath liquid interface (Figure 9.6) the CE capillary is surrounded by a second metal capillary in a coaxial arrangement. Through this second capillary a sheath liquid (typical flow rate, $2-5\,\mu L\,min^{-1}$) is led, which merges with the CE effluent at the capillary outlet. The terminating CE voltage is applied to the sheath liquid through the metal capillary and thereby closes the electrical circuit. Usually, a third coaxial tube delivers a gas flow, which aids solvent nebulization and evaporation. In the second – "sheathless" – CE-ESI-MS approach, the voltage is directly applied to the CE buffer. This can be achieved by applying a metal coating to the end of a tapered separation capillary or by introducing a metal-coated, full metal, or conductive sprayer tip at the end of the CE capillary. Another option is the insertion of a microelectrode through the wall of the capillary end or the introduction of the electrode in a liquid junction setup. Because of the inherently low flow (nanoliter per minute range) emerging from the CE capillary, sheathless CE–MS employs

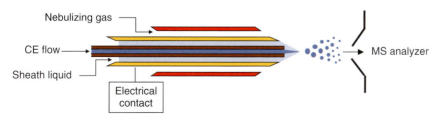

Figure 9.6 Schematic representation of a sheath liquid CE–MS interface.

nanoelectrospray, achieving highly favorable ionization efficiencies. Optimized sheathless CE–MS setups offer improved sensitivity, but until now their practical use has been limited, most likely because they were not commercially available and robustness was often an issue. Recent sheathless designs [42] show improved performance and in 2014 a porous tip CE–MS interface (CESI8000 from Sciex Separations) was introduced to the market.

The composition of the BGE is an important aspect in CE–ESI-MS. Volatile BGEs containing relatively low concentrations of formic acid, acetic acid, and/or ammonium hydroxide are preferred for sensitive ESI-MS [43, 44]. It should be noted that ammonia, often used to adjust the pH of the BGE, might negatively affect the ESI signal of analytes. It has also been shown that occasionally non-volatile BGEs, based on, for example, phosphate, Tris, or borate, can be used for CZE-ESI-MS analysis when their concentration is kept relatively low [45–47]. However, this has not been demonstrated yet for ACE–MS or IA-CE–MS.

In some cases, inductively coupled plasma (ICP) interfacing was used to hyphenate ACE with MS. With ICP ionization, the sample and buffer are passed through a plasma torch (>5000 K) and all compounds are totally degraded to their charged elements, typically in the M^+ state [48, 49]. Elements such as carbon, oxygen, and nitrogen cannot be used for compound detection, since they are usually also abundantly present in the BGE. However, with CE–ICP-MS highly selective detection of, for example, sulfur-, phosphorus-, and metal-containing compounds can be performed. Interestingly, CE–ICP-MS allows the use of even higher concentrations of nonvolatile BGEs, more closely resembling the physiological conditions during separation. Therefore, nonvolatile BGEs are frequently used in ACE–ICP-MS studies.

9.5.2
Specific Requirements for ACE–MS and IA-CE-MS

In ACE-UV molecular interactions may be probed using conditions that resemble the physiological environment, that is, high salinity buffers with a pH between 6.5 and 8.0. These conditions, however, are often not compatible with ESI-MS. Replacing the nonvolatile BGE by a volatile BGE in principle might lead to deviating values for, for example, the K_a. Using NECEEM, it was demonstrated that exchanging nonvolatile Tris–acetate for ammonium acetate, bicarbonate, or formate leads to insignificant differences in the dissociation constants of three pairs of noncovalent protein–ligand complexes [50]. The volatile BGEs did facilitate good separation of free ligands from the protein–ligand complexes. In a similar experiment using FA, both a nonvolatile phosphate BGE and a volatile ammonium acetate BGE were used to determine binding percentages for 13 low molecular weight drugs toward two different proteins [51]. A good correlation was found between binding percentage obtained with both BGEs ($R^2 > 0.980$), with a slope and an intercept value that were close to one and zero, respectively (Figure 9.7a). The ammonium acetate BGE was found to be acceptable for interaction measurements of the drugs and proteins.

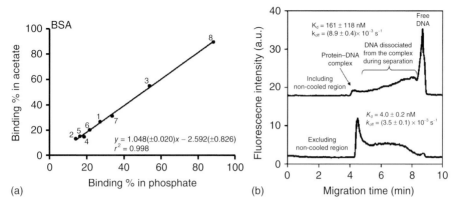

Figure 9.7 Effect of (a) BGE composition and (b) capillary cooling on obtained ACE results. (a) Binding percentages obtained in phosphate buffer (67 mM, pH 7.4) compared to those obtained in ammonium acetate solution (100 mM, pH 7.4) for the interaction of eight drugs with BSA. 1: Carbamazepine, 2: propranolol, 3: warfarin, 4: bupivacaine, 5: lidocaine, 6: L-tryptophan, 7: salicylic acid, and 8: diclofenac [51]. Copyright Wiley-VCH Verlag GmbH & Co. Reproduced with permission. (b) Effect of the noncooled capillary inlet on NECEEM-based determination of rate (k_{off}) and dissociation constants (K_d) of a protein–DNA interaction. Reprinted with permission from Ref. [52]. Copyright 2010 American Chemical Society.

Another parameter that can affect the observed results in ACE is the capillary temperature. Rate constants of complex dissociation will increase with increasing capillary temperature according to the Arrhenius equation

$$k = Ae^{-\frac{E_a}{RT}} \tag{9.13}$$

where A is a pre-exponential factor, E_a is the activation energy of the reaction, T is the temperature, and R is the gas constant. Electric currents in CE inherently lead to heat formation in the capillary (also referred to as *Joule heating*) and may influence the rate constants of complex association/dissociation. Consequently, efficient thermostatting of the CE capillary will be required to obtain reliable results [53]. Active cooling (e.g., by means of a cooling liquid or air) is recommended in ACE [53]. Notably, noncooled portions of the capillary (e.g., near the inlet and outlet) can affect the observed affinity and complexation kinetics. Since temperatures may significantly increase in these parts, which in CE–MS can represent until 50% of the capillary length, rate constant values will deviate [54]. In extreme cases deviations in dissociation constants of more than an order of magnitude have been reported (Figure 9.7b) [52]. This issue can – at least partly – be avoided by minimizing the length of the uncooled parts and by moving the sample through the noncooled inlet into the cooled region by pressure to prevent it from exposure to the elevated temperature (Figure 9.7b).

In ACE–MS, the sheath liquid composition can influence the intensity and stability of the detected signals [51]. Moreover, in order to detect ligand–receptor complexes during ESI-MS, the sheath liquid should not contain high concentrations of organic and acidic/basic components [55]. Nondenaturing conditions are preferred, which means that the sheath liquid should comprise a volatile buffer with near-physiological pH and a low percentage of organic solvent. Besides sheath liquid composition, MS settings such as dry gas flow rate, nebulizer gas pressure, and ion optic voltages have been shown to influence the signal intensity of both free ligand and complex [56]. Therefore, it is recommended to carefully optimize or at least investigate the influence of these parameters when ACE–MS is pursued.

In ACE, adsorption of either ligand or receptor to the capillary wall may cause low precision of binding data, since modification of the inner surface of the capillary can contribute to change in measured electrophoretic mobilities. In order to sufficiently reduce adsorption – especially when dealing with proteins – appropriate rinsing procedures and/or capillary coatings are required [57]. These approaches have been well established in ACE-UV, but might not always be compatible with MS detection. For example, hydrochloric acid or sodium hydroxide are frequently used to flush between runs and remove any adsorbed protein from the capillary wall. However, both are not MS-compatible and their use could lead to contamination of the mass spectrometer and potential corrosion of metal parts. MS compatible alternatives such as ammonium hydroxide [51, 58] or formic and acetic acid [59, 60] have been used in ACE–MS and IA-CE–MS. In cases that MS-incompatible solvents are required for optimal performance, offline flushing by decoupling the CE system from the ion source might be indicated [61–64]. Over time, various capillary coatings for CE–MS have been proposed and their usefulness has been shown [65]. In order to avoid suppression of analyte ionization, background signals, and/or contamination of the ion source and MS optics by coating agents, coatings should be permanently attached to the capillary wall. In ACE–MS a variety of coatings have been successfully used to prevent analyte adsorption and increase the separation window [56, 62, 66–72], providing the migration time stability required for affinity determinations.

9.6
Application of ACE–MS

An overview of reported ACE–MS studies is reported in Table 9.2. One of the first application fields of ACE–MS was the screening of peptide libraries toward their affinity for receptors [66, 67, 69, 73, 74]. CE allows rapid and efficient separation of minute quantities of mixtures of synthesized peptides. Dynamic equilibrium ACE-UV can be used to search for interacting peptides showing mobility shifts. However, no structural information on interacting species is obtained from these experiments. On coupling with MS, ACE is capable of not only selecting but also identifying peptide ligands from a library of peptides. Libraries ranging from 19 [74] until 1000 [66] peptides have been screened by ACE–MS. Predominantly, vancomycin or related glycopeptide antibiotics have been used as receptor. In a typical example, a 19-peptide library was screened against vancomycin [74]. The peptides had the general structure Fmoc-DXYA, where X and Y were random

Table 9.2 Overview of ACE–MS applications ordered based on their kinetics.

Ligand	Receptor	Background electrolyte	log K_a	Remarks	References
Dynamic equilibrium ACE (fast complexation kinetics)					
Dipeptides	Glycopeptide antibiotics	18.5 mM ammonium acetate, pH 5.1	5.5–5.6	Application of home-written software to allow interpretation of data	[73]
Low molecular weight drugs	β-Cyclodextrin	10 mM ammonium acetate	2.8–4.2	Comparison of ACE–MS, ACE–UV, and direct infusion MS	[75]
Oligosaccharides	Stromal cell-derived factor-1	75 mM ammonium acetate, pH 6.5	n.d.	Screening of oligosaccharide binding toward target. Stoichiometry of binding could be determined	[68]
Phosphorylated compounds	Europium	25 mM sodium formate, containing 10% methanol, pH 3.7	2.0–3.9	Also observation of 2 : 1 complex. Detection with inductively coupled plasma MS	[72]
Tetrapeptides	Vancomycin	15 mM Tris–acetate, pH 8.0	3.4–5.4	Library screening; simultaneous determination of K_a of multiple components	[69]
Tetrapeptides	Vancomycin	20 mM Tris–acetate, pH 8.1	4.6–5.3	Library screening; simultaneous determination of K_a of multiple components	[74]
Tetrapeptides	Vancomycin	20 mM Tris–acetate, pH 8.1	4.2–5.2	Library screening; simultaneous determination of K_a of multiple components	[67]
Tri- and tetrapeptides	Vancomycin	20 mM Tris–acetate, pH 8.1	3.8–5.2	Library screening; simultaneous determination of K_a of multiple components	[66]
Pre-equilibrium ACE (slow complexation kinetics)					
Anticancer metal-based drug	Human serum albumin, transferrin	10 mM phosphate buffer, 100 mM sodium chloride, and 10 mM CTAB, pH 7.4	3.7–4.1	Detection with inductively coupled plasma MS	[76]
Chymostatin	α-Chymotrypsin	30 mM ammonium formate, pH 4.0	n.d.	Degree of fragmentation used as a way to describe interaction strength	[56]

(continued overleaf)

Table 9.3 (Continued)

Ligand	Receptor	Background electrolyte	log K_a	Remarks	References
Gallium(III) compounds	Human serum albumin, transferrin	10 mM ammonium bicarbonate, pH 7.4	n.d.	Detection with inductively coupled plasma MS	[77]
Inorganic antimony species	Fish DNA	30 mM ammonium acetate and 50 mM Tris (pH 8.0)	6.1	Detection with inductively coupled plasma MS	[78]
Low molecular weight drugs	α_1-Acid glycoprotein and bovine serum albumin	100 mM ammonium acetate, pH 7.4	4.2–5.2	Method transfer from UV to MS detection. Evaluation of separation and interfacing conditions	[51]
Low molecular weight drugs	Human serum albumin	100 mM ammonium acetate, pH 7.4	2.8–5.9	Method transfer from UV to MS detection	[79]
Neodynium	Fulvic acid	Sodium nitrate, different concentrations and pH	4.0–14.0	Detection with inductively coupled plasma MS	[80]
Phosphotyrosine peptides	src Homology 2 domain	1–2% ampholine pH 3.5–10.0	n.d.	Affinity determinations using capillary isoelectric focusing coupled to ESI-MS	[70]
Platinum-containing compounds	Human serum albumin	15 mM phosphate buffer; pH 7.4	n.d.	Detection with inductively coupled plasma MS	[81]
Rare earth metals	Humic acids	100 mM sodium nitrate, pH 6.0–10.0	9.0–16.0	Detection with inductively coupled plasma MS	[82]
Sulfated pentasaccharide	Antithrombin	30 mM ammonium hydrogen carbonate, pH 8.5	n.d.	Use of nondenaturing interfacing conditions	[55]
β-Endorphin digest	Monoclonal human β-endorphin antibody	50 mM ε-aminocaproic acid–acetic acid, pH 5.1 and 20 mM Tris–acetic acid, pH 8.0	n.d.	Screening for binding epitope of β-endorphin after enzymatic digestion	[71]
Kinetic ACE (intermediate complexation kinetics)					
β-Blockers	α_1-Acid glycoprotein	25 mm Tris–acetate, pH 7.2	4.0–5.7	Offline collection of fractions. Buffer exchanged for methanol before MS analysis	[83]
Lipopolysaccharides	Fatty acid	20 mM ammonium acetate, pH 9.0	n.d.	Potentially the first observation of NECEEM phenomenon with MS detection	[84]
Organotin compounds	Human serum albumin	10 mM phosphate, pH 7.4	5.4–7.4	Comparison mobility-shift ACE and NECEEM. Detection with inductively coupled plasma MS	[85]

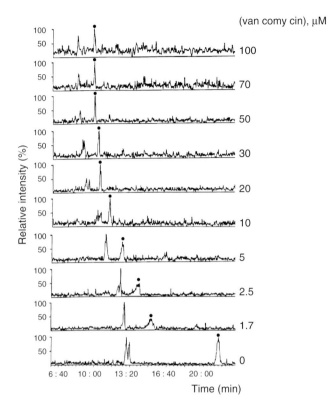

Relative intensity (%)

(van comy cin), μM

100
70
50
30
20
10
5
2.5
1.7
0

6 : 40   10 : 00   13 : 20   16 : 40   20 : 00

Time (min)

Figure 9.8 ACE–MS binding experiments with the tetrapeptide Fmoc-DDYA. The electropherograms are extracted ion traces of $(M + Tris + H)^+$, *m/z* 826. The tetrapeptide is indicated with the black circle. Reprinted with permission from Ref. [74]. Copyright 1998 American Chemical Society.

amino acid residues. On the basis of the mobility shift of the peptides, as exemplified for the peptide Fmoc-DDYA in Figure 9.8, determination of K_d values for the 19 peptides was achieved. K_d values all ranged in the micromolar range and correlated well with ACE-UV experiments done in parallel. However, the simultaneous determination of multiple K_d values by ACE–MS was found to be advantageous over ACE-UV that only allowed determination of one peptide K_d at the time.

The most common field of application for ACE–MS is the study of interactions of ligands with proteins [51, 55, 56, 68, 70, 71, 76, 77, 79, 81, 83, 85]. The investigated ligands range from low molecular weight components [51, 76, 77, 79, 81, 83, 85] to saccharides [55, 68] and peptides [56, 70, 71] using pre-equilibrium (direct separation and FA) and NECEEM approaches coupled to MS. For instance, FACCE–MS was used for the affinity analysis of antithrombin and a sulfated pentasaccharide [55]. As the pentasaccharide is strongly anionic, the complex was more negatively charged than the free protein thereby allowing the separation of both species (Figure 9.9a,b). The FACCE–MS method was compared with direct analysis of the pre-equilibrated sample. It was demonstrated that a higher sensitivity was obtained with FACCE–MS, most likely due to the continuous injection. Protein denaturation and complex dissociation was observed when a sheath liquid containing water–acetonitrile–formic acid was applied. Therefore, ammonium acetate (pH 6.5) was used as a nondenaturing sheath liquid. Under these conditions, good-quality mass spectra were obtained that allowed identifying both the nonbound protein and the complex (Figure 9.9c,d). A similar approach was used to get insight into the dimerization process of stromal cell–derived factor-1 [68]. Using dynamic equilibrium mobility-shift ACE–MS it was observed that in the absence of sulfated oligosaccharides the protein is largely monomeric. However, dimer formation was observed on interaction with heparin-sulfated oligosaccharides.

When one of the two interacting species contains a metal ion, detection with ICP-MS is considered a valuable approach [72, 76–78, 80, 81, 82]. Dynamic equilibrium mobility-shift ACE and NECEEM combined with ICP-MS were, for example, used as tools to determine the binding constants of organotin compounds toward human serum albumin [85]. ACE–ICP-MS allowed the use of nonvolatile BGEs, such as Tris, resembling physiological conditions during separation, while still obtaining good sensitivity. Notably, such conditions might not feasible in CE–ESI-MS. With mobility-shift ACE–MS, measured log K_d values were on average around 6.0 for four different organotin components, which was in agreement with literature values. On the other hand, for NECEEM-MS, log K_d values were around 7.0 (1 order of magnitude higher). Although the authors did not fully investigate the reason for the difference, they suspected protein adsorption to the inner wall of the capillary. Resulting loss of compound and peak tailing would affect accurate determination of the peak areas needed to calculate K_d.

Comparisons between different ACE–MS methods or other methods to determine affinities are scarcely found in literature. Apart from the study described in the previous paragraph, only one other study reported a direct comparison of different methods for affinity determinations [75]. Dynamic equilibrium ACE coupled to either ultraviolet (UV) or MS detection and direct infusion mass spectrometry (DIMS) were compared for their ability to probe the interaction of low molecular weight components with a β-cyclodextrin (Figure 9.10). It was concluded that DIMS provided less reliable K_d values than ACE-UV and ACE–MS. The significant ionization suppression by the cyclodextrin (Figure 9.10c) hampered determination of K_d values for four out of seven interacting ligands. With ACE-UV, affinities of six out of seven interacting ligands were determined with a good accuracy. The seventh component did not show any mobility shift, prohibiting affinity determination by ACE-UV (peak 8 in Figure 9.10a). With ACE-UV, overlapping peaks complicated the measurement of migration times of individual components and, thus, the calculation of the respective affinity constants. ACE–MS resolved comigrating compounds since all ligands had different molecular weights

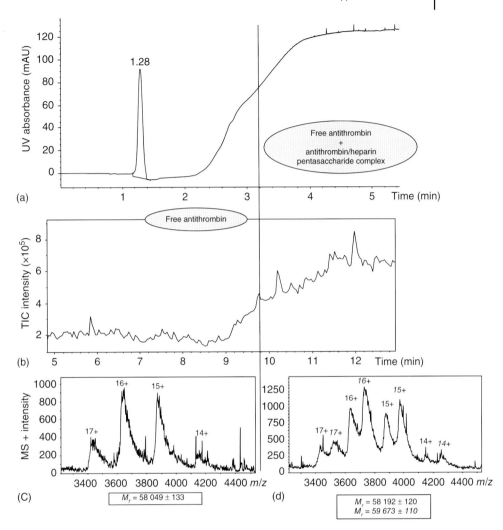

Figure 9.9 (a) UV absorbance electrophero-gram and (b) TIE of the FACCE–MS analysis of the antithrombin–pentasaccharide interaction. Mass spectrum integrated over (c) 9–10 min and (d) 10–12 min time interval shows the presence of free antihrombin and the antithrombin–pentasaccharide complex. Antithrombin molecular mass is indicated in bold and the molecular mass of the antithrombin–pentasaccharide complex in italic. Reprinted with permission from Ref. [55]. Copyright 2007 American Chemical Society.

(Figure 9.10b). Moreover, the affinity of the ligand showing no mobility shift could even be derived based on the change of its ion intensity in MS. Ionization suppression was observed in ACE–MS, as well as some peak broadening compared with the ACE-UV; however, these aspects did not affect the reliability of the affinity determination.

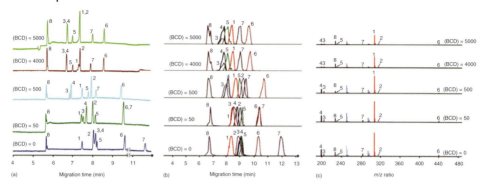

(a) Migration time (min) (b) Migration time (min) (c) m/z ratio

Figure 9.10 Experimental data from (a) mobility-shift ACE-UV, (b) mobility-shift ACE–MS, and (c) direct infusion mass spectrometry to study the interaction between low molecular weight components and β-cyclodextrin. Titration experiments were performed with a fixed concentration of a low molecular weight component (15 μM each) and different concentrations of β-cyclodextrin (BCD) from 0 to 5000 μM. 1: phenylbutazone, 2: diclofenac, 3: ibuprofen, 4: s-flurbiprofen, 5: naproxen, 6: folic acid, 7: 4,4-(propane-1,3-diyl) dibenzoic acid, and 8: resveratrol. Reprinted from Ref. [75] with kind permission from Springer Science and Business Media.

9.7
Applications of IA-CE–MS

IA-CE with optical detection has been widely applied for the analysis of low-abundance biomarkers, drugs, and metabolites. The technology has been pioneered by Guzman and Phillips and, subsequently, they also widely demonstrated the strength of the technology [33, 34]. Key aspects are the selective trapping and subsequent recovery of target compounds using an immobilized affinity phase followed by separation of the enriched compounds. The last few years witnessed growing interest in online IA-CE with MS detection. An overview of reported IA-CE–MS studies is provided in Table 9.3.

Applications of IA-CE–MS have focused on trace analysis of peptides [32, 59, 60, 63] and proteins [61, 64] in complex matrices. In all cases, polyclonal antibodies or antibody fragments were covalently immobilized on sepharose, silica, or magnetic particles. In all reported studies, the preparation of the IA phase and microcolumn was extensively investigated, followed by a demonstration of the selectivity and performance of the IA-CE–MS method.

An IA-CE–MS method for the determination of opioid peptides in plasma was developed [59]. A site-specific antibody immobilization approach was tested based on the covalent attachment of oxidized antibodies through their carbohydrate moieties to hydrazide silica particles. After method optimization, standard solutions of opioid peptides were used in order to establish the IA-CE–MS methodology. Acceptable repeatability, reproducibility, and linearity were obtained for the proposed methodology. Compared to a common sample cleanup method, based on C_{18} material, the selectivity of the IA sorbent against the peptides was significantly better (Figure 9.11a). Most isobaric matrix

Table 9.3 Overview of IA-CE–MS applications.

Ligand	Receptor	Sample matrix	BGE	IA sorbent	Remarks	References
Antibody and milk proteins	Polyclonal antibody	Milk and serum	10% acetic acid, pH 2.0	Magnetic beads	Investigation into components causing milk allergy	[62]
Gonadotropin-releasing hormone	Polyclonal F(ab)$_2$ antibody fragments	Urine and serum	60 mM ammonium bicarbonate, pH 8.0, containing 1% (v/v) acetonitrile	Glass beads retained by frits	First paper on IA-CE–MS	[32]
Opioid peptides	Polyclonal antibody	Aqueous solution and plasma	20 mM ammonium acetate, pH 7.0	Silica particles retained by frits	Comparison with other online cleanup approaches	[59]
Opioid peptides	Polyclonal Fab antibody fragments	Aqueous solution and plasma	20 mM ammonium acetate, pH 7.0	Silica particles retained by frits	Improved affinity capture compared to [20]	[60]
Recombinant human erythropoietin peptides	Polyclonal antibody	Aqueous solution	20 mM ammonium acetate, pH 7.0	Sepharose particles retained by capillary ID	Differences between erythropoietin variants	[63]
Transferrin	Polyclonal IgG antibody	Aqueous solution and serum	25 mM ammonium acetate, pH 8.5	Silica particles retained by frits	First online IA-CE–MS of intact proteins	[64]
α-Lactalbumin, β-Lactoglobulin	Polyclonal antibody	Milk	10% acetic acid, pH 2.0	Magnetic beads	Spotting on MALDI plate	[61]

components retained by the C_{18} sorbent had no retention on the IA phase. Consequently, target peptides could be detected in human plasma down to $1 \, ng \, mL^{-1}$ (Figure 9.11b) representing a 100-fold enhancement with respect to conventional CE–MS. In a subsequent study, Fab antibody fragments were coupled to the IA sorbent [60]. Compared to the full antibody approach, it was shown that an even more selective sample clean-up could be achieved, leading to limits of detection as low as $0.5 \, ng \, mL^{-1}$.

One of the major issues in IA-CE is the nonspecific binding to the IA sorbent. Highly abundant proteins from biological samples (e.g., serum albumin) may bind to IA sorbents and, consequently, disturb the analysis of the target ligand. A simple protein precipitation before analysis can solve this issue when the target compound is not a protein [59, 60]. However, when the target is a protein,

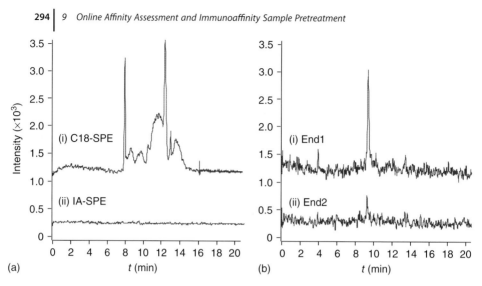

Figure 9.11 (a) Sum of the extracted ion electropherograms of peptides endomorphin 1 (end1) and 2 (end2) for blank plasma samples by (i) C_{18}-SPE-CE–MS and (ii) IA-CE–MS. (b) Extracted ion electropherograms for a plasma sample spiked with 100 ng mL^{-1} of End1 and End2 by IA-CE–MS. Reprinted from Ref. [59] with permission from Elsevier.

such an approach is not an option. This was confirmed by a recent IA-CE–MS study, which showed that albumin was more abundantly captured from serum than transferrin for which the IA phase was designed [64]. The CE step after IA capture allowed separation of transferrin and albumin, both of which were identified by MS detection. A dilution and desalting step was found necessary to prevent IA sorbent saturation, flow blocking, and electrophoretic current breakdown. It has been demonstrated that addition of a surfactant to the sample could also significantly lower or even prevent unspecific binding of proteins [61, 62]. This approach was successfully used to specifically determine the quantity of two major proteins in milk samples by IA-CE–MS [61]. In this study, desorbed proteins were spotted on a matrix-assisted laser desorption ionization (MALDI) plate, before MS determination of their molecular weight. Only the fractions containing the proteins were spotted on the MALDI plate, thereby preventing nonvolatile components, such as the detergent, to impair ionization and enter the mass spectrometer. The same IA-CE–MALDI-MS approach was used to determine the total amount of IgE in blood serum as a marker for allergies [62]. In a subsequent experiment, captured IgE was chemically cross-linked to the IA phase. This IA phase was used to screen potential milk allergens. Obtained results indicated that bovine serum albumin, lactoferrin, and α-casein were allergy sensitizers for the patient whose blood was used to purify the IgE from. The authors conclude that their diagnostic procedure could be useful for the personalized (food) allergy diagnosis including the identification of unusual and unexpected allergens.

9.8
Conclusions

This chapter provides an overview of the various ways ligand–receptor interactions can be probed by CE in conjunction with MS detection. Various ACE modes are available to study affinity binding and kinetics. The application of a particular mode depends on the analytical question posed. Dynamic equilibrium ACE is useful for assessment of interactions that have association/dissociation times that are much faster than the analysis time. Moreover, since the separation is based on CZE principles it provides a possibility to separate multiple components and simultaneously determine their affinity. FA-ACE can be used for pre-equilibrated ligand–receptor complexes that slowly dissociate, implying that the dissociation time should be much longer than the CE run time. NECEEM can be used to cover most interactions in which dissociation occurs within the analysis time. Unfortunately, in the latter two cases often only one ligand–receptor interaction can be probed at a time. Besides the affinity modes discussed in this chapter, a variety of other ACE approaches have been reported in literature [22]. Until now, these have not been combined with MS detection. However, in terms of experimental design they are quite similar to the applied ACE–MS methodologies. Therefore, in principle they should be amenable as well for coupling with MS. Especially in terms of limited ligand and receptor consumption and increased analysis speed, approaches such as multiple-injection ACE [86] and partial filling ACE [87] are interesting methods to explore. The latter approach also has the advantage that it can be used in combination with online sample stacking [88], something that is not feasible with most ACE methods as local concentration differences of receptor can strongly influence the ligand interaction. Online IA has been shown to be an efficient approach for biological sample cleanup before analysis by CE–MS. Using high-affinity IA sorbents, high selectivity and simultaneously sample enrichment can be achieved.

MS takes an important position in contemporary analytical chemistry research. With the maturation of CE–MS interfacing the number of publications in the field of ACE–MS and IA-CE–MS has been growing steadily. Many studies have demonstrated the added value of MS detection over classical UV detection. MS adds selectivity to the analysis, enabling determination of affinities of comigrating components. Moreover, using ESI, the ligand–receptor complex can be kept intact in the gas phase, and MS data can reveal the stoichiometry of the interaction. In some cases, MS information can also be used to calculate ligand–receptor affinities. Only few studies have reported on the latter two aspects yet, as most of the studies merely use the mass spectrometer as a sensitive and selective detector.

From an application point of view, ACE–MS and IA-CE–MS have proven to be useful for a large variety of ligands. They range from low molecular weight components of less than 200 Da to proteins of several tens of kilodalton. A similar diversity is found for the receptors studied until now. One of the fields of application in which ACE–MS and IA-CE–MS can still grow and more studies are required is the study of protein–protein interactions. ACE–MS will allow study of the protein – receptor interaction, or evaluate potential protein inhibitors. Since product diversity is an issue with intact proteins having a large variety of isoforms that can make up one sample, the separation efficiency of CE might be an added value here to allow isoform-selective affinity determinations. IA-CE – MS could be a valuable tool in the bioanalysis of proteins and in studying their faith during biotransformation. Also, here the CE separation after affinity capture can potentially reveal differences between isoforms.

References

1. Pioch, M., Bunz, S.C., and Neususs, C. (2012) *Electrophoresis*, **33**, 1517.
2. Ibanez, C., Simo, C., Garcia-Canas, V., Cifuentes, A., and Castro-Puyana, M. (2013) *Anal. Chim. Acta*, **802**, 1.
3. Pascali, J.P., Bortolotti, F., and Tagliaro, F. (2012) *Electrophoresis*, **33**, 117.
4. Stalmach, A., Albalat, A., Mullen, W., and Mischak, H. (2013) *Electrophoresis*, **34**, 1452.
5. Ramautar, R., Somsen, G.W., and De Jong, G.J. (2013) *Electrophoresis*, **34**, 86.
6. Chu, Y.-H., Avila, L.Z., Biebuyck, H.A., and Whitesides, G.M. (1992) *J. Med. Chem.*, **35**, 2915.
7. Guzman, N.A., Trebilcock, M.A., and Advis, J.P. (1991) *J. Liq. Chromatogr.*, **14**, 997.
8. Krieger, J., Stinson, S., Worthy, W., and Dagani, R. (1991) *Chem. Eng. News*, **69**, 28.
9. Tiselius, A. (1937) *Trans. Faraday Soc.*, **33**, 524.
10. Hjerten, S. (1967) *Chromatogr. Rev.*, **9**, 122.
11. Virtanen, R. (1974) *Acta Polytech. Scand., Chem. Technol. Ser.*, **123**, 67.
12. Mikkers, F.E.P., Everaerts, F.M., and Verheggen, T.P.E.M. (1979) *J. Chromatogr. A*, **169**, 11.
13. Jorgenson, J.W. and Lukacs, K.D. (1981) *Anal. Chem.*, **53**, 1298.
14. Kok, W.T.H. (2000) *Chromatographia*, (51), S1.
15. Tanaka, Y. and Terabe, S. (2002) *J. Chromatogr. B*, **768**, 81.
16. Rundlett, K.L. and Armstrong, D.W. (2001) *Electrophoresis*, **22**, 1419.
17. Guijt-van Duijn, R.M., Frank, J., van Dedem, G.W.K., and Baltussen, E. (2000) *Electrophoresis*, **21**, 3905.
18. Østergaard, J. and Heegaard, N.H.H. (2006) *Electrophoresis*, **27**, 2590.
19. Busch, M.H., Carels, L.B., Boelens, H.F., Kraak, J.C., and Poppe, H. (1997) *J. Chromatogr. A*, **777**, 311.
20. Rundlett, K.L. and Armstrong, D.W. (1997) *Electrophoresis*, **18**, 2194.
21. Chen, Z. and Weber, S.G. (2008) *TrAC, Trends Anal. Chem.*, **27**, 738.
22. Jiang, C. and Armstrong, D.W. (2009) *Electrophoresis*, **31**, 17.
23. Colton, I.J., Carbeck, J.D., Rao, J., and Whitesides, G.M. (1998) *Electrophoresis*, **19**, 367.
24. El-Hady, D., Kuhne, S., El-Maali, N., and Watzig, H. (2010) *J. Pharm. Biomed. Anal.*, **52**, 232.
25. Kraak, J.C., Busch, S., and Poppe, H. (1992) *J. Chromatogr.*, **608**, 257.
26. Avila, L.Z., Chu, Y.-H., Blossey, E.C., and Whitesides, G.M. (1993) *J. Med. Chem.*, **36**, 126.
27. Heegaard, N.H.H. (1994) *J. Chromatogr. A*, **680**, 405.
28. Krylov, S.N. (2007) *Electrophoresis*, **28**, 69.
29. Krylov, S.N. (2006) *J. Biomol. Screen.*, **11**, 115.
30. Berezovski, M. and Krylov, S.N. (2002) *J. Am. Chem. Soc.*, **124**, 13674.
31. Berezovski, M., Nutiu, R., Li, Y., and Krylov, S.N. (2003) *Anal. Chem.*, **75**, 1382.
32. Guzman, N.A. and Stubbs, R.J. (2001) *Electrophoresis*, **22**, 3602.
33. Guzman, N.A., Blanc, T., and Phillips, T.M. (2008) *Electrophoresis*, **29**, 3259.
34. Guzman, N.A. and Phillips, T.M. (2011) *Electrophoresis*, **32**, 1565.
35. Amundsen, L.K. and Sirén, H. (2007) *Electrophoresis*, **28**, 99.

36. Delaunay-Bertoncini, N. and Hennion, M.C. (2004) *J. Pharm. Biomed. Anal.*, **34**, 717.

37. Chen, H.-X., Busnel, J.-M., Peltre, G., Zhang, X.-X., and Girault, H.H. (2008) *Anal. Chem.*, **80**, 9583.

38. Smith, R.D., Olivares, J.A., Nguyen, N.T., and Udseth, H.R. (1988) *Anal. Chem.*, **60**, 436.

39. Olivares, J.A., Nguyen, N.T., Yonker, C.R., and Smith, R.D. (1987) *Anal. Chem.*, **59**, 1230.

40. Hommerson, P., Khan, A.M., de Jong, G.J., and Somsen, G.W. (2011) *Mass Spectrom. Rev.*, **30**, 1096.

41. Maxwell, E.J. and Chen, D.D.Y. (2008) *Anal. Chim. Acta*, **627**, 25.

42. Heemskerk, A.A.M., Deelder, A.M., and Mayboroda, O.A. (2014) *Mass Spectrom. Rev.* doi: 10.1002/mas.21432, accepted for publication.

43. Huber, C.G., Premstaller, A., and Kleindienst, G. (1999) *J. Chromatogr. A*, **849**, 175.

44. Catai, J.R., Sastre Torano, J., de Jong, G.J., and Somsen, G.W. (2007) *Analyst*, **132**, 75.

45. Eriksson, J.H.C., Mol, R., Somsen, G.W., Hinrichs, W.L.J., Frijlink, H.W., and de Jong, G.J. (2004) *Electrophoresis*, **25**, 43.

46. van Wijk, A.M., Muijselaar, P.G., Stegman, K., and de Jong, G.J. (2007) *J. Chromatogr. A*, **1159**, 175.

47. Gottardo, R., Mikšík, I., Aturki, Z., Sorio, D., Seri, C., Fanali, S., and Tagliaro, F. (2012) *Electrophoresis*, **33**, 599.

48. Michalke, B. (2005) *Electrophoresis*, **26**, 1584.

49. Zoorob, G., McKiernan, J., and Caruso, J. (1998) *Microchim. Acta*, **128**, 145.

50. Bao, J. and Krylov, S.N. (2012) *Anal. Chem.*, **84**, 6944.

51. Vuignier, K., Veuthey, J.-L., Carrupt, P.-A., and Schappler, J. (2012) *Electrophoresis*, **33**, 3306.

52. Musheev, M.U., Filiptsev, Y., and Krylov, S.N. (2010) *Anal. Chem.*, **82**, 8637.

53. Berezovski, M. and Krylov, S.N. (2004) *Anal. Chem.*, **76**, 7114.

54. Musheev, M.U., Filiptsev, Y., and Krylov, S.N. (2010) *Anal. Chem.*, **82**, 8692.

55. Fermas, S., Gonnet, F., Varenne, A., Gareil, P., and Daniel, R.g. (2007) *Anal. Chem.*, **79**, 4987.

56. Hoffmann, T. and Martin, M.M. (2010) *Electrophoresis*, **31**, 1248.

57. El Deeb, S., Wätzig, H., and El-Hady, D.A. (2013) *TrAC, Trends Anal. Chem.*, **48**, 112.

58. Guzman, N.A. (2000) *J. Chromatogr. B Biomed. Sci. Appl.*, **749**, 197.

59. Medina-Casanellas, S., Benavente, F., Barbosa, J., and Sanz-Nebot, V. (2012) *Anal. Chim. Acta*, **717**, 134.

60. Medina-Casanellas, S., Benavente, F., Barbosa, J., and Sanz-Nebot, V. (2013) *Anal. Chim. Acta*, **789**, 91.

61. Gasilova, N., Gassner, A.-L., and Girault, H.H. (2012) *Electrophoresis*, **33**, 2390.

62. Gasilova, N. and Girault, H.H. (2014) *Anal. Chem.*, **86**, 6337.

63. Gimenez, E., Benavente, F., de Bolos, C., Nicolas, E., Barbosa, J., and Sanz-Nebot, V. (2009) *J. Chromatogr. A*, **1216**, 2574.

64. Medina-Casanellas, S., Benavente, F., Giménez, E., Barbosa, J., and Sanz-Nebot, V. (2014) *Electrophoresis*, **35**, 2130.

65. Huhn, C., Ramautar, R., Wuhrer, M., and Somsen, G.W. (2010) *Anal. Bioanal. Chem.*, **396**, 297.

66. Chu, Y.-H., Dunayevskiy, Y.M., Kirby, D.P., Vouros, P., and Karger, B.L. (1996) *J. Am. Chem. Soc.*, **118**, 7827.

67. Chu, Y.-H., Kirby, D.P., and Karger, B.L. (1995) *J. Am. Chem. Soc.*, **117**, 5419.

68. Fermas, S., Gonnet, F., Sutton, A., Charnaux, N., Mulloy, B., Du, Y., Baleux, F., and Daniel, R. (2008) *Glycobiology*, **18**, 1054.

69. Lynen, F., Zhao, Y., Becu, C., Borremans, F., and Sandra, P. (1999) *Electrophoresis*, **20**, 2462.

70. Lyubarskaya, Y.V., Carr, S.A., Dunnington, D., Prichett, W.P., Fisher, S.M., Appelbaum, E.R., Jones, C.S., and Karger, B.L. (1998) *Anal. Chem.*, **70**, 4761.

71. Lyubarskaya, Y.V., Dunayevskiy, Y.M., Vouros, P., and Karger, B.L. (1997) *Anal. Chem.*, **69**, 3008.

72. Varenne, F., Bourdillon, M., Meyer, M., Lin, Y., Brellier, M., Baati, R., Charbonniere, L.J., Wagner, A., Doris, E., Taran, F., and Hagege, A. (2012) *J. Chromatogr. A*, **1229**, 280.

73. Machour, N., Place, J., Tron, F., Charlionet, R., Mouchard, L., Morin,

C., Desbène, A., and Desbène,
P.-L. (2005) *Electrophoresis*, **26**,
1466.

74. Dunayevskiy, Y.M., Lyubarskaya,
Y.V., Chu, Y.-H., Vouros, P., and
Karger, B.L. (1998) *J. Med. Chem.*, **41**,
1201.

75. Mironov, G., Logie, J., Okhonin, V.,
Renaud, J., Mayer, P., and Berezovski, M.
(2012) *J. Am. Soc. Mass Spectrom.*, **23**,
1232.

76. Polec-Pawlak, K., Abramski, J.K.,
Semenova, O., Hartinger, C.G.,
Timerbaev, A.R., Keppler, B.K., and
Jarosz, M. (2006) *Electrophoresis*, **27**,
1128.

77. Groessl, M., Bytzek, A., and Hartinger,
C.G. (2009) *Electrophoresis*, **30**, 2720.

78. Li, Y., Liu, J.-M., Han, F., Jiang, Y., and
Yan, X.-P. (2011) *J. Anal. At. Spectrom.*,
26, 94.

79. Wan, H., Östlund, Å., Jönsson, S., and
Lindberg, W. (2005) *Rapid Commun.
Mass Spectrom.*, **19**, 1603.

80. Sonke, J.E. and Salters, V.J.M. (2004) *J.
Anal. At. Spectrom.*, **19**, 235.

81. Timerbaev, A.R., Aleksenko, S.S.,
Polec-Pawlak, K., Ruzik, R., Semenova,
O., Hartinger, C.G., Oszwaldowski, S.,
Galanski, M., Jarosz, M., and Keppler,
B.K. (2004) *Electrophoresis*, **25**, 1988.

82. Stern, J.C., Sonke, J.E., and Salters,
V.J.M. (2007) *Chem. Geol.*, **246**, 170.

83. Bao, J., Krylova, S.M., Wilson, D.J.,
Reinstein, O., Johnson, P.E., and Krylov,
S.N. (2011) *ChemBioChem*, **12**, 2551.

84. Tuffal, G., Tuong, A., Dhers, C.,
Uzabiaga, F., Riviere, M., Picard, C.,
and Puzo, G. (1998) *Anal. Chem.*, **70**,
1853.

85. Sun, J., He, B., Liu, Q., Ruan, T., and
Jiang, G. (2012) *Talanta*, **93**, 239.

86. Chinchilla, D., Zavaleta, J., Martinez,
K., and Gomez, F. (2005) *Anal. Bioanal.
Chem.*, **383**, 625.

87. Brown, A., Silva, I., Chinchilla, D.,
Hernandez, L., and Gomez, F.A. (2004)
LC-GC Eur., **17**, 32.

88. Fukushima, E., Yagi, Y., Yamamoto, S.,
Nakatani, Y., Kakehi, K., Hayakawa, T.,
and Suzuki, S. (2012) *J. Chromatogr. A*,
1246, 84.

10

Label-Free Biosensor Affinity Analysis Coupled to Mass Spectrometry

David Bonnel, Dora Mehn, and Gerardo R. Marchesini

10.1
Introduction to MS-Coupled Biosensor Platforms

Interfacing bioassays with confirmatory methods is not something new; however, it has evolved considerably since the 1970s. Already in 1957, Schneider [1] pioneered a bioassay using a full organ as the most challenging biological interface in a biosensor system. Schneider measured small changes in voltage from the antennal olfactory organ of insects in response to different pheromones. Interfacing such an ultimate bioassay with a separation technique was pioneered by Moorhouse *et al.* in 1969 [2]. Moorhouse added a new dimension to a gas chromatography–flame ionization system by splitting the outlet flow and measuring the voltage changes of the living antenna of an insect exposed to the compounds eluting from the chromatographic column. The work of Moorhouse marked the beginning of several successful studies in the following years coupling biosensors or other bioassay technologies with classical analytical techniques (for a comprehensive review see [3, 4]).

Label-free biosensors are analytical systems aiming for the detection of biological and biochemical targets without the use of coupled enzymes, dyes, and radiolabels. They can be generally divided in two big families: the non-optical techniques, including (but not limited to) microcalorimetry, resonant acoustic profiling, field electron transistors, nanocantilevers, quartz crystal microbalances, electrical impedance, and the optical signal-mediated methods such as Raman microscopy, surface enhanced Raman spectroscopy (SERS), optical waveguides, biolayer interferometry, and last but not least surface plasmon resonance (SPR) [5].

The label-free optical biosensors' field has been dominated by SPR-based biosensor applications during the last decade [6, 7]. These applications exploit biorecognition events of a minimum of two binding partners interacting at the sensor chip surface where one of them is immobilized. Upon binding of the partner in solution, the total mass and concentration of molecules increases at the chip surface resulting in a local refractive index change. The local refractive index change is proportional to the SPR biosensor signal output (Figure 10.1).

Analyzing Biomolecular Interactions by Mass Spectrometry, First Edition.
Edited by Jeroen Kool and Wilfried M.A. Niessen.
© 2015 Wiley-VCH Verlag GmbH & Co. KGaA. Published 2015 by Wiley-VCH Verlag GmbH & Co. KGaA.

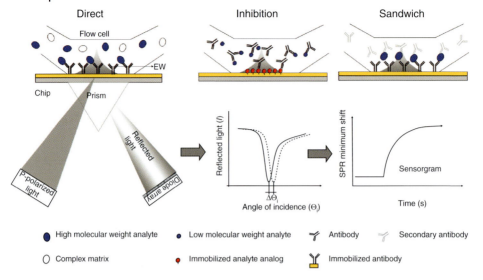

Figure 10.1 Biosensor principle and assay types – direct, inhibition, and sandwich-type assay systems (top), typical SPR equipment configuration, and signal processing (bottom).

The major advantage is that the biomolecular interaction can be monitored in real time enabling the determination of the kinetic parameters including association and dissociation rate constants, and identification of the potential interactions between binding sites. While other conventional binding-based techniques such as radioimmunoassays or enzyme-linked immunosorbent assays (ELISAs) require specific labels (radioactive, fluorescent, or coupled enzyme activity) for the detection, SPR can be considered as intrinsically a label-free method [8]. The concentration of interacting biomolecules can be quantitatively estimated either by the direct binding of the analyte to the chip surface (direct binding assay (DBA)) or by constructing a competitive binding assay (CBA). Since the signal output depends directly on the mass of the interacting analyte, the DBA format is widely used for fishing high molecular weight biomolecules out of complex biological matrices. Nevertheless, with the proper experimental forethought, this approach has also proven to be effective for the determination of kinetic parameters of low molecular weight molecules binding to antibodies or receptors [7, 9]. However, for diagnostics of low molecular weight molecules, the competitive assay format is generally preferred in biosensor immunoassays (BIAs) because of its flexibility, higher response, and robustness. In the competitive BIA format, the analyte, or a structural analog, is immobilized on the sensor surface preserving its affinity for the biorecognition molecule. The sample containing the analyte is mixed with the biorecognition molecule and brought into contact with the analyte derivative or structural analog immobilized on the sensor chip surface. The higher the concentration of analyte in the sample, the lower the number of biorecognition molecules bound to the chip surface. The signal drop due to the competition of the biomolecules binding to the immobilized analyte on the chip surface will be proportional to the analyte concentration in the sample [8].

Besides the above-mentioned strategies, a sandwich-type assay design is another potential solution for detection and mapping (Figure 10.1).

The main drawback of the SPR biosensor screening technology is the lack of information about the identity of the interacting analyte. Hence, a synergistic complement can be obtained by combining SPR biosensor technology based on molecular biorecognition events with analytical techniques based on the physico-chemical properties of the analytes. Thus, its combination with mass spectrometry is a powerful approach that already yielded numerous results in the proteomics field [10–18]. SPR biosensors were combined offline with micropreparative high-performance liquid chromatography (HPLC) [19], online as a detector for IgG when using CE (capillary electrophoresis) for separation [20], in parallel with liquid chromatography with diode array detector–mass spectrometry (LC/DAD–MS) for drug screening [21] and offline and online with MS [10]. The choice of ionization technique amenable for hyphenating SPR biosensor technology with MS greatly depends on the characteristics of the molecule to be analyzed and on the assay format chosen to detect it. When a DBA format is chosen, the analyte is captured on the chip surface and can be eluted for further offline or online LC–MS (liquid chromatography–mass spectrometry) analysis. In contrast, with the CBA format the analyte interacting with the biorecognition element is usually discarded to the waste and a different strategy needs to be implemented.

10.2
Strategies for Coupling Label-Free Analysis with Mass Spectrometry

Biomolecular interaction analysis by mass spectrometry (BIA/MS) was first introduced in 1997 [22, 23]. Two different approaches are generally described and employed: the on-chip and the off-chip configurations. The off-chip approach uses an elution of the analyte from the surface before MS analysis, whereas the on-chip setup is based on the direct analysis of the sample surface.

10.2.1
On-Chip Approaches

10.2.1.1   SPR-MALDI-MS
The emergence of immuno-MS techniques [24–30] illustrates the keen interest of MS for ligand identification. It is therefore logical that coupling between SPR and MS appeared as a major asset in bioengineering (Figure 10.2). Indeed, the association of the two technologies allows multiplex detection in SPR and supplements MS with the spot array selection filter tool. Additionally, the miniaturization of bioassays by high-density printed microarrays provides further advantages such as minimizing reagent consumption while maximizing throughput.

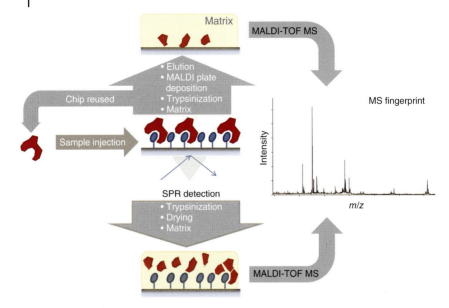

Figure 10.2 SALDI, MALDI-MS interface – protein analysis by identification based on peptide mass fingerprint after trypsinization and elution and deposition on MALDI plate (top) or direct measurement on the SPR chip after drying and matrix deposition (bottom).

The presence of proteins retained during the DBA interaction analysis can be unambiguously confirmed by coupling the SPR biosensor with matrix-assisted laser desorption ionization time-of-flight mass spectrometry (MALDI-TOF-MS) [23].

Among the many sources of MS ionization, the MALDI [31] source has emerged as an attractive choice for the SPR-MS combination. In MALDI, molecular desorption is triggered by a laser beam on a deposited matrix that absorbs the laser energy leading to the sample ablation. Neutral and ionized molecules are then accelerated to the detector and finally the mass over the charge ratio is calculated. Because of its large covered mass range, robustness in instrumentation, and higher tolerance to contaminants including salts, detergents, or buffer components, the MALDI ion source has been widely applied for the SPR sample analysis [10–13, 32–52]. The MALDI source is commonly used today for metabolomics, lipidomics, peptidomics, and proteomics analysis.

The SPR-MALDI-TOF-MS analysis setup has evolved from the direct measurement of the interacting biomolecules from the chip surface to the subsequent deposition of the biomolecules on MALDI plates [23] for analysis by elution [53, 54] to a small elution volume using automated commands [40, 55, 56] and lately to a bifunctional SPR/MS flow cell with removable MS probes [38]. For example, Nelson *et al.* [23] applied a chip-like biosensor for simultaneous monitoring of biomolecular interactions in separate liquid flow cells on the chip. MALDI-TOF analysis was performed directly on the area of the flow cells containing the molecules captured in the DBA, after matrix deposition and drying (Figure 10.2, bottom). In another multiliquid flow cell–based system the peptide myotoxin A from snake venom was captured on an SPR chip and the MALDI matrix was deposited on the SPR chip surface, air-dried, and the presence and identity of the myotoxin was confirmed with MALDI-MS [22]. Nevertheless, the SPR chip is left unusable after a single binding cycle, making this approach cost-inefficient for a large number of samples. In order to overcome this limitation, the elution of the analytes was performed in a small liquid volume between two air bubbles that was later respotted onto a MALDI plate (Figure 10.2, top) [54, 55]. A similar approach was further improved with completely automated commands, including those needed for sample preparation before the MALDI analysis [40]. Moreover, imaging SPR has enabled SPR-MS array platforms for the high-throughput screening of large numbers of interactors, especially in the field of proteomics [11]. For protein analysis by SPR-MALDI, a bottom-up approach based on a proteolytic activity such as trypsinization is applied on the array and then the protein is identified based on its peptide mass fingerprint [12, 51]. This step could be automatized but may induce some experimental bias if the experimental conditions are not standardized.

SPR-MALDI combined with database search tools allows the identification of enzymatically digested proteins by peptide mass fingerprinting, given the 10–50 ppm mass accuracy of MALDI-TOF-MS for small polypeptides [57].

In SPR-MALDI, as a consequence of the matrix-assisted ionization strategy, it is not necessary to use an elution step as in ESI (electrospray ionization, see description later). Nevertheless, the matrix must be deposited on top of the sample and the deposition of a homogenous and reproducible matrix layer is crucial for comparison matters. On this point, the MALDI imaging technology clearly contributed to overcome this limitation since sprayers and spotters are now commercially available and widely used [58–60]. Moreover, routine analysis on high-density chips became possible using lasers working at 1000 Hz frequency. Under this condition, each spot can be analyzed in 0.5 s applying 500 laser shots per position, enabling a throughput of about 7200 spots per h, noting that this setting is under constant improvement by manufacturers.

This throughput capacity opens the door to improved pharmaceutical studies, clinical diagnosis, or research by screening interactions and by identifying the molecules involved. This technique not only provides direct ligand detection on the sensor surface, but also enables the identification of enzyme-induced modifications on the immobilized molecules. In this way, many applications in the field of diagnostics, pharmaceutical, and basic research could be exploited based on the immobilization of antibodies, peptides, proteins, or small molecules (drug, metabolites, lipids) on a MALDI-coupled sensor surface.

In SPR-MALDI-MS, the sensitivity of the MS may be disadvantageous due to the detection of nonspecific contaminants. The use of human plasma causes nonspecific binding on the surface of the chip what necessarily requires a preventive saturation step with blocking agents (bovine serum albumin BSA, casein, etc.)

[61]. Additionally, the incompatibility of certain functionalized surfaces with some polymers such as PEG (polyethyleneglycol) is well described. The molecules of this polymer constitute an important contaminant for this type of approach, since they are readily ionized on the sample surface resulting in a distinctive spectral profile (and pollution of MS signal). Besides the efforts on automation and reproducibility of matrix deposition, elimination, or substitution of these organic molecules would be very advantageous on the MS performance.

10.2.1.2 SPR-LDI-MS

The Laser desorption ionization (LDI) is a soft ionization method allowing the transfer of intact molecule ions from liquid or solid sample to the gas phase. In case of direct SPR-LDI-MS coupling, the surface must be compatible with the SPR analysis to complete both acquisitions on the same chip. The first coupling between SPR and LDI was described in 1998 by Owega *et al.* [62].

Indeed, the LDI ion source offers advantages for the SPR-MS coupling, considering that the main drawback of MALDI is the difficulty to analyze low molecular weight compounds (at $m/z < 1000$) because the matrix allowing ionization interferes in this mass range. Different matrix-free approaches were described to optimize LDI by using nanomaterials [63–68], porous silicon surface [69–72], and, most recently, sol–gel films [73–75]. The optical, chemical, and physical properties of nanomaterials are governed by size, composition, and their assembly into secondary structures on the nano and micro scale. These features make nanomaterials very attractive for a wide array of applications by SPR-LDI. Desorption ionization on porous silicon (DIOS) utilizes etched silicon wafers to fabricate a porous nanostructure that can be used for direct ionization and desorption of analytes from the surface. Nevertheless, one disadvantage of the DIOS surface is poor stability due to its susceptibility to surface oxidation, which could lead to decreased performance and reproducibility [76]. Porous silver nanoparticles (AgNP)-impregnated thin films (sol–gel) were described for LDI applications in separate studies. Thin films impregnated with classical MALDI matrices such as α-cyano-4-hydroxycinnamic acid (CHCA) [77] and 2,5-dihydroxybenzoic acid (DHB) [75] were used to detect a number of molecule classes in sol–gel preparations. Several oligonucleotides were also analyzed by doping sol–gel solutions with 3,4-diaminobenzoic acid (DABA) or 3,5-DABA during the preparation of thin films. Recently, a group described the advantage of incorporating TiO_2 into the sol–gel layer for detection of surfactants, peptides, and tryptic digest proteins [78].

Thus, the LDI methodology offers several distinct advantages such as less noise from the matrix cluster in the low mass range and easier sample preparation process (solvents, pH, etc.) and often requires lower minimum laser fluence for ion production. Nevertheless, further developments are still needed in order to improve and optimize the surface for routine use in laboratories.

10.2.2
Off-Chip Configurations

10.2.2.1 ESI-MS

A different approach for SPR-MS analysis using a DBA format is the elution of the captured analytes and their offline or online analysis using ESI coupled to MS. As in the case of LDI, ESI is a gentle ionization technique that produces singly or multiply charged ions in a gaseous phase directly from a liquid phase (either organic or aqueous). ESI overcomes the use of an interfering matrix and solves in one step the transition of ions to the gas phase needed for MS analysis, simplifying the interfacing as compared to MALDI. Early studies attempted the offline ESI-MS analysis of SPR eluates containing low molecular weight protease inhibitors by a direct infusion into a nano-ESI ion trap MS. However, this strategy was not satisfactory due to the chemical noise arising from the lack of an analytical chromatographic step [79]. Later, a similar offline approach addressed the lack of a chromatographic step using a larger chip surface and nano-ESI-LC–MS/MS analysis of tryptic digests of the SPR eluate and successfully achieved the identification of cyclic nucleotide-binding proteins from cell lysates [80]. An alternative to the laborious handling of the eluates is to hyphenate the fluidic system of the biosensor directly with a reversed-phase trapping column that is subsequently coupled to LC–ESI-MS analysis. This strategy has been successfully tested for MS-based amino acid sequencing after on-chip digestion of proteins using a simple Biacore X instrument [81]. Bouffartigues *et al.* [82] attempted the analysis of nucleoproteins using a Biacore 2000 instrument with a similar system setup. Unfortunately, the ion suppression caused by a buffer detergent prevented MS identification. Finally, it was reported by Hayano *et al.* that proteins purified from a cell lysate were successfully identified using a fully automated online SPR biosensor-nano-LC–MS/MS system [15].

10.2.2.2 In Parallel Approach

A complementary strategy is required for coupling SPR-CBA with MS (Figure 10.3). One option is to use a split-flow strategy for parallel coupling of the screening and MS. Such an approach was proven using a yeast bioassay for the screening of androgens and estrogens in calf urine in combination with LC-ESI quadrupole (Q)-TOF-MS [83, 84]. A similar parallel setup was used for the detection and confirmation of fluoroquinolones (FQs) antibiotics in chicken muscle tissue [85]. Two competitive binding immunoassays (CBIAs) detecting six FQs were used instead of the yeast bioassay and were able to screen among a large number of samples. Only when a sample was considered noncompliant during screening for FQs it was fractionated by LC and the effluent was split toward two 96-well fraction collectors. The immunoactive LC fractions in one of the 96-well plates were immunodiscriminated using the CBIAs creating an immunogram and the positive well's positions were used in the second 96-well plate for immunoactive-oriented identification with ESI-TOF-MS.

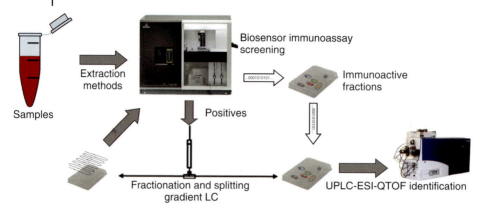

Samples

Extraction methods

Biosensor immunoassay screening

Positives

Immunoactive fractions

Fractionation and splitting gradient LC

UPLC-ESI-QTOF identification

Figure 10.3 Parallel SPR-MS-CBIAs used for detection of small molecules: positive, noncompliant samples are fractionated by LC and the effluent is split toward two 96-well fraction collectors. The immunoactive LC fractions are used in the second 96-well plate for immunoactive-oriented identification with ESI-TOF-MS.

Although successful, the CBIA screening–MS offline strategy has some limitations. First, it requires a relatively large sample volume. Second, because of the amount of sample handling required, the chances of sample contamination are high. Third, the fractionation resolution is limited and this might hamper the identification given that complex samples may contain hundreds of compounds' peaks separated during the chromatographic step, but pooled back together in the same fraction. Fourth, the fractionation process is unfeasible when screening for bioactive compounds whose half-life is short or unknown.

10.2.3
Chip Capture and Release Chromatography – Electrospray-MS

Considering the above limitations, a system requiring minimum sample volume and handling was developed using the antibiotic enrofloxacin as a model analyte [86]. This system was subsequently adapted to PSP (Paralytic Shellfish Poisoning) toxin analysis (Figure 10.4). The SPR-MS system interface was based on a nanoscale affinity chip that linked the CBIA screening with a nano-LC–TOF-MS system by recovering only the relevant analytes. The same biospecific recognition molecule was used in the CBIA and as a biosorbent in the recovery chip, which specifically captured the analyte while the sample matrix flowed through. The operation procedure of the SPR/nano-LC–TOF-MS system comprised four stages. After the sample extracts were prepared (Figure 10.4, frame 1) they were screened with the CBIA (Figure 10.4, frame 2) and only those suspected to be non-compliant were reinjected onto the recovery chip analogous to an affinity chromatographic step on a chip. Once captured, the analyte was eluted into a vial and automatically diluted (Figure 10.4, frame 3) to be later injected on to a loop-type interface. From the loop-type interface, the analytes entered the nano-LC system

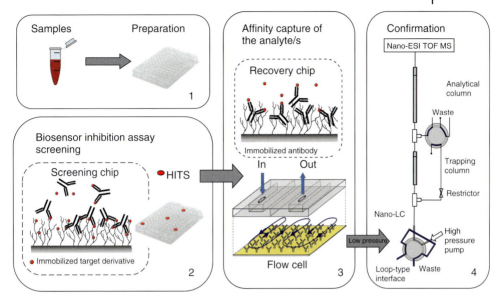

Figure 10.4 Serial SPR-MS sample extraction (frame 1), CBIA (frame 2), affinity capture, and analyte elution from recovery chip (frame 3), nano-LC, trapping column, TOF-MS system (frame 4).

and were retained in a trapping column followed by an analytical column connected to the TOF-MS for mass analysis (Figure 10.4, frame 4). The main features of this strategy were minimum sample volume consumption and the possibility of acquiring in-depth information about the performance of the bioaffinity sorbent before use due to periodical quality control by placing the recovery chip into the SPR biosensor.

10.3
New Sensor and MS Platforms, Opportunities for Integration

10.3.1
Imaging Nanoplasmonics

In the case of prism-based SPR detection, the biosensor surface does not require special fabrication processes: a thin metallic (usually gold) layer is deposited on the prism surface and the change in the reflected light minimum is monitored to generate a sensorgram [87–91]. However, an alternative way to excite surface plasmon polariton (SPP) modes is coupling via a nanoscale diffraction grating that matches the momentum difference between the localized plasmon and the incident light [92–97]. The later setup is based on tuning the resonant wavelength by tailoring the lattice period of the nanoscale grating at a fixed incidence angle. This allows high-resolution sensing at normal incidence with very simplified optical constraints. Recently, the addition of a parallel detector, a charge-coupled device (CCD) or CMOS (Complementary Metal Oxide Silicon) camera, has paved the road to surface plasmon resonance imaging (SPRi) [98–100]. SPRi, combined with the use of microarray platforms and image analysis tools, allows real-time, high-sensitivity multiplexing detection of biomolecules. While earlier the commercial exploitation of such a powerful technology was limited by the high costs of the instrumentation and the noncompetitive cost per assay, with respect to classical ELISA-based assays, recently an affordable cost imaging nanoplasmonics platform was presented using nanostructured gold-polymer chips compatible with a compact stand-alone fully portable reading instrument. The sensing surfaces consisted of a 2D hexagonal lattice of plasma polymerized polyacrylic acid (ppAA) nanopillars embedded in an optically thick gold matrix on a glass substrate [101–103]. The performance of the system was demonstrated using pentaxin 3 (PTX3), a multifunctional molecule, found to represent a possible new prognostic marker in cardiovascular diseases, kidney pathologies, infections, and inflammation [104].

Nanostructured noble metal surfaces play a crucial role also in other plasmonic-based analytical techniques, such as SERS. This method – similarly to SPR – is intrinsically label-free and is based on the enhancement of the fingerprint type but weak Raman scattered signal of molecules situated close to a rough noble metal surface. As this enhancement is shown to reach 10^{4-6} orders of magnitude in practice, SERS is expected by many researchers to evolve into a powerful technique in trace material analysis in the near future. The common need for similar noble metal substrate and light source presents the obvious opportunity of the *in situ* coupling of SPR with SERS. Simultaneous measurement of the Kretschmann configuration-based SPR signal and the SERS spectrum of the captured analyte was demonstrated on commercially available flat metallic surfaces [105] as well as on nanostructured substrates such as gold nanoparticles embedded in a gold-dielectric film [106]. In these integrated setups, SERS completes the quantitative-type SPR detection with structure-related information about the analyte, allowing the identification of the captured molecules based on their characteristic Raman fingerprint.

10.3.2
Evanescent Wave Silicon Waveguides

Evanescent wave silicon waveguide devices are close relatives of the SPR technology providing quantitative, label-free, and real-time detection of analytes based on the phase shift of the propagating optical field due to the accumulation of target molecules on the functionalized sensor surface. They usually consist of a high refractive index, thin waveguide silicon-based core layer on top of a low refractive index cladding or between two cladding layers. In case of planar waveguide evanescent field sensors the propagating field is the evanescent tail of the optical waveguide mode that in principle permits a larger length for the interaction of the light with the biomolecules compared to SPR (where the propagating field is the surface plasmon mode). The resulting high-sensitivity detection system is realized through various geometric waveguide configurations such as Mach–Zehnder-type interferometer sensors or ring resonators. The appropriate submicron features of silicon waveguide chips can be easily fabricated even at the mass production level using the already well-established manufacturing toolset of the semiconductor industry. Planar chips are compatible with microarray formats and can be integrated with microfluidic systems allowing minimization of sample volumes and convenient analyte solution handling including automatization [107]. With the appearance of commercially available chips and instruments for silicon waveguide-based biomolecule detection, the coupling of these platforms to mass spectrometers is expected to evolve similarly to the already existing SPR-MS solutions.

10.3.3
New Trends in MS Matrix-Free Ion Sources

New matrix-free ion sources for MS-based surface analysis have emerged in recent years. Among them, three sources could be adapted to SPR-MS coupling: DESI (desorption electrospray ionization) [108], DART (direct analysis in real time) [109], and LAESI (laser ablation electrospray ionization) [110].

In the DESI method, a fine spray of charged droplets hits the surface of interest. The spray picks up small organic or large biomolecules from the surface, ionizes them, and delivers them into the mass spectrometer.

In the DART method, an electric potential is applied to a gas of high ionization potential to form a plasma of excited-state atoms and ions that trigger the ejection of low molecular weight radicals from the surface of a sample.

Finally, the LAESI operates by firing an infrared laser ($\lambda = 2490\,\text{nm}$) at the sample, which excites the O–H bonds of the water molecules within the sample. This causes phase explosion leading to microscale ablation and ejection of neutral particles. The particles are then intercepted by highly charged liquid droplets from an ESI source. The droplets capture a small amount of the emitted particles converting them into gas-phase ions.

Each of the above-mentioned three ion sources is able to work under normal ambient conditions and atmospheric pressure offering a great advantage in high-throughput applications. These ionization sources also show excellent performance in QMSI (quantitative mass spectrometry imaging) and could be particularly well adapted to SPR-MS coupling [58, 111–118]. QMSI offers the benefit of quantification in addition to the information obtained by SPR, such as association constants or identification of ligand–receptor interactions. One of the QMSI methodologies involves the addition of a pseudo-internal standard in the MALDI matrix for normalization, allowing the correlation of the signal obtained on the surface of the target chip with a calibration curve [111, 117]. Finally, Quantinetix© software designed by ImaBiotech can now handle all data sets obtained and reprocess them [111].

10.3.4
Tag-Mass

As described previously for protein identification by SPR-MALDI, a trypsinization step of each chip spot remains necessary for peptide fingerprint generation. This step needs to be carefully controlled and reproducible (homogeneous spray of the enzyme on the surface, incubation time, temperature and atmospheric conditions) to be positioned on targeting diagnostic applications in the future. However, an encouraging alternative relies on the use of the tag-mass concept: using antibodies labeled with a fixed tag of a known mass [115–122]. Thus, incubation of different tag-mass molecules on the surface of the chip allows identification of the ligands of interest. This quantitative multiplex approach does not require the use of an enzyme and allows targeting some known markers as a diagnostic assay.

10.3.5
Integration

MS technology is becoming a standard instrument in most laboratories and a must in bioanalytical laboratories. The cost-effectiveness of SPR biosensor screening and MS coupling will very likely grow for most laboratories together with the increase of labor costs and the decrease in the cost of SPR biosensor technology. Apart from economic considerations, they have a great potential in more aspects of bioanalytics. With the advent of personalized medicine such hyphenated systems would be of interest for clinical diagnostics and drug discovery, as well as for toxicology and forensics research settings. The SPR-MS approach will need to face the challenge of lowering costs of instrumentation and the development of robust and standardized quality controls for both components in order to allow routine use and establish conformity with laboratories working under a quality control system.

References

1. Schneider, D. (1957) Elektrophysiologische Untersuchungen von Chemo- und Mechanorezeptoren der Antenne des Seidenspinners Bombyx mori L. *Z. Vergl. Physiol.*, **40** (1), 8–41.
2. Moorhouse, J.E. *et al.* (1969) Method for use in studies of insect chemical communication. *Nature*, **223** (5211), 1174–1175.
3. Fishman, H.A., Greenwald, D.R., and Zare, R.N. (1998) Biosensors in chemical separations. *Annu. Rev. Biophys. Biomol. Struct.*, **27** (1), 165–198.
4. Trojanowicz, M. and SzewczyImageskac, M. (2005) Biosensing in high-performance chemical separations. *Trends Anal. Chem.*, **24** (2), 92–106.
5. O'Malley, S. (2008) Recent advances in label-free biosensors applications in protein biosynthesis and HTS screening, in *Protein Biosynthesis* (eds T.E. Esterhouse and L.B. Petrinos), Nova Science Publishers, Inc..
6. Malmqvist, M. (1993) Biospecific interaction analysis using biosensor technology. *Nature*, **361** (6408), 186.

7. Rich, R.L. and Myszka, D.G. (2006) Survey of the year 2005 commercial optical biosensor literature. *J. Mol. Recognit.*, **19** (6), 478–534.

8. Hahnefeld, C., Drewianka, S., and Herberg, F.W. (2004) Determination of kinetic data using surface plasmon resonance biosensors. *Methods Mol. Med.*, **94**, 299–320.

9. Rich, R.L. *et al.* (2002) Kinetic analysis of estrogen receptor/ligand interactions. *Proc. Natl. Acad. Sci. U.S.A.*, **99** (13), 8562–8567.

10. Nedelkov, D. and Nelson, R.W. (2003) Surface plasmon resonance mass spectrometry: recent progress and outlooks. *Trends Biotechnol.*, **21** (7), 301.

11. Nedelkov, D. (2007) Development of surface plasmon resonance mass spectrometry array platform. *Anal. Chem.*, **79** (15), 5987–5990.

12. Rouleau, A. *et al.* (2012) Immuno-MALDI-MS in human plasma and on-chip biomarker characterizations at the femtomole level. *Sensors (Basel)*, **12** (11), 15119–15132.

13. Visser, N.F. and Heck, A.J. (2008) Surface plasmon resonance mass spectrometry in proteomics. *Expert Rev. Proteomics*, **5** (3), 425–433.

14. Buijs, J. and Franklin, G.C. (2005) SPR-MS in functional proteomics. *Briefings Funct. Genomics Proteomics*, **4** (1), 39–47.

15. Hayano, T. *et al.* (2008) Automated SPR-LC-MS/MS system for protein interaction analysis. *J. Proteome Res.*, **7** (9), 4183–4190.

16. de Mol, N.J. (2012) Surface plasmon resonance for proteomics. *Methods Mol. Biol.*, **800**, 33–53.

17. Zheng, P. *et al.* (2012) QUICK identification and SPR validation of signal transducers and activators of transcription 3 (Stat3) interacting proteins. *J. Proteomics*, **75** (3), 1055–1066.

18. Abbady, A.Q. *et al.* (2012) Chaperonin GroEL a Brucella immunodominant antigen identified using Nanobody and MALDI-TOF-MS technologies. *Vet. Immunol. Immunopathol.*, **146** (3-4), 254–263.

19. Nice, E. *et al.* (1994) Synergies between micropreparative high-performance liquid chromatography and an instrumental optical biosensor. *J. Chromatogr. A*, **660** (1–2), 169–185.

20. Whelan, R.J. and Zare, R.N. (2003) Surface plasmon resonance detection for capillary electrophoresis separations. *Anal. Chem.*, **75** (6), 1542–1547.

21. Minunni, M. *et al.* (2005) An optical DNA-based biosensor for the analysis of bioactive constituents with application in drug and herbal drug screening. *Talanta*, **65**, 578–585.

22. Krone, J.R. *et al.* (1997) BIA/MS: interfacing biomolecular interaction analysis with mass spectrometry. *Anal. Biochem.*, **244** (1), 124–132.

23. Nelson, R.W., Krone, J.R., and Jansson, O. (1997) Surface plasmon resonance biomolecular interaction analysis mass spectrometry. 1. Chip-based analysis. *Anal. Chem.*, **69** (21), 4363–4368.

24. Anderson, N.L. *et al.* (2004) Mass spectrometric quantitation of peptides and proteins using Stable Isotope Standards and Capture by Anti-Peptide Antibodies (SISCAPA). *J. Proteome Res.*, **3** (2), 235–244.

25. Kiernan, U.A. *et al.* (2006) Quantitative multiplexed C-reactive protein mass spectrometric immunoassay. *J. Proteome Res.*, **5** (7), 1682–1687.

26. Nelson, R.W. *et al.* (2004) Quantitative mass spectrometric immunoassay of insulin like growth factor 1. *J. Proteome Res.*, **3** (4), 851–855.

27. Trenchevska, O., Kamcheva, E., and Nedelkov, D. (2010) Mass spectrometric immunoassay for quantitative determination of protein biomarker isoforms. *J. Proteome Res.*, **9** (11), 5969–5973.

28. Trenchevska, O., Kamcheva, E., and Nedelkov, D. (2011) Mass spectrometric immunoassay for quantitative determination of transthyretin and its variants. *Proteomics*, **11** (18), 3633–3641.

29. Trenchevska, O. and Nedelkov, D. (2011) Targeted quantitative mass spectrometric immunoassay for human protein variants. *Proteomics Sci.*, **9** (1), 19.

30. Yassine, H. *et al.* (2013) Mass spectrometric immunoassay and MRM

as targeted MS-based quantitative approaches in biomarker development: potential applications to cardiovascular disease and diabetes. *Proteomics Clin. Appl.*, **7** (7-8), 528–540.

31. Karas, M. and Hillenkamp, F. (1988) Laser desorption ionization of proteins with molecular masses exceeding 10,000 daltons. *Anal. Chem.*, **60** (20), 2299–2301.

32. Nedelkov, D. and Nelson, R.W. (2001) Analysis of human urine protein biomarkers via biomolecular interaction analysis mass spectrometry. *Am. J. Kidney Dis.*, **38** (3), 481.

33. Nedelkov, D. and Nelson, R.W. (2000) Practical considerations in BIA/MS: optimizing the biosensor-mass spectrometry interface. *J. Mol. Recognit.*, **13** (3), 140–145.

34. Nedelkov, D. and Nelson, R.W. (2001) Delineation of in vivo assembled multiprotein complexes via biomolecular interaction analysis mass spectrometry. *Proteomics*, **1** (11), 1441–1446.

35. Nelson, R.W., Nedelkov, D., and Tubbs, K.A. (2000) Biosensor chip mass spectrometry: a chip-based proteomics approach. *Electrophoresis*, **21** (6), 1155–1163.

36. Nedelkov, D. and Nelson, R.W. (2003) Detection of Staphylococcal enterotoxin B via biomolecular interaction analysis mass spectrometry. *Appl. Environ. Microbiol.*, **69** (9), 5212–5215.

37. Nedelkov, D. and Nelson, R.W. (2003) Design and use of multi-affinity surfaces in biomolecular interaction analysis-mass spectrometry (BIA/MS): a step toward the design of SPR/MS arrays. *J. Mol. Recognit.*, **16** (1), 15–19.

38. Grote, J. *et al.* (2005) Surface plasmon resonance/mass spectrometry interface. *Anal. Chem.*, **77** (4), 1157–1162.

39. O'Connor, P.B. *et al.* (2004) A high pressure matrix-assisted laser desorption ion source for Fourier transform mass spectrometry designed to accommodate large targets with diverse surfaces. *J. Am. Soc. Mass. Spectrom.*, **15** (1), 128–132.

40. Zhukov, A. *et al.* (2004) Integration of surface plasmon resonance with mass spectrometry: automated ligand fishing and sample preparation for MALDI MS using a biacore 3000 biosensor. *J. Biomol. Tech.*, **15** (2), 112–119.

41. Grasso, G. *et al.* (2006) In situ AP/MALDI-MS characterization of anchored matrix metalloproteinases. *J. Mass Spectrom.*, **41** (12), 1561–1569.

42. Nedelkov, D. and Nelson, R.W. (2006) Surface plasmon resonance mass spectrometry for protein analysis. *Methods Mol. Biol.*, **328**, 131–139.

43. Nedelkov, D., Tubbs, K.A., and Nelson, R.W. (2006) Surface plasmon resonance-enabled mass spectrometry arrays. *Electrophoresis*, **27** (18), 3671–3675.

44. Bellon, S. *et al.* (2009) Hyphenation of surface plasmon resonance imaging to matrix-assisted laser desorption ionization mass spectrometry by on-chip mass spectrometry and tandem mass spectrometry analysis. *Anal. Chem.*, **81** (18), 7695–7702.

45. Konig, S. (2008) Target coatings and desorption surfaces in biomolecular MALDI-MS. *Proteomics*, **8** (4), 706–714.

46. Borch, J. and Roepstorff, P. (2010) SPR/MS: recovery from sensorchips for protein identification by MALDI-TOF mass spectrometry. *Methods Mol. Biol.*, **627**, 269–281.

47. Grasso, G. *et al.* (2009) Enzyme solid-state support assays: a surface plasmon resonance and mass spectrometry coupled study of immobilized insulin degrading enzyme. *Eur. Biophys. J.*, **38** (4), 407–414.

48. Nedelkov, D. (2010) Integration of SPR biosensors with mass spectrometry (SPR-MS). *Methods Mol. Biol.*, **627**, 261–268.

49. Kim, Y.E. *et al.* (2011) Gold patterned biochips for on-chip immuno-MALDI-TOF MS: SPR imaging coupled multi-protein MS analysis. *Analyst*, **137** (2), 386–392.

50. Madeira, A., Vikeved, E., Nilsson, A., Sjögren, B., Andrèn, P.E. and Svenningsson, P. (2011) Identification of Protein-Protein Interactions by Surface Plasmon Resonance followed by Mass Spectrometry. *Current Protocols*

in Protein Science, 65:19.21:19.21.1-19.21.9.

51. Remy-Martin, F. *et al.* (2012) Surface plasmon resonance imaging in arrays coupled with mass spectrometry (SUPRA-MS): proof of concept of on-chip characterization of a potential breast cancer marker in human plasma. *Anal. Bioanal. Chem.*, **404** (2), 423–432.

52. McKee, C.J., Hines, H.B., and Ulrich, R.G. (2013) Analysis of protein tyrosine phosphatase interactions with microarrayed phosphopeptide substrates using imaging mass spectrometry. *Anal. Biochem.*, **442** (1), 62–67.

53. Gilligan, J.J., Schuck, P., and Yergey, A.L. (2002) Mass spectrometry after capture and small-volume elution of analyte from a surface plasmon resonance biosensor. *Anal. Chem.*, **74** (9), 2041–2047.

54. Nedelkov, D. and Nelson, R.W. (2001) Analysis of native proteins from biological fluids by biomolecular interaction analysis mass spectrometry (BIA/MS): exploring the limit of detection, identification of non-specific binding and detection of multi-protein complexes. *Biosens. Bioelectron.*, **16** (9–12), 1071.

55. Borch, J. and Roepstorff, P. (2004) Screening for enzyme inhibitors by surface plasmon resonance combined with mass spectrometry. *Anal. Chem.*, **76** (18), 5243–5248.

56. Lopez, F. *et al.* (2003) Improved sensitivity of biomolecular interaction analysis mass spectrometry for the identification of interacting molecules. *Proteomics*, **3** (4), 402–412.

57. Yates, J.R. 3rd, (1998) Mass spectrometry and the age of the proteome. *J. Mass Spectrom.*, **33** (1), 1–19.

58. Bonnel, D. *et al.* (2011) MALDI imaging techniques dedicated to drug-distribution studies. *Bioanalysis*, **3** (12), 1399–1406.

59. Franck, J. *et al.* (2009) MALDI imaging mass spectrometry: state of the art technology in clinical proteomics. *Mol. Cell Proteomics*, **8** (9), 2023–2033.

60. Schuerenberg, M. and Deininger, S.-O. (2010) *Imaging Mass Spectrometry*, Springer, pp. 87–91.

61. Vaisocherova, H. *et al.* (2008) Ultralow fouling and functionalizable surface chemistry based on a zwitterionic polymer enabling sensitive and specific protein detection in undiluted blood plasma. *Anal. Chem.*, **80** (20), 7894–7901.

62. Owega, S., Lai, E.P., and Bawagan, A.D. (1998) Surface plasmon resonance-laser desorption/ionization-time-of-flight mass spectrometry. *Anal. Chem.*, **70** (11), 2360–2365.

63. Northen, T.R. *et al.* (2008) A nanostructure-initiator mass spectrometry-based enzyme activity assay. *Proc. Natl. Acad. Sci. U.S.A.*, **105** (10), 3678–3683.

64. Northen, T.R. *et al.* (2007) Clathrate nanostructures for mass spectrometry. *Nature*, **449** (7165), 1033–1036.

65. Okuno, S. *et al.* (2005) Requirements for laser-induced desorption/ionization on submicrometer structures. *Anal. Chem.*, **77** (16), 5364–5369.

66. Wen, X., Dagan, S., and Wysocki, V.H. (2007) Small-molecule analysis with silicon-nanoparticle-assisted laser desorption/ionization mass spectrometry. *Anal. Chem.*, **79** (2), 434–444.

67. Piret, G. *et al.* (2012) Surface-assisted laser desorption-ionization mass spectrometry on titanium dioxide (TiO2) nanotube layers. *Analyst*, **137** (13), 3058–3063.

68. Zhou, A. *et al.* (2006) Structural mechanism for the carriage and release of thyroxine in the blood. *Proc. Natl. Acad. Sci. U.S.A.*, **103** (36), 13321–13326.

69. Finkel, N.H. *et al.* (2005) Ordered silicon nanocavity arrays in surface-assisted desorption/ionization mass spectrometry. *Anal. Chem.*, **77** (4), 1088–1095.

70. Go, E.P. *et al.* (2005) Desorption/ionization on silicon nanowires. *Anal. Chem.*, **77** (6), 1641–1646.

71. Mengistu, T.Z., DeSouza, L., and Morin, S. (2005) Functionalized porous silicon surfaces as MALDI-MS substrates for protein identification studies. *Chem. Commun. (Cambridge)*, **45**, 5659–5661.

72. Wei, J., Buriak, J.M., and Siuzdak, G. (1999) Desorption-ionization mass spectrometry on porous silicon. *Nature*, **399** (6733), 243–246.

73. Chen, W.Y. and Chen, Y.C. (2003) Reducing the alkali cation adductions of oligonucleotides using sol-gel-assisted laser desorption/ionization mass spectrometry. *Anal. Chem.*, **75** (16), 4223–4228.

74. Gamez, R.C., Castellana, E.T., and Russell, D.H. (2013) Sol-gel-derived silver-nanoparticle-embedded thin film for mass spectrometry-based biosensing. *Langmuir*, **29** (21), 6502–6507.

75. Lin, Y.S. and Chen, Y.C. (2002) Laser desorption/ionization time-of-flight mass spectrometry on sol-gel-derived 2,5-dihydroxybenzoic acid film. *Anal. Chem.*, **74** (22), 5793–5798.

76. Ostman, P. *et al.* (2006) Minimum proton affinity for efficient ionization with atmospheric pressure desorption/ionization on silicon mass spectrometry. *Rapid Commun. Mass Spectrom.*, **20** (24), 3669–3673.

77. Laughlin, J.B., Cassady, C.J., and Cox, J.A. (1997) Matrix-assisted laser desorption ionization of small biomolecules impregnated in silica prepared by a sol-gel process. *Rapid Commun. Mass Spectrom.*, **11**, 1505–1508.

78. Chen, C.T. and Chen, Y.C. (2004) Molecularly imprinted TiO2-matrix-assisted laser desorption/ionization mass spectrometry for selectively detecting alpha-cyclodextrin. *Anal. Chem.*, **76** (5), 1453–1457.

79. Sönksen, C. *et al.* (2001) Capture and analysis of low molecular weight ligands by surface plasmon resonance combined with mass spectrometry. *Eur. J. Mass Spectrom.*, **7** (4), 385–391.

80. Visser, N.F.C. *et al.* (2007) Surface-plasmon-resonance-based chemical proteomics: efficient specific extraction and semiquantitative identification of cyclic nucleotide-binding proteins from cellular lysates by using a combination of surface plasmon resonance, sequential elution and liquid chromatography-tandem mass spectrometry. *ChemBioChem*, **8** (3), 298–305.

81. Natsume, T. *et al.* (2000) Combination of biomolecular interaction analysis and mass spectrometric amino acid sequencing. *Anal. Chem.*, **72** (17), 4193–4198.

82. Bouffartigues, E. *et al.* (2007) Rapid coupling of Surface Plasmon Resonance (SPR and SPRi) and ProteinChipTM based mass spectrometry for the identification of proteins in nucleoprotein interactions. *Nucleic Acids Res.*, **35**, 1–10.

83. Nielen, M.W.F. *et al.* (2006) Urine testing for designer steroids by liquid chromatography with androgen bioassay detection and electrospray quadrupole time-of-flight mass spectrometry identification. *Anal. Chem.*, **78** (2), 424–431.

84. Nielen, M.W.F. *et al.* (2004) Bioassay-directed identification of estrogen residues in urine by liquid chromatography electrospray quadrupole time-of-flight mass spectrometry. *Anal. Chem.*, **76** (22), 6600–6608.

85. Marchesini, G.R. *et al.* (2007) Dual biosensor immunoassay-directed identification of fluoroquinolones in chicken muscle by liquid chromatography electrospray time-of-flight mass spectrometry. *Anal. Chim. Acta*, **586** (1-2), 259.

86. Marchesini, G.R. *et al.* (2008) Nanoscale affinity chip interface for coupling inhibition SPR immunosensor screening with Nano-LC TOF MS. *Anal. Chem.*, **80** (4), 1159–1168.

87. Homola, J. (2008) Surface plasmon resonance sensors for detection of chemical and biological species. *Chem. Rev.*, **108** (2), 462–493.

88. Smith, E.A. and Corn, R.M. (2003) Surface plasmon resonance imaging as a tool to monitor biomolecular interactions in an array based format. *Appl. Spectrosc.*, **57** (11), 320A–332A.

89. Piliarik, M., Vaisocherova, H., and Homola, J. (2005) A new surface plasmon resonance sensor for high-throughput screening applications. *Biosens. Bioelectron.*, **20** (10), 2104–2110.

90. Piliarik, M. *et al.* (2012) High-resolution biosensor based on localized surface plasmons. *Opt. Express*, **20** (1), 672–680.

91. Shin, Y.-B. *et al.* (2010) A new palm-sized surface plasmon resonance (SPR) biosensor based on modulation of a light source by a rotating mirror. *Sens. Actuators, B*, **150** (1), 1–6.

92. De Leebeeck, A. *et al.* (2007) On-chip surface-based detection with nanohole arrays. *Anal. Chem.*, **79** (11), 4094–4100.

93. Im, H. *et al.* (2012) Nanohole-based surface plasmon resonance instruments with improved spectral resolution quantify a broad range of antibody-ligand binding kinetics. *Anal. Chem.*, **84** (4), 1941–1947.

94. Brolo, A.G. *et al.* (2004) Surface plasmon sensor based on the enhanced light transmission through arrays of nanoholes in gold films. *Langmuir*, **20** (12), 4813–4815.

95. Lindquist, N.C. *et al.* (2012) Ultra-smooth metallic films with buried nanostructures for backside reflection-mode plasmonic biosensing. *Ann. Phys.*, **524** (11), 687–696.

96. Martinez-Perdiguero, J. *et al.* (2012) Enhanced transmission through gold nanohole arrays fabricated by thermal nanoimprint lithography for surface plasmon based biosensors. *Procedia Eng.*, **47**, 805–808.

97. Monteiro, J.P. *et al.* (2013) Effect of periodicity on the performance of surface plasmon resonance sensors based on subwavelength nanohole arrays. *Sens. Actuators, B*, **178**, 366–370.

98. Campbell, C.T. and Kim, G. (2007) SPR microscopy and its applications to high-throughput analyses of biomolecular binding events and their kinetics. *Biomaterials*, **28** (15), 2380–2392.

99. Ouellet, E., Lund, L., and Lagally, E.T. (2013) Multiplexed surface plasmon resonance imaging for protein biomarker analysis. *Methods Mol. Biol.*, **949**, 473–490.

100. Scarano, S. *et al.* (2010) Surface plasmon resonance imaging for affinity-based biosensors. *Biosens. Bioelectron.*, **25** (5), 957–966.

101. Giudicatti, S. *et al.* (2010) Plasmonic resonances in nanostructured gold/polymer surfaces by colloidal lithography. *Phys. Status Solidi A*, **207** (4), 935–942.

102. Giudicatti, S. *et al.* (2012) Interaction among plasmonic resonances in a gold film embedding a two-dimensional array of polymeric nanopillars. *J. Opt. Soc. Am. B*, **29** (7), 1641–1647.

103. Giudicatti, S., Marabelli, F., and Pellacani, P. (2013) Field enhancement by shaping nanocavities in a gold film. *Plasmonics*, **8** (2), 975–981.

104. Bottazzi, B. *et al.* (2014) Multiplexed label-free optical biosensor for medical diagnostics. *J. Biomed. Opt.*, **19** (1), 017006.

105. Meyer, S.A., Le Ru, E.C., and Etchegoin, P.G. (2011) Combining surface plasmon resonance (SPR) spectroscopy with surface-enhanced Raman scattering (SERS). *Anal. Chem.*, **83** (6), 2337–2344.

106. Yih, J.-N. *et al.* (2004) A compact surface plasmon resonance and surface-enhanced Raman scattering sensing device. *Proc. SPIE*, **5327**, 5–9.

107. Densmore, A. *et al.* (2007) *Label-Free Biosensing Using Silicon Planar Waveguide Technology.* website: Proc. SPIE6796, Photonics North 2007, 67962X (October 26, 2007).

108. Takats, Z. *et al.* (2004) Mass spectrometry sampling under ambient conditions with desorption electrospray ionization. *Science*, **306** (5695), 471–473.

109. Cody, R.B., Laramee, J.A., and Durst, H.D. (2005) Versatile new ion source for the analysis of materials in open air under ambient conditions. *Anal. Chem.*, **77** (8), 2297–2302.

110. Nemes, P. and Vertes, A. (2007) Laser ablation electrospray ionization for atmospheric pressure, in vivo, and imaging mass spectrometry. *Anal. Chem.*, **79** (21), 8098–8106.

111. Hamm, G. *et al.* (2012) Quantitative mass spectrometry imaging of propranolol and olanzapine using tissue extinction calculation as normalization factor. *J. Proteomics*, **75**, 4952–4961.

112. Sun, N. and Walch, A. (2013) Qualitative and quantitative mass spectrometry

imaging of drugs and metabolites in tissue at therapeutic levels. *Histochemistry and Cell Biology.* **140**(2) 93–104.

113. Takai, N. *et al.* (2012) Quantitative analysis of pharmaceutical drug distribution in multiple organs by imaging mass spectrometry. *Rapid Commun. Mass Spectrom.*, **26** (13), 1549–1556.

114. Vismeh, R. *et al.* (2012) Localization and quantification of drugs in animal tissues by use of desorption electrospray ionization mass spectrometry imaging. *Anal. Chem.*, **84**, 5439–5445.

115. Bonnel, D. *et al.* (2013) Ionic matrices pre-spotted MALDI plates for patients markers following, drugs titration and MALDI MSI. *Anal Biochem.* **434**(1): 187–98.

116. Goodwin, R.J.A. *et al.* (2011) Qualitative and quantitative MALDI imaging of the positron emission tomography ligands raclopride (a D2 Dopamine Antagonist) and SCH 23390 (a D1 Dopamine Antagonist) in rat brain tissue sections using a solvent-free dry matrix application method. *Anal. Chem.*, **83** (24), 9694–9701.

117. Hamm, G. (2012) Toward quantitative imaging mass spectrometry. *Spectroscopyonline. Special Issues.*.

118. Koeniger, S.L. *et al.* (2011) A quantitation method for mass spectrometry imaging. *Rapid Commun. Mass Spectrom.*, **25** (4), 503–510.

119. Lemaire, R. *et al.* (2007) Tag-mass: specific molecular imaging of transcriptome and proteome by mass spectrometry based on photocleavable tag. *J. Proteome Res.*, **6** (6), 2057–2067.

120. Thiery, G. *et al.* (2008) Improvements of TArgeted multiplex mass spectrometry IMaging. *Proteomics*, **8** (18), 3725–3734.

121. Thiery, G. *et al.* (2012) Targeted multiplex imaging mass spectrometry with single chain fragment variable (scfv) recombinant antibodies. *J. Am. Soc. Mass. Spectrom.*, **23** (10), 1689–1696.

122. Yung-Chi, C. and Prusoff, W.H. (1973) Relationship between the inhibition constant (KI) and the concentration of inhibitor which causes 50 per cent inhibition (I50) of an enzymatic reaction. *Biochem. Pharmacol.*, **22** (23), 3099.

Part IV
Direct Post Column Coupled Affinity Techniques

Analyzing Biomolecular Interactions by Mass Spectrometry, First Edition.
Edited by Jeroen Kool and Wilfried M.A. Niessen.
© 2015 Wiley-VCH Verlag GmbH & Co. KGaA. Published 2015 by Wiley-VCH Verlag GmbH & Co. KGaA.

11

High-Resolution Screening: Post-Column Continuous-Flow Bioassays

David Falck, Wilfried M.A. Niessen, and Jeroen Kool

11.1
Introduction

High-resolution screening (HRS) represents a way of assessing biological properties (affinity or activity) that is especially suited for (complex) mixtures of molecules. In contrast to the more widely used high-throughput screening (HTS), HRS allows differential assessment of biological properties of individual molecules in a mixture by the integration of a separation step. A typical HRS platform consists of four essential parts, as exemplified in Figure 11.1a. At the core, there is liquid chromatography (LC), which separates the different analytes of a mixture in time. Although the analysis of biological properties of (complex) mixtures of molecules always relies on some form of separation, two features distinguish HRS. On the one hand, the biological assay (bioassay) is performed after the separation (post-column) while many other techniques select biological properties before separation (pre-column). On the other hand, the bioassay is operated in the same flow system as the separation step; thus we speak of an on-line or continuous-flow assay [1]. This is in contrast to approaches in which separation and bioassay are connected by a (micro-)fractionation step, the so-called at-line assays, or are completely separate entities, for example, classical effect-directed analysis (EDA) (see Chapter 4; Jonker *et al.*, "Methodologies for Effect-Directed Analysis: applications in environmental chemistry, food quality assessment, and drug discovery"). Although, originally, HRS only measured biological properties, nowadays the parallel acquisition of chemical properties for structure elucidation is so popular that it has become an integral part of HRS platforms [1]. In order to allow two parallel detection techniques, one for structure and one for effect analysis (affinity or activity), the effluent of the separation step has to be divided. This is achieved by a flow split in the second part. The third and fourth parts constitute the two parallel detection techniques. The bioassay measurement reports the effect, by which we mean the biological activity (bioactivity) or the biological affinity (bioaffinity) toward the target protein. This is most commonly achieved by incubating the analyte first with the target protein and then with a competitive binder or substrate, followed by

Analyzing Biomolecular Interactions by Mass Spectrometry, First Edition.
Edited by Jeroen Kool and Wilfried M.A. Niessen.

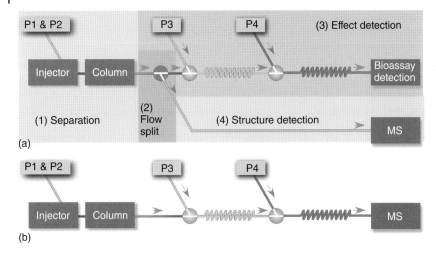

Figure 11.1 (a) A typical HRS platform. Detailed descriptions can be found in Section 11.2. Fluorescence is probably the most widely used bioassay detection in HRS, but other techniques such as fluorescence polarization have been employed as well. Similarly, MS is the most successful technique for gaining chemical information in parallel, but UV/vis spectroscopy is often used instead or in addition. As discussed in Section 11.1.1.2, MS instruments offering both fragmentation and high mass accuracy provide the most useful data for structure elucidation. (b) HRS platforms with MS for both structure and affinity measurements usually use one MS instrument.

spectrometric detection of the bioaffinity or bioactivity. In parallel, structural information is gained by chemical detection, sometimes by ultraviolet/visible (UV/vis) detection or spectroscopy, but most importantly by mass spectrometry (MS). These four stages are further scrutinized in Section 11.2.

Although the term *HRS* can be useful to describe all on-line biological characterization techniques involving the separation of individual analytes, we use it in its original meaning referring to post-column setups only [2]. The discussion is focused on on-line post-column assays, but many of the aspects are interesting to users of at-line assays as well. To provide a framework of reference, we start with a brief overview of HRS variants and further introduce some basic concepts (Section 11.1.1). This is followed by an overview of which targets, samples, and analytes have been investigated (Section 11.1.2). The most obvious requirement for HRS is a suitable, robust, and sufficiently sensitive analytical platform. Therefore, the setup of such a platform is discussed in detail in the second section. It includes practical tips to design a state-of-the-art analytical HRS system (Section 11.2). The third section is dedicated to data analysis, which needs special attention due to the innate complexity of the data obtained in HRS. Furthermore, certain concepts in HRS data analysis, some of which are still under development, are just a bit different than in HTS or in conventional chromatographic detection. These concepts are needed to fully harness the advantages of HRS platforms over HTS platforms, but awareness of sources of error is critical, especially where the concepts are not yet fully developed (Section 11.3). We finish by briefly comparing HRS to other approaches, by highlighting important trends in HRS, by sketching a possible future of HRS, and with some general conclusions (Section 11.4). Screening campaigns at the scale realized in HTS have not been performed with on-line post-column HRS assays. Arguably, the most extensive study in this context was recently published by our laboratory, focusing on the application of HRS to metabolite-like lead libraries for p38α mitogen-activated protein kinase [3]. Therefore, many lessons remain to be learned.

11.1.1
Variants of On-line Post-Column Assays Using Mass Spectrometry

In Tables 11.1 and 11.2, an overview of the currently described HRS platform with MS detection is given. It is clear that MS has become an integral part of state-of-the-art HRS platforms [1]. First and foremost, MS is used as a means of implementing rapid structure elucidation into HRS platforms (see Tables 11.1 and 11.2), that is, in a parallel setting next to bioassay detection mostly by fluorescence

Table 11.1 Overview of HRS platforms featuring MS and bioactivity detection.

Bioassay technique	Protein target	Detected species	Chemical detection	Publication
UV/vis, activity	Acetylcholinesterase	Product[a]	UV, MS^1 (IT)	[4, 5]
UV, activity	Alpha-glucosidase	Product	MS^2 (IT)	[6]
FLD, activity	Phosphodiesterase	Substrate	ELSD, DAD, MS^1 (q-TOF)	[7]
	Angiotensin-converting enzyme	Product	(pseudo)MS^2 (q-TOF)	[8]
	Acetylcholinesterase	Product	MS^1 (IT), DAD	[9]
	Serine proteases	Product	UV, MS^1 (q-TOF)	[10]
	Acetylcholinesterase	Product	UV, MS^2 (q-TOF MS)	[11]
	Soluble epoxide hydrolase	Product	MS^1 (q-TOF)	[12]
FLD, activity, FLE[b]	Alkyline phosphatase	Product[c]	MS^1 (q-TOF)	[13]
MS, (***also MS/MS), activity	Cathepsin B, acetylcholinesterase	Substrate, products, cofactors	MS^1 (*,**quadrupole, ***IT, ****q-TOF)	[27]*, [28]**, [29]***, [30]****

a) Via secondary chemical reaction.
b) Fluorescence enhancement, FLE.
c) Via labeled enzyme binding.

Table 11.2 Overview of HRS platforms featuring MS and bioaffinity detection.

Bioassay technique	Protein target	Detected species	Chemical detection	Publication
FLD, affinity, FLE[a)]	Estrogen receptor	Ligand	MS^2 (IT), (+UV)	[14−16]
	Estrogen receptor	Ligand	DAD, MS^2 (IT-TOF)	[17]
	Acetylcholine-binding protein	Ligand	DAD, MS^2 (IT-TOF)	[18]
	p38 Mitogen-activated protein kinase	Ligand	MS^3 (IT-TOF)	[19, 20]
+ on-line pre-column synthesis[b)]	Estrogen receptor	Ligand	UV, at-line LC−MS/MS via SPE	[21, 22]
	p38 Mitogen-activated protein kinase	Ligand	MS^3 (IT-TOF)	[23]
LIF[c)], affinity, FLE[a)]	Acetylcholine-binding protein	Ligand	MS (nano-ESI-IT-TOF)	[24−26]
MS, affinity	Streptavidin	Ligand	MS^2 (triple quadrupole or q-TOF)	[31, 32]
	Anti-digoxigenin-antibody	Ligand	MS^2 (q-TOF)	[32]

a)   Fluorescence enhancement.
b)   A mixture of related molecules is created from a single substrate by (bio-)chemical compound diversification in the HRS setup before analysis.
c)   Laser-induced fluorescence.

detection (FLD) [4−26]. It additionally allows the detection of analytes that do not show a biological effect. However, MS has also been used for effect detection in bioassays, thereby exploiting its excellent flexibility between general and specific detection [27−32]. These two features of MS in HRS have led to two different types of HRS platforms featuring MS. In one, the bioassay readout is achieved by parallel spectrophotometric detection and MS only serves as the structure readout (see Figure 11.1a). In the other, both the bioassay and the structure readout are performed with the same MS (see Figure 11.1b). While the latter is less complex, the former circumvents the incompatibility of MS with many classical biological buffer systems and bioassay constituents. This is probably the main reason why most HRS systems use MS only as a structure readout, if at all. Consequently, we treat this variant (Figure 11.1a) as typical in Sections 11.2−11.4, mentioning only differences of the fully MS-based systems.

Many of the early HRS platforms did not incorporate MS and are therefore not discussed. However, when it comes to the screening of mixtures of unknown compounds, integrated MS structure analysis is undeniably advantageous.

11.1.1.1 Mass Spectrometry as Readout for Biological Assays

The measurement of bioactivity or bioaffinity involves two stages: a reaction and a detection stage. Bioactivity leads to the formation of a product and the consumption of a substrate by an enzymatic reaction in the reaction stage (Figure 11.2b). The resulting concentration changes of the substrate, the product, or both (reactants) are assessed in the detection stage. Bioaffinity results in the formation of

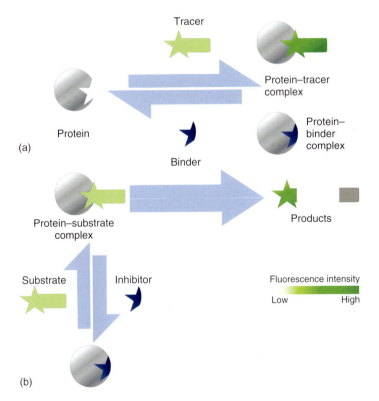

Figure 11.2 Examples of common bioassay principles used in HRS. (a) For bioaffinity, assays based on fluorescence enhancement are most widely used in HRS. Binding to the protein leads to strongly enhanced fluorescence of the tracer. A binder competing for the binding site of the protein reduces the amount of protein–tracer complex and consequently the fluorescence intensity. (b) For bioactivity, enzymatic conversion of a non- or weakly fluorescent substrate to a strongly fluorescent product is most commonly employed. Often, competitive inhibitors are measured in this way, which interferes with product formation and therefore reduces fluorescence intensity. However, the principle is equally suited for agonists and allosteric modulators. A variant of this principle where substrate and products are distinguished based on MS (and not fluorescence) is the basis for most HRS assays featuring MS bioassay readout.

an affinity complex in the reaction stage (Figure 11.2a). A difference in properties between the protein-bound and the free form of a ligand is employed in the detection stage. An exhaustive discussion on bioassay principles is not within the scope of this chapter, so we briefly mention some aspects especially relevant for HRS.

The reaction stage can theoretically consist of any bioactivity or bioaffinity imaginable. An overview of the pharmaceutical targets assessed by HRS in the past is provided in Section 11.1.2.1. Whether a bioactivity- or bioaffinity-based approach is preferred is a complex choice, which depends on many factors. In brief, bioaffinity can also be used to study the inactive form of a protein and it avoids the need to use expensive cofactors. However, only bioactivity assays can distinguish whether the protein activity is inhibited or induced by an analyte. In addition, bioactivity assays usually require a lower concentration of target protein, because instead of providing one bound ligand per protein (binding site), as bioaffinity does, numerous product molecules are formed in iterative reaction cycles by enzymatic catalysis.

The selection of a suitable reaction stage is crucial as it determines how comparable the screening results are to the physiological situation. Often artificial substrates or ligands are needed for sensitive detection by the envisioned spectrometric technique. Artificial substrates can be *de novo* synthesized compounds with the envisioned properties or chemically labeled physiological substrates or ligands that rarely possess the appropriate properties by themselves (see Figure 11.3). However, the equivalence of conversion or binding of these artificial placeholders with the physiological entities has to be established. Therefore, the availability of a suitable substrate or ligand becomes a criterion for choosing a suitable detection stage.

In bioactivity assays, the (artificial) substrate gains or loses a certain spectrometric property, for example, fluorescence, after conversion to the products (see Figure 11.3a,b). Inhibitors, competitors, or allosteric modulators are measured through their influence on the monitored reaction (see Figure 11.2b). In some cases, a secondary reaction has been employed to detect the physiological reaction products. For example, the enzymatic product may undergo a secondary chemical reaction to achieve specific UV/vis absorption or fluorescence [4, 5, 12] or its binding may enhance the fluorescence of a labeled receptor [13]. An example of a secondary chemical reaction is the nucleophilic attack of the artificial acetylcholinesterase product thiocholine on the disulfide known as *Ellman's reagent*, which produces a yellow thiol detectable by UV/vis [35]. Fluorescence labeling of the phosphate-binding protein has been used for a generic phosphate assay as the complex of phosphate and the labeled protein is highly fluorescent [13]. Although these solutions present a way of employing physiological substrates, if it results in added complexity with regard to the setup and the data analysis, the desirability of this solution may be limited. Early HRS systems used UV/vis for detection of bioactivity [4, 5]. However, this was rapidly replaced by the more selective and sensitive FLD yielding more robust assays. Because generally fluorescence is a more specific molecular property than UV/vis absorption, there are fewer interferences but it takes more development time. Although there are examples where

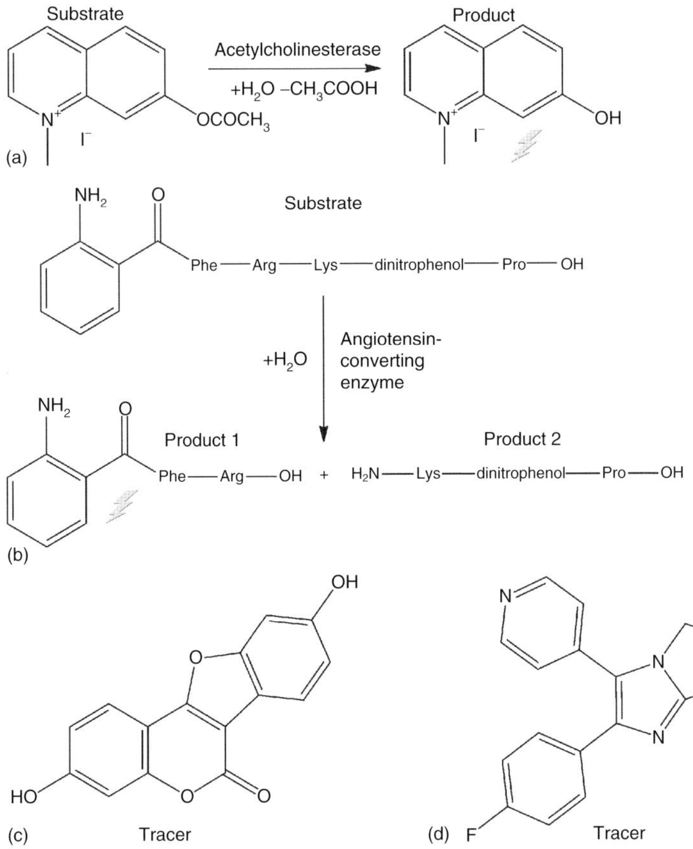

Figure 11.3 A number of examples for artificial substrates and ligands used in HRS assays are shown. (a) The artificial substrate 7-acetoxy-1-methylquinolinium iodide is enzymatically converted into 7-hydroxy-1-methylquinolinium iodide, which absorbs light and emits it at a higher wavelength. Thus, the product can be monitored by FLD in a homogeneous assay without interference from the substrate [33]. (b) Next to small molecules, peptides have been employed as artificial substrates. Here, a peptide is labeled with a fluorophore and a quencher. When these are separated by amide hydrolysis, the fluorophore can be detected [8]. (c) Coumestrol [34] and (d) SKF86002 [19] are ligands that show fluorescence enhancement upon protein binding. Coumestrol is a phytoestrogen that binds the estrogen receptors α and β. SKF86002 is a synthetic inhibitor of p38 mitogen-activated protein kinase.

a loss in fluorescence by substrate conversion is observed [7], the more common approach by far is having a fluorescent product. When probing enzyme activity directly, the signal-to-background ratio is far higher using a fluorescent product than with a fluorescent substrate, because the background is obviously much higher in the latter case. Even with assays probing modulation of enzyme activity, product detection is advantageous. Conversion rates have to be kept low (usually below 10%) for compatibility with Michaelis–Menten kinetics and negative peaks greatly reduce the number of false positives due to auto-fluorescence of the analytes when screening for inhibitors, as is most common. MS, being a label-free detection method, circumvents the need for synthesis of artificial substrates [30]. Although it is not always trivial to achieve efficient ionization of the physiological substrate and products, the need to exclude biases resulting from the difference of physiological and artificial substrates is avoided. Furthermore, with MS, substrate and products are easily distinguished and measured in parallel. This significantly increases the reliability of the assay, mainly by uncovering interferences, such as side reactions. In addition, cofactors can be monitored, revealing interesting additional information about the enzymatic reaction, such as uncoupling. The main drawbacks of MS detection are related to ionization suppression effects and instrument contamination. MS requires the use of volatile buffers to prevent performance deterioration due to salt precipitation. As classical biological buffer systems contain nonvolatile buffers and additives, designing a buffer system that is MS-compatible while retaining physiologically relevant conditions is challenging. Even volatile buffers should not be used in high concentrations as that decreases the sensitivity by ionization suppression. Related to this is the misinterpretation of ionization suppression effects caused by analyte elution as an inhibitory signal on the product formation. Fortunately, the distinction can be made either by looking at the substrate signal, which increases in inhibition and decreases with ionization suppression, or by using system-monitoring compounds whose signals should remain unchanged [27, 30]. Thus, like in FLD, where interferences due to auto-fluorescence are detected as a distortion of the peak shape, the analyte-caused ionization suppression effects are reliably detected because the analysis is based on multiple reactants. An advantage of MS-based bioaffinity assays lies in the simultaneous detection of biological and chemical information as opposed to the parallel detection in FLD/MS systems. As a consequence, the correlation of both data sets is even more robust (see Section 11.3.1.5).

In HRS, bioaffinity is almost exclusively detected with FLD. This detection is mostly based on a poorly understood phenomenon called *fluorescence enhancement*. Some weakly fluorescent ligands show great enhancement of their fluorescence intensity upon binding to a protein (see Figure 11.2a). The difference in fluorescence between bound and free ligand is often large enough to allow excellent homogeneous assays [19]. However, some early bioaffinity HRS assays suffered from the fact that this difference was not pronounced enough [2]. Consequently, they had to employ a sample preparation step in between the reaction and detection step (heterogeneous assay). Owing to the highly dynamic nature of bioaffinity, this separation is not trivial. Additionally, heterogeneous assays are usually much more labor-intensive and error-prone than their homogeneous counterparts [34]. Therefore, state-of-the-art HRS assays are usually homogeneous assays. As it would be impractical to design a fluorescent analog for every analyte, one labeled compound is used as a tracer to measure the interaction of many analytes, which cannot be directly detected (see Figure 11.3c,d). In bioaffinity assays, binders compete with the tracer for the monitored binding site, resulting in reduced fluorescence, because a smaller fraction of the tracer is subjected to fluorescence enhancement. Direct detection of bioaffinity by MS is extremely challenging as it requires the protein–ligand complex to stay intact during ionization under the prevailing conditions, even if it is not detected itself. This is the likely reason why post-column bioaffinity measurement by MS has only been reported for the relatively manageable, high-affinity model systems streptavidin/biotin, and anti-digoxigenin-antibody/digoxigenin + digoxin [31, 32]. Therefore, MS-based bioaffinity measurement most often relies on affinity selection, separation of the bound and free analytes, and release and detection of the analytes with affinity [36–38]. As the pre-column separation of analytes additionally presents an effective desalting/buffer exchange step, the advantages of a heterogeneous post-column assay are limited in comparison. MS-based continuous-flow bioaffinity assays are thus predominantly pre-column assays.

A technique that shows many similarities to HRS is the measurement of antioxidant properties [39]. Three different categories exist, which are all chemical in nature. In the first, oxidation of a substrate is monitored using oxidizing reagents that are based on reactive oxygen species. The second measures radical scavengers by one-electron reduction of a radical chromophore, which loses its optical properties after reaction. The third relies on electrochemical detection of redox potentials. Although the antioxidant properties are often referred to as *bioactivity*, they are fundamentally different from the bioactivity or bioaffinity measured in HRS systems. The antioxidant measurement lacks the level of complexity associated with protein–ligand interaction and (bio-)catalysis. However, the instrumental setups, especially of the second variant, are often quite comparable to HRS.

11.1.1.2 Structure Elucidation by Mass Spectrometry

In modern HRS platforms, the assessment of biological information is often complemented by the acquisition of chemical information. This is usually achieved in a parallel way through (see flow splitting Section 11.2.2) or simultaneously in the case of MS-based bioassays. In the simplest case, it is done by single (unselective) wavelength UV/vis detection, which mainly provides a means of additionally detecting inactive compounds [40]. However, the main strength of additional acquisition of chemical information lies in structural elucidation of the analytes. Two main methods, full spectrum UV/vis and MS, have been employed. A UV/vis spectrum can be collected by means of a diode array detector (DAD) to extract information on functional groups, and conjugated and aromatic systems. MS provides different information, whose density can be greatly enhanced by high-resolution mass spectrometry (HR-MS) and fragmentation data. For example, the elemental composition can be calculated from HR-MS measurements and fragmentation gives hints to substructures and structural connections. Because the two methods give complementary information, UV/vis is often used in addition to MS (see Tables 11.1 and 11.2). In addition, the response factors of different analytes are generally more similar in UV/vis, when using short (unspecific) UV wavelengths, than they are in MS, affording a more reliable detection of inactive compounds and a more reliable estimation of relative abundances by UV/vis. MS is one of the most sensitive techniques for structure elucidation and fast enough for real-time collection of structural data in LC−MS^n (liquid chromatography and mass spectrometry) combinations. The MS instruments used for structure elucidation in HRS were quadrupole, ion-trap (IT), quadruple-time-of-flight (q-TOF), and ion-trap-time-of-flight (IT-TOF) instruments, which were used in full spectrum mode. On MS instruments that allowed fragmentation (IT, q-TOF, and IT-TOF), MS^2 and MS^3 (IT-TOF only) were used for structure elucidation, mostly in data-dependent acquisition mode. A detailed discussion on the strategies and instrumentation of MS structure elucidation in HRS is found in Chapter 1.

11.1.2
Targets and Analytes

This Section 11.1.2 is concerned with the application of HRS to different target proteins and (complex mixtures of) analytes. Again, the discussion is strictly limited to those HRS platforms with integrated MS, and is thus not a comprehensive overview for HRS targets and applications in general.

11.1.2.1 Targets

Every individual protein target presents new challenges for the development of a suitable HRS platform, even when an HTS assay for that target is already available. Limited stability of the protein or the substrate, cofactors causing interferences, and extreme hydrophilicity or hydrophobicity of the ligands are only a few examples of the many challenges faced in assay development. In addition, like any *in vitro* bioassay, HRS requires the target protein in a sufficiently purified form. Therefore, it is not surprising that the number of protein targets for which HRS assays have been established is still limited and that HRS-like platforms are most popular in antioxidant assays that do not involve proteins.

HRS has been particularly popular in the area of neurological disease research. Several assay formats have been developed for acetylcholinesterase in the context of Alzheimer's disease research [4, 5, 9, 11, 30], possibly resulting from a combination of the success of the natural product Galantamine as anti-Alzheimer drug and the particular suitability of HRS for natural product screening [41]. Another anti-Alzheimer target protein studied by HRS is the nicotinic acetylcholine receptor (nAChR) analog acetylcholine-binding protein (AChBP) [18, 24, 25]. In terms of enzymes, hydrolysis catalyzing enzymes such as esterases, proteases, phosphatases, epoxidases, and glycosidases have been in the focus of HRS development. HRS assays have been developed for the serine proteases trypsin and thrombin [10], for cathepsin B [27−29], and for angiotensin-converting enzyme [8]. Phosphodiesterase (PDE) [7] and alkaline phosphatase [13] have also been investigated with HRS (see also Figure 11.4b). Angiotensin-converting enzyme and PDEs are targets related to cardiovascular diseases. This is also true for soluble epoxide hydrolase (sEH), which is in addition a target for analgesic and anti-inflammatory drugs [12]. Furthermore, the antidiabetic target α-glucosidase was employed in an HRS study [6]. Membrane receptors present a serious challenge for HRS assay development that has yet to be truly overcome. The two receptors that have been extensively studied by HRS, nAChR, and estrogen receptor (ER), present fortunate exceptions, because suitable soluble analogs are available in the form of AChBP and the free ligand-binding domain of ER, respectively. ER is a contraception target [14–16, 21, 22]. Receptors necessarily rely on bioaffinity measurement, but bioactivity measurement seems to be strongly preferred for enzymes (see also Section 11.1.1.1). The activity/affinity hybrid assay developed for alkaline phosphatase slightly deviates from this trend [13], but the only true exception is presented by a p38 mitogen-activated protein kinase assay (p38) [19] (see Figure 11.4a). For p38, the (less active) non-phosphorylated form has been proposed as a drug target in its own right, which, in combination with better availability of the non-phosphorylated form, resulted in bioaffinity measurement being the preferable measurement strategy for this particular enzyme.

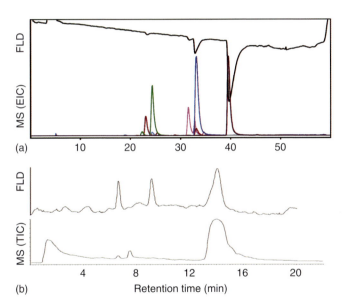

Figure 11.4 HRS chromatograms of complex mixtures. (a) A crude synthesis mixture of drug-like molecules generated by a biocatalytic approach was analyzed in an HRS system featuring a fluorescence enhancement-based bioaffinity assay and parallel IT-TOF-MS detection [20]. The FLD chromatogram (top) shows several binders (negative peaks). Several extracted ion chromatograms (EIC; below) represent the MS data that were used for structure identification of the synthesis products. (b) An HRS analysis of a plant extract spiked with the two PDE inhibitors, theophylline and papaverine, is shown [7]. Bioactivity in this PDE assay is detected as a positive peak in the FLD chromatogram (top). Parallel chemical analysis is represented by the MS total ion chromatogram (TIC). The second of the three bioactivity signals is caused by auto-fluorescence.

11.1.2.2 Analytes and Samples

Next to the target, the final composition of an HRS platform is also heavily influenced by the type of sample that is analyzed. Although HRS can screen pure compounds [10, 18], its real strength lies in the analysis of (complex) mixtures. Natural extracts include some of the most complex mixtures imaginable and are therefore a favored application area of HRS [41]. Plant extracts (see Figure 11.4b), for example, in the form of teas, have received the most attention [4–7, 9, 13, 27–30]. The active compounds discovered were mainly alkaloids and flavonoids, but also mycotoxins and small, drug-like molecules. Furthermore, extracts from fungi [27] and foodstuff [8], and more recently from animal venoms, more specifically snake venoms [24, 25], have been analyzed by HRS. HRS has also found application in *in vitro* drug metabolism [11, 12, 16] and pharmaceutical degradation studies [11]. Next to an identification of the metabolites and degradation products, HRS also hints to whether individual compounds contribute to the pharmacological effect of the drug. Lastly, HRS can also be found in the context of lead optimization. Here, shotgun synthetic approaches using chemical [3], photochemical [3], electrochemical [23], and biosynthetic [17, 20–22] conversion of lead molecules produce mixtures of related molecules that can be used as lead libraries (see Figure 11.4a). This concept has the potential to significantly speed up lead optimization as it rapidly provides chemical and pharmacological information on multiple molecules. Because it allows an initial structure–activity relationship of a crude synthesis mixture, laborious synthesis optimization and purification efforts can be limited to promising molecules without much compromise on the information available for deselected molecules.

11.2
The High-Resolution Screening Platform

In this section, instrumental aspects of the HRS setup (see Figure 11.1) are discussed in detail: (1) the separation of complex mixtures by LC, (2) the flow splitting to achieve a parallel bioassay and MS setup, (3) the bioassay, and (4) the parallel MS detection.

11.2.1
Separation

The purpose of the separation step is for the separation of different analytes in a mixture on a time scale. An efficient separation means obtaining as much resolution in as little time as possible. Thus, the utilization of standard state-of-the-art chromatographic equipment is essential. While this point is of course true for any separation, there are significant differences to separations with more classical detectors such as UV/vis, which can mainly be attributed to the very specialized bioassay measurement. We focus on high-performance liquid chromatography

(HPLC) as the use of other separation techniques, for example, incorporating elec-trophoretic separation, is still rare [42].

The most important challenge to the use of LC in on-line combination with bioassays is the innate incompatibility of many classical LC mobile-phase con-stituents (and flow rates) with the bioassay measurement. The main reason for this incompatibility is the unfolding/denaturation of the target protein that is triggered by excessive concentrations of organic modifiers after post-column mixing of effluent and protein solution [1]. Additionally, extreme acid, salt, or base content or immiscibility of nonpolar solvents with the aqueous buffers used in the bioassay measurement could interfere when using separation by, for example, ion-exchange LC [43] or normal-phase LC, respectively. Fortunately, standard reversed-phase (RP-)LC solvents like acetonitrile and especially methanol usually only produce unfolding at higher concentrations. The exact number is strongly dependent on the protein, incubation time, incubation temperature, and other additives. In a serine protease assay (trypsin and thrombin), for example, the influence of acetonitrile and methanol only became visible above 20% and 30% (initial assay concentrations), respectively [10]. Thus, compatibility can often be achieved by suitable dilution of the eluent with the bioassay reagent solutions. If several reagent addition steps are involved, the compatibility has to be achieved in the step in which the target protein is added, which is usually the first, in order to prevent the (generally irreversible) unfolding of the protein molecules. For example, an sEH bioassay features an about $1:11$ dilution in the first enzyme mixing step, but only a $1:0.15$ dilution when the substrate is subsequently mixed in [12]. In contrast, in a p38 bioassay, the influence of the gradient on the baseline is much more pronounced, because the first mixing step is only $1:5$, although the total mixing dilution (D_M) in both the sEH and p38 assay is in the range of $1:9$ [19]. We would like to point out that in the case of the p38 assay the baselines were straightened by background subtraction. The background is determined by a blank run (injection of solvent without analytes).

However, although dilution assists in solving compatibility issues, it is often an important challenge in HRS. Dilution reduces the final concentration at the point of detection and thereby reduces the sensitivity of the HRS platform. There are two inevitable dilution steps involved in any bioassay measurement by HRS (see also Figure 11.5 and Section 11.3.1.3). The first one is chromatographic dilution (D_C). It is a result of diffusion and convection, which transforms the initially homogeneous injection plug into a two-sided concentration gradient by mixing with the solvent flow (see also Figures 11.8 and 11.10). The result is the series of (nearly) Gaussian signal distributions so typical for chromatographic separations. Depending on the interplay between injection volume (V_i), flow rate (u), and void volume, the analyte is diluted and the maximum concentration and thus the signal height is reduced. Because this additionally results in broader peaks, the phenomenon is also known as *chromatographic band broadening*. Although we can consider increasing the ratio of V_i to u, this has a negative influence on the separation due to the consequential increase in peak width. The second dilution D_M has already been mentioned earlier. It is exactly the same dilution

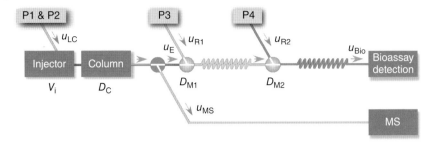

Figure 11.5 A typical HRS platform. The variables describing flows u and dilutions D are linked to their position in the platform. The correlation between flows and dilutions is examined in Section 11.3.1.3 and Equations 11.3–11.5. Platforms with MS-based bioassays (compare Figure 11.1b) sometimes contain a flow split as well, because optimal flow rates may differ largely between u_{LC} and u_E. u_{MS} is directed to waste in that case.

that appears when sample and reagents are mixed in a beaker or well plate, for example, in HTS. D_M is mainly relevant for the bioassay measurement. Chemical detection by electrospray ionization (ESI) MS does not usually need reagents or suffers as much from incompatibility with LC (see later). If a makeup flow is desired, the reduced ionization efficiency at higher flow rates can be more than compensated by suitable additives and/or an increased organic modifier concentration in the makeup flow. We provide a more quantitative look at dilution in the respective data analysis section (Section 11.3.1.3). In general, the total dilution is usually on the order of 50–100 times, thus significantly impairing the HRS measurements of low concentration samples. As a result, the analysis of samples from clinical or animal studies, for example, has not been achieved by HRS so far.

We have discussed the compatibility of separation with biochemical detection, but the compatibility with MS is equally important, although usually easier to achieve. Surely, it is superfluous to recap here the hyphenation issues of LC and MS as there are excellent books and reviews on that topic [44, 45]. However, when it comes to the compatibility of MS with a bioassay, one has to be aware of the many mobile-phase additives used to enhance ionization efficiency (which may have additional functions in the separation). For example, formic acid (FA) is often used in concentrations of 0.1–1% to enhance ESI$^+$ efficiency. FA, like any other acid, salt, or base, can result in two major interferences with the bioassay. Firstly, the pH of the bioassay, which is crucial for the protein–ligand interaction, can be affected. Secondly, interaction of any kind with the protein, for example, ionic or hydrophobic, can denature the protein structure. This adds to the influence of the organic modifier and might lead to more unfolding. In a p38 HRS assay, for example, even 0.1% (v/v) FA in the mobile phase in combination with methanol caused unacceptable protein unfolding [19]. In this case, reducing the mobile-phase content of FA to 0.01% (v/v) resulted in a sufficient compromise between bioassay compatibility and enhancement of ESI$^+$ efficiency.

The influences on the bioassay described are less critical with isocratic elution LC, because they are at least constant. However, in the analysis of complex mixtures, gradient elution is more favorable from an LC point of view. When gradient elution is used in HRS, this leads to changing assay conditions during the chromatographic run, due to the changing modifier content as is demonstrated in Figure 11.6. Mainly, the assay window, being the difference between the assay baseline and the maximum assay signal, is reduced with increasing protein unfolding. Furthermore, in assays based on signal reduction, the high signal baseline decreases as a consequence of protein unfolding, which complicates peak detection and integration (see also Figure 11.4a). On the one hand, this can be prevented by increasing D_M to the point where the change in influence is negligible compared to the noise and in return accepting the accompanying decrease in sensitivity. On the other hand, the changing assay window, as long as it is sufficiently large over the whole gradient, neither prevents effect detection nor significantly changes measured IC_{50} values. The trade-off here is the increased protein concentration, which ensures a sufficient assay window at higher modifier concentrations. Issues with (manual and automatic) peak detection and integration can be minimized by subtracting a blank gradient run from every bioassay chromatogram [19]. Another possible interference of the gradient that has to be taken into account during assay development is a change in detection properties of the analyte or tracer (in the bioassay detection), such as fluorescence wavelength and/or quantum yield, or the ESI efficiency.

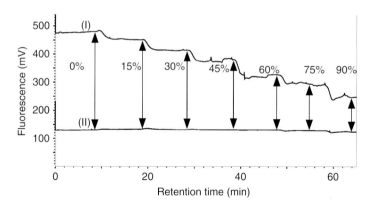

Figure 11.6   Typical example of the influence of the organic modifier concentration in gradient LC on baseline and assay window. (I) In this example, the methanol concentration in the LC eluent was increased stepwise from 0% to 90% (the concentration experienced by the protein was one-fifth due to the mixing dilution) and the baseline was recorded [19]. (II) Response to the same step gradient was recorded at maximum competition using a high concentration of a known strong inhibitor. Therefore, the arrows indicate the assay window at each methanol concentration, respectively. In Figure 11.4a the influence of a linear gradient on the assay baseline can be seen.

To eliminate the mentioned drawbacks, two technological solutions have been investigated to combine gradient chromatography with bioassays: countergradient systems and high-temperature LC. A countergradient system provides a constant organic modifier concentration to the bioassay by dilution of the LC eluent. The simplest version adds a gradient makeup flow between LC and the bioassay, which is the inverse of the LC gradient [22]. The resulting twofold analyte dilution is rarely an issue. However, this countergradient requires an additional gradient LC system, increasing the complexity of the HRS platform. In order to integrate a countergradient without the additional gradient pumps, half of the eluent from one run is stored in a reservoir without mixing and added to the other half of the next run in inverse order. This version does not require the additional gradient system, but is not necessarily less complex than the other countergradient solution [10]. High-temperature liquid chromatography (HTLC) presents a more elegant solution [28]. It employs isocratic solvent elution in combination with a temperature gradient. As solvents change their polarity with temperature, for example, water reaches the polarity of methanol at around 200 °C, HTLC can be very effectively used to replace organic modifier gradients in RP-LC [46]. Because the effluent is cooled post column, there is no influence of the temperature gradient on the bioassay. Although temperature stability of the analytes can be an issue, HTLC offers great advantages in combination with bioassays [28]. In some cases, the solvent can be simplified down to pure water without compromising the separation, allowing minimization of D_M [47]. However, adsorption of hydrophobic analytes in the bioassay and MS sensitivity can become issues at these very low modifier concentrations.

In conclusion, the composition of the mobile phase is a compromise between separation efficiency, ionization efficiency, and compatibility with the bioassay. The choice of the stationary phase is thus often limited by these restrictions. The flow rate and column dimensions are discussed in the following section as this is, via D_M, closely linked to the concept of flow splitting.

11.2.2
Flow Splitting

Flow splitting has two main functions: (i) to allow the parallel use of several detectors and (ii) to control the flow rate at which the eluent is provided to either detector. We first discuss principles behind flow splitting and the attributes of a good flow split. Later, we go into detail on how to optimize the split ratio.

Flow splitting is achieved by providing more than one exit for the eluent after the separation step. The total flow rate applied during the separation (u_{LC}) then divides itself according to the resistance encountered (see Figure 11.5). This means that the flow rate toward each exit (e.g., u_{Bio} and u_{MS}) is such that the resistance or, in LC terms, the backpressure experienced as a consequence of this flow rate is the same in all lines. One way to make a simple split is to take a connector with

$1 + N$ connections where one is connected to the LC column and N is the number of desired exits. The exits are then fitted with tubing of various dimensions. When calculating these dimensions, the first decision to be made is the operating pressure of the split. On the one hand, it should be high enough to compensate for post-split pressure differences and variations. Especially ESI capillaries can be an unexpected source of backpressure in this context. For example, at an operating pressure of 50 bar, a difference of 1 bar in two lines already causes 2% change in the split ratio while at an operating pressure of 5 bar the same interference renders the split effectively useless because of a 20% error. On the other hand, the pressure from the split adds to the demand on the LC pumping system and should therefore not be excessive. While especially with analytical and microbore LC columns the optimal flow rate for separation is often only reached at backpressures over 100 bar, standard LC equipment has an upper limit between 200 and 400 bar. Pressures of 40–60 bar in the flow split have proven to be a sensible compromise.

Of course, the pressure is dependent not only on split dimensions but also on solvent viscosity and total flow rate. Solvent viscosity is especially tricky to handle as it changes during a chromatographic gradient. However, this change does not influence the split ratio. For ease and safety of calculating and testing a split, it is useful to employ water as the solvent. However, it is always necessary to check that the pressure is still sufficiently high for a stable split ratio at the lowest viscosity encountered in the gradient and that at the highest viscosity the combined backpressure of column and split does not exceed the system limits. Like all connections in the LC system, the splitter should not introduce significant additional void volume (see discussion on void volumes in Section 11.2.3).

The u_{LC} is determined mainly by the separation and by MS compatibility, the latter because the majority of the flow is usually directed toward this detector. Although it would be possible to add a waste line to the split in order to divert excess flow, the split is the part of the HRS system most liable to failure, the possibility of which increases with the number of exits. Additionally, splits with a larger number of exits are also increasingly complicated to prepare. Therefore, it is often preferred to match the column dimensions to the MS ion source to achieve compatibility without additional splitting. The influence of the split ratio on the decision for u_{LC} is limited. The bioassay is usually conducted with a concentration-dependent detector, for example, a fluorescence detector. However, the split ratio has no direct influence on the concentration of the analyte in the fraction of eluent that enters the bioassay. The fraction of total injected analyte mass entering the bioassay, which is equivalent to the split ratio, is of interest merely in samples considered mass-limited with respect to the high absolute sensitivity of both HRS detectors [24].

There are also commercial solutions available for flow splitting, which operate by the same principles. In some variants, the split ratio is adjustable within a limited range. However, when using a costly commercial flow splitter, the integration of a sufficiently fine particle filter between column and split is advised as clogging of the split is a frequent problem, especially in matrix-intensive samples.

11.2.3
Bioassay

In the introduction, especially in Section 11.1.1.1, we discussed general aspects of the bioassay choice. In this section, we focus on the technical aspects of an HRS bioassay.

After flow splitting, the column effluent carrying the analytes enters the HRS bioassay. The flow rate directed toward the bioassay (u_E) is a result of the desired D_M and the flow rates at which the bioassay reagents are mixed in ($u_{R1} + u_{R2}$) (see Equation 11.1 and Figure 11.5). This is exemplified with values taken from an sEH HRS platform in Equation 11.1 [12]. It is preferred that u_{Bio} be as low as possible to minimize reagent consumption, while the ratio of u_{R1} to u_E be large enough to prevent protein unfolding (see Section 11.2.1). Both are important as the purified protein is often the most precious reagent in the HRS platform. There are two factors that limit minimization of the reagent flow. The flow rate needs to be exceptionally stable, which becomes an increasing technical challenge the lower the flow rate is. This also sets a limit to the achievable ratio of u_{R1} to u_{R2}. Additionally, u_{Bio} has to match the dimensions of the optical detector used in the bioassay to avoid extensive peak broadening in the detector cell. Thus, miniaturization of the bioassay by employing micro- or even nanofluidic instrumentation promises a significant reduction in reagent consumption [48].

$$u_E = \frac{u_{Bio}}{D_M} = \frac{u_{R1} + u_{R2}}{D_{M-1}} = \frac{155 + 30\,\mu l\,min^{-1}}{12,\overline{3}} = 15\,\mu l\,min^{-1} \quad (11.1)$$

One strategy frequently employed to stabilize the reagent flow is the use of pulse dampeners (see Figure 11.7). A main contribution to flow instability is the sinus-shaped fluctuation resulting from the work cycle of the pump pistons. This is true provided the appropriate measures (solvent degassing and regular seal maintenance) have been taken to prevent the formation of air bubbles in the solvent delivery system. The pulses are smoothened by passing the solvent over an oil-supported, impermeable membrane in the pulse dampener. For its function, it is

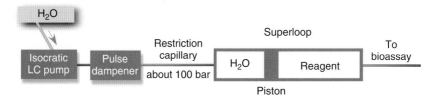

Figure 11.7 Solvent delivery setup for on-line bioassays. An isocratic LC pump provides deionized water to a superloop. In between, a pulse dampener and restriction capillary steady the solvent flow. The influx of water pushes the piston in the superloop, which in turn expels the reagent (enzyme or substrate or other) toward the bioassay. The setup is important for good flow stability and low reagent consumption (see text).

necessary to apply high pressure to the solvent, which is achieved by a restriction capillary providing at least 100 bar backpressure.

We can gain distinct advantages by delivering the reagents indirectly instead of pumping them through an LC pump. Indeed, the disadvantages (see later) of direct reagent delivery via an LC pump are so grave that we strongly discourage it. Indirect delivery can be achieved by running the pump on deionized water that is passed through a pulse dampener and restriction capillary to generate a stable solvent flow, which in turn operates a superloop (see Figure 11.7). A superloop is a plastic-covered glass cylinder. Both ends are sealed watertight but provide a standard LC connection. The cylinder also hosts a movable, watertight piston. On the inlet side, the water enters the superloop and pushes the piston. On the outlet side, the reagents are stored. By the movement of the piston, the reagents are pushed out at the same flow rate at which the water enters. The reagents are then mixed with the LC eluent. By using this setup, we save the reagent solution as we only need to fill the transfer tubing from superloop to mixer with the reagent and do not have to purge the significant void volume of an LC pump. By keeping the void volume of the transfer tubing low, the equilibration time, needed to achieve a stable assay baseline, is also shortened. In addition to saving expensive reagents, the setup reduces maintenance and increases the lifetime of the pump as contact with chemical and/or biological reagents is avoided. If necessary, the superloop is also easily cooled by placing it in an ice bucket. This allows an HRS system to run for days without interruption, depending on the stability of the protein or reagents.

There are a number of measures we should consider to achieve an optimal performance of the superloop. The piston should move as freely as possible within the superloop. Thus, regular greasing is essential. While filling the superloop with reagent solution and water, air bubbles should be avoided. Air bubbles increase fluctuations in the outlet flow as they compress and decompress every time the piston movement is slightly hindered. Because prevention of air bubbles is not always possible, the superloop should be stored with the inlet slightly elevated. This at least prevents artifacts from release of the bubbles into the reactors during the analysis. The specific described superloop type has a pressure limit of around 10 bar.

Mixing of the eluent with the reagents is achieved with a T-piece. The mixing in an on-line post-column bioassay is highly efficient, especially when coiled reactors or hairpin curve containing microfluidic chips are used as discussed later. This is one of the major advantages over assays in well plates or test tubes and the reason why shorter incubation times can be achieved in HRS at similar reagent concentrations [12].

Although they are essentially only a simple piece of polyetheretherketone (PEEK), polytetrafluoroethylene (PTFE), or fused silica tubing, much attention has to be devoted to the reactors in post-column bioassays. One reason for this lies in the fact that we purposefully introduce additional void volume between separation and detection. This necessarily leads to post-column band broadening. As there are other sources of band broadening, we call this effect physical band broadening for distinction. The efforts described in the following examples are only justified if other prominent sources of void volume are also meticulously managed. Minimal void volume connectors and T-pieces should be utilized and special attention has to be directed to avoid ill-fitted connections. In addition, excessively large spectrophotometric detection flow cells compared to u_{Bio} are a main cause of peak broadening [49]. From post-column derivatization, the coiled open tubular reactor (coiled reactor) was adopted to reduce peak broadening. It is simply a specially "knitted" PTFE tubing that induces secondary flow patterns, thus increasing radial mass transfer and reducing peak broadening [50]. By increasing the radial mass transfer, the coiled reactor also contributes to the mixing of eluent and reagents.

A second kind of band broadening results from the interaction between the reactor material and the analytes (chemical band broadening). In PEEK and PTFE reactors, these are mainly hydrophobic interactions. We therefore recently proposed to use chemically modified fused silica tubing for highly lipophilic analytes [19]. Chemical modification is necessary, as the use of untreated fused silica results in strong dipole–dipole or ionic interactions of the free silanol groups with nitrogen-containing groups often found in drug-like molecules. By chemical modification, the fused silica surface properties can be tuned to minimize interactions with the respective analyte. For example, polyethylene glycol (PEG)-modified fused silica tubing has been successfully used for highly lipophilic drug-like molecules [19]. A disadvantage of the fused silica tubing is that its low flexibility does not allow knitting it in the same way as the PTFE tubing. Therefore, depending on whether physical or chemical band broadening is the main contributor, a choice must be made between coiled PTFE and straight fused-silica reactors, respectively, until fused-silica reactors with the knitted-coil geometry become commercially available.

Bioassay peaks also increase in width when their height is close to the maximum signal intensity assay signal. However, this biochemical band broadening is different from the physical and chemical band broadening, because it does not influence the actual concentration of the analyte at the top of the peak or any other data point for that matter. The reasons and results are comparable to overloading of a classical detector. We discuss this aspect in more detail in Section 11.3.1.2.

Miniaturized assay formats can tackle chemical band broadening quite effectively, despite their larger surface-to-volume ratio, by the use of glass microfluidic chips that can be deactivated by chemical modification [29, 48]. Moreover, microfabrication of chips offers a lot of freedom in designing the flow paths. The introduction of hairpin curves prevents a laminar flow profile [51], and thus enables efficient mixing and avoids excessive physical peak broadening.

For optimal performance, the inner diameter (I.D.) of the tubing should match the flow rate applied, although smaller I.D. is generally preferred. Although it promotes chemical band broadening (see earlier), the large surface-to-volume ratio of the coiled reactors is favorable for the heat transfer [50]. This tight and rapid temperature control is another advantage of the HRS over a classical HTS system and a second reason why shorter reaction times can be achieved [12]. Furthermore, incubation in HTS, especially at elevated temperatures, results in solvent evaporation, which is a major source of errors in HTS. Due to the (virtually) closed system, this is not an issue in HRS. Of course, minimizing the I.D. is limited in practice by the pressure limits of the superloop and the cost of longer coiled reactors. An I.D. around 250 μm presents the most common compromise.

We have now discussed the delivery and mixing of a reagent with the column effluent. However, usually, there are at least two different reagents needed: on the one hand, the enzyme or receptor whose interaction with or manipulation by the analytes is studied, and on the other hand, the substrate or tracer indicating inhibition of or affinity with the target protein. These two reagents might be delivered in one [48] or two separate steps [19]. If a two-step delivery is chosen, the protein is delivered first in order to increase the sensitivity of analyte measurement (see discussion about kinetics later). Delivering protein and tracer with the same superloop reduces the complexity of the system. Whether the reagents can be premixed depends on the bioassay principle. For example, it is obviously not possible for bioactivity assays. Due to the short incubation times typical in HRS (usually ≤5 min), affinity assays can be negatively influenced as well. While binding to the free protein is usually quite fast, competition between a ligand and a tracer for a binding site significantly delays the binding of the analyte [52]. Therefore, it is often favorable for the sensitivity of the affinity/activity measurement to make a first mixing and incubation step of the protein with the eluent, thus allowing fast association with the analytes, and then to introduce the substrate or tracer. Nonetheless, for binders with slow association kinetics (small k_{on}), which do not reach equilibrium in the first incubation step, concentration sensitivity of the affinity measurement is lowered and IC_{50} values are underestimated. This issue may also be observed in bioactivity assays. Further details can be found in Section 11.3.1.1.

Of course, as in any biological assay, the post-column bioassay should be buffered and contain the appropriate cofactors, for example, NADPH or Mg^{2+}. Sometimes, reagents like bovine serum albumin (BSA) are added to improve protein stability. In addition, chemicals may be added that help to prevent adsorption of analytes, proteins, and substrate/tracer to the coiled reactors. This presents a major challenge, especially due to the large surface-to-volume ratio of reactors in HRS (see earlier). Often, reagents that are classically employed to counter unspecific binding are used to this end, such as enzyme-linked immunosorbent assay (ELISA) blocking reagent, PEG, or lactic acid. These have the additional advantage of preventing unspecific binding to the target protein, but their influence on protein stability has to be carefully monitored.

Detection of substrate conversion or tracer competition is most often achieved with a chromatographic fluorescence detector. When using a spectrophotometric LC detector, the main source of interference is air bubbles in the detector cell. These can be effectively prevented by creating backpressure at the detector outlet, either by sufficiently narrow I.D. waste tubing or by using commercially available backpressure regulators. Herein, the pressure limits of the detector cell as well as of the superloops have to be carefully observed to prevent damage to the system. However, usually a pressure excess of only 1–4 bar is sufficient.

In most cases, FLD (usually of a tracer) is preferred because it is superior to other detectors in combining sensitivity, selectivity, and linearity, while still being applicable to a large number of molecules. Sensitivity is very important as it is directly related to reducing reagent consumption. Selectivity is a prerequisite for any homogeneous assay, as the detection should not be disturbed by other bioassay constituents present [34]. For example, in an activity assay with fluorescence product detection, the substrate should not show similar fluorescence properties [12]. Linearity is also important, but much more likely a hurdle for MS-based bioassay detection than for FLD- or UV/vis-based platforms. The linear range of the detector should extend at least over a reasonable part of the pseudolinear range of the bioassay response (see Section 11.3.1.2). When no analyte is eluting, the stable baseline of tracer binding or substrate conversion can be considered as 100% bioassay response, although both tracer binding and substrate conversion are usually far from complete. A ligand or inhibitor decreases binding or conversion, resulting in a bioassay signal. We should remind the reader here that the direction of the signal, meaning increase or decrease of the detector response, depends on the detected species (see Figure 11.4). An allosteric modulator may also increase binding or substrate conversion, thus resulting in a bioassay signal with the opposite sign than the ligand or inhibitor. For a full binder or inhibitor, the pseudolinear range is usually between 10% and 90% bioassay response. Following this logic, a positive allosteric modulator yields a bioassay response above 100%. However, the discussion of allosteric modulators in the context of HRS is somewhat theoretical, as they have not been measured by HRS yet. The pseudolinear range depends on the amount of binding/conversion at 100% bioassay response. In bioaffinity assays, 50% tracer binding is often employed, which would result in a pseudolinear range between 110% and 190% for a positive allosteric modulator. However, positive allosteric modulators are not all equally effective, so often 200% will not be the target value. In bioactivity assays, the conversion ratio is traditionally low (10% or even less), in order to ensure the applicability of Michaelis–Menten kinetics. Therefore, the bioassay response could exceed 200%, but that might also compromise the applicability of Michaelis–Menten kinetics. Consequently, matching the detector linear range to the detection of 5–200% bioassay response should provide good IC_{50} curves for most ligands, inhibitors, and allosteric efforts. Extreme cases of agonism are probably better measured in a separate assay.

11.2.4
MS Detection

We have dwelled briefly on chemical detection (or structure elucidation) by MS in Section 11.1.1.2. In this section, we take a closer look at what aspects are (especially) important in the context of HRS. We first introduce instrumentation and then data acquisition and analysis strategies that have been combined with HRS.

The ionization methods of choice in HRS platforms are ESI [27] and, in some cases, atmospheric pressure chemical ionization (APCI) [15], because their coupling to LC is relatively straightforward and because they allow the determination of the protonated (or deprotonated) molecules from (complex) mixtures. The general bias toward the use of positive ionization mode in pharmaceutical MS is reflected in HRS, but examples with negative ionization exist as well [15].

Mainly two developments have enabled structure elucidation by MS. Firstly, the high mass accuracy of the high-resolution mass analyzers, in HRS currently mostly TOF (see Section 11.1.1.2 and Tables 11.1 and 11.2), allows the determination of the elemental composition of ions [53]. Secondly, fragmentation reveals functional groups and residues by deducing specific neutral losses and allowing interpretation of their structural connections. When using soft ionization techniques, fragmentation of the protonated molecule can be achieved by tandem mass spectrometry (MS–MS). MS–MS instruments employed in HRS are either q-TOF [11] or IT-TOF combinations, the latter providing IT MS^n [17]. IT instruments were also frequently used [16]. Obviously, the combination of high mass accuracy and fragmentation makes the HR–MS–MS instruments ideally suited for structure elucidation.

The hyphenation of MS to LC, which is part of HRS systems, does not only enable the detection of individual isobaric and isomeric compounds, but it also significantly reduces the occurrence of ionization suppression issues in atmospheric pressure ionization (API). Fortunately, modern RP-LC and API-MS are very compatible. In addition, LC–HR-MS can resolve isobaric compounds that have not been chromatographically separated, but issues with isomers and ionization suppression cannot be elevated by HR-MS alone. In some cases, fragmentation can help in the discrimination of isomers.

MS–MS and IT MS^n can be applied with various data acquisition strategies, but only product-ion analysis mode has been used for HRS. Herein, a precursor ion is selected, fragmented, and a mass spectrum of the fragments collected. The reason is probably that it is the most effective and unbiased method for the screening of unknown compounds, which is one of the key applications of HRS. This is certainly true nowadays as product-ion analysis is becoming increasingly fast and sensitive, for example, in q-TOF instruments [54]. Data-dependent acquisition is a powerful strategy, frequently used in HRS for the monitoring of unknowns [23]. First, a full spectrum without fragmentation is collected and processed in real time. Based on preselected criteria, the software selects the most abundant ions to be fragmented and collects their MS–MS spectra. In ITs, this can even be an iterative process (MS^n). In this way, the process of collecting fragmentation data for structure elucidation can be greatly automated.

The challenging analysis of the collected data is discussed in Section 11.3.1.4.

11.3
Data Analysis

11.3.1
Differences between HRS and HTS

11.3.1.1 Influence of Shorter Incubation Times

In HTS, incubation times are chiefly governed by enzyme stability and work flow practicality. Several hours up to a day are not exceptional, especially when cell-based assays are involved. In contrast, incubation time in HRS is generally limited to a maximum of ~5 min by physical peak broadening (see Section 11.2.3). As a consequence, incubation times are often much shorter in on-line bioassays than in HTS. Still, HRS assays achieve comparable assay quality at the same or even lower enzyme concentrations due to advantages in mixing and temperature control, as discussed earlier (see Section 11.2.3), but also because of advantages in data analysis (see Section 11.3.1.2). For pure compounds, especially when larger sample numbers are involved, HTS can more than compensate for the longer incubation times, because parallel incubation is effortlessly possible. We can attempt an indirect comparison of throughput between HRS and HTS for mixture analysis using EDA, which is in many ways comparable to HTS, except that it is able to deal with complex mixtures. Most EDA protocols involve preparative separation, which then becomes the bottleneck for multiplexing of samples. Owing to this limitation of EDA, HRS should show at least a similar throughput.

One of the consequences of the short incubation times in HRS is that competition between the analytes and the tracer or substrate in on-line bioassays is more likely to be under kinetic rather than under thermodynamic control (compare Equation 11.2). This can have an influence on the determined IC_{50} values and the sensitivity of the assay. In the most common two-superloop setup (see Section 11.2.3), where the protein is first incubated alone with the column effluent, the association kinetics (k_{on}) of the analyte with the target is fast, because no competition with the tracer or substrate is experienced in the first incubation step. Most binders/inhibitors reach a significant amount of binding in 30 s of pre-incubation. The sensitivity of the assay is therefore only negatively influenced for analytes with a very slow k_{on}. In the second step, the tracer can only associate with free protein or protein that is freed by dissociation of the analyte. This creates a bias toward the analyte over the tracer or substrate as compared to competition under equilibrium conditions. Especially, binders or inhibitors with a slow dissociation kinetic (k_{off}) will not dissociate to a large extent before the detection, thus largely preventing competition by the tracer. However, as current opinion sees a slow k_{off} as a favorable property for lead compounds, this bias seems acceptable [55]. A less common variant of the two-superloop setup mixes protein, substrate/tracer, and analyte at the same time, so that competition starts immediately and is unbiased [56]. However, this setup probably does not reach equilibrium either. Therefore, the ratio of tracer to analyte binding will be governed chiefly by the ratio of the respective association kinetics (k_{on}). In a single-superloop setup, protein and tracer are mixed in the superloop before the analyte competition starts. Therefore, the k_{off} of the tracer will be more influential than under equilibrium conditions. Although this means that there is a bias in the absolute IC_{50} values, as this bias is the same for all analytes, relative to each other the IC_{50} values should be unbiased. However, slow k_{on} binders will be even more discriminated than in both two-superloop setups.

Thus, all three setups are biased, but as k_{off} is pharmacologically seen as the more important parameter, the first two-superloop setup produces the most acceptable bias. In addition, the bias toward analyte binding in the first two-superloop setup also leads to a significantly higher sensitivity. Consequently, the single-superloop setup shows the lowest sensitivity, because it is biased toward tracer binding.

$$K_D = \frac{k_{off}}{k_{on}} \qquad (11.2)$$

Because of the limitations in incubation time, conversion of assay principles requiring very long incubation in HTS to an on-line format may not always be feasible. In such a case, an at-line format may be attractive. In contrast to the on-line format, where the bioassay reaction is directly detected in a chromatographic detector, at-line formats use microscale fractionation to transfer the bioassay into well plates [1]. In some at-line assays, the HPLC eluent is directly fractionated [57]. The parallel chemical detection, for example, by MS, and the high time resolution of the fractionation still distinguish these at-line assays from EDA, which mainly relies on multiple orthogonal separation methods. All advantages of the on-line format over HTS with respect to mixing and the closed system are lost, but the advantages for data analysis are retained, because a bioassay chromatogram can be reconstructed. One advantage is that the fractions can be dried and reconstituted eliminating organic modifier influence. However, analyte degradation has to be avoided and small amounts of organic modifier may be needed for reconstitution of lipophilic analytes. Other at-line assays add the bioassay reagents before fractionation, thus retaining the efficient mixing of HRS [56]. This has even been demonstrated for living bacteria [58]. The most important advantage of all at-line formats is that they set a limit to post-column band broadening, by restricting diffusion to the boundaries of the fraction. In this respect, it is very important that each fraction contain only a small time window of the chromatogram (2.5–12 s), leading to a sufficient data point frequency in the reconstructed bioassay chromatogram. This allows comparison of separation and peak shapes between the MS and bioassay measurement. As a consequence, the at-line format can be used in combination with long incubation times (up to 24 h were reported) without a loss in chromatographic resolution for the bioassay measurement. An additional advantage is the good compatibility with standard HTS assays. However, at sufficient resolution one chromatogram can easily fill a whole well plate. Therefore, throughput is lower than in on-line formats, unless robotics is employed for post-fractionation liquid handling.

11.3.1.2 The Assay Signal

One difference in data analysis between HRS and HTS is the calculation of the signal that is directly linked to the origin of the reference. In HTS, the reference is represented by the control sample, which gives the assay readout without any observable effect. Replicates produce a mean and standard deviation (SD) of the control, μ_C and σ_C, respectively. This is essentially an external reference. In another experiment, mean and SD of the sample are obtained, μ_S and σ_S, respectively. The signal is then calculated by subtracting the means $|\mu_S - \mu_C|$. In HRS, no control sample needs to be measured, as the baseline of the on-line bioassay readout provides an easily accessible and highly relevant internal reference for each individual sample. Thus, the signal in HRS is identified as the peak height ($|\mu_H|$) as shown in Figure 11.8. Therefore, contrary to HTS, changes in the background do not influence the signal, which results in a lower SD of the peak height (σ_H) compared to using σ_S. However, changes in the assay window are equally uncorrected. The variation of the reference, also called the *noise*, is usually calculated differently in HRS than in HTS (see σ_C). Instead of calculating the SD of the baseline, the maximum deviation of the baseline signal is calculated, which is more easily accessible. This is halved to account for positive and negative deviation from the baseline. Although the noise is slightly overestimated in this way, this method is more robust in dealing with baseline changes due to gradient elution. This discussion is highly relevant for the assay quality and therefore we come back to it in Section 11.3.2 where we also need the variables introduced here for Equations 11.6 and 11.7.

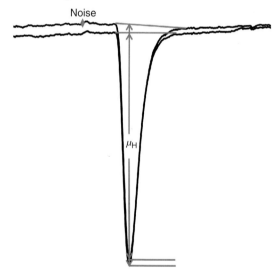

Figure 11.8 Peak height ($|\mu_H|$) and noise in the bioassay. These two values are applied in HRS instead of the external control used in HTS. The picture clearly shows the benefit of this approach. While the absolute value of the signal changes due to the shift of the baseline, $|\mu_H|$ is unaffected in the replicate. This correction is only possible in HRS, because the detection of the baseline allows the extraction of an internal control value.

In bioassays, there is always a (spectrometric) signal that indirectly detects the underlying reaction stage. For correct measurement of this signal, the chromatographic detector has to be within its linear range (see Section 11.2.3). However, the resulting concentration–signal dependence is not linear due to the underlying biochemical interaction. Direct detection of biocatalytical conversion of the analyte follows Michaelis–Menten kinetics in most cases. The much more common detection of affinity or inhibition results in the typical sigmoidal dose–response curves when plotting $|\mu_H|$ against the logarithm of the concentration (if the detector response is linear; see Figure 11.9). However, the influence is not limited to the peak height. In any chromatographic peak, the analyte molecules are distributed according to a Gaussian distribution, because of the longitudinal diffusion of the initially homogeneous injection plug (V_i), so each data point has its own analyte concentration. In the bioassay detection, this results in a distortion of the ideal Gaussian shape. Fortunately, the sigmoidal curve has a significant pseudolinear part, usually between 10% and 90% of the assay window (see Figure 11.9), thus allowing the calculation of important peak parameters like the full width at half maximum (FWHM) (see Figure 11.11). However, if $|\mu_H|$ is outside this pseudo-linear range, the consequences are much the same as when a classical signal is outside the linear range. For example, if $|\mu_H|$ is above 90% of the assay window, the FWHM increases as the flanks of the peak are still growing linearly while the peak is not, the so-called overloading effect. We also referred to this earlier as biochemical band broadening.

While the use of signal intensity (height) is natural to a pharmacologist, as HTS and other batch assays simply do not show peaks, separation scientists prefer working with peak areas, as these are less sensitive to fluctuations in the parameters controlling the separation than the peak heights are. The most prominent assumption in classical chromatographic detection is the presence of an underlying (pseudo-)linear phenomenon. However, in order to obtain reliable IC$_{50}$ values, we would like to acquire sufficient data points, in the nonlinear parts of the

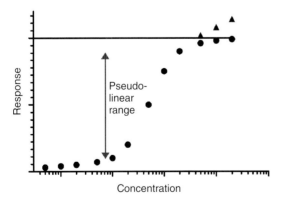

Figure 11.9 Theoretical dose–response curve. The dots represent the curve that would be expected when using peak height as readout. The triangles at high concentrations present the systematic *y*-value error that would occur when using peak area instead of peak height.

sigmoidal dose–response curve as well, as this increases the quality of the data fitting. Thus acquiring data to determine IC_{50} values automatically violates the detection–linearity assumption. As a consequence, employing the peak area of the bioassay signal would create a relatively large, systematic y-error in the asymptotic region of the dose–response curve (see Figure 11.9). Owing to the fact that the curve is asymptotic toward a y-value, the x-value error caused through the dilution calculation by distortion of the FWHM is far less influential (see Section 11.3.1.3). More importantly, while the y-value error in the peak area increases continuously with increasing injected concentration, the x-value error in the calculated concentration becomes increasingly meaningless with increasing x-values. In addition, the FWHM of the pseudolinear range might be used for the nonlinear range, thus reducing the x-error for the most affected data points close to the pseudolinear range. The determination of IC_{50} values from the dose–response curve is discussed in more detail in the next section.

11.3.1.3 Dilution Calculations

As explained earlier, there are two types of dilution, D_M and D_C, whose product forms the total dilution D. While D_M represents essentially the same step as the mixing of the analyte solution with the assay reagents necessary in HTS, D_C is unique to HRS. D_M is simply calculated by dividing the total flow rate through the bioassay detector (u_{Bio}) by the flow rate at which the analyte containing the effluent enters the bioassay (u_E) (for variables also see Figure 11.5). D_C results from the transformation of the injection volume (V_i) into the near-Gaussian distribution (see Figure 11.10) and can therefore be calculated by Equation 11.3. With D_M and D_C, we can calculate the final concentration (c_F) at μ_H from the injected concentration (c_i), using Equation 11.5. It has to be remembered, however, that

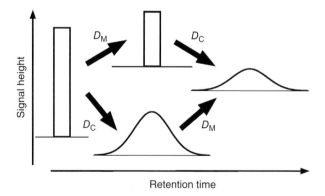

Figure 11.10 Schematic representation of the influence of the dilutions on the peak parameters. During the mixing, dilution of the concentration, and accordingly of the peak height, is reduced by the value of D_C (depicted as a factor of 2). D_M results from diffusion and convection along the flow direction. These lead to transformation of the homogeneous injection plug into a concentration gradient described (ideally) by a Gaussian distribution.

Equation 11.5 is derived for an ideal Gaussian distribution, which means that the results grow less accurate the more the shape of the bioassay peak deviates from the ideal Gaussian distribution (see Section 11.3.1.2). Therefore, this concept should be further developed in the future to include deviations from the ideal distribution, for example, by including the peak tailing factor in a revised equation. In the publication that introduced these calculations, u was identified with u_{BIO} [10]. However, the initial width of the injection plug is determined by u_{LC}. If we had, for example, simple UV detection and no split, we would have to use u_{LC} for the calculation of D_C. The splitting and mixing theoretically cannot influence FWHM and therefore D_C should not change upon entering the bioassay. Even though there is band broadening while the analyte passes through the bioassay, which changes both variables, this does not result from the change in flow rate. Although this calculation is, among other things, extremely useful for the calculation of IC_{50} values, it might be prudent to determine D_C experimentally, until these calculations have been experimentally supported.

$$D_C = \frac{FWHM}{2} \times \frac{u}{V_i} \times \sqrt{\frac{\pi}{\ln 2}} \tag{11.3}$$

$$D_M = \frac{u_{Bio}}{u_E} \tag{11.4}$$

$$c_F = \frac{c_i}{D_M \times D_C} \tag{11.5}$$

11.3.1.4 MS Structure Elucidation

Most often HRS is used for the analysis of mixtures of related molecules, that is, molecules that have the same or similar core structures. They can be derived, for example, from chemical or biochemical diversification approaches [3] or they can be different molecules from one class or type of natural products [24, 30]. The structure of individual compounds is identified by a combination of accurate mass measurements and collision-induced dissociation (CID) fragmentation experiments. As HRS often deals with unknown compounds, data-dependent acquisition (involving automatic switching between MS and MS^n mode) is of key importance. MS instrument settings should be optimized using an available model compound from within the compound class targeted at expected concentrations. However, if the bioassay is MS-based, the sensitive detection of the reactants often takes precedence. If data-dependent precursor selection fails, the afflicted experiments have to be repeated with manual precursor selection after data processing of the MS^1 data. Peptides and small proteins might be elucidated (partly) by database comparison [24] or by top−down proteomics approaches [26]. For small molecules, the first step is deducing the elemental composition of the parent ion. To this end, a range of elemental compositions preselected from prior knowledge of the analyte class, modification pathways, and so on, is matched to the accurate mass. On the basis of the stringency in matching accurate and exact mass and the range of elemental compositions allowed, one or more elemental compositions are yielded. For the fragmentation analysis, the elemental compositions can be limited to that of the precursor ion. Specific neutral losses can then be used to identify functional groups and elucidate their structural connection (see later). If several possibilities for the precursor ion exist, some wrong elemental compositions might be uncovered through proof by contradiction, that is, if no sensible neutral losses connect the precursor ion with the product ions. Software procedures are under development to assist in these calculations [59].

There are a number of data interpretation methods that were specifically developed for the analysis of unknown mixtures of related compounds such as they occur, for example, in metabolism studies or natural extracts. The profile group approach [60, 61] assumes that the identity of one compound in the mixture is known and that this compound is related to the other compounds in the mixture by simple chemical modifications, for example, a substrate and its metabolites [61] or an active ingredient and its multiple products from a shotgun synthesis approach [58]. Firstly, the fragmentation of the standard (parent drug) is thoroughly analyzed with the MS instrument intended for the study, the data are interpreted, and a fragmentation tree is established. A number of structurally informative fragment ions are selected as profile groups. Then, the fragmentation trees of all analytes are measured and compared to that of the standard, following mainly the profile groups. The difference in elemental composition between the analyte and the standard is found in some profile groups, which is evidence that the molecule is modified in the corresponding part of the molecule. The difference depends on the modification, for example, oxygenation or dehydrogenation result in $+O$ ($+15.995$ Da) or $-H_2$ (-2.016 Da), respectively. Thus, the site of modification can be narrowed down. In addition, changes in (relative) retention time in RP-LC can be used to distinguish the isomeric modifications of common metabolic reactions, such as hydroxylation and N-oxidation [62]. There are also a large number of software solutions available that predict either fragmentation of analytes or a specific set of modifications to a template, such as metabolic reactions. Although these can speed up the data interpretation process by providing inspiration and probabilities, their prediction has to be critically examined as they are far from fail-safe.

When analyzing bioactive peptides, proteomics tools can be applied for structure elucidation. The mass of the precursor and possibly of fragment ions are extracted and used for database searches [8, 24, 25]. In contrast to most proteomics applications, the native peptides (no reduction/alkylation) are measured in HRS. Therefore, the detected masses possibly have to be corrected for the missing modifications before the database search [24].

Structure elucidation by MS is an integral part of HRS nowadays. HR-MS instruments are preferred, as accurate analyte masses and superior full-spectrum sensitivity can be obtained. The powerful combination of full-spectrum acquisition, data-dependent acquisition and the profile group approach or proteomics strategies for structure analysis of mixtures of unknown compounds presents a milestone in the maturation of HRS. In case the bioassay is also based on MS detection, the combination of structure and affinity measurements within one MS instrument offers an attractive synergy [32].

11.3.1.5 Structure–Affinity Matching

Matching the structure and the affinity of a molecule by matching its MS and its bioaffinity response starts by aligning the extracted ion currents (EICs) of the MS detection with the bioaffinity chromatogram, for example, from FLD. As both chromatograms result from the same chromatographic run, one chromatogram can simply be time-adjusted for the stable retention time delay (Δ_r), which is a consequence of differences in post-column void volume. This is depicted in Figure 11.11. Variations in flow rate could cause Δ_r to vary, but this variation is not easy to determine. Therefore, the SD of the total retention time could be used as a boundary [23]. All MS peaks with retention times ($t_{r,MS}$) within this SD around the bioaffinity peak with $t_{r,BIO}$ are accepted as peaks potentially having bioaffinity. This is evidently an overcautious approach, especially considering all measures taken to ensure flow stability in the bioassay, because the analytes usually spend significantly more time in the separation part than post column. Additionally, variations in temperature and solvent composition influence only the chromatographic retention time and not the additional time spent post column. Overall, other factors are usually more crucial for an accurate peak matching (see later). In most HRS setups, the FWHM of the bioaffinity peaks is

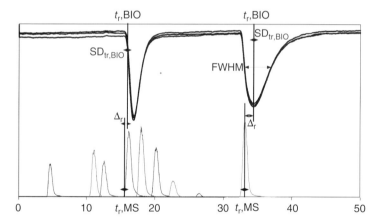

Figure 11.11 An unmatched HRS chromatogram. Chemical detection via MS (bottom; several EICs) and bioassay detection via FLD (top; triplicates). First, the chromatograms are matched by adjusting the retention times of one of the traces by the stable retention time difference (Δ_r, see text). Then, the candidate masses for the bioactive compound are identified, considering all MS peaks with a retention time ($t_{r,MS}$) within a margin around the bioassay retention time ($t_{r,BIO}$). In most cases, the deviation in peak width at different concentrations Δ_w is the largest contribution to matching insecurity and consequently $0.5 \cdot \Delta_w$ is identified as the margin. The deviation of t_r is almost identical for MS and bioassay, as both detect the same separation. Therefore, the deviation of Δ_r is usually insignificant.

larger than the SD of t_r, let alone the deviation of Δ_r. Especially when dealing with overloading (see Section 11.3.1.2) in either of the responses, peak maxima and consequently t_r values can be difficult to identify. Therefore, we often mark both MS and bioaffinity peaks at their onset rather than at their top, because the onset is better defined. Nonetheless, this may result in different marks for the same compound depending on its concentration and thus in unstable retention time differences for different compounds, because the onset is further away from t_r in a broad peak than it is in a sharp peak. Consequently, the variation in peak width (Δ_w) is another criterion for the accuracy in peak matching and overloaded peaks have to be even more carefully assigned. If peaks are assigned on the onset, it is more logical to use the width at 10% rather than FWHM to determine Δ_w. Δ_w can be deduced, for example, from IC_{50} measurement curves during the validation and then $0.5 \cdot \Delta_w$ can be used in the same way as described for the SD of t_r defined earlier. The alignment of the two chromatograms and the matching via Δ_r can also be tested and possibly adjusted if a known affinity compound is present to function as a marker. However, co-elution of other affinity compounds and/or overloading of either the MS or the bioaffinity response can influence this matching as well.

In HRS setups where the same MS is used for structure and bioassay detection, Δ_r is zero by definition. However, the uncertainty due to Δ_w fully applies.

11.3.2
Validation

Validation is the final stage of every assay development, not excluding HRS systems. In HRS, we usually start by validation of the bioassay and add the MS (structure elucidation) later for further validation of the whole setup. A good first validation step is the measurement of a number of specific assay parameters, which give a quick first quality assessment of the assay and if necessary show potential for improvement. Simple measurements, such as those depicted in Figure 11.12, reveal all contributions to the background signal as well as the assay window [7, 12]. An assay window (for calculation see later) with a signal-to-noise ratio (S/N) between 50 and 100 is an indication of a good-quality bioassay, because it allows measurement (S/N $\geq$ 3) of the flat part of the sigmoidal curve between 5% and 10% signal and quantitation (S/N $\geq$ 9) of (most of) the pseudolinear part of the sigmoidal curve between 10% and 90% signal [12, 19]. The former greatly enhances sensitivity of the bioassay; the latter is essential to obtain good-quality IC_{50} values. If sufficient S/N is not yet reached, the individual background contributions provide a clue to which part still has to be optimized. For example, if by far the largest contribution comes from the tracer, decreasing the tracer concentration or increasing the enzyme concentration is an obvious starting point. It should be mentioned here that sometimes lower S/N for the assay window are preferred due to reagent consumption or other limitations, accepting the resulting loss in assay quality [48, 63].

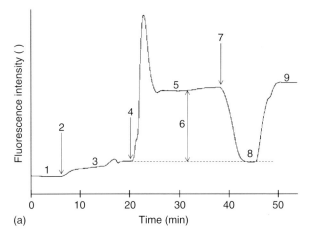

(a)

Figure 11.12 Assay quality parameters. A typical initial validation for a two-superloop bioactivity assay based on detection of a fluorescent product is depicted [12]. A similar figure can be obtained for other bioactivity and also bioaffinity assays. In the beginning, both superloops contain the assay buffer (1). After addition of the substrate (or tracer) to the second superloop (2), the contribution of the substrate (or tracer) to the background can be assessed (3). The substrate (or tracer) itself may show weak fluorescence and for bioactivity assays autolysis of the substrate may contribute as well. Then, the enzyme is added to the first superloop (4) and after some equilibration time, the stable baseline response is reached (5). Under the (often) reasonable assumption that the enzyme does not contribute to the fluorescent background (8), the assay window (6) can be estimated as the difference between baseline (5) and substrate (or tracer) background (3). By putting the reaction coils on ice (7), the biocatalytic reaction is stopped. If the signal goes back to the level of the substrate background, this confirms, firstly, that the baseline response is indeed caused by the biocatalytic reaction; secondly, that there is no contribution of the enzyme to the fluorescence background; and, thirdly, the estimation of the assay window. However, binding is not affected in the same way as catalysis, so this specific experimental estimation of the assay window is not accessible in bioaffinity assays. When the assay is brought back to the assay temperature, the baseline response should recover (9).

We quickly explain how two important benchmarks of bioassay quality can be calculated in an HRS setup: the assay window and the Z' factor [64]. For both of them, we need μ_H at full inhibition or binding (μ_{Hmax}) as well as the baseline noise. For activity assays, μ_{Hmax} can often be found by comparing the signal at the incubation temperature and at $0\,°C$ (where virtually no enzymatic reaction is observed) as shown in Figure 11.12. Injection of an analyte concentration that results in complete inhibition or binding is a way to find μ_{Hmax}, which works for both activity and affinity assays. To calculate the S/N of the assay window, we simply divide $|\mu_{Hmax}|$ by the noise. The Z' factor as defined for HTS is given in Equation 11.6 [64]. Like the S/N, it reflects the ratio between the noise and the assay window. However, in contrast to S/N, the Z' factor accounts for the deviation in both the signal and the baseline, and thus for the likelihood of both false negatives and false positives. To measure the Z' factor, the mean (μ), and SD of positive (subscript C+; maximum signal) and negative (subscript C–; no signal) control have to be determined.

$$Z' = 1 - \frac{(3 \cdot SD_{C+} + 3 \cdot SD_{C-})}{|\mu_{C+} - \mu_{C-}|} \tag{11.6}$$

With the adjustments discussed in Section 11.3.1.2, we derived the Z'_{Chrom} factor for HRS in Equation 11.7 [19]. $\mu_{C+} - \mu_{C-}$ is identified with μ_{Hmax} and, consequentially, SD_{C+} changes to SD_{Hmax}. As discussed, the baseline represents the internal (negative) control and therefore SD_{C-} becomes the noise.

$$Z'_{Chrom} = 1 - \frac{(3 \cdot SD_{H\,max} + 3 \cdot noise)}{|\mu_{H\,max}|} \tag{11.7}$$

HRS platforms have been successfully used to quantify the affinity/activity of analytes by measuring their IC_{50} values. However, this is not a core strength of HRS platforms as to this end they still rely on purified compounds as much as HTS platforms, due to the as yet missing capacity of HRS platforms to quantify unknown compounds. Nonetheless, it is valuable to discuss the measurement of IC_{50} values in HRS platforms. In the context of validation, good sigmoidal curves confirm that signal changes observed are indeed due to the interaction of the analytes with the target and do not result from (nonspecific) interferences or artifacts. If literature values measured with the same assay principle, are available, the actual IC_{50} values provide a validation. However, great caution is advised when comparing IC_{50} values as there are many pitfalls. Even seemingly identical assay formats can show major differences. A classic example is the use of different substrates in an activity assay, but sometimes changes in buffer constituents might impact the interaction of analyte and target as well. Transforming IC_{50} values into K_i values [65, 66] can account for some differences between assays and thus deliver a more solid basis for quantitative dose–response comparison. However, K_i values are not always reported and often not all the necessary factors are available, in order to convert reported IC_{50} values to K_i values. Also, K_i values certainly cannot account for all differences between assays. Additionally, kinetics might impact IC_{50} values when comparing assays with largely different incubation times (see Section 11.3.1.1) [19, 52]. However, these are by no means issues unique to HRS, although the kinetics issue is more pronounced than in HTS. In terms of application, measuring IC_{50} values with HRS sometimes offers an advantage as well. Due to the influences discussed in Section 11.3.1.2, HRS assays are often either of higher quality at the same protein concentration or necessitate a lower protein concentration to achieve the same quality. While the influence of the former on IC_{50} value quality is evident, the latter allows a better distinction of high-affinity binder/inhibitors [19]. The reason is that half of the protein concentration (c_p) is the theoretical lower limit of measurable IC_{50} values, because below that limit the enzyme is simply titrated. Assume, for example, that we have c_p values of 90 and 450 nM in an HRS assay and in an HTS assay, respectively, and an analyte with an IC_{50} of 45 nM. If we add the analyte at 45 nM, it interacts with 50% of the protein in the HRS assay, thus correctly giving an IC_{50} value of 45 nM. Note that also any tested analyte with a higher affinity yields this value. If we add the same analyte concentration to the HTS assay, it only occupies 10% of the protein-binding sites. Only with 225 nM of the analyte, we reach 50% interaction; thus the apparent IC_{50} in the HTS is 225 nM ($0.5 \cdot c_p$). We should add that this advantage is more often observed in affinity assays, because in activity assays c_p can be reduced by increasing the incubation time.

Similarly to HTS, IC_{50} values are determined by measuring a series of concentrations and fitting the resulting dose–response curve (see Figure 11.9). To this end, the individual concentrations of the pure compound are injected into the HRS platform. This can be done either in flow injection analysis (FIA) mode accounting for the injection peak by subtraction of a blank injection or by isocratic elution with minimal retention, thus separating the injection peak. Then, the critical step is the calculation of c_F by dividing the injected concentration c_i by the two dilution events (see Section 11.3.1.3, Equation 11.5). The models and algorithms used for fitting the dose–response curve and calculating the IC_{50} are the same as in HTS. Theoretically, it is possible to deduce individual IC_{50} values by injecting a dilution series of a mixture. The repeatability of IC_{50} values in HRS systems seems to be quite good, as can be judged by almost identical values obtained for TAK-715 in different studies (81 nM (71–92 nM) [19] and 71 nM (36–140 nM) [20]).

A good approach to test the complete HRS platform, and especially the structure–affinity correlation, is the analysis of an artificial mixture of pure compounds, preferably containing both effectors (inhibitors, ligands, etc.) and nonbinders [19, 48]. The first indication is that the correct number of compounds is detected in the MS response and that only as many signals as effectors are detected in the bioassay response. Then, the MS and bioassay responses are matched blindly and the compounds are identified from the MS response. Only the effectors should show bioaffinity.

11.4
Conclusions and Perspectives

11.4.1
The Relation of On-line Post-Column Assays to Other Formats

In this chapter, we mostly used HTS as a reference, for example, to work out the unique properties of on-line post-column HRS assays, because it is the most common approach. Therefore, it is useful to dwell shortly on the differences with other separation-based formats.

Post-column assays are open to both bioaffinity and biochemical measurements while pre-column assays rely entirely on bioaffinity (see, for example, Chapters 8–10) [67]. In addition, saturation of the binding assay is less likely to occur in post-column assays due to the separation before analyte–target interaction. In general, post-column assays require more specific affinity measurement (see Section 11.1.1.1), which, compared to pre-column assays, has advantages in reduced nonspecific binding interferences, but disadvantages for novel targets or novel binding sites. Additionally, pre-column assays do not allow the measurement of analytes without biological effect as those are discarded before the separation step. This reduces information content, but also interferences.

In on-line assays, the (virtually) closed system and the retention of a peak shape for the effect signal result in many advantages, compared to HTS (see Sections 11.2.4 and 11.3.1.2). The (hybrid) at-line methods lose the former advantages, but retain the latter [1]. At-line assays need more hands-on time and lose the excellent temperature control of their on-line counterparts, but can be made to retain the excellent mixing properties. Because they decouple separation and bioassay, at-line assays allow for extended incubation time with minor compromises in chromatographic resolution (see Section 11.3.1.1). Mainly due to this advantage, at-line assays can be used for functional assays, a mode that has not been demonstrated for on-line assays.

11.4.2
Trends in High-Resolution Screening

Over the years, many variations of the typical HRS platform (see Figure 11.1) have been described. We take a look at interesting solutions and key developments in this section.

While the typical HRS platform provides high data density and quality for interaction with a single protein, physiologically and pharmacologically the relative affinity/activity toward multiple proteins is often more relevant for the actions of an analyte. The assessment of this selectivity has been introduced into HRS platforms by the addition of a second on-line bioassay (see Figure 11.13a) [17]. Because the bioassay responses result from the same separation, relative affinity of unknown analytes for the two targets can be estimated without prior quantification. This dual affinity setup has found another interesting application in FLD-based HRS bioassays. If the second line is operated with the same enzyme, but without the tracer, interferences due to auto-fluorescence or even fluorescence enhancement of the analytes can be quantitatively detected and the bioassay response corrected for the interference [18]. The pharmacological selectivity assessed by the two-enzyme dual-HRS assay provides clues to which compound might finally possess a better therapeutic window (difference between dose resulting in therapeutic effect and dose causing serious side effects). So far, HRS has been used to measure selectivity of closely related enzymes such as the ER α and β [17]. It may be interesting in the future to combine the pharmacological assessment of the target with the toxicological assessment of notorious off-targets like the cytochromes P450 [40] in parallel assays to gather more hints on the potential therapeutic window. Although HRS is certainly far away from tackling whole enzyme panels, such as are commercially available for protein kinases, for example, combining multiple assays in one HRS platform is certainly a valuable long-term goal. This will require further integration and possibly simplification, but especially miniaturization (see later).

On-line pre-column synthesis presents another intriguing trend in HRS platforms aimed at the further integration of drug discovery/development

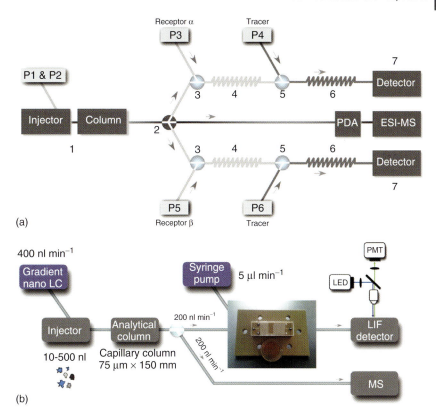

Figure 11.13 Trends in HRS. (a) An HRS platform for the screening of mixtures from biocatalytic lead diversification for estrogen receptors is shown [17]. The remarkable feature of this setup is that it combines two parallel bioassays: one for the α- and one for the β-isoform. This allows a direct assessment of the selectivity of the analytes for two different isoforms. (b) The depicted HRS platform was miniaturized in all relevant aspects. The separation of peptides and small molecules is achieved in a nano-LC system, chemical detection is accomplished by nano-ESI-MS and the bioassay featured a microfluidic chip as reactor and laser-induced fluorescence (LIF) to achieve sufficiently sensitive detection in a very small-diameter and small-volume detector cell. Because of the possibility to inject very small sample amounts, the system was successfully applied to the screening of mass-limited animal venom extracts [24, 25].

workflows. It is enabled by the mixture analysis feature of HRS. By combining a novel synthetic and a novel analytical approach, namely, flow chemistry and HRS, on-line pre-column synthesis promises more efficient drug discovery not in the least by stimulating the integration of synthetic, analytical, and pharmaceutical sciences. In practice, an initial structure–activity relationship is gained by injection of a single compound that presents an excellent starting point for further pharmacological and toxicological evaluation. This is achieved without the need for laborious synthesis optimization and purification efforts for compounds that are not pharmacologically interesting. The two systems that have been developed employ two different out-of-the-box synthetic approaches: biocatalysis and electrochemical conversion (EC). Biocatalysis offers superior regio- and stereoselectivity, but requires purification. This can be solved by the integration of an SPE (solid phase extraction) step into the HRS platform with on-line biocatalysis, increasing the complexity of the system to such an extent that MS analysis was performed at-line as LC−MS/MS [21, 22]. The prominent feature of EC with regard to HRS platforms is the relatively simple matrix required for the reaction. This allows EC to be integrated directly into HRS systems without any additional purification steps [3, 23]. One of the strengths of this approach lies in the detection of reactive/unstable products.

A powerful addition to HRS platforms would be the on-line quantitation of unknown mixtures without an individual molecular standard for each analyte. In combination with structure and affinity/activity analysis, biochemical/pharmaceutical significance of the HRS data would be greatly enhanced, for example, by direct delivery of quantitative structure−activity relationships (QSARs). However, this is extremely challenging. Although UV/vis and evaporative light-scattering detection (ELSD) are sometimes claimed to show the same response for similar molecules, the influence of small structural modifications on absorption and aggregation, respectively, can be quite significant. Thus, so far, quantitation has to be achieved by the use of standards as demonstrated by van Elswijk *et al.* [8], who used MS for quantitation in addition to structure elucidation. The combination of quantitation and structure elucidation from precursor-ion analysis combining the strength of HR-(tandem)MS in structure elucidation with the quantitative interpretation of CID reactions or neutral losses has received much attention recently [68]. In the future, detection techniques based on atomic rather than molecular properties may present a way to still achieve the formulated goal. Inductively coupled plasma mass spectrometry (ICP-MS) has been implemented successfully for the quantitation of pharmaceuticals containing metal ions [69] or other heavy atoms, such as bromine and iodine [70], without molecular standards. With advances in sensitivity for a wider array of atoms, especially chlorine and sulfur, ICP-MS may become a prime candidate for implementation in HRS. Reaction cells and triple quadruple ICP-MS instruments are examples of two recent improvements [71]. Nuclear magnetic resonance (NMR) spectrometry features a much wider applicability than ICP-MS, because hydrogen and carbon can be detected. However, certainly within the relevant chromatographic timescale, the sensitivity of NMR is much lower than standard molecular property-based detection techniques such as ESI-MS or FLD. Additionally, hyphenation to LC is far less developed than LC−ESI-MS or even LC−ICP-MS, which presents a serious hurdle to the implementation into HRS systems. On the upside, recent advantages in the miniaturization of NMR flow probes are rapidly closing the sensitivity gap [72, 73] and a synergy with NMR structure elucidation could be especially useful (see later).

As described in Section 11.1.1.2, MS is currently the key technique to structure elucidation in HRS [74]. It provides an elemental composition and information about connections of structural elements. This can be supplemented by UV/vis spectrometry using DAD, which reveals functional groups and conjugated and aromatic systems. In the future, the hyphenation of ion mobility spectrometry (IMS) and MS, which has made important progress resulting in new commercial instrumentation, may find its way into HRS structure elucidation [75]. Because the drift time is usually on the timescale of the MS experiment or even shorter and separation is based on size and shape rather than on the physicochemical properties employed in LC, IMS might offer an interesting secondary separation dimension. However, in contrast to separation techniques, IMS does not reduce ionization suppression as it is positioned after the ionization step (with the exception of high-field asymmetric-waveform IMS). Nowadays, more in-depth structural information, for example, the exact configuration of regional isomers, *E*/*Z*-isomerism, and so on, is often obtained by NMR after purification by, for example, preparative LC. If NMR structure elucidation could be directly integrated into HRS platforms, this would greatly increase confidence in and refinement of HRS structure elucidation. Hyphenated LC–MS–NMR or even LC–DAD–MS–NMR systems are already used for various applications [76, 77].

Probably one of the most important developments in HRS at the moment is miniaturization, which applies to all parts of the HRS platform (see Figure 11.13b). Miniaturization of the separation reduces the required amount of sample, which is a huge advantage, for example, in natural extract screening [24]. In a miniaturized bioassay, the reagent consumption is greatly reduced, which could prove critical to the competitiveness of HRS. The use of chip-based on-line reactors also provides an opportunity for more freedom in reactor design with regard to material and geometry (flow patterns; see Section 11.2.3). The reduced sample volumes obviously necessitate more sensitive detection of the bioassay. This has been successfully demonstrated using LIF and (nano-)ESI-MS [24, 29]. The key vulnerability of HRS miniaturization lies in the robustness of the systems, much like it used to lie in the robustness of the analytical-scale HRS setups. However, in the last 5–10 years, the latter have matured into very reliable systems that provide excellent data quality. Therefore, considering the promises they hold, miniaturized HRS systems are likely to see the same maturation process. The reduction in sample amounts could open entirely new application fields, for example, the HRS analysis of clinical samples.

11.4.3
Conclusions

The strong link between a molecular structure and an individual biological effect, which is created by the integration of separation, bioassay, and structure elucidation, is undoubtedly the greatest benefit of HRS. In addition, the (virtually) closed system, the efficient mixing and heat exchange, and the internal control offered by the baseline hold the possibility of achieving high-quality bioassays with this on-line post-column format. Although throughput is not a primary attribute of HRS, the data density intrinsic to the approach is so high that it presents both a potential and a challenge. Similarly to any hyphenated system, another challenge is presented by finding a useful compromise between the complex requirements of each individual analytical technique. However, such a compromise has been reached in various ways and sometimes even synergies become evident. Dilution is a key factor to compatibility, but at the same time one of the biggest hurdles for HRS sensitivity (with regard to the amount and concentration limit of detection (LOD)). We have shown that a great number of the (data analysis) concepts, generally applied to bioassays, can be adapted to HRS, although many of these derived concepts would surely benefit from further refinement. Exploring the potentials of HRS can certainly lead to routine application in the future, especially if accompanied by continued technical development of the systems.

Currently, HRS is mostly applied to the screening of mixtures, among them natural extracts [24, 41] and lead libraries from shotgun synthesis approaches [3, 17], where its strengths are fully utilized. Due to the large developmental head start of HTS over HRS, full competitiveness will still require much progress. Miniaturization is a key to reducing reagent consumption in HRS, while robustness increases with maturation of the integrated systems. In spite of these obvious prejudices, the bigger hurdle toward industrial implementation of HRS is probably knowledge. Because the whole system is integrated, the operator must command a solid knowledge of all analytical and pharmacological disciplines involved in the HRS platform. Although training efforts can certainly contribute, widespread use will only be possible if the hurdles are lowered by commercial availability of robust and fully integrated systems and automated data analysis. In addition, HRS is a flow approach to chemistry that is in contrast to the majority of the chemical and pharmaceutical industry, which still uses batch approaches. This incompatibility puts HRS at another (apparent) disadvantage. However, there is much pressure on the pharmaceutical industry to improve the effectiveness of their processes [78]. When the potential of flow chemistry, for example, in terms of flexibility, is recognized, HRS may become an even more interesting alternative to HTS.

References

1. Kool, J., Giera, M., Irth, H., and Niessen, W.M. (2011) Advances in mass spectrometry-based post-column bioaffinity profiling of mixtures. *Anal. Bioanal. Chem.*, **399** (8), 2655–2668.
2. Irth, H., Oosterkamp, A.J., Vanderwelle, W., Tjaden, U.R., and Vandergreef, J. (1993) Online immunochemical detection in liquid-chromatography using fluorescein-labeled antibodies. *J. Chromatogr.*, **633** (1-2), 65–72.
3. Falck, D., Rahimi Pirkolachachi, F., Giera, M., Honing, M., Kool, J., and Niessen, W.M. (2014) Comparison of (bio-)transformation methods for the generation of metabolite-like compound libraries of p38alpha MAP kinase inhibitors using high-resolution screening. *J. Pharm. Biomed. Anal.*, **88**, 235–244.
4. Ingkaninan, K., de Best, C.M., van der Heijden, R., Hofte, A.J., Karabatak, B., Irth, H., Tjaden, U.R., van der Greef, J., and Verpoorte, R. (2000) High-performance liquid chromatography with on-line coupled UV, mass spectrometric and biochemical detection for identification of acetylcholinesterase inhibitors from natural products. *J. Chromatogr. A*, **872** (1-2), 61–73.
5. Ingkaninan, K., Hazekamp, A., de Best, C.M., Irth, H., Tjaden, U.R., van der

Heijden, R., van der Greef, J., and Verpoorte, R. (2000) The application of HPLC with on-line coupled UV/MS-biochemical detection for isolation of an acetylcholinesterase inhibitor from narcissus 'Sir Winston Churchill'. *J. Nat. Prod.*, **63** (6), 803–806.

6. Li, D.Q., Qian, Z.M., and Li, S.P. (2010) Inhibition of three selected beverage extracts on alpha-glucosidase and rapid identification of their active compounds using HPLC-DAD-MS/MS and bio-chemical detection. *J. Agric. Food. Chem.*, **58** (11), 6608–6613.

7. Schenk, T., Breel, G.J., Koevoets, P., van den Berg, S., Hogenboom, A.C., Irth, H., Tjaden, U.R., and van der Greef, J. (2003) Screening of natural products extracts for the presence of phosphodiesterase inhibitors using liquid chromatography coupled online to parallel biochemical detection and chemical characterization. *J. Biomol. Screening*, **8** (4), 421–429.

8. van Elswijk, D.A., Diefenbach, O., van der Berg, S., Irth, H., Tjaden, U.R., and van der Greef, J. (2003) Rapid detection and identification of angiotensin-converting enzyme inhibitors by on-line liquid chromatography-biochemical detection, coupled to electrospray mass spectrometry. *J. Chromatogr. A*, **1020** (1), 45–58.

9. Rhee, I.K., Appels, N., Hofte, B., Karabatak, B., Erkelens, C., Stark, L.M., Flippin, L.A., and Verpoorte, R. (2004) Isolation of the acetylcholinesterase inhibitor ungeremine from Nerine bowdenii by preparative HPLC coupled on-line to a flow assay system. *Biol. Pharm. Bull.*, **27** (11), 1804–1809.

10. Schebb, N.H., Heus, F., Saenger, T., Karst, U., Irth, H., and Kool, J. (2008) Development of a countergradient parking system for gradient liquid chro-matography with online biochemical detection of serine protease inhibitors. *Anal. Chem.*, **80** (17), 6764–6772.

11. Marques, L.A., Kool, J., de Kanter, F., Lingeman, H., Niessen, W., and Irth, H. (2010) Production and on-line acetyl-cholinesterase bioactivity profiling of chemical and biological degradation products of tacrine. *J. Pharm. Biomed. Anal.*, **53** (3), 609–616.

12. Falck, D., Schebb, N.H., Prihatiningtyas, S., Zhang, J., Heus, F., Morisseau, C., Kool, J., Hammock, B.D., and Niessen, W.M. (2013) Development of on-line liquid chromatography-biochemical detection for soluble epoxide hydrolase inhibitors in mix-tures. *Chromatographia*, **76** (1-2), 13–21.

13. Schenk, T., Appels, N.M., van Elswijk, D.A., Irth, H., Tjaden, U.R., and van der Greef, J. (2003) A generic assay for phosphate-consuming or -releasing enzymes coupled on-line to liquid chromatography for lead finding in natural products. *Anal. Biochem.*, **316** (1), 118–126.

14. Schobel, U., Frenay, M., van Elswijk, D.A., McAndrews, J.M., Long, K.R., Olson, L.M., Bobzin, S.C., and Irth, H. (2001) High resolution screening of plant natural product extracts for estro-gen receptor alpha and beta binding activity using an online HPLC-MS bio-chemical detection system. *J. Biomol. Screening*, **6** (5), 291–303.

15. van Elswijk, D.A., Schobel, U.P., Lansky, E.P., Irth, H., and van der Greef, J. (2004) Rapid dereplication of estro-genic compounds in pomegranate (Punica granatum) using on-line bio-chemical detection coupled to mass spectrometry. *Phytochemistry*, **65** (2), 233–241.

16. Kool, J., Ramautar, R., van Liempd, S.M., Beckman, J., de Kanter, F.J., Meerman, J.H., Schenk, T., Irth, H., Commandeur, J.N., and Vermeulen, N.P. (2006) Rapid on-line profiling of estrogen receptor binding metabolites of tamoxifen. *J. Med. Chem.*, **49** (11), 3287–3292.

17. de Vlieger, J.S., Kolkman, A.J., Ampt, K.A., Commandeur, J.N., Vermeulen, N.P., Kool, J., Wijmenga, S.S., Niessen, W.M., Irth, H., and Honing, M. (2010) Determination and identification of estrogenic compounds generated with biosynthetic enzymes using hyphen-ated screening assays, high resolution mass spectrometry and off-line NMR.

J. Chromatogr. B Anal. Technol. Biomed. Life Sci., **878** (7-8), 667–674.

18. Kool, J., de Kloe, G.E., Bruyneel, B., de Vlieger, J.S., Retra, K., Wijtmans, M., van Elk, R., Smit, A.B., Leurs, R., Lingeman, H., de Esch, I.J., and Irth, H. (2010) Online fluorescence enhancement assay for the acetylcholine binding protein with parallel mass spectrometric identification. *J. Med. Chem.*, **53** (12), 4720–4730.

19. Falck, D., de Vlieger, J.S., Niessen, W.M., Kool, J., Honing, M., Giera, M., and Irth, H. (2010) Development of an online p38alpha mitogen-activated protein kinase binding assay and integration of LC-HR-MS. *Anal. Bioanal. Chem.*, **398** (4), 1771–1780.

20. Rea, V., Falck, D., Kool, J., de Kanter, F.J.J., Commandeur, J.N.M., Vermeulen, N.P.E., Niessen, W.M.A., and Honing, M. (2013) Combination of biotransformation by P450 BM3 mutants with on-line post-column bioaffinity and mass spectrometric profiling as a novel strategy to diversify and characterize p38 alpha kinase inhibitors. *MedChemComm*, **4** (2), 371–377.

21. van Liempd, S.M., Kool, J., Niessen, W.M., van Elswijk, D.E., Irth, H., and Vermeulen, N.P. (2006) On-line formation, separation, and estrogen receptor affinity screening of cytochrome P450-derived metabolites of selective estrogen receptor modulators. *Drug Metab. Dispos.*, **34** (9), 1640–1649.

22. Van Liempd, S.M., Kool, J., Meerman, J.H., Irth, H., and Vermeulen, N.P. (2007) Metabolic profiling of endocrine-disrupting compounds by on-line cytochrome p450 bioreaction coupled to on-line receptor affinity screening. *Chem. Res. Toxicol.*, **20** (12), 1825–1832.

23. Falck, D., de Vlieger, J.S., Giera, M., Honing, M., Irth, H., Niessen, W.M., and Kool, J. (2012) On-line electrochemistry-bioaffinity screening with parallel HR-LC-MS for the generation and characterization of modified p38alpha kinase inhibitors. *Anal. Bioanal. Chem.*, **403** (2), 367–375.

24. Heus, F., Vonk, F., Otvos, R.A., Bruyneel, B., Smit, A.B., Lingeman, H., Richardson, M., Niessen, W.M., and Kool, J. (2013) An efficient analytical platform for on-line microfluidic profiling of neuroactive snake venoms towards nicotinic receptor affinity. *Toxicon*, **61**, 112–124.

25. Otvos, R.A., Heus, F., Vonk, F.J., Halff, J., Bruyneel, B., Paliukhovich, I., Smit, A.B., Niessen, W.M., and Kool, J. (2013) Analytical workflow for rapid screening and purification of bioactives from venom proteomes. *Toxicon*, **76**, 270–281.

26. Heus, F., Otvos, R.A., Aspers, R.L., van Elk, R., Halff, J., Ehlers, A.W., Dutertre, S., Lewis, R.J., Wijmenga, S.S., Smit, A.B., Niessen, W.M., and Kool, J. (2014) Miniaturized bioaffinity assessment coupled to mass spectrometry for guided purification of bioactives from toad and cone snail. *Biology*, **3**, 139–156.

27. de Boer, A.R., Letzel, T., van Elswijk, D.A., Lingeman, H., Niessen, W.M., and Irth, H. (2004) On-line coupling of high-performance liquid chromatography to a continuous-flow enzyme assay based on electrospray ionization mass spectrometry. *Anal. Chem.*, **76** (11), 3155–3161.

28. de Boer, A.R., Alcaide-Hidalgo, J.M., Krabbe, J.G., Kolkman, J., van Emde Boas, C.N., Niessen, W.M., Lingeman, H., and Irth, H. (2005) High-temperature liquid chromatography coupled on-line to a continuous-flow biochemical screening assay with electrospray ionization mass spectrometric detection. *Anal. Chem.*, **77** (24), 7894–7900.

29. de Boer, A.R., Bruyneel, B., Krabbe, J.G., Lingeman, H., Niessen, W.M., and Irth, H. (2005) A microfluidic-based enzymatic assay for bioactivity screening combined with capillary liquid chromatography and mass spectrometry. *Lab Chip*, **5** (11), 1286–1292.

30. de Jong, C.F., Derks, R.J., Bruyneel, B., Niessen, W., and Irth, H. (2006) High-performance liquid chromatography-mass spectrometry-based acetylcholinesterase assay for the screening of inhibitors in natural extracts. *J. Chromatogr. A*, **1112** (1-2), 303–310.

31. Hogenboom, A.C., de Boer, A.R., Derks, R.J., and Irth, H. (2001) Continuous-flow, on-line monitoring of biospecific interactions using electrospray mass spectrometry. *Anal. Chem.*, **73** (16), 3816–3823.

32. Derks, R.J., Hogenboom, A.C., van der Zwan, G., and Irth, H. (2003) On-line continuous-flow, multi-protein biochemical assays for the characterization of bioaffinity compounds using electrospray quadrupole time-of-flight mass spectrometry. *Anal. Chem.*, **75** (14), 3376–3384.

33. Rhee, I.K., Appels, N., Luijendijk, T., Irth, H., and Verpoorte, R. (2003) Determining acetylcholinesterase inhibitory activity in plant extracts using a fluorimetric flow assay. *Phytochem. Anal.*, **14** (3), 145–149.

34. Oosterkamp, A.J., Villaverde Herraiz, M.T., Irth, H., Tjaden, U.R., and van der Greef, J. (1996) Reversed-phase liquid chromatography coupled on-line to receptor affinity detection based on the human estrogen receptor. *Anal. Chem.*, **68** (7), 1201–1206.

35. Ellman, G.L., Courtney, K.D., Andres, V. Jr., and Feather-Stone, R.M. (1961) A new and rapid colorimetric determination of acetylcholinesterase activity. *Biochem. Pharmacol.*, **7**, 88–95.

36. Derks, R.J., Letzel, T., De Jong, C.F., van Marle, A., Lingeman, H., Leurs, R., and Irth, H. (2006) SEC–MS as an approach to isolate and directly identifying small molecular GPCR-ligands from complex mixtures without labeling. *Chromatographia*, **64** (7/8), 379–385.

37. Jonker, N., Kretschmer, A., Kool, J., Fernandez, A., Kloos, D., Krabbe, J.G., Lingeman, H., and Irth, H. (2009) Online magnetic bead dynamic protein-affinity selection coupled to LC-MS for the screening of pharmacologically active compounds. *Anal. Chem.*, **81** (11), 4263–4270.

38. Annis, D.A., Athanasopoulos, J., Curran, P.J., Felsch, J.S., Kalghatgi, K., Lee, W.H., Nash, H.M., Orminati, J.A., Rosner, K.E., Shipps, G.W., Thaddupathy, G.R., Tyler, A.N., Vilenchik, L., Wagner, C.R., and Wintner, E.A. (2004) An affinity selection–mass spectrometry method for the identification of small molecule ligands from self-encoded combinatorial libraries Discovery of a novel antagonist of E. coli dihydrofolate reductase. *Int. J. Mass Spectrom.*, **238**, 77–83.

39. Niederlander, H.A., van Beek, T.A., Bartasiute, A., and Koleva, I.I. (2008) Antioxidant activity assays on-line with liquid chromatography. *J. Chromatogr. A*, **1210** (2), 121–134.

40. Kool, J., van Liempd, S.M., van Rossum, H., van Elswijk, D.A., Irth, H., Commandeur, J.N., and Vermeulen, N.P. (2007) Development of three parallel cytochrome P450 enzyme affinity detection systems coupled on-line to gradient high-performance liquid chromatography. *Drug Metab. Dispos.*, **35** (4), 640–648.

41. Potterat, O. and Hamburger, M. (2013) Concepts and technologies for tracking bioactive compounds in natural product extracts: generation of libraries, and hyphenation of analytical processes with bioassays. *Nat. Prod. Rep.*, **30** (4), 546–564.

42. Chen, J. and Lee, C.S. (2002) On-line post-capillary affinity detection of immunoglobulin G for capillary zone electrophoresis. *J. Chromatogr. B Anal. Technol. Biomed. Life Sci.*, **768** (1), 105–111.

43. Schebb, N.H., Vielhaber, T., Jousset, A., and Karst, U. (2009) Development of a liquid chromatography-based screening methodology for proteolytic enzyme activity. *J. Chromatogr. A*, **1216** (20), 4407–4415.

44. Niessen, W.M.A. (2006) *Liquid Chromatography – Mass Spectrometry*, Chromatographic Science Series, 3rd edn, CRC Press.

45. Kostiainen, R. and Kauppila, T.J. (2009) Effect of eluent on the ionization process in liquid chromatography-mass spectrometry. *J. Chromatogr. A*, **1216** (4), 685–699.

46. Teutenberg, T. (2009) Potential of high temperature liquid chromatography for the improvement of separation

efficiency--a review. *Anal. Chim. Acta*, **643** (1-2), 1–12.

47. Causon, T.J., Shellie, R.A., and Hilder, E.F. (2009) High temperature liquid chromatography with monolithic capillary columns and pure water eluent. *Analyst*, **134** (3), 440–442.

48. Heus, F., Giera, M., de Kloe, G.E., van Iperen, D., Buijs, J., Nahar, T.T., Smit, A.B., Lingeman, H., de Esch, I.J., Niessen, W.M., Irth, H., and Kool, J. (2010) Development of a microfluidic confocal fluorescence detection system for the hyphenation of nano-LC to on-line biochemical assays. *Anal. Bioanal. Chem.*, **398** (7-8), 3023–3032.

49. Schebb, N.H., Falck, D., Faber, H., Hein, E.M., Karst, U., and Hayen, H. (2009) Fast method for monitoring phospholipase A2 activity by liquid chromatography-electrospray ionization mass spectrometry. *J. Chromatogr. A*, **1216** (27), 5249–5255.

50. Engelhardt, H. and Neue, U.D. (1982) Reaction detector with three dimensional coiled open tubes in HPLC. *Chromatographia*, **15** (7), 403–408.

51. Yamaguchi, Y., Takagi, F., Yamashita, K., Nakamura, H., Maeda, H., Sotowa, K., Kusakabe, K., Yamasaki, Y., and Morooka, S. (2004) 3-D simulation and visualization of laminar flow in a microchannel with hair-pin curves. *AlChE J.*, **50** (7), 1530–1535.

52. Pargellis, C., Tong, L., Churchill, L., Cirillo, P.F., Gilmore, T., Graham, A.G., Grob, P.M., Hickey, E.R., Moss, N., Pav, S., and Regan, J. (2002) Inhibition of p38 MAP kinase by utilizing a novel allosteric binding site. *Nat. Struct. Biol.*, **9** (4), 268–272.

53. Kind, T. and Fiehn, O. (2006) Metabolomic database annotations via query of elemental compositions: mass accuracy is insufficient even at less than 1 ppm. *BMC Bioinf.*, **7**, 234.

54. Ramanathan, R., Jemal, M., Ramagiri, S., Xia, Y.Q., Humpreys, W.G., Olah, T., and Korfmacher, W.A. (2011) It is time for a paradigm shift in drug discovery bioanalysis: from SRM to HRMS. *J. Mass Spectrom.*, **46** (6), 595–601.

55. Copeland, R.A., Pompliano, D.L., and Meek, T.D. (2006) Drug-target residence time and its implications for lead optimization. *Nat. Rev. Drug Discovery*, **5** (9), 730–739.

56. Giera, M., Heus, F., Janssen, L., Kool, J., Lingeman, H., and Irth, H. (2009) Microfractionation revisited: a 1536 well high resolution screening assay. *Anal. Chem.*, **81** (13), 5460–5466.

57. Kool, J., Rudebeck, A.F., Fleurbaaij, F., Nijmeijer, S., Falck, D., Smits, R.A., Vischer, H.F., Leurs, R., and Niessen, W.M. (2012) High-resolution metabolic profiling towards G protein-coupled receptors: rapid and comprehensive screening of histamine H(4) receptor ligands. *J. Chromatogr. A*, **1259**, 213–220.

58. Giera, M., de Vlieger, J.S., Lingeman, H., Irth, H., and Niessen, W.M. (2010) Structural elucidation of biologically active neomycin N-octyl derivatives in a regioisomeric mixture by means of liquid chromatography/ion trap time-of-flight mass spectrometry. *Rapid Commun. Mass Spectrom.*, **24** (10), 1439–1446.

59. Pelander, A., Tyrkko, E., and Ojanpera, I. (2009) In silico methods for predicting metabolism and mass fragmentation applied to quetiapine in liquid chromatography/time-of-flight mass spectrometry urine drug screening. *Rapid Commun. Mass Spectrom.*, **23** (4), 506–514.

60. Kerns, E.H., Volk, K.J., Hill, S.E., and Lee, M.S. (1995) Profiling new taxanes using LC/MS and LC/MS/MS substructural analysis techniques. *Rapid Commun. Mass Spectrom.*, **9** (15), 1539–1545.

61. Kerns, E.H., Rourick, R.A., Volk, K.J., and Lee, M.S. (1997) Buspirone metabolite structure profile using a standard liquid chromatographic mass spectrometric protocol. *J. Chromatogr. B*, **698** (1-2), 133–145.

62. Holcapek, M., Kolarova, L., and Nobilis, M. (2008) High-performance liquid chromatography-tandem mass spectrometry in the identification and determination of phase I and phase II

drug metabolites. *Anal. Bioanal. Chem.*, **391** (1), 59–78.

63. Retra, K., Geitmann, M., Kool, J., Smit, A.B., de Esch, I.J., Danielson, U.H., and Irth, H. (2010) Development of surface plasmon resonance biosensor assays for primary and secondary screening of acetylcholine binding protein ligands. *Anal. Biochem.*, **407** (1), 58–64.

64. Zhang, J.H., Chung, T.D., and Oldenburg, K.R. (1999) A simple statistical parameter for use in evaluation and validation of high throughput screening assays. *J. Biomol. Screening*, **4** (2), 67–73.

65. Oosterkamp, A.J., Irth, H., Villaverde Herraiz, M.T., Tjaden, U.R., and van der Greef, J. (1997) Theoretical concepts of on-line liquid chromatographic-biochemical detection systems. I. Detection systems based on labelled ligands. *J. Chromatogr. A*, **787** (1-2), 27–35.

66. Cheng, Y. and Prusoff, W.H. (1973) Relationship between the inhibition constant (K1) and the concentration of inhibitor which causes 50 per cent inhibition (I50) of an enzymatic reaction. *Biochem. Pharmacol.*, **22** (23), 3099–3108.

67. Jonker, N., Kool, J., Irth, H., and Niessen, W.M. (2011) Recent developments in protein-ligand affinity mass spectrometry. *Anal. Bioanal. Chem.*, **399** (8), 2669–2681.

68. Campbell, J.L. and Le Blanc, J.C. (2012) Using high-resolution quadrupole TOF technology in DMPK analyses. *Bioanalysis*, **4** (5), 487–500.

69. Telgmann, L., Faber, H., Jahn, S., Melles, D., Simon, H., Sperling, M., and Karst, U. (2012) Identification and quantification of potential metabolites of Gd-based contrast agents by electrochemistry/separations/mass spectrometry. *J. Chromatogr. A*, **1240**, 147–155.

70. de Vlieger, J.S., Giezen, M.J., Falck, D., Tump, C., van Heuveln, F., Giera, M., Kool, J., Lingeman, H., Wieling, J., Honing, M., Irth, H., and Niessen, W.M. (2011) High temperature liquid chromatography hyphenated with ESI-MS and ICP-MS detection for the structural characterization and quantification of halogen containing drug metabolites. *Anal. Chim. Acta*, **698** (1-2), 69–76.

71. Ammann, A.A. (2007) Inductively coupled plasma mass spectrometry (ICP MS): a versatile tool. *J. Mass Spectrom.*, **42** (4), 419–427.

72. Falck, D., Oosthoek-de Vries, A.J., Kolkman, A., Lingeman, H., Honing, M., Wijmenga, S.S., Kentgens, A.P., and Niessen, W.M. (2013) EC-SPE-stripline-NMR analysis of reactive products: a feasibility study. *Anal. Bioanal. Chem.*, **405** (21), 6711–6720.

73. Gokay, O. and Albert, K. (2012) From single to multiple microcoil flow probe NMR and related capillary techniques: a review. *Anal. Bioanal. Chem.*, **402** (2), 647–669.

74. Falck, D., Kool, J., Honing, M., and Niessen, W.M. (2013) Tandem mass spectrometry study of p38alpha kinase inhibitors and related substances. *J. Mass Spectrom.*, **48** (6), 718–731.

75. Kanu, A.B., Dwivedi, P., Tam, M., Matz, L., and Hill, H.H. (2008) Ion mobility-mass spectrometry. *J. Mass Spectrom.*, **43** (1), 1–22.

76. Corcoran, O. and Spraul, M. (2003) LC-NMR-MS in drug discovery. *Drug Discovery Today*, **8** (14), 624–631.

77. Tang, H., Xiao, C., and Wang, Y. (2009) Important roles of the hyphenated HPLC-DAD-MS-SPE-NMR technique in metabonomics. *Magn. Reson. Chem.*, **47** (Suppl. 1), S157–162.

78. Crommelin, D., Stolk, P., Besancon, L., Shah, V., Midha, K., and Leufkens, H. (2010) Pharmaceutical sciences in 2020. *Nat. Rev. Drug Discovery*, **9** (2), 99–100.

12
Conclusions

Jeroen Kool

There is a wide spectrum of methodologies to study bioaffinity interactions with mass spectrometry (MS), as demonstrated in the various chapters of this book. The methodologies to study direct ligand–protein affinity interactions can be separated into two generic categories. The first category is based on immobilization of target protein on a solid support (see Chapters 4, 5, and 8). The second category utilizes in-solution ligand binding to target protein followed by a rapid separation step of bound ligands and unbound molecules (see Chapters 4 and 5). Ultrafiltration and size exclusion methodologies are solution-based (see Chapters 4 and 5). This also holds true for dynamic protein affinity selection (see Chapter 4) and for affinity capillary electrophoresis (CE) methods (see Chapter 9). Other methodologies used for ligand fetching followed by identification with MS belong to the first category and most of them are based on frontal affinity, zonal elution, or affinity capture (see Chapter 8). Furthermore, most CE-based methods and all surface plasmon resonance (SPR)-based (see Chapter 10) ones use immobilized proteins or other target macromolecules.

When assessing the current state of the art on using immobilized proteins, it can be stated that frontal affinity chromatography (FAC) will remain relevant as it is cost-effective and straightforward for many target proteins to assess ligand binding. For more complex drug targets, that is, membrane-bound proteins, given their intrinsic complexity, instability, and environment needed for maintaining their activity, in-solution based methods are commonly more difficult to implement than FAC methodologies. SPR-MS methodologies (discussed in detail later in this conclusion) can produce similar affinity data as FAC. However, SPR in many cases is limited by its inability to measure small ligands, although advances in sensitivity allow analysis of small ligands for a few membrane-bound targets (e.g., GPCRs (G-protein coupled receptors)). Further advances in this field are slowly allowing the more difficult target proteins to be used for analysis of small molecule binding. Solution-based methodologies have the advantage of not suffering from impaired binding properties due to protein immobilization. Furthermore, binding interactions occur in solution. Ultrafiltration approaches benefit from the possibility of reusing target protein and throughput. CE affinity selection will continue to be important despite issues with nonspecific adsorption of proteins to capillary walls.

Analyzing Biomolecular Interactions by Mass Spectrometry, First Edition.
Edited by Jeroen Kool and Wilfried M.A. Niessen.
© 2015 Wiley-VCH Verlag GmbH & Co. KGaA. Published 2015 by Wiley-VCH Verlag GmbH & Co. KGaA.

Specific coating approaches can overcome these issues. Dynamic protein affinity selection is very sensitive for weak binders, and would therefore be well suited for fragment-based drug discovery programs. Size exclusion chromatography (SEC) methodologies, such as the Automated Ligand Identification System (ALIS), have successfully been used in the pharmaceutical industry for screening campaigns (see Chapters 4 and 5). They are straightforward to implement and to automate for 96-well plate processing. Such approaches are also significantly faster than similar methodologies discussed. Overall, much of the research on affinity selection MS in the pharmaceutical industry uses SPR or SEC technologies. The other methodologies discussed form a smaller niche for specific protein targets and/or library types to be screened.

Membrane-bound proteins comprise the most important drug targets for drug discovery and development. Usually, screening affinity interactions of ligands for these receptors are undertaken using traditional approaches such as ligand binding assays, functional assays, and calcium flux assays. These methods, however, are laborious and do not allow the analysis of mixtures for individual ligands. Furthermore, these assays are not used or are difficult to use when it comes to drug discovery programs focusing on orphan receptors. One alternative is the use of bioaffinity chromatography to study receptor ligand interactions (main topic of Chapter 8). In this approach, the target protein (e.g., receptor or enzyme) is immobilized onto a chromatographic support. As a result, chromatographic stationary phases are created that allow for retaining ligands based on their affinity for the target protein. This label-free technique allows for screening orphan receptors as well as analyzing mixtures. Protein-based stationary phases are versatile, and there have been many successful applications known in the drug discovery area. Stationary phases for transmembrane proteins are very difficult to develop due to the labile receptors and the undefined membrane surrounding them. But several groups have succeeded in the production of such stationary phases in which the immobilized receptor retained its functional activity. For identification of unknown ligands in mixtures, the detector of choice is MS coupled online after affinity chromatography. The order of ligand elution is a direct measure of the relative affinities of the respective ligands' eluting. In this specific setup of FAC, a ligand or mixture is continuously infused and breakthrough times of ligands are an indication of their affinity. In mixtures, individual breakthrough times give the same indication of relative affinities of individual ligands judged by their breakthrough times.

Radioligand binding assays are widely used for assessment of ligand affinity toward a target protein. For this, the competition between a radioligand (i.e., a ligand labeled with a radioisotope) and ligands under study is a means of measuring affinity by performing these experiments in a concentration–response manner. Sensitivities using these assays are usually extremely high and measurement takes place after separation of radiolabel bound to target receptor and radiolabel displaced by the test ligand and thus present in solution. After this separation, the fraction with the radiolabel bound to the target receptor is measured. High costs of radiolabels and the evident problems associated with radioactivity are drawbacks of this approach. Although assays based on fluorescent tracer ligands are feasible alternatives, a tracer ligand with fluorophore or an intrinsically fluorescent tracer ligand must be available and most often these assays are by far not as sensitive as radiolabel-based assays.

The use of MS as a label-based, or label-free, alternative is highly attractive since the readout is generic (main topic of Chapter 5). In other words, any ligand can be used since targeted or untargeted analysis of its molecular mass is the principle of MS. Using current-day mass spectrometers, this can be done in a highly versatile, specific, and sensitive manner, and at the same time structural information of detected binders can be acquired. Also, the improvements and options in the area of ionization helped putting MS-based binding assays as a well-accepted alternative to traditional radioligand binding assays. Nowa-days, many examples exist of homogenous assay formats coupled with direct measurement of target−ligand complexes by means of electrospray ionization (ESI)-MS analysis. Using MS readout for assays requiring filtration steps (e.g., membrane-bound receptor assays that comprise the most important drug targets of today), strategies are required with a more complex setup since they require a separation step. Of these, FAC, ultrafiltration, and SEC are options that are especially suitable for analyzing mixtures with ligands.

Filtration-based assays can be adapted to MS readout in a straightforward man-ner. Here, MS quantification can be attractive as it benefits from the same simple working principle as filtration-based radioligand binding, a specific strength of this method, while simultaneously circumventing the use of radioactivity. In this approach, quantitation of nonlabeled ligands is simply done with another also unlabeled reporter ligand, called *native marker*, using MS, and after the filtration step of unbound and bound ligand. MS binding assays share the advantages of radioligand binding assays, while avoiding all disadvantages intrinsically involved in using radiolabels such as the necessity of synthesizing these ligands and the problems associated with radioactivity in general.

CE is a very suitable measurement technique for a variety of molecules ranging from small ions to large biomolecules where their charge and shape are impor-tant determinants for their behavior during separation. CE can also be used to assess biomolecular affinity interactions (main topic of Chapter 9). Furthermore, using affinity materials for selective online extraction before CE analysis can be mentioned. Currently, both CE-based approaches have proved to be successful in many different applications. In recent times, MS has also been coupled to CE as a detection technique for measuring affinity interactions. Chapter 9 provides an overview where both small molecules and proteins are studied for their affinity interactions.

Label-free biosensors commonly aim at detection of biological and biochemi-cal targets without the indirect use of coupled enzymes or dyes for readout, such as, for example, employed in traditional enzyme-linked immunosorbent assays (ELISAs). Most widespread optical biosensors are SPR based. A great advantage of SPR is that the interactions under study can be monitored in real time. When the analytes under study are known, the concentration of these affinity analytes can be quantified since SPR is a mass sensitive detector (see Chapter 10). Besides label free analysis, one can also choose for competitive binding assays. This can be advantageous for analyzing small molecular analytes, as the signal output depends directly on the mass of the interacting analyte. A major drawback of SPR, however, is that chemical information on the identity of unknown analytes is lacking. Therefore, it is advantageous to couple SPR in series, or in parallel, to chemical detectors of choice. In this regard, MS is by far the most logical option. In the proteomics field, numerous applications of this are known. Borch and Roepstorff elaborately discussed SPR protocols for protein capture followed by elution to ESI-MS. This is usually done after a digestion step for efficient identification by proteomics approaches. Another review by Nedelkov describes protocols and know-how for generic coupling of SPR to ESI-MS. SPR – matrix-assisted laser desorption ionization (MALDI) – MS has also been done, where proteins trapped on the SPR chip are subsequently analyzed by MALDI-MS, usually after an on-chip digestion step. When it comes to analyzing small molecular affinity analytes, SPR has also been used in combination with MS, in many cases with implementation of offline or online liquid chromatography (LC) or CE as a separation step. The additional separation allows for more complex samples to be handled and identification of individual affinity analytes. Localized surface plasmon resonance (LSPR) sensors in combination with MALDI-MS have been described. They have the added advantage of fewer interferences from the bulk refractive index. For example, Anker *et al.* used this combination to study amyloid beta oligomers that have important roles in Alzheimer's disease. Possible ionization techniques for hyphenating SPR with MS depend on the characteristics of the molecule, but, for the major part, on bioassay compatibility. In case of a direct binding assay, for example, after analyte capture on the chip, elution toward offline LC – MS (liquid chromatography – mass spectrometry) analysis gives more freedom of buffer and/or ionization technique than subsequent online LC – MS analysis. In the last case, the solvent composition for elution has to be compatible with the immobilized target on the chip, if the chip is to be reused again, as well as the ionization technique used.

Different methodologies can be distinguished in which postcolumn bioaffinity is performed for identity assessment. Where online and at-line screening approaches comprise real integrated systems that combine biochemical and biological assays to be performed in parallel with MS-based ligand identification (see Chapters 11 and 5), more traditional effect direct analysis (EDA) approaches are employed for determining the correlation of bioactivity to identity in an indirect manner (see Chapter 5). This is also the case for antivenomics approaches (see Chapter 7). Many different online postcolumn bioaffinity screening methodologies have been developed (see Chapter 11). Compared to traditional offline approaches that are comprehensive and elaborate, online approaches are fast and straightforward. A disadvantage of online approaches is that they are only applicable for fast bioassays with soluble target proteins. Important membrane-bound drug targets such as GPCRs and ion channels are therefore out of reach. One reason for this is that a compromise has to be made between the separation and the bioassay conditions as they have to be combined and thus have to be compatible with each other. Furthermore, traditional strategies are directly implemented in traditional drug discovery workflows. Deconvolution of bioactive mixtures and correlation of identity to bioactivities with these offline methodologies, however, is very cumbersome and often does not allow all bioactives to be identified. Especially the minor bioactives are often not easily identified. Therefore, hyphenated postcolumn strategies have their niche and once such a methodology is developed, bioactive mixtures can be screened rapidly and all bioactive peaks found can be correlated to their MS spectra in a straightforward way. Although commercial bioassays cannot be implemented directly, after adaptation to online screening, they comprise strong profiling platforms for analysis of bioactive mixtures. Binding assays with nuclear receptors, protein kinases, proteases, phosphodiesterases, phosphatases, and metabolic enzymes are most conveniently modified to online screening assays. Enzymes with high turnover rates and soluble receptors with a ligand showing fluorescence enhancement in the binding pocket are most suitable. Target enzymes with low turnover rates and membrane-bound drug targets are better suited for at-line or nanofractionation approaches (Chapter 4), thereby overcoming the drawbacks mentioned for traditional offline screening approaches. These approaches are also more easily implemented in traditional drug discovery programs. Finally, online and at-line strategies for bioactive mixture screening are strong tools for deconvolution of the bioactives. They allow rapid and straightforward correlation of bioactives for their identity. Before usage, evidently a thorough method development per drug target/bioassay is pivotal.

One can conclude that a wide variety of techniques are used for EDA-like approaches. Amongst the different research fields, these methodologies are applied in many different ways but all with the same aim to identify biologically active compounds from a complex mixture. Unfortunately, thus far, the majority of these approaches are labor-intensive and not always successful in identifying the bioactive compounds. The latter is especially the case for environmental samples. Furthermore, broad knowledge is necessary for the use of both biological and chemical techniques. Both the labor-intensive methodologies and the necessary expertise have thus far prevented the use of EDA as a technique in routine analysis and monitoring. It would, however, be beneficial if EDA-like approaches could be implemented in general environmental and water laboratories. The use of EDA in, for example, water monitoring laboratories could aid in the identification of newly appearing contaminants. In addition, there are many thinkable applications in food monitoring, for example, in the screening for new hormones that may be used as growth stimulators in livestock and for the identification of new kinds of doping agents in sports doping control. In order to implement EDA, it is necessary that robust systems are developed. It is, however, important to keep in mind that, depending on the obtained results, adjustments have to be made to the method in order to isolate specific compounds. Finally, if a new bioactive compound is identified, faster, more sensitive, and robust traditional techniques have to take over for quantitation and subsequent monitoring. This means that challenges have to be overcome in sample preparation, separation, and both chemical and biological detection. Recent developments such as accelerated membrane-assisted cleanup (AMAC) for lipid-rich matrices, SPE (solid phase extraction) for the extraction of a broad range of contaminants from plasma samples, nanospotting technologies, and, importantly, recent developments in hyphenation of biological and chemical detection are promising tools for future research and might encourage laboratories to use EDA-like approaches as more standard analytical methodologies.

The development of techniques and strategies for elucidation of snake venoms using proteomics approaches has gained widespread use because of its success and comprehensive eventual study results. These studies are called *venomics studies*, whereas antivenomics is venomics in a translational manner (see Chapter 7). In this regard, the quantification of the cross-reactivity of antivenoms against venoms is being referred to here. The analytical methods used for this approach are straightforward and might help in the design and manufacture of improved therapeutic antivenoms. It is in a way complementary to more traditional immunoassays and Western blots for the assessment of immunoreactivity of antibodies.

Plant metabolomics approaches are positioned in the field of chemical identification of the metabolome, which can then be correlated to diverse biological properties of plants in general (see Chapter 6). Plant metabolomics plays important roles in different research fields including drug discovery, crop improvement, food safety, and also for the elucidation of the plant metabolome. There are huge challenges in terms of analytical methods that allow full assessment of metabolites present. Challenges to be named include the differences in concentrations of metabolites and their different chemical properties. MS is the analytical detection method of choice when it comes to elucidation of plant metabolomes of which two specific approaches are distinguished: targeted and nontargeted. Whereas targeted approaches allow for higher sensitivity as metabolites are known, nontargeted approaches can be an initial step in generating biomarker discovery data. Evidently, for metabolic profiling, chromatographic separation techniques before detection play a pivotal role. Within this combination, the often applied possibility of chemical fingerprinting can be used, which has to be supported by modern software for statistical analysis, most often in combination with database searches.

Over the last two decades, MS has become a powerful tool for investigating large biomolecules and their interactions. Traditional top-down MS approaches have included the study of intact proteins under denaturing conditions (acidic solution and/or high concentration of organic solvent) to determine accurate protein masses or map posttranslational modifications (PTMs) using MS/MS fragmentation. Protein scientists are becoming increasingly interested in the use of "native" approaches, where conformations of individual proteins and noncovalent complexes are preserved throughout the ionization process and analysis inside the mass spectrometer. This interest is because most biological processes are regulated by protein–protein interactions. Furthermore, probably the majority of proteins exist as multi-subunit assemblies that comprise the actual functional protein complexes within a cell. To understand cellular functioning, it is crucial to know these cellular protein assemblies, and map their interactions. Classically, this is done using X-ray crystallography, nuclear magnetic resonance, as well as electron microscopy. Unfortunately, however, many protein assemblies are not well compatible with these techniques due to their low solubility, weak stability, large size, inability to form crystals, for example, when they reside in membranes. Furthermore, these methodologies do not allow one to dynamically study protein assemblies or to study them in solution. Different alternative low-resolution methods can provide complementary structural data to form a complete picture of what is happening in cells. Of the many different methods to be used, structural MS and chemical cross-linking MS can be applied to reveal structural features of protein assemblies. Advances in mass spectrometric hardware over the last decade have opened new ways to study noncovalent protein complexes and assemblies. Analysis speed and the analysis of intact complexes in solution are advantages. These relatively new MS-based technologies allow for the characterization of protein–protein, protein–DNA, protein–RNA, protein–metal ion, protein–drug, protein–carbohydrate, and protein–lipid interactions (see Chapters 2 and 3). Whereas membrane-bound proteins are now slowly entering the arena of these MS-based approaches, also sometimes called *native MS approaches*, many examples dealing with non-membrane-bound protein interactions including even whole virus capsids directly being analyzed in the MS have already been described in the literature. As formation of biologically active protein structures is commonly mediated by non-covalent-binding interactions, soft ionization techniques are evidently crucial. It is thus desirable to study these systems under preservative conditions. As the complexes are large in size, modified mass spectrometers should be used in order to cope with these complexes. For example, the flight time of these native complexes in time-of-flight (TOF) mass analyzers is relatively long.

Native ion mobility MS is one of the newest possibilities being introduced in the field of structural biology (main topic of Chapter 3). It is an add-on complementarity to the existing native MS approaches. The ion mobility cell has the unique feature to discriminate ions of the same m/z value by their shape. As this shape tells something about the 3D structure of a protein complex, this addition is a very valuable one that has entered the native MS field. All slightly different native MS-based approaches are complementary in themselves, and each of these methods has its advantages and limitations. In all native MS approaches, however, artificially elevated concentrations of proteins have to be used for a study, but in comparison with other structural biology methods (e.g., crystallography), native MS approaches use only marginal amounts of sample. These direct analysis approaches facilitated studying many protein assemblies with a wide variety of biological functions, but *in vitro* (using crude cell extracts) or even *in vivo* (e.g., from cells or tissues) sampling for direct analysis with native MS approaches is not feasible yet. Still, the buffer conditions of the samples to be injected into an MS do mimic *in vitro* conditions (in contrary to the often elevated concentrations of the proteins under study). Although this is the case, actual analysis is performed on ions in the gas phase and thus the results obtained have to be looked at critically and many different control experiments should be included. In most cases, it was found that results obtained from studying protein–protein and protein–ligand interactions with native MS resemble results from traditional biochemical approaches. In this regard, the mass spectrometric results are also complementary to many traditional biochemical approaches used to study protein complexes with the added advantage that protein complexes can be studied in time and parameters can be changed while studying the sorted effects on the complexes under study. Native MS is also capable of studying extremely large and diverse protein assemblies. High-resolution images of the shapes of protein complexes are evidently not possible with native MS approaches. For this, other complementary technologies such as scanning electron microscopy and atomic force microscopy have to be used.

Interactomics or pull-down proteomics is able to unravel whole networks of interacting proteins on a protein, DNA, small molecule, or any other target molecule (shortly described in Chapter 4). This field of research is expanding rapidly, and many studies unraveling biological functions of protein interactions and networks are being published nowadays. Critical evaluation of the data obtained is very important taking into account nonspecific binding processes during the pull-down experiments. For the pull-down or "protein fishing" process, thorough optimization and validation should be conducted before the actual study. After eventual identification of the proteins identified as binders, traditional biochemical approaches should be performed for confirmation purposes. For chemical proteomics, usually an immobilized lead compound or drug is studied. Here, in-solution lead compounds or drugs under study are used in different concentrations, thereby binding to proteins for which they have affinity. This in turn prevents these proteins from binding to the immobilized lead compound or drug resulting in eventual detection of these proteins by MS in a lower abundance. Using this approach, proteins with high affinity (for the in solution lead compound or drug under study) are displaced first. This methodology allows selectivity analysis of many different binding proteins in one experiment. Especially for protein kinase drug discovery, this approach is very valuable. The methodology enables inhibitory profiles to be measured toward whole panels of (hundreds) of proteins at once and addresses the issue of selectivity profiling of ligands for protein kinase drug discovery programs. The interactomics and chemical proteomics studies are shortly discussed in this book as they are covered elaborately in other books focusing at proteomics in general and interactomics and/or chemical proteomics specifically.

To summarize, the different techniques discussed in this book nicely show that MS can be used efficiently in combination with many biochemical and biological approaches. In several chapters, the implementation of separation strategies is shown and comprises an efficient way to study biologically active mixtures. The different techniques and methodologies discussed are often complementary to each other in their approach and results obtained. Some of them are suited for a few specific research questions, while others are more broadly applicable. In many cases, using several of the techniques discussed to study the same samples will result in similar end results that confirm each other. Surely the exiting field of MS for measurement of biological interactions will expand in the coming decades allowing more specific and in-depth studies toward biological events and biochemical interactions occurring in life.

Index

a

accelerated membrane assisted clean-up (AMAC)   117
acetonitrile   331
7-acetoxy-1-methylquinolinium iodide   325
acetylcholinesterase   328
affinity capillary electrophoresis (ACE)
– dynamic equilibrium   276
– kinetic   280
– pre-equilibrium   279
affinity capillary electrophoresis – mass spectrometry (ACE–MS)
– application   286, 287
– specific requirements   284
affinity ligands   281
agonists and allosteric modulators   323
air bubbles   337
alkaline phosphatase   328
allosteric modulator   340
α-lactalbumin   92
amyloid beta oligomers   368
amyloidogenic proteins   96
analogue acetylcholine binding protein (AChBP)   328
androgen chemical activated luciferase expression (AR-CALUX)   117
angiotensin-converting enzyme   328
anti-Alzheimer drug   328
anti-Alzheimer target protein   328
antibiotic enrofloxacin   306
anti-digoxigenin-antibody/digoxigenin + digoxin   327
antivenomics   370
ArMet   215
assay quality parameters   351
assay signal   344
association constants   276, 278
association kinetics   342

b

background electrolyte (BGE)   271
base-peak chromatogram (BPC)   2
benzophenones   66
β-isoform   355
Biacore 2000 instrument   305
binding affinity
– frontal affinity chromatography   246, 264
– staircase method   247
– zonal chromatography   255
bioactive peptides   348
bioaffinity   323
– analysis   131, 133
– chromatogram   349
– chromatography   243, 366
bioassay   336
– chromatogram   333, 343
– detector   346
bioassay guided fractionation (BGF)   124
biocatalysis   351, 356
biochemical band broadening   345
bioengineering   301, 302
biological assays   323
biological buffer systems   326
biosensor immunoassays (BIAs)   300
bottom-up approach   40
bulk refractive index   368

c

atmospheric pressure chemical ionization (APCI)   10, 341
automated ligand identification system (ALIS)   135

CALUX assays   138
capillary electrophoresis (CE)   272, 273, 275, 367

Analyzing Biomolecular Interactions by Mass Spectrometry, First Edition.
Edited by Jeroen Kool and Wilfried M.A. Niessen.
© 2015 Wiley-VCH Verlag GmbH & Co. KGaA. Published 2015 by Wiley-VCH Verlag GmbH & Co. KGaA.

capillary electrophoresis – mass
   spectrometry (CE – MS) 207
– ACE-MS 284
– general requirements 283
– IA 284
capillary zone electrophoresis (CZE) 275
cell reproduction tests 137
charged residue model (CRM) 8, 86
chemical cross linking proteomics 136
chemical cross-linking MS 64
chemical proteomics approach 136
chip capture and release chromatography
   306
chromatographic detector 343
chromatographic dilution 331
cluster analysis 263
collisional cooling 89
collisional heating 89
collision-induced dissociation 20
collision induced unfolding 98
competition assay
– GAT1 176
– SERT 186
competitive binding assay (CBA) 300
competitive binding immunoassays (CBIAs)
   305
conformational mobility 264
Coumestrol 325
crude synthesis mixture 330
cyclodextrin 290, 292
cytochrome P450 354

d
data acquisition 5, 30
data analysis 68
data-dependent acquisition (DDA) 143
data interpretation methods 348
deconvolution 144
de novo synthesized compounds 324
desalting 83
desorption electrospray ionization (DESI)
   309
desorption electrospray ionization mass
   spectrometry (DESI-MS) 212
desorption ionization on porous silicon
   (DIOS) 304
dextromethorphan 265
dialysis 83, 117
3,4-diaminobenzoic acid (DABA) 304
diazirines 66
dilution calculations 346
diode array detector (DAD) 327
direct analysis in real time (DART) 309
direct binding assay (DBA) 300

dissociation constant 280
dopamine receptors 188
dose – response curve 353
double-bond equivalent 7
DriftScope 95
drug metabolites 33
drug – protein interactions 254
dynamic combinatorial chemistry 181
dynamic equilibrium ACE 276
dynamic protein affinity selection 365

e
effect directed analysis (EDA) 111, 342, 368
– automated ligand identification system
   (ALIS) 135
– bioassay testing 136
– fractionation 126
– identification and confirmation 141
– principles 113
– RapidFire system 134
– sample preparation 113
– size exclusion chromatographic (SEC)
   approaches 133
– SpeedScreen technology 135
– ultrafiltration 134
– workflow of 127
electron capture dissociation 21, 90
electron ionization 206
electron multiplier 18
electron transfer dissociation 21, 90
electroosmotic flow (EOF) 274, 275
electroosmotic mobility 274
electrophoresis 273
electrophoretic mobility 274
electrospray ionization (ESI) 8, 84, 283, 332
electrospray ionization mass spectrometry
   (ESI-MS) 305
ELISA blocking reagent 339
Ellman's reagent 324
elution time 255
energy-sudden techniques 7
enhanced product ion analysis 25
Envimass 147
enzymatic catalysis 324
enzyme assisted extraction (EAE) 202
enzyme-linked immunosorbent assays
   (ELISAs) 282, 300
equilibrium dissociation constant 168
ESI capillaries 335
estrogen receptor (ER) 329, 354, 355
ethoxyresorufin-O-deethylase (EROD) assay
   116
evanescent wave silicon waveguides 308

evaporative light scattering detection (ELSD)   356
extracted ion chromatograms (EIC/XIC)   2, 329
extracted ion currents (EIC)   349

f

filtration based assays   367
first generation antivenomics   229, 230
FlashQuant MS system quantitation   179
flavor and fragrance analysis   123
flow injection analysis (FIA)   353
flow splitting   334
fluorescence   320
fluorescence detection (FLD)   322
fluorescence detector   335
fluorescence enhancement based bioaffinity assay   329
fluorophore   325, 367
fluoxetine   184–186
formic acid (FA)   332
Fourier transform ion cyclotron resonance mass spectrometry (FT-ICR-MS)   16, 210
fragmentation   20
frontal affinity chromatography (FAC)   244, 245, 365
frontal analysis continuous capillary electrophoresis (FACCE)   279
fused-silica capillaries   274
fused-silica reactors   338

g

Galantamine   328
γ-aminobutyric acid (GABA)   171
gas chromatography olfactometry (GC-O)   122
gas chromatography–mass spectrometry (GC–MS)   128, 141
Gaussian distribution   345
gel electrophoresis   42
gel permeation chromatography (GPC)   119
glass microfluidic chips   338
Global Snakebite Initiative (GSI)   228
GLUT1   244
glycopeptide antibiotics   286
GPR17   249
GraphPad Prism   278
GroEL   84

h

headspace analysis   122
heat dissipation   273

heterobifunctional cross-linker   70
high-efficiency free-zone electrophoresis   272
high-field asymmetric waveform ion mobility spectrometry   29
high performance liquid chromatography (HPLC)   131, 331
high-precision nano-FAC system   247
high-resolution screening (HRS)
– bioassay   336
– flow splitting   334
– fluorescence   320
– *vs.* HTS   342
– MS detection   340
– on-line post-column assays   321
– separation   330
– trends   354
high-temperature liquid chromatography (HTLC)   334
high throughput affinity screening   252
high-throughput screening (HTS)   124, 319
homobifunctional cross-linker   69
hydrazone libraries   181
hydrolysis catalyzing enzymes   328
7-hydroxy-1-methylquinolinium iodide   325

i

IC$_{50}$ measurement curves   350
ideal Gaussian distribution   347
immobilized artificial membrane (IAM) stationary phase   244
immobilized proteins   244
immunoaffinity capillary electrophoresis (IA-CE)   271, 272, 281
immunoaffinity capillary electrophoresis–mass spectrometry (IA-CE–MS)
– applications   292
– specific requirements   284
immunoaffinity chromatography   233, 235
incubation times   342
inductively-coupled plasma (ICP)   284
in silico tools   145
interactomics   136, 372
intermediate complexation kinetics, *see* kinetic ACE
internal energy   7
in vitro bioassays   137
ion detection   18
ion dissociation   20
ion electrophoretic mobility   273
ion-evaporation model   8
ionization suppression effects   326

ion mobility 90
ion-mobility mass spectrometry (IMMS) 81, 210
ion mobility spectrometry (IMS) 28, 357
ion-trap instrument 328
ion-trap mass analyzer 15
ion trap mass spectrometers 209
ion-trap-time-of-flight (IT-TOF) 328
isocratic solvent 334

j
Joule heating 285

k
kinase inhibitors 98
kinetic ACE 280
kinetic assay
– GAT1 175
– SERT 186

l
label-free biosensors 299, 367
label free techniques 366
laser ablation electrospray ionization (LAESI) 213, 309
laser desorption imaging (LDI)-MS 213
ligand binding 88
linear-acceleration high-pressure collision cell 22
linear discriminant analysis (LDA) 215
linear ion trap quadrupole (LTQ) 209
linear regression analysis 249
linear solvent energy relation (LSER) model 147
linear two-dimensional ion traps 15
lipophilic analytes 343
liquid chromatography (LC) separation method 61
liquid chromatography with diode array detector–MS (LC/DAD–MS) 301
liquid chromatography–mass spectrometry (LC–MS) 142, 203, 206
localized surface plasmon resonance (LSPR) 368
logarithm of thermodynamic retention factors 262
low-resolution methods 371

m
mass analyzer 12
– ion-trap 15
– Orbitrap 17
– quadrupole 13
– time-of-flight 15

mass defect 4
mass spectrum 2
matrix effect 11
matrix-assisted laser desorption ionization (MALDI) 10, 211
matrix-free ion sources 309
mecamylamine 261
membrane bound proteins 366
Merck index 145
metabolic reactions 348
Metabolomics Standards Initiative (MSI) 215
META-PHOR 216
methanol 331
METLIN 145
Michaelis–Menten kinetics 325, 340
micro-affinity column 247
microchannel plate detectors 18
microfluidic chips 337
micro/nano-fluidic instrumentation 336
Microsoft Excel 278
microwave-assisted extraction (MAE) 201
middle-down approach 43
miniaturized bioassay 357
minimal void volume connectors 338
mobility-shift ACE 276, 277
MOLGEN-MS 146
MS Binding Assays 167
– analytics 170
– competitive 176
– dopamine receptors 188
– γ-aminobutyric acid (GABA) 171
– GAT1 179
– kinetic 175
– library screening 181, 183
– *vs.* radioligand binding assays 167, 168
– saturation assay 174
– serotonin transporter (SERT) 183
MS-coupled biosensor platforms 299
MS detection 340
MS structure elucidation 347
multiple reaction monitoring 22
multiprotein stationary phase 252

n
nano-electrospray 284
nanoelectrospray ionization (nano-ESI) 9, 86
nanoflow-electrospray ionization source 60
nanofractionation 129
nanoplasmonics 307
native ion mobility 371
native MS analysis 58, 60

natural extracts   330
N-hydroxysuccinimide (NHS) esters   64
nicotinic acetylcholine receptor (nAChR)
   328
nicotinic receptors   261
nitrogen rule   7
noise   344
nominal mass   4
noncovalent protein–ligand complexes
   284
non-equilibrium CE   276
non-equilibrium CE of equilibrium mixtures
   (NECEEM)   280
non-linear chromatography (NLC)   261
non-volatile buffers   326
nonvolatile phosphate BGE   284
NORMAN network   145
novel screening method   249
nucleoproteins   305

o
olfactometric analysis   130
oligonucleotides   304
on-chip approaches
– SPR-LDI-MS   304
– SPR-MALDI-MS   301
online affinity analysis   141
online bioluminescence   139
on-line post-column assays   353
on-line post-column bioaffinity screening
   methodology   368
on-line pre-column synthesis   354
opioid peptides   292
Orbitrap   17, 143, 209
orphan receptors   366
orthogonal partial least squares discriminant
   analysis (OPLS-DA)   215
orthogonal separation methods   343
overlayed frontal chromatograms   251

p
paperspray ionization   11
passive sampler device   113
pentasaccharide   289
pentaxin 3 (PTX3)   308
peptides   347
P-glycoprotein transporter   259, 260
phosphate-binding protein   324
phosphodiesterase (PDE)   328
photoreactive cross-linkers   65
plant extracts   330
plant metabolomics approach   370
plasma polymerized poly-acrylic acid
   (ppAA)   308

PLE extraction   120
p38 mitogen-activated protein kinase   321,
   325, 329
polyethylene glycol (PEG)   338
porous silver nanoparticles (AgNP)   304
post-column bioaffinity   368
– measurement   327
post-fractionation liquid handling   343
pre-equilibrium ACE   279
precursor-ion analysis   356
pressurized hot water extraction (PHWE)
   201
pressurized liquid extraction (PLE)   117,
   201
principal component analysis (PCA)
   214
projection approximation   96
protein analysis   58
protein concentration   352
protein fishing   372
protein–ligand interaction   332
protein molecular-weight determination
   37
protein–protein interactions, *in vivo*
   strategy   71
protein target   328
protein unfolding   333
proteomics   39, 368
protonation   38
pseudostatic libraries   182
PSP toxins analysis   306, 307
PubChem   145
pulse dampeners   336
pulsed ultrafiltration   135

q
QTAX approach   70
quadrupole mass analyzer   13
quadrupole mass spectrometer   208
quadrupole-time-of-flight (q-TOF)   328
– mass spectrometer   208
qual/quant strategy   32
quantitative analysis   31
quantitative structure–retention
   relationships (QSRRs)   257

r
radioflow detector system   245
radioimmunoassays   300
radiolabeled thyroxin–transthyretin
   (T4*-TTR) binding assay   117
radioligand binding assays   366
RapidFire system   134
reconstitution methods   59

resolving power   2
retention factor   255, 256

s

sandwich type assay design   300, 301
saturation assay
– GAT1   174
– SERT   185
Scatchard analysis   279
second generation antivenomics   232, 234
secondary ion mass spectrometry (SIMS)
   212
selected-reaction monitoring   22
semi permeable membrane device   115
semi-quantitative determination   88
separation   330
serine protease assay   331
serotonin transporter (SERT)   183
sheath liquid interface   283
sigmoidal dose–response curve   345
signal-to-noise ratio   350
size exclusion chromatography (SEC)   133,
   366
size exclusion methodology   365
SKF86002   325
slow association kinetics   339
slow complexation kinetics, *see*
   pre-equilibrium ACE
snake venomics approach   231
solid-phase extraction (SPE)   114, 115
solid-phase microextraction (SPME)   115,
   123
soluble epoxide hydrolase (sEH)   328
solvent assisted flavor extraction (SAFE)
   122
solvent composition   257
solvent viscosity   335
Soxhlet extraction   125
spectrophotometric LC detector   339
SpeedScreen technology   135
split ratio   335
SPR-LDI-MS   304
SPR-MALDI-MS   301
staircase method   247, 248
stir-bar sorptive extraction (SBSE)   115, 122
streptavidin/biotin   327
stromal cell-derived factor-1   290
structure-affinity matching   349
supercritical fluid extraction (SFE)   201
superloop   337
surface-enhanced laser desorption
   ionization   10
surface-induced dissociation   101
surface plasmon polariton (SPP)   307

surface plasmon resonance imaging (SPRi)
   308
switching function   261

t

tag-mass concept   310
tandem mass spectrometry   18
– hybrid instruments   25
– ion dissociation   20
– ion mobility spectrometry   28
– ion-trap instrument   23
– MS–MS configuration   21
– Orbitrap-based hybrid systems   27
– TOF–TOF instruments   23
target antigen   281, 282
target protein immobilization   365
theoretical dose response curve   345
thin layer chromatography (TLC)   124
thiocholine   324
tiagabine   172
time-of-flight (TOF)
– mass analyzer   15
– mass spectrometer   208
time-to-digital converter   18
top-down approach   39
total-ion chromatogram (TIC)   2
toxicity identity evaluation (TIE)   112
transient protein–protein interactions   70
transthyretin   190
travelling-wave IMS   92
tribrid Orbitrap-based system   27
two-enzyme dual-HRS assay   354

u

ultrafiltration   134, 365
ultrahigh performance liquid
   chromatography system   191
ultrasound assisted extraction (UAE)   201
unmatched HRS chromatogram   349

v

vancomycin   286
vinblastin   260
volatile ammonium acetate BGE   284
volatile BGE   284

y

yeast (anti)androgen assay   116

z

zeta potential   274, 275
Z′-factor   351
zonal affinity chromatography   253
zonal elution   258
zonal elution experiment   255, 256